Pitman Research Notes in Mathematics Series

Main Editors

H. Brezis, Université de Paris

R.G. Douglas, Texas A&M University

A. Jeffrey, University of Newcastle upon Tyne *(Founding Editor)*

Editorial Board

H. Amann, University of Zürich

R. Aris, University of Minnesota

G.I. Barenblatt, University of Cambridge

A. Bensoussan, INRIA, France

P. Bullen, University of British Columbia

S. Donaldson, University of Oxford

R.J. Elliott, University of Alberta

R.P. Gilbert, University of Delaware

R Glowinski, University of Houston

D. Jerison, Massachusetts Institute of Technology

K. Kirchgässner, Universität Stuttgart

B. Lawson, State University of New York at Stony Brook

B. Moodie, University of Alberta

S. Mori, Kyoto University

L.E. Payne, Cornell University

G.F. Roach, University of Strathclyde

I. Stakgold, University of Delaware

W.A. Strauss, Brown University

Submission of proposals for consideration

Suggestions for publication, in the form of outlines and representative samples, are invited by the Editorial Board for assessment. Intending authors should approach one of the main editors or another member of the Editorial Board, citing the relevant AMS subject classifications. Alternatively, outlines may be sent directly to the publisher's offices. Refereeing is by members of the board and other mathematical authorities in the topic concerned, throughout the world.

Preparation of accepted manuscripts

On acceptance of a proposal, the publisher will supply full instructions for the preparation of manuscripts in a form suitable for direct photo-lithographic reproduction. Specially printed grid sheets can be provided and a contribution is offered by the publisher towards the cost of typing. Word processor output, subject to the publisher's approval, is also acceptable.

Illustrations should be prepared by the authors, ready for direct reproduction without further improvement. The use of hand-drawn symbols should be avoided wherever possible, in order to maintain maximum clarity of the text.

The publisher will be pleased to give any guidance necessary during the preparation of a typescript, and will be happy to answer any queries.

Important note

In order to avoid later retyping, intending authors are strongly urged not to begin final preparation of a typescript before receiving the publisher's guidelines. In this way it is hoped to preserve the uniform appearance of the series.

Addison Wesley Longman Ltd
Edinburgh Gate
Harlow, Essex, CM20 2JE
UK
(Telephone (0) 1279 623623)

Laurent Véron

University of Tours, France

Singularities of solutions of second order quasilinear equations

 LONGMAN

Addison Wesley Longman Limited
Edinburgh Gate, Harlow
Essex CM20 2JE, England
and Associated Companies throughout the world.

Published in the United States of America
by Addison Wesley Longman Inc.

First published 1996

AMS Subject Classifications: (Main) 35J, 35K, 35B
                               (Subsidiary) 49F, 53C, 34E

ISSN 0269-3674

ISBN 0 582 03539 2

British Library Cataloguing in Publication Data

A catalogue record for this book is
available from the British Library

Library of Congress Cataloging-in-Publication Data

Veron, Laurent.
    Singularities of solutions of second order quasilinear equations /
L. Veron.
    p. cm. -- (Pitman research notes in mathematics series, ISSN
0269-3674 ;)
    1. Differential equations, Elliptic--Numerical solutions.
2. Differential equations, Parabolic--Numerical solutions.
3. Differential equations, Nonlinear--Numerical solutions.
4. Singularities (Mathematics) I. Title. II. Series.
QA377.V46 1995
515'.353--dc20   94-33324
                                                       CIP

Printed and bound in Great Britain
by Biddles Ltd, Guildford and King's Lynn

Contents

6 Singularities of nonlinear parabolic equations

Preface

This book is an attempt to present in a unified way the development of the theory of singularities of second order nonlinear elliptic or parabolic equations starting from the linear equations and the pionneering work of J. Serrin on quasilinear equations. Emphasis is put on the very rich singularity theory of second order equations with a superlinear perturbation.

I would like to express my deep thanks to Haim Brezis who encouraged me to write this book and James Serrin who is at the origin of many of the problems which are discussed here.

My sincere thanks also go to many of my friends with whom I spent so many working periods; part of the material contained in this volume comes from these cooperations. I don't also forget the valuable help of those have accepted to read and criticise parts of the manuscript. Among the people I want to thank, I would like to mention Ph. Bénilan, M.F. Bidaut-Véron, P. Chrusciel, J.I. Diaz, E.B. Dynkin, A. Friedman, S. Kamin, V. Kondratiev, M. Marcus, H. Matano, Y. Pinchover, J. Spruck, J.L. Vazquez. My thanks also go to my former PhD students who gave me the opportunity to explore new problems in the singularity theory. Among them I want to mention M. Bouhar, J. Fabbri, A. Gmira, M. Guedda, B. Guerch, M. Grillot, J.R. Licois, I. Moutoussamy and Y. Richard.

Finally, I would like to thank the staff at Addison Wesley Longman for their patience with my repeated delays in delivering my work and for the time spent in checking a preliminary version of it.

Tours, June 1996   Laurent Véron

Introduction

The singularity problem for a quasilinear equation can be stated as follows. We are given an open subset Ω in $\mathbb{R}^N$ or in some N-dimensional Riemannian manifold M, and a subset Σ in Ω. We are also given a function u defined in $\Omega \backslash \Sigma$ satisfying some quasilinear equation in $\Omega \backslash \Sigma$. The problem is either to extend u to whole Ω in a natural way in order that the new function $\tilde{u}$ satisfies the same quasilinear equation in Ω, or, if no reasonable extension of u is possible, to describe the behaviour of u near Σ; such a behavior is called admissible. The two types of equations which will be considered in this book are the second order quasilinear elliptic or parabolic equations, most often in divergence form, that is

$$\mathrm{div}A(x,u,\nabla u) + B(x,u,\nabla u) = 0, \tag{1}$$

$$u_t - \mathrm{div}A(t,x,u,\nabla u) - B(t,x,u,\nabla u) = 0, \tag{2}$$

where A and be are respectively vector valued and real valued functions. In connection with the problem stated above, it is also a natural question to wonder whether there truly exist solutions of (1) or (2) with some admissible singular behaviour near Σ. If we take the simple example of an harmonic function u in $\mathbb{R}^N \backslash \Sigma$, with $\Sigma = \{O\}$, a single point, then the behaviour of u is obtained by an expansion in series of spherical harmonics under the following form

$$u(x) = u(r,\sigma) = \sum_{n=0}^{\infty} \mu^{(n)}(r)\psi_n(\sigma) + \sum_{n=0}^{\infty} r^n \tilde{\psi}_n(\sigma), \tag{3}$$

where (r,σ) are the spherical coordinates in $\mathbb{R}^N \backslash \{O\}$, μ the fundamental solution of the Laplacian and ψ_n and $\tilde{\psi}_n$ spherical harmonics of degree n. The question of the removability of the possible singularity at $\{O\}$ is clearly conditioned by the growth of u near this point. If, for example, $u(x) = o(\mu(|x|))$, then u is a regular harmonic function in whole $\mathbb{R}^N$ but, if $u(x) = O(\mu^{(k)}(|x|))$ for some nonnegative integer k, then u admits an asymptotic expansion of the following form

$$u(x) = u(r,\sigma) = \sum_{n=0}^{k} \mu^{(n)}(r)\psi_n(\sigma) + \sum_{n=0}^{\infty} r^n \tilde{\psi}_n(\sigma). \tag{4}$$

Therefore, a crucial step to study singularity problem is the knowledge of an *a priori* blow-up estimate of u near the singular set Σ. The size of Σ in the sense of its topological dimension is of importance and, more generally, the singularity problems will involve a balance between the growth of u and the size of Σ.

The different models of elliptic operators which will be considered in the sequel are the following:

$$
\operatorname{div}A(x,u,\nabla u) = \Delta u = \sum_{j} u_{x_j x_j}, \tag{5}
$$

which is the ordinary Laplacian,

$$
\operatorname{div}A(x,u,\nabla u) = \Delta_p u = \operatorname{div}\left(|\nabla u|^{p-2}\nabla u\right) = \sum_{j}\left(|\nabla u|^{p-2} u_{x_j}\right)_{x_j}, \tag{6}
$$

which is called the p-Laplacian $(1 < p < \infty)$, and

$$
\operatorname{div}A(x,u,\nabla u) = \operatorname{div}\left(\frac{\nabla u}{\sqrt{1+|\nabla u|^2}}\right) = \sum_{j}\left(\frac{u_{x_j}}{\sqrt{1+|\nabla u|^2}}\right)_{x_j}, \tag{7}
$$

which is the mean curvature operator. The types of phenomenon we shall find will be strongly dependent of the lower order tem B. If we just consider a nonlinearity with the same order of growth as A, then its effect is, in some sense, negligible. But if the order of growth of B is bigger than the one of A, then the situation is extremely complicated and many parameters are interacting in the singularity problem. For example, let us consider the two following equations

$$
-\Delta u = u^q \tag{8}
$$

$$
\Delta u = u^q \tag{9}
$$

with $q > 1$. If we look for radial power-like solutions under the form $r \mapsto \alpha r^\beta$, then always $\alpha = -2/(q-1)$, and $\beta = \left[(2/(q-1))(N-2q/(q-1))\right]^{1/(q-1)}$ for equation (8) but β exists if and only if $N > 2$ and $q > (N/(N-2)$; and $\beta = \left[(2/(q-1))(2q/(q-1)-N)\right]^{1/(q-1)}$ for equation (9), and this β exists if and only if $1 < q < N/(N-2)$, with no restriction when $N = 2$. Therefore it is clear that the value of q with respect to $N/(N-2)$ is crucial for the study of solutions of (8) or (9) with a singularity at $\{O\}$. This simple observation was done by Emden [Em] who published in 1897 the first work of radial solutions of (8). Later on, in the begining of the Century, Fowler [Fo], carried on the first almost complete study of the radial solutions of (8) and (9).

For a long time the only study of singularity problems dealt with linear equations or radial solutions of equations of type (8) and (9). Among the people who obtained significant results on the linear equations are Picard, Gevrey [Ge], Bôcher [Bô] and more recently, Gilbarg and Serrin [GS]. As for the differential equations associated with (8) or (9), they have been studied, for example, by Sommerfeld [So], Chandrasekhar [Ch], Hille [Hi1], [Hi2] and many others. In fact the first breakthrough is due to Serrin [Se1], [Se2] who obtained the first general results on quasilinear equations of type (1) under the assumption that the perturbation term has its growth dominated by the one of A. Around 1980 the sharp developpment of nonlinear partial differential equations theory allowed another breakthrough in the study of nonradial singular solutions of (8) and (9). This was initiated by Gidas and Spruck [GSk1], Lions [Ls] and Véron [Ve1] [Ve2]. After this first period, many articles have been published taking into account the different aspects of the singularity problem for equations of type (8) or (9) and also for parabolic equations such that

$$u_t - \Delta u - u^q = 0, \tag{10}$$

$$u_t - \Delta u + u^q = 0. \tag{11}$$

Among the people who published significative results in this direction are Aviles [Av1] [Av2] [Av3], Brezis and Véron [BV], Caffarelli, Gidas and Spruck [CGS], Loewner and Nirenberg [LN], Baras and Pierre [BP1], [BP2], Schoen [Sc], Schoen and Yau [SY], Richard and Véron [RV], Vazquez [Va1], [Va2], Vazquez and Véron [VV1], [VV2], [VV3], Friedman and Véron [FV], Chen, Matano and Véron [CMV], Mazzeo and Smale [MS] or Paccard[MP], Paccard [Pc1], [Pc2], [Pc3], [Pc4] , Bidaut-Véron and Véron [B.VV] and Véron [Ve3], [Ve4], [Ve5], [Ve6] for the semilinear elliptic equation; as for the nonlinear parabolic equations, the first significative results about the singularity problem are due to Brezis and Friedman [BF], and then after came Kamin and Peletier [KP], Brezis, Peletier and Terman [BPT], Galaktionov, Kurdyumov and Samarskii [GKS], Baras and Pierre [BP2], Oswald [Os] ,Weissler [We1], [We2], Giga and Kohn [GK1], [GK2], Escobedo and Kavian [EK1], Véron [Ve7], Moutoussamy and Véron [MV1], [MV2].

Presentation of the contents

Chap. 1. After a brief introduction dealing with the simplest linear cases, we present Gilbarg and Serrin's results on linear elliptic equations and Serrin's theory on the general quasilinear equations (1) with a restricted growth assumption on the lower order term B.

Chap. 2. We treat semilinear equations of the following type:

$$-\Delta u + g(u) = 0 \tag{12}$$

where g is essentially a nondecreasing function.The first section deals with the description of the possible behaviour of a solution near an isolated singularity. One tool for that is the isotropy theorem due to Vazquez and Véron; the classification theorems of Véron and Chen-Matano and Véron in the particular case where $g(u) = |u|^{q-1}u$ (q > 1), are derived. The second Section deals with the various removability results due to Brezis and Véron, Baras and Pierre, Vazquez and Véron and Kondratiev and Landis.

Chap. 3. We study semilinear equations of the following type

$$-\Delta u = g(u) \tag{13}$$

where g is nonnegative and, most often, $g(u) = u^q$ (q > 1). The starting point is Brezis-Lions Lemma on isolated singularities of positive superharmonic functions, and Richard and Véron's applications to nonlinear inequalities. The description of the isolated singularities of positive solutions of (8) when $1 < q < N/(N-2)$ or of (13) with $g(u) = e^u$ is derived from these results. A much more difficult case for (8) deals with the range $N/(N-2) \leq q < (N+2)/(N-2)$; we present Gidas and Spruck, Aviles results with the various extensions given by Bidaut-Véron and Véron. Many of these problems are linked to the study of some semilinear elliptic equations on compact manifolds, in which problems the curvature plays a fundamental role; we present some results in that direction. Another crucial step in that study is the understanding of the asymptotics of the bounded solutions of the following equation

$$v_{tt} + av_t + \Delta_s v + f(v) = 0 \tag{14}$$

on $[0,\infty) \times S^{N-1}$ where Δ_s is the Laplace-Beltrami operator on S^{N-1} a some real number and f a real valued function; a breaktrough in that direction is due to Simon. The last sections is devoted to the construction of solutions with more complicated singular set (finite number of points or submanifold).

Chap. 4. We present some boundary singularity problems for equations of type (9) when q > 1. The first results in a nonlinear framework are due to Gmira and Véron who studied the removability of boundary singularities, the construction of solutions with measure boundary data and the description of the possible blow-up at an isolated boundary singularity. Another series of results due to Marcus and Véron deals with the construction and the uniqueness of solutions of (9) with uniform blow-up on $\partial\Omega$; the regularity of the boundary being crucial for these problems. The last section is devoted to the presentation of Marcus and Véron's boundary trace theory of positive solutions of (9).

Chap. 5. This chapter is devoted to the singularity problem for quasilinear equations such as

$$-\text{div}A\left(|\nabla u|\nabla u\right) + g(u) = 0, \tag{16}$$

when $A(r) = 1/\sqrt{1+r^2}$ or $A(r) = |r|^{p-2}r$ with p > 1 and when the growth of g is bigger than the one of A. We first present Kichenassamy and Véron's results on isolated singularities of p-harmonic functions, then Vazquez and Véron's theorem on removability of isolated singularities for (16) when $A(r) = |r|^{p-2}r$ and $g(u) = |u|^{q-2}u$, as well as Friedman and Véron classifications of the singularities of the solutions of such equation when they exist. In the second section we give Bidaut-Véron's extension of Brezis and Lions Lemma to positive p-superharmonic functions. As a consequence a description of isolated singularities of positive solutions of

$$-\text{div}\left(|\nabla u|^{p-2}\nabla u\right) = u^q, \tag{17}$$

is derived in the subcritical case $p-1 < q < N(p-1)/(N-p)$. The radial study due to Guedda and Véron and completed by Bidaut-Véron is given in the general case $q > p-1$. For capillarity type equations

$$-\mathrm{div}\left(\frac{\nabla u}{\sqrt{1+\left|\nabla u\right|^2}}\right) = f(u), \tag{18}$$

where $f(u) = \lambda u$ $(\lambda > 0)$ in the original physical problem, or is an increasing function of u. Existence and uniqueness result of singular radial solutions due to Concus and Finn or Bidaut-Véron are presented.

Chap. 6. We present Aronson's results concerning isolated singularities of linear parabolic equations of second order. The model example we treat in details is the semilinear heat equation with absorption.

$$u_t - \Delta u + u|u|^{q-1} = 0. \tag{19}$$

For this equation we study the measure initial data problem as well as the removability question, two types of results which were initiated by Brezis and Friedman and then developped following different techniques by Baras and Pierre. Next we present Oswald, and Moutoussamy and Véron's results dealing with the description of the local behaviour of a solution of (19) near a isolated initial singularity. For the heat equation with forcing term below

$$u_t - \Delta u - u^q = 0, \tag{20}$$

similar questions are studied. The remaining part of this Chapter is devoted to the extension of such problems to some of the most classical equations arising in the theory of nonlinear filtration;

$$u_t - \Delta u^m = 0, \tag{21}$$

the porous media equation;

$$u_t - \mathrm{div}\left(|\nabla u|^{p-2}\nabla u\right) = 0, \tag{22}$$

the p-hamonic flow; and the pertubation of the two preceeding equations:

$$u_t - \Delta u^m + u|u|^{q-1} = 0, \tag{23}$$

$$u_t - \mathrm{div}\left(|\nabla u|^{p-2}\nabla u\right) + u|u|^{q-1} = 0. \tag{24}$$

Concerning these type of equations the problems are the existence of fundamental solutions as well as the existence of very singular solutions and the connection between those two types of solutions.

1 Singularities of linear and quasilinear elliptic equations

1-1 Linear equations

1-1-1 Some basic notations in Riemannian geometry

We define a n-dimensional Riemannian manifold (M, g) as a differential manifold M endowed with a symmetric positive definite tensor $g = (g_{ij})$ called the metric tensor, with inverse matrix $g^{-1} = (g^{ij})$. If (x^j), $j=1,..., n$, is a system of local coordinates on M, we define the length L of a curve C given by its parametric representation

$$x^j = x^j(t), \qquad a \le t \le b, \tag{1.1.1}$$

as being the following quantity

$$L = \int_a^b \left(\sum_{ij} g_{ij} \frac{dx^i}{dt} \frac{dx^j}{dt} \right)^{1/2} dt. \tag{1.1.2}$$

If we write $|g| = \det(g_{ij})$, then the volume element on (M, g) is

$$dv = \sqrt{|g|} dx^1 ... dx^n, \tag{1.1.3}$$

and the scalar product on the tangent bundle TM of M is defined for some x on M and $\mathbf{X}$ and $\mathbf{Y}$ belonging to $T_x M$ by

$$\mathbf{X} . \mathbf{Y} = \sum_{ij} g_{ij}(x) X^i Y^j, \tag{1.1.4}$$

where (X^i) and (Y^i) are the components of $\mathbf{X}$ and $\mathbf{Y}$. The Christoffel symbols are

$$\Gamma^i_{jk} = \frac{1}{2} \sum_l \left(\frac{\partial g_{kl}}{\partial x^j} + \frac{\partial g_{lj}}{\partial x^k} - \frac{\partial g_{jk}}{\partial x^l} \right) g^{il}. \tag{1.1.5}$$

With the Christoffel symbols it is possible to express the Riemann curvature tensor of the metric $\mathrm{Riem}(M, g)$ by

$$\mathrm{Riem}(M, g) = \left(R^i_{\ jkl} \right) = \frac{\partial}{\partial x^k} \Gamma^i_{lj} - \frac{\partial}{\partial x^l} \Gamma^i_{kj} + \sum_m \left(\Gamma^i_{km} \Gamma^m_{lj} - \Gamma^i_{lm} \Gamma^m_{kj} \right). \tag{1.1.6}$$

From this it is natural to define two curvature elements which will play some role in the sequel: the Ricci tensor $\mathrm{Ricc}\,(M,g)$, which is

$$\mathrm{Ricc}(M,g) = \left(R_{ij}\right) = \left(\sum_{l} R^{l}{}_{ijl}\right), \tag{1.1.7}$$

and the scalar curvature which is

$$R_{g} = \sum_{ij} g^{ij} R_{ij}. \tag{1.1.8}$$

If u is a C^{1} function defined on M, the gradient of u, quoted by ∇u, is the vector field on M with covariant components $\nabla_{i} u = \dfrac{\partial u}{\partial x^{i}}$ and contravariant components $\nabla^{i} u = \sum_{k} g^{ik} \dfrac{\partial u}{\partial x^{k}}$. Therefore

$$|\nabla u|^{2} = \nabla u . \nabla u = \sum_{ij} g^{ij} \frac{\partial u}{\partial x^{i}} \frac{\partial u}{\partial x^{j}}. \tag{1.1.9}$$

The divergence of a C^{1} vector field $\mathbf{X} = \left(\mathbf{X}^{i}\right)$ is defined by

$$\mathrm{div}_{g}(\mathbf{X}) = \frac{1}{\sqrt{|g|}} \sum_{k} \frac{\partial}{\partial x^{k}}\left(\sqrt{|g|}\mathbf{X}^{k}\right), \tag{1.1.10}$$

and the Laplacian of a $C^{2}(M)$-function u is the divergence of the covariant gradient of u, that is

$$\Delta_{g} u = \frac{1}{\sqrt{|g|}} \sum_{ij} \frac{\partial}{\partial x^{i}}\left(\sqrt{|g|}\,g^{ij}\,\frac{\partial u}{\partial x^{j}}\right). \tag{1.1.11}$$

The following formulas are straightforward:

$$\Delta_{g}(uv) = u\Delta_{g}v + v\Delta_{g} + 2\nabla u . \nabla v, \tag{1.1.12}$$

$$\Delta_{g} f(u) = \frac{df}{du} \Delta_{g} u + \frac{d^{2}f}{du^{2}} |\nabla u|^{2}, \tag{1.1.13}$$

for any C^{2} real function f . The second covariant derivatives of u are the following

$$\nabla_{ij} u = \frac{\partial^{2} u}{\partial x^{i}\partial x^{j}} - \sum_{k} \Gamma^{k}_{ij} \frac{\partial u}{\partial x^{k}}, \tag{1.1.14}$$

and the Hessian of u is the covariant 2-tensor

$$\left(\text{Hess}(u)\right)_{ij} = \nabla_{ij}u. \tag{1.1.15}$$

With these notations we can recall the Bochner-Lichnerowicz-Weitzenböck formula which will be of a great use in Chapter 3:

THEOREM 1.1. *Assume* u *is a* C^3 *function on a Riemannian manifold* (M, g). *Then*

$$\frac{1}{2}\Delta_g|\nabla u|^2 = |\text{Hess}(u)|^2 + \nabla(\Delta_g u).\nabla u + \text{Ricc}(\nabla u, \nabla u). \tag{1.1.16}$$

For the proof see [BGM].

If (M, g) is a compact manifold without boundary (resp. with a non-empty boundary) then $-\Delta_g$ has a nonnegative (resp. positive) spectrum in the Sobolev space $W^{1,2}(M)$ (resp. $W_0^{1,2}(M)$). Let $0 = \lambda_0 < \lambda_1 \leq ... \leq \lambda_k \leq ...$ (resp. $0 < \lambda_1 \leq \lambda_2 \leq .... \leq \lambda_k \leq ....$) be the infinite sequence of eigenvalues of $-\Delta_g$ in $W^{1,2}(M)$ (resp. $W_0^{1,2}(M)$ and H_k the kernel of $-\Delta_g - \lambda_k I$, then H_k is finite dimensional and $W^{1,2}(M)$ (resp. $W_0^{1,2}(M)$ is the Hilbertian sum of the H_k.

1-1-2   Spherical harmonics

The full spectrum of the Euclidean n-sphere S^n is known for a long time and we recall it [BGM]:

THEOREM 1.2. *The eigenvalues of* $-\Delta_g$ *in the space* $W^{1,2}(S^n)$ *are the* $\lambda_k = k(k+n-1)$, $k = 0,1,2,....$ *The corresponding eigenspace* H_k *is the space of the restrictions to* S^n *of the harmonic polynomials of degree k, and the dimension of* H_k *is*

$$d_k = \frac{(n+k-2)(n+k-3)...(n+1)n}{k!}(n+2k-1). \tag{1.2.1}$$

For example, the first eigenspace H_0 is reduced to constant functions; the second eigenspace H_1 is generated by the restriction to S^n of the coordinate functions in $\mathbb{R}^{n+1}$, $\left(x_i\right)_{i=1,2,...,n+1}$, and any of these functions is positive on one hemisphere and negative on the other; the third eigenspace H_2 is generated by the restrictions to S^n of $\left(x_i x_j\right)_{1 \leq i < j \leq n+1}$ and of $\left(x_i^2 - x_j^2\right)_{1 \leq i < j \leq n+1}$.

With this result, it is easy to construct singular harmonic functions in $\mathbb{R}^N \setminus \{0\}$: Let μ be the fundamental solution of the Laplacian in $\mathbb{R}^N$, that is

$$\mu(x) = \begin{cases} \ln(1/|x|) & \text{if } N = 2, \\ |x|^{2-N} & \text{if } N \geq 3, \end{cases} \tag{1.2.2}$$

then if ψ_k is any spherical harmonic of degree k on the (N-1)-sphere, the function

$$u(x) = u(r, \sigma) = \frac{d^k \mu}{dr^k}(r) \psi_k(\sigma) \tag{1.2.3}$$

is a singular harmonic function in $\mathbb{R}^N \setminus \{O\}$. Moreover, due to the fact that any harmonic function u in $\mathbb{R}^N \setminus \{O\}$ admits an expansion in series of spherical harmonics such as

$$u(x) = u(r, \sigma) = \sum_{n=0}^{\infty} \mu^{(n)}(r) \psi_n(\sigma) + \sum_{n=0}^{\infty} r^n \tilde{\psi}_n(\sigma), \tag{1.2.4}$$

where the ψ_n and $\tilde{\psi}_n$ are spherical harmonics of degree n , it is clear that an *a priori* bound on u near O of the following type: $u(x) = O(|x|^{\alpha})$ for some $\alpha < 0$, implies that the expansion of u is reduced to

$$u(x) = u(r, \sigma) = \sum_{n=0}^{[\alpha]+1} \mu^{(n)}(r) \psi_n(\sigma) + \sum_{n=0}^{\infty} r^n \tilde{\psi}_n(\sigma). \tag{1.2.5}$$

1-1-3 Sobolev spaces and capacities

We recall here the basic notations concerning Sobolev spaces on an open subset Ω of $\mathbb{R}^N$. Let $C_0^\infty(\Omega)$ be the space of C^∞ functions with compact support in Ω, m a nonnegative integer and $1 \leq p \leq \infty$. We denote

$$W^{m,p}(\Omega) = \left\{ u \in L^p(\Omega): D^\alpha u \in L^p(\Omega), \forall \alpha / |\alpha| \leq m \right\}, \tag{1.3.1}$$

and

$$\|u\|_{m,p} = \left(\sum_{|\alpha| \leq m} \int_\Omega |D^\alpha u|^p dx \right)^{1/p}, \tag{1.3.2}$$

with the usual modification when $p = \infty$. The space $W_0^{m,p}(\Omega)$ is the closure of $C_0^\infty(\Omega)$ in $W^{m,p}(\Omega)$. The definitions of the Sobolev spaces for a compact Riemannian manifold follows the same ideas excepted that the differentiation has to be replaced by the covariant differentiation. For the properties of the Sobolev spaces one could consult [Ad], [Ma1] or [GT] and [Au2] for the Riemannian manifold case.

If K is any compact subset of Ω we define the (elliptic) capacity of order (m,p) by

$$c_{m,p}(K) = \inf\left\{\|v\|_{m,p}^p : v \in W_0^{m,p}(\Omega)\,/\,v \geq 1 \text{ in some neibourhood of K}\right\}. \quad (1.3.3)$$

If ω is any open subset relatively closed in Ω, we set

$$c_{m,p}(\omega) = \sup\left\{c_{m,p}(K): \text{K compact, } K \subset \omega\right\}, \quad (1.3.4)$$

and, for any subset E relatively closed in Ω,

$$c_{m,p}(E) = \inf\left\{c_{m,p}(\omega): \omega \text{ open}, E \subset \omega \subset \overline{\omega} \subset \Omega\right\}. \quad (1.3.5)$$

In the sequel we shall only consider the cases where m = 1 or 2. There are many other types of capacities (see [Fr] for example). The elliptic capacities are useful objects to study the regularity of solutions of partial differential equations. A natural extension of the classical notion of dimension is the Hausdorff measure [Be1]. If α is any positive real number, we define the α-Hausdorff measure of some subset E in $\mathbb{R}^N$ by

$$H_\alpha(E) = \liminf_{\rho \to 0}\left\{\sum_i r_i^{\,\alpha} : E \subset \bigcup_i \overline{B}(a_i, r_i), 0 < r_i \leq \rho\right\}, \quad (1.3.6)$$

where $B(a_i, r_i)$ is the open ball of center a_i and radius r_i . Since it is clear that

$$H_\alpha(E) \leq H_{\alpha'}(E) \qquad \text{if } \alpha' \leq \alpha, \quad (1.3.7)$$

the Hausdorff dimension of E is defined by

$$\dim_H(E) = \inf\left\{\alpha > 0: H_\alpha(E) = 0\right\}. \quad (1.3.8)$$

Among the relations between capacities and Hausdorff dimension we mention the fact that a compact subset K of Ω is of $c_{1,2}$-capacity zero if and only if its (N-1)- Hausdorff measure is zero, which is, for example, the case if K lies on any submanifold of Ω with codimension d > 1. The following results concerning capacities will be used in the sequel:

PROPOSITION 1.1. *Let* K *be a compact subset of* Ω *such that* $c_{m,p}(K) = 0$ *for some* $m \geq 1$ *and* $p < \infty$. *Then there exists a sequence of* C_0^∞ *functions* $\{\eta_n\}$ *with* $0 \leq \eta_n \leq 1$, *taking the value 1 in some neighborhood of* K *and such that* $\eta_n(x) \to 0$ *a.e. in* Ω *and* $\|\eta_n\|_{m,p} \to 0$.

The proof can be found in [AP] for the general case and Serrin [Se1] in the case m = 1 (in his proof η_n is replaced by $1 - \eta_n$). As a consequence any subset with $c_{m,p}$-capacity zero has measure zero. The $c_{1,s}$-capacity of a ball can be easily computed by means of the s-Laplacian operator.

PROPOSITION 1.2. *Let* B_R, $R > 0$, *be a ball of radius* R *and* $1 < s \leq N$. *Then*

$$c_{1,s}(K) = \begin{cases} \left|S^{N-1}\right| \left(\dfrac{N-s}{s-1}\right)^{s-1} R^{N-s} & \text{if } 1 < s < N, \\[2mm] \left|S^{N-1}\right| \left(\ln(R_0/R)\right)^{1-N} & \text{if } s = N, \end{cases} \tag{1.3.9}$$

where R_0 *is any number larger than* R.

1-1-4 Linear equations in non-divergence form

We consider a second order linear elliptic operator in non-divergence form, namely

$$Lu = \sum_{ij} a_{ij} \frac{\partial^2 u}{\partial x_i \partial x_j} + \sum_i b_i \frac{\partial u}{\partial x_i} + cu, \tag{1.4.1}$$

where the coefficients a_{ij}, b_i, c are defined in some open subset Ω of $\mathbb{R}^N$ and satisfy the strong ellipticity condition, that is

$$\lambda^{-1} \sum_i \xi_i^2 \leq \sum_{ij} a_{ij}(x)\xi_i\xi_j \leq \lambda \sum_i \xi_i^2 , \tag{1.4.2}$$

for some $\lambda > 0$, independently of x in Ω, and for any $\xi = (\xi_1, ..., \xi_N)$ in $\mathbb{R}^N$. The regularity assumption on the coefficients of L is the Dini continuity for the a_{ij}.

Definition. A function f is Dini continuous at some point x_0 if

$$\left|f(x) - f(x_0)\right| \leq \varphi(\left|x - x_0\right|), \tag{1.4.3}$$

where φ, the Dini modulus of continuity of f at x_0, is an increasing function satisfying

$$\int_0^1 \frac{\varphi(r)}{r} dr < \infty. \tag{1.4.4}$$

It is well known that if the coefficients of L are continuous, then the maximum principle holds in the sense that if c(x) is zero, any non-constant solution u of the differential inequality

$$Lu \geq 0 \tag{1.4.5}$$

11

cannot take on a maximum value in the interior of its domain of definition (and if $c \leq 0$, any non-constant solution of (1.4.5) cannot take on a nonnegative maximum in the interior of its domain of definition). The removability result of Gevrey ([Ge],[GS]), shows that this maximum principle still holds if u is defined in $B^*(O,R) = B(O,R) \setminus \{O\}$ and satisfies a growth condition at O, provided the coefficients a_{ij} be Dini continuous at O.

THEOREM 1.3. *Let the coefficients* a_{ij} *be Dini continuous at* O, *the coefficients* b_i *bounded in* B(O, R) *for some* $R > 0$, c *be zero and* u *be a* C^2 *solution of* (1.4.5) *in* $B^*(O,R)$ *which satisfies*

$$\lim_{|x| \to 0} u(x) / \mu(x) = 0, \tag{1.4.6}$$

where $\mu(x) = |x|^{2-N}$ *if* $N \geq 3$ *or* $\mu(x) = \ln(1/|x|)$ *if* $N = 2$. *Then* $u(y) < M_\rho = \max_{|x|=\rho} u(x)$, *for any* $\rho \in (0,R)$ *and* $y \in B^*(O,\rho)$. *Moreover*

$$\limsup_{x \to 0} u(x) < M_\rho. \tag{1.4.7}$$

If $c \leq 0$, *the same conclusion holds with* $M_\rho = \max_{|x|=\rho}(0, u(x))$.

Proof. Let us define

$$h(r) = \int_0^{g(r)} \exp K\left(r + \int_0^r s^{-1}\varphi(s)ds \right) g'(r)dr, \tag{1.4.8}$$

where K is some positive constant and g(r) is either r^{2-N} if $N > 2$ or $\ln(\rho / r)$ if $N = 2$. If K is taken large enough, it is easy to check that

$$Lh \leq 0 \quad \text{in} \quad B^*(O,\rho), \tag{1.4.9}$$

and, from (1.4.4), $h(r) \approx g(r)$ near 0. Defining $v_\varepsilon = u - \varepsilon h - M_\rho$, then $Lv_\varepsilon \geq 0$ in $B^*(O,\rho)$. Moreover $v_\varepsilon < 0$ near O and for $|x| = \rho$. Applying the maximum principle in $B^*(O,\rho) \setminus B(O,2^{-n}\rho)$ implies $v_\varepsilon < 0$. Letting n tend to infinity and ε to 0 yields $u \leq M_\rho$ in $B^*(O,\rho)$. From the strong maximum principle equality cannot occur in $B^*(O,\rho)$. This implies the strict inequality and (1.4.7) holds.

Remark 1.1. When $N > 2$ it is possible to drop the Dini continuity assumption over the coefficients a_{ij} by replacing the condition (1.4.6) by the more restrictive one

$$u(x) = O\left(|x|^{2-N+\delta}\right), \tag{1.4.10}$$

near O, for some $\delta > 0$.

It is clear that Theorem 1.3 implies the boundedness of the solutions u of

$$Lu = f \tag{1.4.11}$$

in $B^*(O,\rho)$, provided (1.4.6) or (1.4.10) be satisfied and f be bounded for simplicity (replace u by $u + A|x|^2$ for suitable A). A natural second step is the behaviour of bounded solutions of (1.4.11) near 0. The following result of Liouville type deals with such a problem

THEOREM 1.4. *Let* u *be a solution of* (1.4.11) *in the punctured ball* $B^*(O,R)$ *and f be bounded. If* $N = 2$ *assume that* u *is bounded below; if* $N > 2$ *suppose that the coefficients* a_{ij} *are continuous at* O *and that* u *is bounded in* $B^*(O,R)$. *Then* u *tends to a limit (not necessarily finite if* $N = 2$) *at* O.

Proof. By changing u into $(1 - A|x|^2)u$ for suitable A, we can assume $c \leq 0$.

Case I. $N = 2$, let u_0 be the lim inf of u at O. It is clear that $u_0 > -M = \inf_{x \in B^*(O,R)} u(x)$. If $u_0 = \infty$, then $\lim u(x) = \infty$ and the result follows. Therefore we may suppose that u_0 is finite and there exist $\varepsilon > 0$ and $R' \in (0,R)$ such that

$$v_\varepsilon = u - u_0 + \varepsilon > 0 \quad \text{in} \quad B^*(O,R'), \tag{1.4.12}$$

moreover v_ε satisfies

$$Lv_\varepsilon = f + c(\varepsilon - u_0) = f', \tag{1.4.13}$$

and f' is also bounded, say $|f'| \leq B'$. Since $\liminf_{x \to 0} v_\varepsilon(x) = \varepsilon$, there exists a sequence $\{x_n\}$ going to O such that $v_\varepsilon(x_n) < 2\varepsilon$. Let us assume that $|x_n| < \min(1, R'/2)$; we can apply an Harnack type inequality due to Serrin [Se1],[GS] and obtain

$$v_\varepsilon(x) \leq K\left(v_\varepsilon(x_n) + B'|x_n|^2\right) \leq K\left(2\varepsilon + B'|x_n|^2\right) \tag{1.4.14}$$

for $|x| = |x_n|$, in which formula K is positive and depends only on the ellipticity constant λ and on the upper bound B of the coefficients of L. Now for sufficiently large A and sufficiently small ρ we have

$$L(v_\varepsilon + A|x|^2) \geq 0 \quad \text{in} \quad B^*(O,\rho) . \tag{1.4.15}$$

If n_0 is so large that $|x_n| \leq \rho$ for $n \geq n_0$, we may apply the maximum principle to $v_\varepsilon + A|x|^2$

in the annuli $B(O,|x_n|) \setminus B(O,|x_{n+1}|)$, for $n \geq n_0$. Using (1.4.14) we see that

$$v_\varepsilon(x) + A|x|^2 \leq K\left(2\varepsilon + B'|x_n|^2\right) + A|x_n|^2 \tag{1.4.16}$$

in these annuli, yielding

$$v_\varepsilon(x) \leq 2K\varepsilon + (A + KB')|x_n|^2 \ , \tag{1.4.17}$$

for $0 < |x| \leq |x_n|$, $n \geq n_0$. Letting x, hence x_n, and finally ε go to 0 implies the existence of a finite limit at 0, since we have

$$\limsup_{x \to 0} u(x) \leq u_o = \liminf_{x \to 0} u(x). \tag{1.4.18}$$

Case II. N > 2. The preceding argument carries over to the present case with minor change arising from the Harnack type inequality (still due to Serrin). In that case there exist a constant $R_0 \in (0, R)$ and a continuous increasing function Φ defined on $\mathbb{R}^+$ vanishing at 0 and depending on the ellipticity constant λ, on the bounds of the coefficients of L , on B', on the bounds of u in $B^*(O, R)$ and on the modulus of continuity of the a_{ij} at O, such that

$$u(x) \leq \Phi\left(u(x') + B'|x|^2\right), \tag{1.4.19}$$

for $0 < |x| = |x'| \leq R_0$. Proceeding as above we still obtain (1.4.18).

Remark 1.2. It is possible to state another type of limit theorem by replacing the boundedness assumption on u by an estimate on the gradient of u : if $|\nabla u(x)| = O(1/|x|)$ near 0 then u admits a limit at O. If there exists a sequence $\{x_n\}$ tending to O with $u(x_n)$ bounded, we can see that u(x) also remains bounded for $|x| = |x_n|$. Therefore u is bounded in the punctured ball and we can apply Theorem 1.4, although a refined analysis would show that the continuity of the a_{ij} is not needed in the case N > 2, since a Harnack type inequality similar to (1.4.19) is valid under the assumption on the gradient of u. If $\limsup_{x \to O} u(x) = \infty$, then the previous argument proves that $\lim_{x \to 0} u(x) = \infty$.

Similarly to Theorem 1.5 the asymptotic behaviour of a bounded solution of Lu = 0 in an exterior domain is derived:

THEOREM 1.5. *Let* u *be a solution of* Lu = 0 *in the exterior domain* $\{x: |x| \geq R\}$ *with* c = 0 *and* $b_i(x) = O(1/|x|)$ *at infinity. If* N = 2 *we assume that* u *is bounded below and, if* N > 2, *that* u *is bounded and the coefficients* a_{ij} *admit a limit at infinity. Then* u *admits a limit at infinity.*

For N > 2, this result can be made more precise and the following counterpart of Theorem 1.3 holds

THEOREM 1.6. *Let* $N > 2$, u *be a bounded solution of* $Lu = 0$ *in an exterior domain and* L *satisfy the same assumptions as in Theorem 1.5. Then*

$$u(x) - \lim_{|y| \to \infty} u(y) = O\left(|x|^{2-N+\delta}\right), \qquad (1.4.20)$$

for every $\delta > 0$. *If we assume moreover that the coefficients* a_{ij} *are Dini continuous at infinity then*

$$u(x) - \lim_{|y| \to \infty} u(y) = O\left(|x|^{2-N}\right). \qquad (1.4.21)$$

If u admits an isolated singularity at O the most interesting problem is to obtain a description of it. The following theorems due to Gilbarg and Serrin generalise Bôcher result.

THEOREM 1.7. *Suppose* $N = 2$ *and that the equation*

$$Lu = 0 \qquad (1.4.22)$$

possesses a solution $V(x)$ *defined in the punctured ball* $B^*(O.R)$, *which tends to infinity at* O. *Then every nonnegative solution* u *of* (1.4.22) *in* $B^*(O.R)$ *can be written under the form*

$$u(x) = kV(x) + w(x), \qquad (1.4.23)$$

where k *is some nonnegative constant and* w *is a* C^1 *function defined in the whole ball* $B(O,R)$.

Proof. As in the proof of Theorem 1.4, we can suppose $c \leq 0$. Taking the constant V_0 large enough we can assume that the function $V^* = V - V_0$ is nonpositive for $|x| = R$ since we can always infer that u and V are continuous up to the boundary. Let α be the largest number such that $u - \alpha V^*$ is nonnegative in $B(O,R)$; this number exists since the set of all values β for which $u - \beta V^* \geq 0$ in $B^*(O.R)$ is closed, bounded above and non empty as O belongs to it. If we set $W = u - \alpha V^*$, then W is nonnegative in the punctured ball. For a continuous function f on $B^*(O.R)$ we define for $0 < r \leq R$

$$\overline{f}(r) = \max_{|x|=r} f(x) \quad \text{and} \quad \underline{f}(r) = \min_{|x|=r} f(x). \qquad (1.4.24)$$

We suppose now that the relation

$$\underline{W}(x) = o\left(\overline{V}(x)\right) \qquad (1.4.25)$$

near O is not true. Then there exist $\eta > 0$ and a sequence $\{r_n\}$ tending to 0 such that

$$\underline{W}(r_n) \geq \eta \overline{V}(r_n) \tag{1.4.26}$$

for n = 1, 2. From (1.4.26) we have $W(x) \geq \eta V^*(x)$ for $|x| = r_n$, while $W(x) - \eta V^*(x)$ is nonnegative for $|x| = R$ as we can take η small enough. But $L(W - \eta V^*) = (\alpha + \eta)cV_0$ is nonpositive; then, applying the maximum principle in the annuli $\{x: r_n < |x| < R\}$ and letting n go to infinity yields $W - \eta V^* \geq 0$ in $B^*(O.R)$. But this means that $u - (\alpha + \eta)V^*$ is nonnegative in the same domains, contradicting the maximality of α; therefore (1.4.25) holds. The function W is a positive solution of

$$LW = c\alpha V_0, \tag{1.4.27}$$

and $c\alpha V_0$ is bounded in absolute value by some constant B'. From Harnack inequality we deduce that the oscillation of W on any circle of radius $r \in (0, \min(1, R_0 / 2))$ is controled as we have

$$\overline{W}(r) \leq K\left(\underline{W}(r) + r^2 B'\right). \tag{1.4.28}$$

The same holds with $\underline{V}$ (without the perturbation term $r^2 B'$) since we can assume that V is positive on some small enough punctured disk $B^*(O, R_1)$. It follows that

$$u - \alpha V = o(V), \tag{1.4.29}$$

near O and $w = u - \alpha V$ remains bounded, hence continuous at O, by the same arguments as the ones of Theorems 1.3 and 1.4. From the maximum principle w is equal to the C^1 solution w^* of the equation $Lw^* = 0$ in the whole disk, which coincides with w on the boundary (the existence of w^* is derived from [GT], [Ni1]).

THEOREM 1.8. *Let* $N \geq 2$, *c be nonpositive and the coefficients of* L *Hölder continuous in* $B(O, R)$. *Then every positive solution of* (1.4.22) *takes the following form*

$$u(x) = kG(x) + w(x), \tag{1.4.30}$$

where G is the Green function for the ball $B(O, R)$ *with singularity at* O, *w is the unique solution of the Dirichlet problem in the ball with prescribed boundary data on* $\partial B(O, R)$ *and k is a nonnegative constant.*

Proof. The proof follows the techniques of Theorem 1.7 where V has to be replaced by the Green function G the existence of which is derived from the classical Schauder theory. If α is the largest number such that $u - \alpha G$ is nonnegative in $B^*(O, R)$, we deduce from an Harnack type inequality (valid from the Hölder regularity of the coefficients) that the oscillations of $w = u - \alpha G$ on any small sphere with center O are bounded. This finally implies

$$w = o(G), \tag{1.4.31}$$

near O. According to [Ge1], w is a regular solution in the whole ball.

Remark 1.3. When the coefficients of L are Hölder continuous it is known that there exists a positive singular solution of (1.4.22) which, near O, behaves like $\ln(1/|x|)$, if N = 2, or $|x|^{2-N}$ if $N > 2$; when the coefficients of L are merely continuous there may not exist any singular solutions (with or without the previous behaviour [GS]). Let us define

$$L^*u = \sum_{ik}\left(\delta_{ik} + g(r)\frac{x_i x_k}{r^2}\right)\frac{\partial^2 u}{\partial x_i \partial x_k} = 0. \tag{1.4.32}$$

Then for every g there exists a radial solution $u = u(r)$ as (1.4.32) reduces to

$$\frac{u''}{u'} = \frac{1-N}{r(1+g)}, \tag{1.4.33}$$

and u is explicitely given by

$$u(r) = \int_a^r \left[\exp\left(\int_a^s \frac{1-N}{1+g(\rho)}\frac{d\rho}{\rho}\right)\right]ds. \tag{1.4.34}$$

As for the ellipticity condition, it is satisfied provided

$$0 < \min(1+g) \le \max(1+g) < \infty. \tag{1.4.35}$$

If we take $N = 2$ and $g(r) = -2/(2+\ln r)$, then $g(0) = 0$,

$$u(r) = a + b/\ln r; \tag{1.4.36}$$

it is clear that L^* is uniformly elliptic near O, its coefficients are continuous (and regular except at O) and u is continuous but not C^1 at O; moreover, the maximun principle does not hold for such a function u.

If we take $g(r) = -2/(\ln r - 2)$, then

$$u(r) = \ln^3(1/r), \tag{1.4.37}$$

which is singular but not of the order of $\ln(1/r)$. For $N > 2$, if we take g as being defined by $g(r) = -1/(1+(N-1)\ln r)$, we find that (1.4.34) gives rise to

$$u(r) = a(1+o(1))r^{2-N}/\ln r. \tag{1.4.38}$$

The coefficients of the elliptic operator are merely continuous at O and u(r) is $o(r^{2-N})$ but not $O(r^{2-N+\delta})$ for any $\delta > 0$. If we take now $N > 2$ and $g(r) = ((N-2)\ln r - 2)/(\ln r + 2)$

and $g(0) = N - 2$, then (1.4.34) implies that $u(r) = a + b / \ln r$. It is clear that the coefficients of L^* are discontinuous at O and the maximum principle does not hold.

Remark 1.4. The previous results have been widely extended since the last ten years in particular to Fuchsian operators which are operators of type (1.4.1) defined in a domain Ω of $\mathbb{R}^N$ which either an exterior domain or a bounded domain containing the origin O and such that

$$|x|\sum_i |b_i(x)| + |x|^2 |c(x)| \le M, \tag{1.4.39}$$

for any x in Ω or $\Omega \setminus \{O\}$ according to the situation. A typical result in that direction due to Pinchover ([Pi]) is the following:

Let Ω be a bounded domain containing O, L an elliptic Fuschian type operator with Hölder continuous coefficients a_{ij} in Ω and b_i and c in $\Omega \setminus \{O\}$. Assume also that

$$\limsup_{x \to O} |x|^2 c(x) \le -m < 0. \tag{1.4.40}$$

Suppose that $(x,r) \mapsto f(x,r)$ is a continuous function in $\Omega \setminus \{O\} \times \mathbb{R}^+$ which satisfies, for some $\delta \in (0,1)$ and M and $C > 0$,

$$|x|^2 |f(x,u)| \le Cu^\delta \qquad (\forall x \in \Omega \setminus \{O\}, \ \forall u \ge M). \tag{1.4.41}$$

If u and v are two positive solutions of

$$Lu + f(x,u) = 0 \qquad and \qquad Lv = 0, \tag{1.4.42}$$

in $\Omega \setminus \{O\}$ which tends to infinity at O. Then there exists a positive constant A such that

$$\lim_{x \to O} u(x) / v(x) = A. \tag{1.4.43}$$

1-1-5 Linear equations in divergence form

Although most of the results of this Section are valid for nonlinear elliptic equations as we shall see it in the next chapter, we present here the techniques of differential inequalities applied to Dirichlet type integrals for second order elliptic equations in divergence form. Such operators are

$$Du = \sum_{ij} \frac{\partial}{\partial x_i}\left(a_{ij}\frac{\partial u}{\partial x_k}\right) + \sum_i b_i \frac{\partial u}{\partial x_i}, \tag{1.5.1}$$

where the coefficients a_{ij} and b_i are merely supposed to be bounded and measurable in the ball $B(O,R)$, and we assume also the ellipticity of D, which means

$$\sum_{ij} a_{ij}(x)\xi_i\xi_j > 0 \tag{1.5.2}$$

for any $\xi = (\xi_1,...,\xi_N)$ in $\mathbb{R}^N$ and any x in $B(O,R)$. The following result is an extended maximum principle to be compared with Theorem 1.3 and Remark 1.1.

THEOREM 1.9. *Let* u *be a non-constant function belonging to* $C^2\big(B^*(O,R)\big)$ *satisfying the inequality*

$$Du \geq 0 \tag{1.5.3}$$

in $B^*(O,R)$. *We assume that the coefficients* b_i *of* D *are identically* 0, *and that* D *is elliptic but not necessarily uniformly elliptic, and we set* $M_\rho = \max_{|x|=\rho} u(x)$ *for* $\rho \in (0,R)$. *Then if*

$$u(x) = \begin{cases} O\big(|\ln|x||^{1-\delta}\big) & \text{if } N = 2 \, , \\ O\big(r^{2-N+\delta}\big) & \text{if } N \geq 3 \, , \end{cases} \tag{1.5.4}$$

for some $\delta > 0$ *and* x *small enough, it follows that* $u \leq M_\rho$.

Proof. Let us suppose that there exists some bounded open connected component G of $B^*(O,\rho)$ where $u > M_\rho$. The function $v = u - M_\rho$ is a positive solution of (1.5.3) in G and vanishes on ∂G except may be at O. Setting $\varepsilon = \delta / N$, we apply the divergence theorem in the domain $G_\sigma = G \cap \{x:|x| > \sigma|\}$ for $\sigma \in (0,\rho)$; if $S_\sigma = \partial G_\sigma \cap \{x:|x| = \sigma\}$, we have

$$\int_{G_\sigma} \sum_{ij} \frac{\partial}{\partial x_i}\left(v^\varepsilon a_{ij}\frac{\partial v}{\partial x_j}\right) dx = \int_{S_\sigma} \sum_{ij} v^\varepsilon a_{ij}\frac{\partial v}{\partial x_j} n_i dS, \tag{1.5.5}$$

where the n_i are the direction cosines of the outward normal unit vector. Developping the derivative and using (1.5.3) yields

$$\varepsilon \int_{G_\sigma} v^{\varepsilon-1}\sum_{ij} a_{ij}\frac{\partial v}{\partial x_i}\frac{\partial v}{\partial x_j} dx \leq \int_{S_\sigma} \sum_{ij} a_{ij}\frac{\partial v}{\partial x_j} n_i dS. \tag{1.5.6}$$

If $G = G_\sigma$ for some $\sigma \in (0,\rho)$, then the boundary integral vanishes and v is zero in G, contradiction. Therefore, for any $\sigma \in (0,\rho)$, $G \neq G_\sigma$ and O lies on ∂G. From Schwarz inequality, we get

$$\int_{S_\sigma} v^\varepsilon \sum_{ij} a_{ij} \frac{\partial v}{\partial x_j} n_i \, dS \le \int_{S_\sigma} v^\varepsilon \left(\sum_{ij} a_{ij} \frac{\partial v}{\partial x_i} \frac{\partial v}{\partial x_j} \right)^{1/2} \left(\sum_{ij} a_{ij} n_i n_j \right)^{1/2} dS,$$

$$\le K \int_{S_\sigma} v^\varepsilon \left(\sum_{ij} a_{ij} \frac{\partial v}{\partial x_i} \frac{\partial v}{\partial x_j} \right)^{1/2} dS, \tag{1.5.7}$$

for some positive K. Therefore

$$\int_{S_\sigma} v^\varepsilon \left(\sum_{ij} a_{ij} \frac{\partial v}{\partial x_j} \frac{\partial v}{\partial x_i} \right)^{1/2} dS \le \left(\int_{S_\sigma} v^{\varepsilon+1} dS \int_{S_\sigma} v^{\varepsilon-1} \sum_{ij} a_{ij} \frac{\partial v}{\partial x_j} \frac{\partial v}{\partial x_i} dS \right)^{1/2}, \tag{1.5.8}$$

and finally

$$\left(\int_{G_\sigma} v^{\varepsilon-1} \sum_{ij} a_{ij} \frac{\partial v}{\partial x_j} \frac{\partial v}{\partial x_i} dx \right)^2 \le K_1 \int_{S_\sigma} v^{\varepsilon+1} dS \int_{S_\sigma} v^{\varepsilon-1} \sum_{ij} a_{ij} \frac{\partial v}{\partial x_j} \frac{\partial v}{\partial x_i} dS. \tag{1.5.9}$$

From (1.5.4) we have

$$\int_{S_\sigma} v^{\varepsilon+1} dS = O(\sigma^{1+\varepsilon}) \qquad \text{if } N > 2, \tag{1.5.10}$$

$$\int_{S_\sigma} v^{\varepsilon+1} dS = O(|\ln \sigma|^{1-\varepsilon} \sigma) \qquad \text{if } N = 2. \tag{1.5.11}$$

Therefore (1.5.9) becomes

$$\left(\int_{G_\sigma} v^{\varepsilon-1} \sum_{ij} a_{ij} \frac{\partial v}{\partial x_j} \frac{\partial v}{\partial x_i} dx \right)^2 \le K_2 \Theta(\sigma) \int_{S_\sigma} v^{\varepsilon-1} \sum_{ij} a_{ij} \frac{\partial v}{\partial x_j} \frac{\partial v}{\partial x_i} dS, \tag{1.5.12}$$

where $\Theta(\sigma) = \sigma^{1+\varepsilon}$ if $N > 2$ and $\Theta(\sigma) = \sigma|\ln \sigma|^{1-\varepsilon}$ if $N = 2$. Denote

$$Q(\sigma) = \int_{G_\sigma} v^{\varepsilon-1} \sum_{ij} a_{ij} \frac{\partial v}{\partial x_j} \frac{\partial v}{\partial x_i} dx = \int_\sigma^{\sigma_0} d\tau \int_{S_\tau} v^{\varepsilon-1} \sum_{ij} a_{ij} \frac{\partial v}{\partial x_i} \frac{\partial v}{\partial x_j} dS, \tag{1.5.13}$$

then Q satisfies a differential inequality, namely

$$\frac{dQ}{d\sigma} = -\int_{S_\sigma} v^{\varepsilon-1} \sum_{ij} a_{ij} \frac{\partial v}{\partial x_j} \frac{\partial v}{\partial x_i} dS \le -\frac{Q^2(\sigma)}{K_2 \Theta(\sigma)}. \tag{1.5.14}$$

By integration we get

$$Q^{-1}(\sigma_0) - Q^{-1}(\sigma) \geq K_3(\sigma^{-\varepsilon} - \sigma_0^{-\varepsilon}),\tag{1.5.15}$$

for $0 < \sigma < \sigma_0 \leq \rho$, if $N > 2$, and

$$Q^{-1}(\sigma_0) - Q^{-1}(\sigma) \geq K_3(|\ln \sigma|^{\varepsilon} - |\ln \sigma_0|^{\varepsilon}),\tag{1.5.16}$$

if $N = 2$. If we let now σ go to 0 the right-and side of (1.5.16) becomes infinite as the left hand side remains finite and bounded by $Q^{-1}(\sigma_0)$. Therefore $u \leq M_\rho$ in $B^*(O,\rho)$.

Remark 1.5. The previous theorem remains true if we just assume that u is C^1 in $B^*(O,R)$. However in assuming the C^2 regularity on u and that the coefficients of D are C^1 in $B^*(O,R)$, we can infer a more precise result, namely that $u < M_\rho$ in $B^*(O,\rho)$ and $\limsup_{x \to O} u(x) < M_\rho$.

Remark 1.6. When $N = 2$ the previous method gives an extended maximum principle for solutions of (1.5.3) in an exterior domain of the plane: if u is C^2 in $\{x:|x| > R\}$ and satisfies

$$u(x) = O(|\ln|x||^{1-\delta}),\tag{1.5.17}$$

for some $\delta > 0$, then $u \leq M_\rho$ for any $\rho > R$

The analogues of Theorems 1.4 and 1.7 when $N = 2$ are proved in a very similar way

THEOREM 1.10. *Assume* $N = 2$ *and* u *is a nonnegative solution of the equation*

$$Du = 0\tag{1.5.18}$$

in the punctured ball $B^*(O,R)$. *Then* u(x) *tends to a limit (possibly infinite) when* x *goes to* O *and the same conclusion at infinity holds if the ball is replaced by an exterior domain .*

THEOREM 1.11. *Assume* $N = 2$, *and* (1.5.18) *possesses a solution* V(x) *in the punctured ball* $B^*(O,R)$ *such that* V(x) *tends to infinity when* x *tends to* O; *then every positive solution of* (1.5.18) *in* $B^*(O,R)$ *takes the following form*

$$u = kV + w,\tag{1.5.19}$$

where k *is nonnegative and* w *is a solution in* $B^*(O,R)$ *which is Hölder continuous at* O.

In the 2-dimensional case L. Nirenberg [Nr1] proved the following result in assuming uniform ellipticity.

THEOREM 1.12. *Assume* $N > 2$, D *is uniformly elliptic with* $b_i = 0$ *and the* a_{ij} *are continuous at* O. *Then any bounded solution of* (1.5.18 *in* $B^*(O,R)$ *admits a limit at* O; *in fact the limit is achieved Hölder continuously. If* u *is a bounded solution of the same uniformly elliptic equation in an exterior domain and if the coefficients* a_{ij} *converges to some constant limit at infinity, then there exist* u_0 *and* $\alpha > N/2 - 1$ *such that*

$$|u(x) - u_0| = O(|x|^{\alpha}) \tag{1.5.20}$$

at infinity.

1-2 Quasilinear equations in divergence form

1-2-1 General framework for quasilinear equations

This section is devoted to the study of Serrin's celebrated results concerning the singularity problem for quasilinear equations in divergence form. We are given a N-dimensional Euclidean space $\mathbb{R}^N$, a connected open subset Ω of $\mathbb{R}^N$, a vector valued function A and a scalar function B such that

$$A: \Omega \times \mathbb{R} \times \mathbb{R}^N \mapsto \mathbb{R}^N \quad \text{and} \quad B: \Omega \times \mathbb{R} \times \mathbb{R}^N \mapsto \mathbb{R}, \tag{2.1.1}$$

both depending on the variables (x, r, q). We assume that A and B are measurable in x and continuous in r and q. The key-hypothesis on the growth of A and B is

$$\begin{cases} |A(x,r,q)| \le a|q|^{p-1} + b|r|^{p-1} + e, \\ |B(x,r,q)| \le c|q|^{p-1} + d|r|^{p-1} + f, \\ q.A(x,r,q) \ge |q|^p - d|u|^p - g, \end{cases} \tag{2.1.2}$$

for some $p > 1$, where a is a positive constant and the b, ..., g are measurable functions of x defined in Ω and belonging to some Lebesgue class, and more precisely

$$b, e \in L^{N/(p-1)}(\Omega), \quad c \in L^{N/(1-\varepsilon)}(\Omega), \quad d, f, g \in L^{N/(p-\varepsilon)}(\Omega) \quad \text{if } 1 < p < N, \tag{2.1.3}$$

$$b, e \in L^{N/(N-1-\varepsilon)}(\Omega), \quad c \in L^{N/(1-\varepsilon)}(\Omega), \quad d, f, g \in L^{N/(N-\varepsilon)}(\Omega) \quad \text{if } p = N, \tag{2.1.4}$$

for some $\varepsilon > 0$, and

$$b, e \in L^{p/(p-1)}(\Omega), \quad c \in L^p(\Omega), \quad d, f, g \in L^1(\Omega) \quad \text{if } p > N. \tag{2.1.5}$$

The equation which is under the consideration is

$$\operatorname{div}(A(x,u,\nabla u)) = B(x,u,\nabla u), \tag{2.1.6}$$

in weak sense in Ω, which means that u belongs to the Sobolev space $W^{1,p}_{loc}(\Omega)$ and that

$$\int_\Omega A(x,u,\nabla u).\nabla\varphi \; dx + \int_\Omega B(x,u,\nabla u)\varphi \; dx = 0, \tag{2.1.7}$$

holds for any function φ belonging to the space $C_0^\infty(\Omega)$. By using Moser's iterative methods as in [Se1] or [Tr] the following *a priori* estimates are derived:

THEOREM 1.13. *Assume* $1 < p < N$, *the structural conditions* (2.1.2) *and* (2.1.3) *are fulfilled and* u *is weak solution of* (2.1.6) *in some ball* $B_{2R} \subset \Omega$, *then*

$$\|u\|_{L^\infty(B_R)} \le CR^{-N/p}\left(\|u\|_{L^p(B_{2R})} + kR^{N/p}\right), \tag{2.1.8}$$

$$\|\nabla u\|_{L^1(B_R)} \le CR^{-1}\left(\|u\|_{L^p(B_{2R})} + kR^{N/p}\right), \tag{2.1.9}$$

where C *and* k *are constants depending only on* (2.1.2) *and* (2.1.3); *in particular*

$$C = C\left(p, N, \varepsilon, a, \|b\|_{L^{N/(p-1)}}, R^\varepsilon\|c\|_{L^{N/(1-\varepsilon)}}, R^\varepsilon\|d\|_{L^{N/(p-\varepsilon)}}\right), \tag{2.1.10}$$

$$k = \left(\|e\|_{L^{N/(p-1)}} + R^\varepsilon\|f\|_{L^{N/(p-\varepsilon)}}\right)^{1/(p-1)} + \left(R^\varepsilon\|g\|_{L^{N/(p-\varepsilon)}}\right)^{1/p}. \tag{2.1.11}$$

In the case $p = N$ this result is the following:

THEOREM 1.14. *Assume* $p = N$, *the structural conditions* (2.1.2) *and* (2.1.4) *hold and* u *is a weak solution of* (2.1.6) *in some ball* $B_{2R} \subset \Omega$, *then*

$$\|u\|_{L^\infty(B_R)} \le CR^{-1}\left(\|u\|_{L^N(B_{2R})} + kR\right), \tag{2.1.12}$$

$$\|\nabla u\|_{L^N(B_R)} \le CR^{-1}\left(\|u\|_{L^N(B_{2R})} + kR\right), \tag{2.1.13}$$

where C *and* k *are nonnegative constants depending only on* (2.1.2) *and* (2.1.4); *in particular*

$$C = C\left(N, \varepsilon, a, \|b\|_{L^{N/(\alpha-1-\varepsilon)}}, R^\varepsilon\|c\|_{L^{N/(1-\varepsilon)}}, R^\varepsilon\|d\|_{L^{N/(N-\varepsilon)}}\right), \tag{2.1.14}$$

$$k = \left(\|e\|_{L^{N/(N-1-\varepsilon)}} + R^\varepsilon\|f\|_{L^{N/(N-\varepsilon)}}\right)^{1/(N-1)} + \left(R^\varepsilon\|g\|_{L^{N/(N-\varepsilon)}}\right)^{1/N}. \tag{2.1.15}$$

Remark 1.7. The previous results hold if u is a weak subsolution of (2.1.6), that is

$$\int_{B_{2R}} A(x,u,\nabla u).\nabla\varphi\ dx + \int_{B_{2R}} B(x,u,\nabla u)\varphi\ dx \le 0, \tag{2.1.16}$$

for any $\varphi \in C_0^\infty(B_{2R})$, $\varphi \ge 0$. Moreover they also hold if (2.1.6) is replaced by

$$\mathrm{div}\big(A(x,u,\nabla u)\big) \ge B(x,u,\nabla u) + C(x,u), \tag{2.1.17}$$

where $rC(x,r) \ge 0$, $\forall (x,r) \in B_{2R} \times \mathbb{R}$.

As a consequence of Theorems 1.13 and 1.134 a maximum principle valid for weak subsolutions is derived.

THEOREM 1.15. *Assume* $p < N$ *and* u *is a weak subsolution of* (2.1.6) *in a domain* D $\subset \Omega$. *Suppose that for any* $\gamma > 0$, *there exists a neighborhood of the boundary of* D *where* $u \le M + \gamma$, *almost everywhere. Assume also that the conditions* (2.1.2), (2.1.3), (2.1.4) *hold. Then*

$$\max u \le M + C'\,|M| + C\big(|D|^{-1/p}\|\tilde{u}\|_{L^p(D)} + k\big) + M, \tag{2.1.18}$$

where

$$\tilde{u} = \max(0, u - M) \quad \text{and} \quad C = C\big(p, N, \varepsilon, |D|^{\varepsilon/N}\|c\|_{L^{N/(1-\varepsilon)}}, |D|^{\varepsilon/N}\|d\|_{L^{N/(p-\varepsilon)}}\big), \tag{2.1.19}$$

and

$$k = \big(|D|^{\varepsilon/N}\|f\|_{L^{N/(p-\varepsilon)}}\big)^{1/(p-1)} + \big(|D|^{\varepsilon/N}\|g\|_{L^{N/(p-\varepsilon)}}\big)^{1/N}, \tag{2.1.20}$$

while C', *which depends on the structural constants, tends to zero with* $\|d\|_{L^{N/(N-\varepsilon)}}$.

A key step for the regularity or the singularity study of quasilinear equations is the estimate of the local oscillation of the solutions and this is obtained via Harnack inequality as in the linear case; Serrin's classical result asserts that the oscillation can be controled since the following is proved:

THEOREM 1.16. *Let* u *be a nonnegative solution of* (2.1.6) *in some open ball* $B_{3R} \subset \Omega$, $p \le N$ *and the conditions* (2.1.2)-(2.1.4) *hold. Then*

$$\max_{x \in B_R} u(x) \le C\Big(\min_{x \in B_R} u(x) + k\Big), \tag{2.1.21}$$

where C *and* k *are as in Theorems* 1.13-1.14.

In the case $p > N$ the situation is somewhat different since by Morrey's imbedding Theorem any $W^{1,p}$ function is Hölder continuous. However the following form of Harnack inequality is also proved by Serrin,

24

THEOREM 1.17. *Let* $p > N$ *and* u *be a nonnegative solution of* (2.1.6) *in some open ball* $B_{3R} \subset \Omega$; *assume also that the structural conditions* (2.1.2) *and* (2.1.5) *are satisfied; then for any* (x,y) *in* B_R *there holds*

$$u(x) \le \left(u(y) + k\right)\exp\left(C\left(|x - y| / R\right)^{1 - N/p}\right), \tag{2.1.22}$$

where

$$C = C\left(p, N, a, R^{(p-N)/p'}\|b\|_{L^{p/(p-1)}}, R^{(p-N)/p}\|c\|_{L^p}, R^{p-N}\|d\|_{L^1}\right), \tag{2.1.23}$$

and

$$k = \left(R^{(p-N)/p'}\|e\|_{L^1} + R^{p-N}\|f\|_{L^1}\right)^{1/(p-1)} + \left(R^{p-N}\|g\|_{L^1}\right)^{1/p}. \tag{2.1.24}$$

In some cases, the equations do not fall into the scope of assumptions (2.1.3) or (2.1.4); it is in particular the case if we replace (2.1.2) by

$$\begin{cases} |A(x,r,q)| \le a|q|^{p-1} + b|r|^{(p-1)/\theta} + e, \\ |B(x,r,q)| \le c|q|^{p-\theta} + d|r|^{(p-\theta)/\theta} + f, \\ q.A(x,r,q) \ge |q|^p - a(|u|^{p/\theta} + g), \end{cases} \tag{2.1.25}$$

where $\theta = (N - p + p\varepsilon) / N$ (we assume $p < N$). For a solution u we can write

$$b(x) = a|u|^{(p-1)(1-\theta)/\theta}, \ c(x) = a|\nabla u|^{1-\theta}, \ d(x) = a|u|^{p(1-\theta)/\theta}, \tag{2.1.26}$$

and, assuming that u is a locally $W^{1,p}$-function, we obtain similar results as those contained in Theorems 1.12 or 1.14, except that the coefficients C and k take into account the $W^{1,p}$-norm of u. Similar remark is valid when $p \ge N$.

1-2-2 Removable singularities

If we just consider the p-Laplacian operator in the N-dimensional space, with $p > N$; then it is straightforward to check that the function $\omega : x \mapsto \omega(x) = |x|^{(p-N)/(p-1)}$ satisfies

$$\int \nabla \varphi . \nabla \omega |\nabla \omega|^{p-2} dx = \left(\frac{p - N}{p - 1}\right)^{p-1} |S^{N-1}|\varphi(0). \tag{2.2.1}$$

Therefore there exist p-harmonic bounded singular functions in the case $p > N$. Consequently we shall restrict our attention to the case $1 < p \le N$. The removability results are expressed in terms of capacity by the following result due to Serrin [Se1]. Moreover it is important to notice that the assumption of continuity on the weak solutions is not restrictive since any such solution is almost everywhere equal to some Hölder continuous function.

THEOREM 1.18. *Let* K *be a compact subset of a domain* D *included in* Ω *such that* $c_{1,s}(K) = 0$ *for some* $1 < p \le s \le N$. *Assume that* (2.1.2)-(2.1.5) *are verified and that* u *is a continuous weak solution of* (2.1.6) *in* $D \setminus K$ *satisfying*

$$u \in L^{\theta(1+\delta)}(D), \quad \theta = s(p-1)/(s-p). \tag{2.2.2}$$

Then u *can be defined on the set* K *so that the resulting function* u *is a continuous weak solution of* (2.1.6) *on whole* D. *As a special case, if the* $c_{1,p}$- *capacity of* K *is zero, then* K *is a removable singularity for any bounded solution of* (2.1.6).

Proof. We shall prove that u can be continuously extended in the neighborhood of any point of D. Let P be a point in D and $R > 0$ such that $B_{2R} = B(P, 2R)$ is included in D. By scaling, we can assume that $R = 1$, and we set

$$k = 1 + \left(\|e\|_{L^{N/(p-1)}} + \|f\|_{L^{N/(p-\varepsilon)}} \right)^{1/(p-1)} + \left(\|g\|_{L^{N/(p-\varepsilon)}} \right)^{1/p}, \tag{2.2.3}$$

for $p < N$, with the modification (2.1.4) on the norm of e when $p = N$. Let $\tilde{u}$ be $|u| + k$ and η, $\tilde{\eta}$ two nonnegative C^∞ functions η with compact support in B_2 and $\tilde{\eta}$ vanishing in a neighborhood of K. As a test function for (2.1.6), we take

$$\varphi = (\eta\tilde{\eta})^p \operatorname{sign} u \left(\tilde{u}^\beta - k^\beta \right) \quad \text{for some } \beta > 0. \tag{2.2.4}$$

Since u is locally a $W^{1,p}$ function and $\tilde{\eta}$ vanishes in a neighborhood of K, such a φ is admissible. A lengthy calculation yields the following estimates (in the case $p < N$, the case $p = N$ requiring only minor changes is left to the reader).

$$\left\| \eta\tilde{\eta}\nabla v \right\|_{L^p} \le Cq^{p/\varepsilon} \left(1 + \beta^{-1} \right)^{1/\varepsilon} \left(\left\| \eta\tilde{\eta}v \right\|_{L^p} + \left\| v\nabla(\eta\tilde{\eta}) \right\|_{L^p} \right), \tag{2.2.5}$$

$$\left\| \eta\tilde{\eta}v \right\|_{L^{p*}} \le Cq^{p/\varepsilon} \left(1 + \beta^{-1} \right)^{1/\varepsilon} \left(\left\| \eta\tilde{\eta}v \right\|_{L^p} + \left\| v\nabla(\eta\tilde{\eta}) \right\|_{L^p} \right), \tag{2.2.6}$$

where $p* = Np / (N-p)$, $q = (p+\beta-1)/p$ and $v = \tilde{u}^q$. We now consider the term $\left\| v\nabla\tilde{\eta} \right\|_{L^p}$ which appears in (2.2.5) and (2.2.6) and get, for $s > p$,

$$\left\| v\nabla\tilde{\eta} \right\|_{L^p}^p = \int |\nabla\tilde{\eta}|^p \tilde{u}^{pq}dx \le \left\| \nabla\tilde{\eta} \right\|_{L^s}^p \left(\int \tilde{u}^{pqs/(s-p)}dx \right)^{1-p/s}. \tag{2.2.7}$$

Choosing $\beta = \beta_0 = \delta(p-1)$, we have $pqs / (s-p) = (p+\beta-1)s / (s-p) = \theta(1+\delta)$. Using (2.2.2) gives us

$$\left\|v\nabla\tilde{\eta}\right\|_{L^p} \leq \text{Const.}\left\|\nabla\tilde{\eta}\right\|_{L^s}, \tag{2.2.8}$$

with $\theta = \infty$ if If $s = p$. Therefore u is bounded and (2.2.8) also holds. We now replace $\tilde{\eta}$ by a sequence $\{\tilde{\eta}_n\}$ such that $\tilde{\eta}_n = 0$ in some neighborhood of K, $\tilde{\eta}_n$ tending to 1 a. e. in D and $\lim\limits_{n\to\infty}\left\|\nabla\tilde{\eta}_n\right\|_{L^s} = 0$. The followings inequalities are derived from (2.2.5) and (2.2.6)

$$\left\|\eta\nabla v\right\|_{L^p} \leq Cq^{p/\varepsilon}\left(\left\|\eta v\right\|_{L^p} + \left\|v\nabla\eta\right\|_{L^p}\right), \tag{2.2.9}$$

$$\left\|\eta v\right\|_{L^{p*}} \leq Cq^{p/\varepsilon}\left(\left\|\eta v\right\|_{L^p} + \left\|v\nabla\eta\right\|_{L^p}\right), \tag{2.2.10}$$

(we recall that K has zero measure, and there is no change in taking the norms over B_2 instead of $B_2 \setminus K$)) . The method is now to derive a uniform estimate by Moser's iterative scheme: we set

$$\Phi(r,h) = \left(\int\limits_{B_h} \tilde{u}^r dx\right)^{1/r}, \tag{2.2.11}$$

and get for $0 < h' < h \leq 2$

$$\Phi(\kappa r_0, h') = \left(C(r_0 / p)^{p/\varepsilon}(h - h')^{-1}\right)^{p/r_0} \Phi(r_0, h), \tag{2.2.12}$$

with $p*/p = N/(N-p)$, $r_0 = p + \beta_0 - 1 = (p-1)(1+\delta)$. This is obtained by taking a function η with value 1 on $B_{h'}$, compactly supported in B_h, $0 \leq \eta \leq 1$ and $\left\|\nabla\eta\right\|_{L^\infty} = 2/(h - h')$. The main point is to obtain a series of inequalities of type (2.2.12) with r_0 replaced by r, for r arbitrary and larger than r_0. We define the following test function

$$F(\tilde{u}) = \begin{cases} \tilde{u}^q & \text{if } k \leq \tilde{u} \leq 1, \\ q_0^{-1}\left(ql^{q-q_0}\tilde{u}^{q_0} + (q - q_0)l^q\right) & \text{if } 1 \leq \tilde{u}, \end{cases} \tag{2.2.13}$$

for $q \geq q_0 = r_0 / p$ and $l > k$ and

$$G(u) = \text{sign}\, u\left(F(\tilde{u})(F'(\tilde{u}))^{p-1} - q^{p-1}k^\beta\right), \tag{2.2.14}$$

where $pq = p + \beta - 1$. F is C^1 and G is piecewise C^1. Moreover

$$F(\tilde{u}) \leq (q / q_0)l^{q-q_0}\tilde{u}^{q_0}, \qquad \tilde{u}F'(\tilde{u}) \leq qF(\tilde{u}), \tag{2.2.15}$$

$$G'(u) \geq \begin{cases} q^{-1}\beta\left(F'(\tilde{u})\right)^p & \text{if } |u| < 1-k, \\ q_0^{-1}\beta_0\left(F'(\tilde{u})\right)^p & \text{if } |u| > 1-k. \end{cases} \tag{2.2.16}$$

We take now $\varphi(x) = (\eta\tilde{\eta})^p G(u)$ and plug it into (2.1.7); since $|G| \leq F(F')^{p-1}$ it gives

$$\left\| v\eta\tilde{\eta} \right\|_{L^{p*}} \leq Cq^{p/\varepsilon}\left(\left\| v\eta\tilde{\eta} \right\|_{L^p} + \left\| v\nabla(\eta\tilde{\eta}) \right\|_{L^p} \right) \tag{2.2.17}$$

as above, with $v = F(\tilde{u})$. Since $v \leq \text{Const.} \tilde{u}^{q_0}$, we also get , as in the first part of the proof, that $\left\| v\nabla\tilde{\eta} \right\|_{L^p} \leq \text{Const.} \left\| \nabla\tilde{\eta} \right\|_{L^p}$. Replacing η by η_n and letting n go to infinity, implies

$$\left\| v\eta \right\|_{L^{p*}} \leq Cq^{p/\varepsilon}\left(\left\| v\eta \right\|_{L^p} + \left\| v\nabla\eta \right\|_{L^p} \right). \tag{2.2.18}$$

We let l tend to infinity, use Lebesgue theorem and obtain

$$\left\| \tilde{u}^{q_0}\eta \right\|_{L^{p*}} \leq Cq^{p/\varepsilon}\left(\left\| \tilde{u}^{q_0}\eta \right\|_{L^p} + \left\| \tilde{u}^{q_0}\nabla\eta \right\|_{L^p} \right), \tag{2.2.19}$$

which in turn implies

$$\Phi(\kappa r, h') = \left(C(r/p)^{p/\varepsilon}(h-h')^{-1} \right)^{p/r_0} \Phi(r,h), \tag{2.2.20}$$

with $r = pq \geq r_0$ and $0 < h' < h \leq 2$. Iterating now these inequalities, with initial $r = r_0$, we deduce

$$\left\| \tilde{u} \right\|_{L^\infty(B_1)} \leq C\left\| \tilde{u} \right\|_{L^{r_0}(B_2)} \leq C\left\| \tilde{u} \right\|_{L^{\theta(1+\delta)}(B_2)}, \tag{2.2.21}$$

which is finite from hypothesis. Therefore u is essentially uniformly bounded on B_1. It remains to prove that u can be continuously extended to whole B_1 and that the resulting function is a solution of (2.1.6) in B_1. Choosing η to be 1 in B_1 and using (2.2.9) yields

$$\left\| \nabla v \right\|_{L^p(B_1)} = C_1\left(\int_{B_1} \tilde{u}^{\beta_0-1}|\nabla u|^p \, dx \right)^{1/p} \leq C_2\left\| v \right\|_{L^p(B_2)} \leq C_3\left(\int_{B_2} \tilde{u}^{pq_0} \, dx \right)^{1/p} \leq C_4. \tag{2.2.22}$$

Consequently u is locally a $W^{1,p}$ function. We take now a smooth test φ function with compact support in B_1 and get

$$\int_{B_1} u\left(\varphi\nabla\eta_n + \eta_n\nabla\varphi \right) dx = -\int_{B_1} \varphi\eta_n\nabla u \, dx. \tag{2.2.23}$$

If we let n go to infinity and use Lebesgue's theorem we get

$$\int_{B_1} u\nabla\varphi \, dx = -\int_{B_1} \varphi\nabla u \, dx, \tag{2.2.24}$$

which means that u is strongly differentiable in B_1. From the above estimates and assumptions (2.1.2)-(2.1.4) it is clear that $A(x,u,\nabla u) \in L^{p/(p-1)}(B_1)$ and $B(x,u,\nabla u) \in L^{p*/(p*-1)}(B_1)$. Taking the same test function $\varphi\eta_n$ and letting n tend to infinity we deduce

$$\int_{B_1} A(x,u,\nabla u).\nabla\varphi \, dx + \int_{B_1} B(x,u,\nabla u)\varphi \, dx = 0. \tag{2.2.25}$$

From the regularity theory of quasilinear equations [Se1] we know that u satifies a Hölder continuity criterion in $B_1 \setminus K$; therefore it satisfies the Cauchy criterion in $B_1 \setminus K$. Since $B_1 \setminus K$ is dense in B_1, the function u can be continuously extended to B_1 and the resulting function, still quoted u, is Hölder continous in B_1 and satisfies (2.2.25), which completes the proof.

Remark 1.8. The exponent $\theta = s(p-1)/(s-p)$ is optimal, at least when s is an integer, as the following example shows it. We consider the p-Laplace operator and take for K the hyperplane $\{x: x_1 = x_2 = ... = x_s = 0\}$. Clearly $c_{1,s}(K) = 0$. Let $\rho(x)$ be the distance to K and $u(x)$ be defined by

$$u(x) = \begin{cases} \rho^{(p-s)/(p-1)} & \text{if } p > s, \\ \ln(1/\rho) & \text{if } p = s. \end{cases} \tag{2.2.26}$$

Then we have

$$u \in \begin{cases} L^{\theta-\varepsilon}(B_1) & \text{for any } \varepsilon > 0, \quad \text{if } p < s, \\ L^s(B_1) & \text{for any } \varepsilon < \infty, \quad \text{if } p = s. \end{cases} \tag{2.2.27}$$

Thus, with a smaller exponent of Lebesgue space θ, such a solution has a non removable singularity on K.

If we want to apply theorem 1.18 in the framework of the Hausdorff dimension, we obtain

COROLLARY 1.1. *Let K be a compact subset of Ω of Hausdorff dimension m and D a subdomain of Ω. Suppose m < N-p and that u is a continuous solution of (2.1.6) in $D \setminus K$ such that, for some $\delta > 0$, one has*

$$u \in L^{\theta(1+\delta)}, \quad \theta = (N-m)(p-1)/(N-m-p). \tag{2.2.28}$$

Then u *can be extended by continuity to* K *so that the resulting function is a continuous solution of* (2.1.6) *in whole* D.

Proof. If $H_\alpha(K)$ is the α-Hausdorff measure of K , then we have seen in Section 1.3 that $H_{m+\gamma}(K) = 0$ for any $\gamma > 0$. This implies that $c_{1,s}(K) = 0$ with s = N-m-γ. From (2.2.28) one has

$$u \in L^\kappa(D), \quad \kappa = s(p-1)(1+\delta')/(s-p), \tag{2.2.29}$$

for some $\delta' > 0$, provided γ be small enough. Therefore K is a removable singularity.

A slightly different version of theorem 1.18 is obtained by replacing the integrability assumption on u by a growth assumption near the singularity. In that case, the singular set has to be replaced by a C^1 compact submanifold of $\mathbb{R}^N$, say K, with the property that there exists a C^1 diffeomorphism $x \mapsto y$ from $\mathbb{R}^N$ into itself such that the image of the manifold lies in the hyperplane $\{x: x_1 = x_2 = ... = x_s = 0\}$.

THEOREM 1.19. *Let* K *be a* C^1 *submanifold of* $\mathbb{R}^N$ *of codimension* s *where* $p \leq s \leq N$ *and let* D *be a domain included in* Ω. *Suppose that* u *is a continuous solution of* (2.1.6) *in* $D \setminus K$, *and that, for some* $\delta > 0$, *one has*

$$u(x) = \begin{cases} O\left(\xi^{(p-s)/(p-1)+\delta}\right) & \text{if} \quad s > p, \\ O\left(\left|\ln(1/\xi)\right|^{1-\delta}\right) & \text{if} \quad s = p, \end{cases} \tag{2.2.30}$$

where $\xi = \xi(x)$ *is the distance from* x *to* K. *Then* u *can be extended by continuity on* K *so that the resulting function is a solution of* (2.1.6) *in whole* D.

Proof. The case s > p is a consequence of Corollary 1.1. Therefore we shall assume s = p. Moreover there exists a change of variables leaving the structural hypotheses (2.1.2)-(2.1.4) invariant such that K is included into the set

$$H = \left\{x: x_1 = x_2 = ... = x_s = 0, \; |x| \leq 3\right\} \tag{2.2.31}$$

and D, in turn, contains the ball B_2. We may even suppose that K is H. We now proceed as in Theorem 1.18. As a test function, we take

$$\varphi = (\eta\tilde{\eta})^p \text{sign}(u)\left(\tilde{u}^\beta - k^\beta\right) \quad \text{for some} \; \beta > 0, \tag{2.2.32}$$

and get

$$q^{-1}\beta\|\eta\tilde{\eta}\nabla v\|^p_{L^p} \le pa\int |v\nabla\eta|\,|\eta\tilde{\eta}\nabla v|^{p-1}dx + pq^{p-1}\int \overline{b}|v\nabla\eta|(\eta\tilde{\eta}v)^{p-1}dx$$

$$+ \int c\eta v|\eta\tilde{\eta}\nabla v|^{p-1}dx + (1+\beta)q^{p-1}\int d(\eta\tilde{\eta}v)^p\,dx \qquad (2.2.33)$$

$$+ p\int |v\nabla\tilde{\eta}|\big(a|\eta\nabla v|^{p-1} + q^{p-1}\overline{b}(\eta v)^{p-1}\big)dx,$$

where $\eta, \tilde{\eta}$ are as in Theorem 1.18 except that they are supposed to take value in $[0,1]$, $v = \tilde{u}^q$, $pq = p + \beta - 1$, k is as in (2.2.23) and $\overline{b} = b + k^{1-p}e$. Let ρ be the distance to K in the new variables; ρ behaves like ξ, and for $\sigma > 0$, we take $\tilde{\eta}$ as

$$\tilde{\eta}(x) = \begin{cases} 0 & \text{if} \quad \rho \le \sigma, \\ 1 & \text{if} \quad \rho \ge \sigma + h, \end{cases} \qquad (2.2.34)$$

and linear if $\sigma \le \rho \le \sigma + h$. If we let h go to 0 keeping σ fixed, we have

$$\lim_{h \to 0}\int |v\nabla\tilde{\eta}|\big(a|\eta\nabla v|^{p-1} + q^{p-1}\overline{b}|\eta v|^{p-1}\big)dx = \oint_{\rho=\sigma} v\big(a|\eta\nabla v|^{p-1} + q^{p-1}\overline{b}|\eta v|^{p-1}\big)ds \quad (2.2.35)$$

for almost all σ, since $|\nabla\tilde{\eta}| = h^{-1}$. Therefore, (2.2.33) becomes

$$q^{-1}\beta\|\eta\nabla v\|^p_{L^p} \le pa\int |v\nabla\eta|\,|\eta\nabla v|^{p-1}dx + pq^{p-1}\int \overline{b}|v\nabla\eta|(\eta v)^{p-1}dx$$

$$+ \int c\eta v|\eta\nabla v|^{p-1}dx + (1+\beta)q^{p-1}\int d(\eta v)^p\,dx \qquad (2.2.36)$$

$$+ p\oint_{\rho=\sigma} |v\nabla\eta|\big(a|\eta\nabla v|^{p-1} + q^{p-1}\overline{b}(\eta v)^{p-1}\big)ds,$$

for almost all value of σ, all the volume integrals being taken on the set where $\rho > \sigma$. If we assume that $p < N$ for simplicity, we deduce that

$$\|\eta\nabla v\|_{L^p} \le C\big(1+\beta^{-1}\big)^{1/\varepsilon} q^{p/\varepsilon}\big(\|\eta v\|_{L^p} + \|v\nabla\eta\|_{L^p}\big) + C\beta^{-1/p}qI(\sigma), \qquad (2.2.37)$$

where

$$I(\sigma) = \oint_{\rho=\sigma} v\big(|\eta\nabla v|^{p-1} + \overline{b}|\eta v|^{p-1}\big)ds . \qquad (2.2.38)$$

From Hölder inequality we obtain

$$I(\sigma) \le \left(2\oint_{\rho=\sigma} v^p ds\right)^{1/p}\left(\oint_{\rho=\sigma}\big(|\eta\nabla v|^p + \overline{b}^{p/(p-1)}|\eta v|^p\big)ds\right)^{(p-1)/p}, \qquad (2.2.39)$$

and, from the hypothesis,

$$\oint_{\rho=\sigma} v^p ds = \oint_{\rho=\sigma} \tilde{u}^{pq} ds \le \tilde{C}\sigma^{s-1}(\ln\ \sigma)^{(p+\beta-1)/(1-\delta)}, \tag{2.2.40}$$

for some positive constant $\tilde{C}$. Since $s = p$ we take $\beta = \beta_0 = \delta(p-1), \quad \varepsilon' = \delta^2$ and get

$$\oint_{\rho=\sigma} v^p ds \le \tilde{C}\left(\sigma\left|\ln\ \sigma\right|^{1-\varepsilon'}\right)^{p-1}. \tag{2.2.41}$$

Let us define J and Ψ by

$$J = J(\sigma) = \int_{\rho>\sigma} \left|\eta\nabla v\right|^p dx\ ,\quad \Psi = \Psi(x) = b^{p/(p-1)}\left|\eta v\right|^p, \tag{2.2.42}$$

then $J'(\sigma) = -\oint_{\rho=\sigma}\left|\eta\nabla v\right|^p ds$. As the two terms in (2.2.37) which do not involve the derivative of v are bounded when σ tends to 0, we can set

$$K = C^p q^{p^2/\varepsilon}\left(\left\|\eta v\right\|_{L^p} + \left\|v\nabla\eta\right\|_{L^p}\right)^p, \tag{2.2.43}$$

where the norm is taken on $\rho > 0$ and get

$$(J(\sigma) - K)^{p/(p-1)} \le \overline{C}\sigma\left|\ln\ \sigma\right|^{1-\varepsilon}\left(-J' + \oint_{\rho=\sigma}\Psi ds\right). \tag{2.2.44}$$

The main point is to prove that $J(\sigma) \le K$ for all $\sigma > 0$. Let us assume that this is not true and that there exists some $\sigma_0 > 0$ such that $J(\sigma_0) > K$. Then $J(\sigma) > K$ on $\left(0, \sigma_0\right]$. If we set $H(\sigma) = J(\sigma) - K$ in (2.2.44), we obtain

$$\frac{1}{\sigma\left|\ln\ \sigma\right|^{1-\varepsilon'}} \le -\frac{\overline{C}H'}{H(\sigma)^{p/(p-1)}} + \frac{\overline{C}}{H(\sigma_0)^{p/(p-1)}}\oint_{\rho=\sigma}\Psi ds \tag{2.2.45}$$

almost everywhere on $\left(0, \sigma_0\right]$, which implies

$$\frac{1}{\varepsilon'}\left(\left|\ln\ \sigma\right|^{\varepsilon'} - \left|\ln\ \sigma_0\right|^{\varepsilon'}\right) \le (p-1)\left(\frac{1}{H(\sigma_0)^{1/(p-1)}} - \frac{1}{H(\sigma)^{1/(p-1)}}\right)$$
$$+ \frac{\overline{C}}{H(\sigma_0)^{p/(p-1)}}\int_{\sigma<\rho<\sigma_0}\Psi dx\ , \tag{2.2.46}$$

and, because

$$\int \Psi dx \leq \left(\int \overline{b}^{N/(p-1)} dx\right)^{p/N} \left(\int |\eta v|^{pn/(N-p)} dx\right)^{1-p/N} \quad \text{and} \quad u(x) = O\left(\left|\ln \rho\right|^{1-\delta}\right), \quad (2.2.47)$$

we get a contradiction when σ goes to 0. Therefore $J(\sigma) \leq K$ for $\sigma > 0$. Going back to (2.2.37) yields

$$\|\eta \nabla v\|_{L^p} \leq Cq^{p/\varepsilon}\left(\|v\nabla\eta\|_{L^p} + \|\eta v\|_{L^p}\right), \quad (2.2.48)$$

where C depends on the structural constants and δ. By using Sobolev inequality we obtain

$$\|\eta v\|_{L^{p^*}} \leq C' q^{p/\varepsilon}\left(\|v\nabla\eta\|_{L^p} + \|\eta v\|_{L^p}\right), \quad (2.2.49)$$

which is the same as (2.2.10). The end of the proof is similar to the one of Theorem 1.18.

1-2-3 Equations of mean curvature type

A classical result due to L. Bers [Be] on singularities of minimal surface equation asserts that, if u is a solution of

$$\operatorname{div}\left(\frac{\nabla u}{\sqrt{1+|\nabla u|^2}}\right) = 0 \quad (2.3.1)$$

in a punctured domain $\Omega' = \Omega \setminus \{P\}$, where P is a point of Ω, then it can be extended as a continuous function which is a solution of (2.2.50) in whole Ω. Later this result was extended by De Giorgi and Stampacchia [DGS] who proved that any compact subset D of Ω with (N-1)- Hausdorff measure zero is a removable singularity for equation (2.2.50). Serrin's result is a generalisation of the ones of De Giorgi and Stampacchia . We still consider equation (2.2.16) but introduce the following hypotheses on A and B

$$\begin{cases} |A(x,r,q)| \leq a , \\ |B(x,r,q)| \leq f , \\ q.A(x,r,q) \geq |q| - d|r| - g , \end{cases} \quad (2.3.2)$$

where a is a positive constant, f and d belong to $L^{N+\varepsilon}(\Omega)$ for some $\varepsilon > 0$ and g is integrable .

THEOREM 1.20. *Let* D *be a compact subset of* Ω *with (N-1)-Hausdorff measure zero and* u *be an element of* $W^{1,1}_{loc}(\Omega \setminus D)$, *which is a weak solution of (2.2.16) in* $\Omega \setminus D$. *Then* u *can be defined in* Ω *so that the resulting function is a solution of (2.2.16) in whole* Ω.

Proof. Step 1: We claim that $u \in L^{N/(N-1)}(\Omega \backslash D)$. From the analysis of Theorem 1.18, u is locally bounded in $\Omega \backslash D$ and we can assume that it stays bounded in absolute value by some $M > 0$ in a neighborhood of $\partial \Omega$. If k and l are such that $l > k > M$, we define

$$\tilde{u} = \begin{cases} k & \text{if} \quad u \leq k, \\ u & \text{if} \quad k < u < l, \\ l & \text{if} \quad u \geq l, \end{cases} \tag{2.3.3}$$

and set $\varphi = (\tilde{u} - k)\tilde{\eta}$ where η is a smooth function taking its values in $[0,1]$ and vanishing in a neighborhood of D. From the equation and (2.3.2) we get

$$\nabla \varphi . A(x,u,\nabla u) + \varphi B(x,u,\nabla u) \geq \tilde{\eta}\left(|\nabla \tilde{u}| - d|\tilde{u}| - g\right) - a|\nabla \tilde{\eta}|(\tilde{u} - k) - \tilde{\eta}f. \tag{2.3.4}$$

Since $|\tilde{u}| = (\tilde{u} - k) + k$, we deduce from (2.2.16) that

$$\left\|\tilde{\eta}\nabla \tilde{u}\right\|_{L^1} \leq \left\|\tilde{\eta}(kd + g)\right\|_{L^1} + a(1-k)\left\|\nabla \tilde{\eta}\right\|_{L^1} + \left\|d + f\right\|_{L^N(u>k)}\left\|\tilde{\eta}(\tilde{u} - k)\right\|_{L^{N/(N-1)}}. \tag{2.3.5}$$

Since Sobolev's inequality reads as $\left\|\tilde{\eta}(\tilde{u} - k)\right\|_{L^{N/(N-1)}} \leq K\left(\left\|\tilde{\eta}\nabla \tilde{u}\right\|_{L^1} + (1-k)\left\|\nabla \tilde{\eta}\right\|_{L^1}\right)$ where K=K(N), we can take k large enough so that $\left\|d + g\right\|_{L^N(u>k)} \leq 1/(2K)$, and obtain

$$\left\|\tilde{u} - k\right\|_{L^{N/(N-1)}} \leq 2K\left\|kd + g\right\|_{L^1}. \tag{2.3.6}$$

If we let l go to infinity we deduce that $u^+ \in L^{N/(N-1)}(\Omega \backslash D)$, and the same holds for the negative part of u.

Step 2: End of the proof. The iteration scheme of the proof of Theorem 1.17 still applies. Therefore u remains essentially bounded in $\Omega \backslash D$ and the conclusion of the proof is as in Theorem 1.17.

Remark 1.9. Under the mere hypothesis (2.2.51) there is no reason that u could be extended as a continuous solution of (2.2.16) in Ω. However, if we assume that $A = A(q)$, $B = B(r)$ and some ellipticity and monotonicity conditions like the following

$$\begin{cases} (q - q').(A(q) - A(q')) > 0 & \text{if} \quad q \neq q' \\ (r - r')(B(r) - B(r')) \geq 0 & \text{if} \quad r \neq r' \end{cases} \tag{2.3.7}$$

then any solution u of (2.2.16) in $\Omega \backslash D$ where D is a compact subset of the domain Ω with $(N-1)$-Hausdorff measure zero can be extended as a continuous solution of the same equation in whole Ω.

1-2-4 Characterization of isolated singularities

In this paragraph Ω is a domain of $\mathbb{R}^N$ containing O. Serrin's results assert that when the isolated singularity of a solution of (2.2.16) is not removable and when $1 < p \leq N$ then it can be described by mean of some fundamental singular solution of the same equation.

THEOREM 1.21. *Let the structural assumptions (2.1.1)-(2.1.5) be satisfied with* $1 < p \leq N$ *and* u *be a solution of (2.1.16) in* $\Omega' = \Omega \setminus \{O\}$ *such that* $u \geq L$ *for some* L. *Then either the singularity at* O *is removable, or there exist* $\varepsilon_0 > 0$, $C > 0$ *such that*

$$C \leq u(x) / |x|^{(p-N)/(p-1)} \leq 1/C, \ \textit{if} \ 1 < p < N, \tag{2.4.1}$$

$$C \leq u(x) / \ln(1/|x|) \leq 1/C \ , \ \textit{if} \ p = N, \tag{2.4.2}$$

for $0 < |x| \leq \varepsilon_0$.

By changing the size of the domain we can replace it by a closed ball $\overline{B}(O,R)$ where R is chosen in order to make the Lebesgue norms of b, c, d, f, g, e small enough.

LEMMA 1.1. *There exists a constant* $\tilde{K}$ *such that for any function* φ *belonging to the space* $W^{1,p}(B(O,R))$, *with value 1 in a neighborhood of O and compact support in* $B(O,R)$, *there holds*

$$\int (\nabla\varphi . A(x,u,\nabla u) + (\varphi - 1)B(x,u,\nabla u))dx = \tilde{K}. \tag{2.4.3}$$

Proof. If φ and $\tilde{\varphi}$ are two test functions with the above properties, we have

$$\int ((\nabla\varphi - \nabla\tilde{\varphi}) . A(x,u,\nabla u) + (\varphi - \tilde{\varphi})B(x,u,\nabla u))dx = 0, \tag{2.4.4}$$

and because φ and $\tilde{\varphi}$ are identically 1 near O, (2.4.4) can be written as

$$\begin{aligned}
\int (\nabla\varphi . A(x,u,\nabla u) + (\varphi - 1)B(x,u,\nabla u))dx = \\
\int (\nabla\tilde{\varphi} . A(x,u,\nabla u) + (\tilde{\varphi} - 1)B(x,u,\nabla u))dx,
\end{aligned} \tag{2.4.5}$$

which is (2.4.3).

LEMMA 1.2. *If* u *has a non-removable singularity at* O, *then*

$$\lim_{x \to 0} u(x) = \infty. \tag{2.4.6}$$

Proof. By assumption there exists a sequence of points $\{x_n\}$ tending to O such that

$$\lim_{n \to \infty} u(x_n) = \infty. \tag{2.4.7}$$

From Theorem 1.15 (Harnack inequality) and by a simple scaling argument, we get

$$\lim_{n \to \infty} \min_{|x|=|x_n|} u(x) = \infty. \tag{2.4.8}$$

Applying Theorem 1.14 in the spherical shell $\left\{ x : |x_{n+1}| < |x| < |x_n| \right\}$, we obtain

$$u(x) \geq (1 - C') \min\left\{ \min_{|y|=|x_n|} u(y), \min_{|y|=|x_{n+1}|} u(y) \right\} - Ck \tag{2.4.9}$$

for $|x_{n+1}| < |x| < |x_n|$, where C' is small, which means that (2.4.6) holds.

We can therefore assume that $u(x) \geq 1$ for $0 < |x| \leq \sigma_0$. Let us denote by $m(\sigma)$ the minimum of u on the sphere $B(O, \sigma)$, $0 < \sigma \leq \sigma_0$. On the shell $\sigma < |x| < R$ we define

$$v = v(x, \sigma) = \begin{cases} 0 & \text{if } u(x) \leq 0, \\ u(x) & \text{if } 0 < u(x) < m(\sigma), \\ m(\sigma) & \text{if } u(x) \geq m(\sigma). \end{cases} \tag{2.4.10}$$

LEMMA 1.3. *There exist some positive constants* R *and* C *such that*

$$m(s)^{p-1} \leq C\left(\tilde{B}(s) + 1\right) \begin{cases} s^{p-N} & \text{if } \quad p < N, \\ \left(\ln(R\,/\,s)\right)^{N-1} & \text{if } \quad p = N, \end{cases} \tag{2.4.11}$$

where

$$\tilde{B}(\sigma) = \int_{\sigma < |x| < R} |B(x, u, \nabla u)|\, dx. \tag{2.4.12}$$

Proof. If we extend v to be $m(\sigma)$ for $|x| \leq \sigma$, it satisfies the assumptions of Lemma 1.1 and we have

$$m(\sigma)\tilde{K} = \int \left(\nabla v \cdot A(x, u, \nabla u) + (v - m)B(x, u, \nabla u)\right) dx$$
$$= \int \left(\nabla v \cdot A(x, v, \nabla v) + (v - m)B(x, v, \nabla v)\right) dx \tag{2.4.13}$$

since $v = u$ and $\nabla u = \nabla v$ on the set where $\nabla v \neq 0$. From (2.1.2) we have

$$\int \nabla v . A(x, v, \nabla v) dx \geq \int \left(|\nabla v|^p - dv^p - g \right) dx, \tag{2.4.14}$$

and, from Hölder and Sobolev inequalities,

$$\int dv^p dx \leq \|d\|_{L^{N/p}} \|v\|_{L^{p^*}}^p \leq C\|d\|_{L^{N/p}} \|\nabla v\|_{L^p}^p, \tag{2.4.15}$$

with $C = C(p, N)$ (we assume $p < N$ for simplicity, the case $p = N$ requiring only minor changes). Since the norm of d is taken over the ball $B(O,R)$, we can take R small enough in order to have $C\|d\|_{L^{N/p}} \leq 1/3$. Therefore, by (2.4.14), (2.4.15),

$$\int \nabla v . A(x, v, \nabla v) dx \geq \frac{2}{3} \int |\nabla v|^p dx - C. \tag{2.4.16}$$

By the capacity estimate (1.3.9), we obtain

$$\int |\nabla v|^p dx \geq |S^{N-1}| m(s)^p \begin{cases} \left(\dfrac{N-p}{p-1} \right)^{p-1} s^{N-p} & \text{if} \quad p < N, \\[2ex] \left(\ln(R/s) \right)^{1-N} & \text{if} \quad p = N. \end{cases} \tag{2.4.17}$$

From the definition of $\tilde{B}(\sigma)$ and using the fact that $|v - m| \leq m$, we get

$$\int (v - m(\sigma)) B(x, u, \nabla u) dx \geq -m(\sigma)\tilde{B}(\sigma), \tag{2.4.18}$$

and, by (2.4.13),

$$m(s)^p \leq C' \left(\left(\tilde{B}(s) + \tilde{K} \right) m(s) + C \right) \begin{cases} s^{p-N} & \text{if} \quad p < N, \\[1ex] \left(\ln(R/s) \right)^{N-1} & \text{if} \quad p = N. \end{cases} \tag{2.4.19}$$

Since $m(\sigma) \geq 1$, we finally obtain (2.4.11).

LEMMA 1.4. *The function* $B(x, u, \nabla u)$ *is integrable on* $B(O,R)$ *and*

$$\int_{|x| \leq s} |B(x, u, \nabla u)| dx \leq C \begin{cases} s^\varepsilon & \text{if} \quad p < N, \\[1ex] s^{\varepsilon/2} & \text{if} \quad p = N. \end{cases} \tag{2.4.20}$$

Proof. From Theorem 1.16, the following holds for $\sigma \leq \sigma_0 / 4$

$$\max_{\sigma/2 \leq |x| \leq 3\sigma} u(x) \leq C \left(\min_{\sigma/2 \leq |x| \leq 3\sigma} u(x) + k \right), \tag{2.4.21}$$

37

where C and k are structural constants. Since $\min\limits_{\sigma/2\le|x|\le3\sigma} \le m(\sigma)$ and $m(\sigma)\ge1$, inequality (2.4.21) implies

$$\max_{\sigma/2\le|x|\le3\sigma} u(x) \le C' m(\sigma). \qquad (2.4.22)$$

Therefore

$$\int_{\sigma\le|x|\le2\sigma}\left(du^{p-1}+f\right)dx \le \int_{\sigma\le|x|\le2\sigma}(d+f)u^{p-1}dx,$$

$$\le \max_{\sigma\le|x|\le2\sigma} u^{p-1}(x)\,\|d+f\|_{L^{N/(N-\varepsilon)}}\left(\int_{\sigma\le|x|\le2\sigma}dx\right)^{N/(N-p+\varepsilon)}, (2.4.23)$$

$$\le Cs^{N-p+\varepsilon}m(\sigma)^{p-1}.$$

Similarly

$$\int_{\sigma\le|x|\le2\sigma} c|\nabla u|^{p-1}dx \le C\sigma^{(N-p+\varepsilon p)/p}\left(\sigma^{-1}\|u\|_{L^p(\sigma/2\le|x|\le4\sigma)} + \sigma^{(N-p)/p}\right)^{p-1}, \qquad (2.4.24)$$

from Hölder's inequality and estimate (2.1.9) in Theorem 1.13. Since

$$\|u\|_{L^p(\sigma/2\le|x|\le4\sigma)} \le C\sigma^{N/p}\max_{\sigma/2\le|x|\le4\sigma} u(x) \le C'\,\sigma^{N/p}m(\sigma), \qquad (2.4.25)$$

we finally get

$$\int_{\sigma\le|x|\le2\sigma} c|\nabla u|^{p-1}dx \le C\sigma^{N-p+\varepsilon}m(\sigma)^{p-1}. \qquad (2.4.26)$$

Moreover $|B(x,u,\nabla u)|\le c|\nabla u|^{p-1}+du^{p-1}+f$ and $\int_{\sigma\le|x|\le2\sigma}|B(x,u,\nabla u)|dx = \tilde{B}(\sigma)-\tilde{B}(2\sigma)$, we deduce

$$\tilde{B}(\sigma)-\tilde{B}(2\sigma)\le \beta\sigma^\varepsilon\left(\tilde{B}(\sigma)+1\right), \qquad (2.4.27)$$

from 2.4.25) , (2.4.26) and (2.4.11) in the case $p<N$ and

$$\tilde{B}(\sigma)-\tilde{B}(2\sigma)\le \beta'\,\sigma^\varepsilon\left(\tilde{B}(\sigma)+1\right)\left(\ln(R/\sigma)\right)^{N-1} \le \beta\sigma^{\varepsilon/2}\left(\tilde{B}(\sigma)+1\right) \qquad (2.4.28)$$

in the case $p=N$, where β and β' are some appropriate positive constants. We shall restrict ourself to treat the difference inequality (2.4.25), the case (2.4.26) being the same. We write (2.4.25) under the form $(1-\beta\sigma^\varepsilon)\tilde{B}(\sigma)\le\beta\sigma^\varepsilon+\tilde{B}(2\sigma)$ and denote $\tilde{\sigma}=(2\beta)^{-\varepsilon}$. Since $(1-\beta\sigma^\varepsilon)^{-1}\le 2\le\exp(2\beta\sigma^\varepsilon)$, for $\sigma\le\tilde{\sigma}$, we have $\tilde{B}(\sigma)\le 2\beta\sigma^\varepsilon+\exp(2\beta\sigma^\varepsilon)\tilde{B}(2\sigma)$ and by iterations we deduce

38

$$\tilde{B}(2^{-k}\tilde{\sigma}) \le 2^{-k\varepsilon} + \exp(2^{-k\varepsilon})\tilde{B}(2^{-k+1}\tilde{\sigma}) \le \ldots$$

$$\le \left(\sum_1^k 2^{-j\varepsilon} + \tilde{B}(\tilde{\sigma})\right)\exp\left(\sum_1^k 2^{-j\varepsilon}\right) \le C\big(B(\tilde{\sigma})+1\big) \tag{2.4.29}$$

since $\displaystyle\sum_1^\infty 2^{-j\varepsilon} < \infty$. Then $\tilde{B}(\sigma)$ remains bounded when σ approaches 0 which means that $\displaystyle\int_{B(O,R)} |B(x,u,\nabla u)|dx < \infty$. Moreover

$$\int_{|x|\le\sigma} |B(x,u,\nabla u)|dx = \sum_1^\infty \left(\tilde{B}(2^{-j}\sigma) - \tilde{B}(2^{-j+1}\sigma)\right) \le \beta\sigma^\varepsilon(B(0)+1)\sum_1^\infty 2^{-j\varepsilon} \tag{2.4.30}$$

from (2.4.27), which is (2.4.20).

As a consequence of Lemmas 1.3 and 1.4, the first upper estimate on u is derived:

$$m(s) \le C \begin{cases} s^{(p-N)/(p-1)} & \text{if } p < N, \\ \ln\,(R/s) & \text{if } p = N. \end{cases} \tag{2.4.31}$$

From the integrability of $B(x,u,\nabla u)$ in $B(O,R)$, the following extension of Lemma 1.1 is valid:

LEMMA 1.5. *There exists a constant K such that for any function φ belonging to the space $W^{1,p}(B(O,R))$, with value 1 in a neighborhood of O and compact support in $B(O,R)$, there holds*

$$\int (\nabla\varphi.A(x,u,\nabla u) + B(x,u,\nabla u)\varphi)dx = K. \tag{2.4.32}$$

LEMMA 1.6. K *is positive.*

Proof. We proceed by contradiction in assuming $K \le 0$. From Lemma 1.4, we have

$$0 \ge \int(\nabla v.A(x,u,\nabla u) + vB(x,u,\nabla u))dx,$$
$$= \int \nabla v.A(x,v,\nabla v)dx + \int_{S_1} vB(x,v,\nabla v)dx + m(\sigma)\int_{S_2} B(x,u,\nabla u)dx, \tag{2.4.33}$$

with $S_1 = \{x : \sigma \le |x| \le R$ and $u(x) < m(\sigma)\}$, $S_2 = \{x : 0 < |x| < \sigma\}\cup\{x : u(x) \ge m(\sigma)\}$. In the same way as in Lemma 1.3, there holds

$$\int (\nabla v . A(x,v,\nabla v)) dx \geq \frac{2}{3}\int |\nabla v|^{p} dx - C, \tag{2.4.34}$$

and

$$\int_{S_1} v|B(x,v,\nabla v)| dx \leq \int \left(cv|\nabla v|^{p-1} + dv^{p} + vf \right) dx. \tag{2.4.35}$$

Moreover, if R is small enough,

$$\int cv|\nabla v|^{p-1} dx \leq \|c\|_{L^{N}} \|v\|_{L^{p*}} \|\nabla v\|_{L^{p}}^{p-1} \leq C\|c\|_{L^{N}} \|\nabla v\|_{L^{p}}^{p} \leq \frac{1}{12}\|\nabla v\|_{L^{p}}^{p} \tag{2.4.36}$$

and

$$\int dv^{p} dx \leq \frac{1}{3}\|\nabla v\|_{L^{p}}^{p}. \tag{2.4.37}$$

Finally

$$\int vf dx \leq \|f\|_{L^{pN/(pN-p-N)}} \|v\|_{L^{p*}} \leq C\|\nabla v\|_{L^{p}} \leq \frac{1}{12}\|\nabla v\|_{L^{p}}^{p} + C. \tag{2.4.38}$$

Combining (2.5.34)-(2.4.36) yields

$$\int_{S_1} v|B(x,v,\nabla v)| dx \leq \frac{1}{2}\|\nabla v\|_{L^{p}}^{p} + C. \tag{2.4.39}$$

Therefore, from (2.4.33)-(2.4.34) and (2.4.39) and (1.3.9), we deduce the estimates

$$m(s)^{p} \leq C' \left(m(s)\int_{S_2} |B(x,u,\nabla u)| dx + C \right) \begin{cases} s^{p-N} & \text{if} \quad p < N, \\ (\ln(R/s))^{N-1} & \text{if} \quad p = N, \end{cases} \tag{2.4.40}$$

which is similar to (2.4.19). The main point is now to prove that

$$m(s) \leq M \begin{cases} s^{\tau(1-\delta)} & \text{if} \quad p < N, \\ (\ln R/s)^{1-1/N} & \text{if} \quad p = N, \end{cases} \tag{2.4.41}$$

where $\tau = (p-N)/(p-1)$ and $\delta = \varepsilon/(N-p+\varepsilon)$, for some M larger than 1. If such an inequality holds, it will imply that the singularity at O is removable by using Harnack inequality and Theorem 1.19, a contradiction.

We first consider the case $p < N$. Since (2.4.41) is trivially verified if $m(\sigma) \leq \sigma^{\tau(1-\delta)}$, we have only to consider the σ such that $m(\sigma) > \sigma^{\tau(1-\delta)}$. For θ to be chosen large enough later on, we set $\Gamma_{\sigma,\theta} = \left\{ x\colon \theta\sigma^{1-\delta} \le |x| \le \sigma_0 / 2 \right\}$. From Harnack inequality and (2.4.31), we have for any x in $\Gamma_{\sigma,\theta}$

$$u(x \le Cm(|x|)) \le C' |x|^\tau \le C' \,\theta^\tau \sigma^{\tau(1-\delta)} \le C' \,\theta^\tau m(\sigma) = m(\sigma), \qquad (2.4.42)$$

provided $C' \,\theta^\tau = 1$. As a consequence of the definition of S_2

$$S_2 \subset \left\{ x\colon 0 < |x| < \sigma \right\} \cup \left\{ x\colon 0 < |x| < \theta\sigma^{1-\delta} \right\} \cup \left\{ x\colon \sigma_0 / 2 < |x| < R \right\}.$$

As $\lim_{\sigma \to 0} m(\sigma) = \infty$, $S_2 \cap \left\{ x\colon \sigma_0 / 2 < |x| < R \right\}$ is empty for $\sigma \le \sigma_1$, small enough, and it can also be chosen such that $\left\{ x\colon 0 < |x| < \sigma \right\} \subset \left\{ x\colon 0 < |x| < \theta\sigma^{1-\delta} \right\}$. Finally $S_2 \subset \left\{ x\colon 0 < |x| < \theta\sigma^{1-\delta} \right\}$. From Lemma 1.4, we have

$$\int_{S_2} |B(x,u,\nabla u)|dx \le \int_{|x| \le \theta\sigma^{1-\delta}} |B(x,u,\nabla u)|dx \le C\sigma^{\varepsilon(1-\delta)}. \qquad (2.4.43)$$

for $\sigma \le \sigma_1$. Pluging this estimate in (2.4.39) yields

$$m(\sigma)^p \le C\!\left(m(\sigma)\sigma^{p-N+\varepsilon(1-\delta)} + \sigma^{p-N} \right), \qquad (2.4.44)$$

and

$$m(\sigma) \le C'\left(\sigma^{\tau(1-\delta)} + \sigma^{(p-N)/p} \right) \le 2C'\,\sigma^{\tau(1-\delta)}, \qquad (2.4.45)$$

as $(p - N)/p \ge \tau(1-\delta)$. Therefore (2.4.31) holds for $\sigma \le \sigma_1$. As it clearly holds for some M larger than 1 on $\sigma > \sigma_1$, consequently it holds in all the cases when $p < N$.

In the case $p = N$, we have only to consider the σ for which $m(\sigma) > \left(\ln(R/\sigma) \right)^{1-1/N}$. We consider again some θ and set $\Gamma_{\sigma,\theta} = \left\{ x\colon Re^{-2y} \le |x| \le \sigma_0 / 2 \right\}$ where $y = y(\sigma) = \left(\ln(R/\sigma) \right)^{1-1/N}$. For θ small enough we have, as in (2.4.42),

$$u(x \le Cm(|x|)) \le C' |x|^\tau \le C' \ln(R/\sigma) \le C' \,\theta y < C' \,\theta m(\sigma) = m(\sigma) \qquad (2.4.46)$$

on $\Gamma_{\sigma,\theta}$. In the same way as above, the set S_2 is included into $\left\{ x\colon 0 < |x| < Re^{-2\theta y} \right\}$ for $\sigma \le \sigma_1$, small enough. By Lemma 1.4, we get

$$\int_{S_2} |B(x,u,\nabla u)|dx \le \int_{|x| \le Re^{-2\theta y}} |B(x,u,\nabla u)|dx \le Ce^{-\varepsilon\theta y}, \qquad (2.4.47)$$

and finally

41

$$m(\sigma) \le C(\ln(R/\sigma))^{1-1/N}\left(y(\sigma)^{1/(N-1)}e^{-\varepsilon\theta y/(N-1)}+1\right). \tag{2.4.48}$$

Since $y(\sigma)^{1/(N-1)}e^{-\varepsilon\theta y/(N-1)}+1$ is uniformly bounded for $y(\sigma)\in[0,\infty)$ we have (2.4.41) for $\sigma\le\sigma_1$. By changing M, we have it also for $\sigma>\sigma_1$, which ends the proof of Lemma 1.6.

We now define $M=M(\sigma)$ as being the maximum of $u(x)$ for $|x|=\sigma$, $0<|x|\le\sigma_0$. Theorem 1.21 will be a mere consequence of the following lemma and Harnack inequality.

LEMMA 1.7. *There exists a positive constant* L *such that*

$$M(s)\ge L\begin{cases} s^{(p-N)/(p-1)} & \text{if } p<N, \\ \ln(R/\sigma) & \text{if } p=N. \end{cases} \tag{2.4.49}$$

Proof. Let V be the function defined by

$$V(x)=\begin{cases} \mathrm{Max}(0,u(x)) & \text{for } s\le|x|\le R, \\ \mathrm{Min}(M(s),u(x)) & \text{for } 0<|x|<s, \\ M(s) & \text{for } x=O. \end{cases} \tag{2.4.50}$$

V belongs to $W^{1,p}(B(O,R))$, has compact support and $V\equiv M(\sigma)$ near O. We choose σ_2 so that $\displaystyle\int_{0<|x|\le\sigma_2}|B(x,u,\nabla u)|dx\le K/2$ and define Θ by

$$\Theta=\Theta(x,\sigma)=\begin{cases} 0 & \text{for } \sigma_2<|x|\le R, \\[2mm] \dfrac{|x|^\tau-\sigma_2^\tau}{\sigma^\tau-\sigma_2^\tau} & \text{for } \sigma<|x|\le s_2, \\[2mm] 1 & \text{for } \sigma>|x|. \end{cases} \tag{2.4.51}$$

for $\sigma<\sigma_2$ (we restrict the proof to the case $p<N$ and recall that $\tau=(p-N)/(p-1)$). The function V satisfies the assumption of Lemma 1.5, consequently

$$\begin{aligned} K &= \int(\nabla\Theta.A(x,u,\nabla u)+\Theta B(x,u,\nabla u))dx, \\ &\le \|\nabla\Theta\|_{L^p}\|A(x,u,\nabla u)\|_{L^{p/(p-1)}}+K/2. \end{aligned} \tag{2.4.52}$$

Because $\|\nabla\Theta\|_{L^p}^p=|S^{N-1}||\tau|^{p-1}(\sigma^\tau-\sigma_2^\tau)^{1-p}$, this yields

$$K'(\sigma^\tau-\sigma_2^\tau)\le\int_{\sigma\le|x|\le\sigma_2}|A(x,u,\nabla u)|^{p/(p-1)}dx, \tag{2.4.53}$$

where K' is some positive constant. On the set of x where $\sigma \le |x| \le \sigma_2$, we have V = u . If we call Δ the whole set where V = u, then the integral term in (2.4.53) is increased if we take Δ as a new domain of integration, which implies

$$\int_{\Delta}|A(x,u,\nabla u)|^{p/(p-1)}dx \le C\int_{\Delta}\left(|\nabla u|^p + b^{p/(p-1)}u^p + e^{p/(p-1)}\right)dx,$$

$$= C\int_{\Delta}\left(\left(|\nabla u|^p - 2du^p - 2g\right) + \left(2d + b^{p/(p-1)}\right)u^p + \left(2g + e^{p/(p-1)}\right)\right)dx. \tag{2.4.54}$$

By Sobolev inequality

$$\int_{\Delta}\left(2d + b^{p/(p-1)}\right)u^p dx \le \int_{\Delta}\left(2d + b^{p/(p-1)}\right)V^p dx,$$

$$\le C\left(\|d\|_{L^{N/p}} + \|b\|^{p/(p-1)}_{L^{N/(p-1)}}\right)\int_{\Delta}|\nabla u|^p dx, \tag{2.4.55}$$

as $\nabla V = \nabla u$ almost everywhere in Δ and $\nabla V = 0$ on the complement of Δ. If R is taken small enough, we deduce from (2.4.55) that

$$\int_{\Delta}\left(2d + b^{p/(p-1)}\right)u^p dx \le \int_{\Delta}|\nabla u|^p dx. \tag{2.4.56}$$

From (2.4.54) and the structural assumptions, we also get

$$\int|A(x,u,\nabla u)|^{p/(p-1)}dx \le C'\left(1 + \int_{\Delta}\nabla u. A(x,u,\nabla u)dx\right),$$

$$= C'\left(1 + \int_{\Delta}\nabla V. A(x,u,\nabla u)dx\right). \tag{2.4.57}$$

and, by Lemma 1.5,

$$\int\nabla V. A(x,u,\nabla u)dx = M(\sigma)K - \int VB(x,u,\nabla u)dx,$$

$$\le M(\sigma)K + \left(\underset{B(O,R)}{\text{Max }} V\right)\int_{B(O,R)}|B(x,u,\nabla u)|dx. \tag{2.4.58}$$

Taking R small enough, we deduce from Theorem 1.15 that

$$\underset{0<|x|\le R}{\text{Max }} u(x) \le CM(\sigma), \tag{2.4.59}$$

since we can always assume that $M(\sigma) \ge 1$, and finally

$$\underset{0<|x|\le R}{\text{Max }} V(x) \le CM(\sigma). \tag{2.4.60}$$

With this estimate and (2.4.74)-(2.4.79) we deduce

$$\int |A(x,u,\nabla u)|^{p/(p-1)} dx \leq CM(\sigma). \tag{2.4.61}$$

Pluging this estimate into (2.4.53) yields

$$M(\sigma) \geq 2C''\left(\sigma^{\tau} - \sigma_2^{\tau}\right), \tag{2.4.62}$$

for $0 < \sigma < \sigma_2$. Taking σ small enough implies

$$M(\sigma) \geq C'' \sigma^{(p-N)/(p-1)}, \tag{2.4.63}$$

for $1 < p < N$. In the same way

$$M(\sigma) \geq C'' \ln(R/\sigma), \tag{2.4.64}$$

if $p = N$, which ends the proof of the lemma and of Theorem 1.21 as well.

Remark 1.10. The existence of solutions of (2.1.6) satisfying the singularity estimates of Theorem 1.21 is not clear under the mere structural assumptions. In [Se2] the existence of such solutions is obtained under the two following hypotheses

P1. For any domain D with small enough diameter the Dirichlet problem

$$\begin{cases} -\text{div} A(x,u,\nabla u) + B(x,u,\nabla u) = 0 & \text{in } D, \\ u = \psi & \text{on } \partial D, \end{cases} \tag{2.4.65}$$

is uniquely solvable for ψ smooth enough and the mapping $\psi \mapsto u$ is continuous in the uniform convergence topology.

P2. Let Γ denote a smooth annular domain in D and v_m be the solution of the Dirichlet problem in Γ, with $\psi_0(x)$ on the outer boundary and $\psi_1(x) + m$ on the inner boundary. Then

$$\lim_{m \to \infty} v_m(P) = \infty, \tag{2.4.66}$$

for any $P \in \Gamma$.

For such a smooth domain D containing O, and assuming ψ is a smooth function on the boundary, there exists a family of solutions $w(x)$ of (2.5.67) in $D' = D \setminus \{O\}$ instead of D such that

$$C^{-1} \leq w(x)|x|^{(N-p)/(p-1)} \leq C \qquad \text{if } p < N,$$

$$C^{-1} \leq w(x)\left(\ln(1/|x|)\right)^{-1} \leq C \qquad \text{if } p = N, \tag{2.4.67}$$

for x in a neighborhood of O and C > 0. Moreover, for any P belonging to D', the range of values of w covers $\left(w_0(P),\infty\right)$ where w_0 is the regular solution of (2.4.65) in D.

2 Singularities of solutions of elliptic equations with absorption term

2-1 Isolated singularities

2-1-1 Isotropic singularities for solutions of equations

Let Ω be an open subset of $\mathbb{R}^N$ containing O, $\Omega^* = \Omega \setminus \{O\}$, and f and g be continuous real valued functions defined respectively on Ω and $\mathbb{R}$. If u is a solution of

$$-\Delta u + g(u) = f \tag{1.1.1}$$

in Ω^*, then it may happen that u is isotropic near O, that is $u(x)$ behaves like its spherical average $\bar{u}(|x|) = \fint_{|y|=|x|} u(y) ds(y)$ in the sense that $u(x) - \bar{u}(|x|) = o(u(x))$ near O. In that case the singular behaviour of u can be described my the mean of the associated ordinary differential equation

$$-y'' - \frac{N-1}{r} y' + g(y) = 0. \tag{1.1.2}$$

The first isotropy results are due to Vazquez and Véron ([VV2] and [VV3]) and deal with a nondecreasing nonlinear term g (also called an absorption term). We introduce the spherical coordinates $(r, \sigma) \in \mathbb{R}_*^+ \times S^{N-1}$.

THEOREM 2.1. *Let* g *be a non-decreasing continuous function and* u *a* $C^2(\Omega^*)$*-solution of* (1.1.1) *in* Ω^* *such that*

$$\liminf_{r \to 0} r^{N-1} \left(\int_{S^{N-1}} |u(r, \sigma) - \bar{u}(r)|^2 d\sigma \right)^{1/2} = 0. \tag{1.1.3}$$

Then there exists $c \in \mathbb{R} \cup \{\pm\infty\}$ *such that*

$$\lim_{x \to 0} u(x)/\mu(x) = c \tag{1.1.4}$$

where μ *is the fundamental solution of the Laplacian in* $\mathbb{R}^N$. *If* c *is zero, then* u *can be extended as a* C^1*-solution of* (1.1.1) *in* Ω.

The notion of isotropy is described via the following estimate

PROPOSITION 2.1. *Under the assumptions of Theorem 2.1, there holds*

$$\lim_{r \to 0} \left\| u(r,.) - \overline{u}(r) \right\|_{L^\infty(S^{N-1})} = 0. \tag{1.1.5}$$

Proof. Step 1. The case $N \geq 3$. We write equation (1.1.1) in spherical coordinates $(r,\sigma) \in (0,R] \times S^{N-1}$, for some $R > 0$. Then u satisfies

$$-\frac{\partial^2 u}{\partial r^2} - \frac{N-1}{r}\frac{\partial u}{\partial r} - \frac{1}{r^2}\Delta_{S^{N-1}} u + g(u) = f. \tag{1.1.6}$$

We introduce now the following change of variable and unknown

$$s = r^{N-2}/(N-2) \ , \ u(r,\sigma) = r^{2-N}v(s,\sigma) \ , \ f(r,\sigma) = \varphi(s,\sigma). \tag{1.1.7}$$

The function v satisfies

$$s^2\frac{\partial^2 v}{\partial s^2} + \frac{1}{(N-2)^2}\Delta_{S^{N-1}} v = A_N s^{N/(N-2)}\left(g\left(\frac{v}{s(N-2)}\right) - \varphi \right) \tag{1.1.8}$$

where $A_N = (N-2)^{(4-N)/(N-2)}$. If $\overline{v}(s)$ (resp. $\overline{\varphi}(s)$) is the average of $v(s,.)$ (resp. $\varphi(s,.)$) on S^{N-1}, then

$$-\int_{S^{N-1}} (v-\overline{v})\Delta_{S^{N-1}}(v-\overline{v})d\sigma \geq \lambda_1 \int_{S^{N-1}} (v-\overline{v})^2 d\sigma = (N-1)\int_{S^{N-1}} (v-\overline{v})^2 d\sigma \tag{1.1.9}$$

and

$$\int_{S^{N-1}} (v-\overline{v})\left(g\left(\frac{v}{s(N-2)}\right) - \overline{g\left(\frac{v}{s(N-2)}\right)} \right)d\sigma$$

$$= \int_{S^{N-1}} (v-\overline{v})\left(g\left(\frac{v}{s(N-2)}\right) - g\left(\frac{\overline{v}}{s(N-2)}\right) \right)d\sigma \geq 0 \tag{1.1.10}$$

because g is nondecreasing. Therefore

$$s^2\int_{S^{N-1}} (v-\overline{v})\frac{\partial^2}{\partial s^2}(v-\overline{v})d\sigma - \frac{N-1}{(N-2)^2}\int_{S^{N-1}} (v-\overline{v})^2 d\sigma$$

$$\geq -\int_{S^{N-1}} (v-\overline{v})(\varphi-\overline{\varphi})d\sigma \tag{1.1.11}$$

If we set $X(s) = \left\| v(s,.) - \overline{v}(s) \right\|_{L^2(S^{N-1})}$, there exists some $M > 0$ such that the following inequality holds in the sense of distributions

$$s^2\frac{d^2 X}{ds^2} - \frac{N-1}{(N-2)^2}X + Ms^{N/(N-2)} \geq 0, \tag{1.1.12}$$

since $\int_{S^{N-1}} (v - \bar{v}) \frac{\partial^2}{\partial s^2} (v - \bar{v}) d\sigma \leq X \frac{d^2 X}{ds^2}$ on the set where $X(s) > 0$. From the maximum principle and (1.1.3) it follows from (1.1.12) that there exists some positive constant β such that

$$X(s) \leq \beta s^{(N-1)/(N-2)} \tag{1.1.13}$$

on $(0, S] = \left(0, R^{N-2} / (N-2)\right]$. This L^2-isotropy estimate is not sufficient and has to be strengthened into a uniform estimate. For that, we first consider the following linear problem

$$\begin{cases} s^2 \dfrac{\partial^2 \omega}{\partial s^2} + \dfrac{1}{(N-2)^2} \Delta_{S^{N-1}} \omega = 0 & \text{in} \quad (\rho, \tau) \times S^{N-1}, \\[2mm] \omega(\rho, .) = \alpha, \qquad \omega(\tau, .) = \gamma & \text{in} \quad S^{N-1}. \end{cases} \tag{1.1.14}$$

where α and γ are two $L^2(S^{N-1})$-nonnegative functions. In fact, up to the transformation (1.1.7), ω is a harmonic function and it can be decomposed into $\omega_1 + \omega_2$ where

$$\begin{cases} s^2 \dfrac{\partial^2 \omega_1}{\partial s^2} + \dfrac{1}{(N-2)^2} \Delta_{S^{N-1}} \omega_1 = 0 & \text{in} \quad (0, \tau) \times S^{N-1}, \\[2mm] \omega_1(\tau, .) = \gamma & \text{in} \quad S^{N-1}, \end{cases} \tag{1.1.15}$$

and

$$\begin{cases} s^2 \dfrac{\partial^2 \omega_2}{\partial s^2} + \dfrac{1}{(N-2)^2} \Delta_{S^{N-1}} \omega_2 = 0 & \text{in} \quad (\rho, \infty) \times S^{N-1}, \\[2mm] w_2(r, .) = g & \text{in} \quad S^{N-1}. \end{cases} \tag{1.1.16}$$

Moreover $\omega < \omega_1 + \omega_2$. From Poisson's formula (see [VV3]) there exists a positive constant $C = C(N)$ such that

$$\omega_1(s, \sigma) \leq C \frac{s}{\tau} \left(1 + \frac{1}{\ln(\tau/s)} \right)^{(N-1)/2} \|\gamma\|_{L^2(S^{N-1})}, \tag{1.1.17}$$

and

$$\omega_2(s, \sigma) \leq C \left(1 + \frac{1}{\ln(s/\rho)} \right)^{(N-1)/2} \|\alpha\|_{L^2(S^{N-1})}, \tag{1.1.18}$$

for any (s, σ) in $(\rho, \tau) \times S^{N-1}$. Therefore

$$\omega(s,\sigma) \leq C\left(1 + \frac{1}{\ln(s/\rho)}\right)^{(N-1)/2}\left(\|\alpha\|_{L^2(S^{N-1})} + \frac{s}{\tau}\|\gamma\|_{L^2(S^{N-1})}\right). \tag{1.1.19}$$

For $0 < \rho < \tau < S$, we set $a = \overline{v}(\rho)$, $b = \overline{v}(\tau)$ and let y be the solution of

$$\begin{cases} s^2 \dfrac{d^2y}{ds^2} - A_N s^{N/(N-2)} g\left(\dfrac{y}{s(N-2)}\right) = 0 & \text{in } (\rho,\tau), \\[2mm] \qquad\qquad y\rho) = a\,, \quad y(\tau) = b. \end{cases} \tag{1.1.20}$$

The function $w = v - y$ satisfies

$$s^2 \frac{\partial^2 w}{\partial s^2} + \frac{1}{(N-2)^2}\Delta_{S^{N-1}} w = A_N(hw - \varphi) \tag{1.1.21}$$

where h is nonnegative since being defined by

$$h(s,s) = \begin{cases} \big(g(v/s(N-2)) - g(y/s(N-2))\big)/(v-y) & \text{if } v \neq y, \\[2mm] 0 & \text{if } v = y. \end{cases} \tag{1.1.22}$$

We fix the function ω solution of (1.1.14) in $(\rho,\tau) \times S^{N-1}$ by the boundary values $\alpha(.) = \big(v(\rho,.) - a\big)^+$ and $\gamma(.) = \big(v(\tau,.) - b\big)^+$ and we call z the solution of

$$\begin{cases} s^2 \dfrac{\partial^2 z}{\partial s^2} + \dfrac{1}{(N-2)^2}\Delta_{S^{N-1}} z = -A_N|\varphi| & \text{in } (0,S)\times S^{N-1}, \\[2mm] \qquad\qquad z(S,.) = 0 & \text{in } S^{N-1}. \end{cases} \tag{1.1.23}$$

Then $\omega^* = \omega + z$ is a nonnegative supersolution for (1.1.21) in $(\rho,\tau) \times S^{N-1}$. Since ω^* dominates w on the boundary, then $w \leq \omega^*$. In the same way we can minorize w. If we use the fact that $0 \leq z(s,\sigma) \leq Cs$ for some positive C' depending on $\|\varphi\|_{L^\infty((0,S)\times S^{N-1})}$, we deduce from (1.1.19) that

$$\|v(s,.) - y(s)\|_{L^\infty(S^{N-1})} \leq C\left(s + \left(1 + \frac{1}{\ln(s/\rho)}\right)^{(N-1)/2}\|v(\rho,.) - a\|_{L^2(S^{N-1})}\right.$$
$$\left. + \frac{s}{\tau}\left(1 + \frac{1}{\ln(\tau/s)}\right)^{(N-1)/2}\|v(\tau,.) - b\|_{L^2(S^{N-1})}\right). \tag{1.1.24}$$

Since y is radially symmetric, this inequality still holds if we replace v by $\overline{v}$, which finally yields

$$\left\|v(s,.) - \overline{v}(s)\right\|_{L^{\infty}(S^{N-1})} \le 2C\left(s + \left(1 + \frac{1}{\ln(s/\rho)}\right)^{(N-1)/2} \left\|v(\rho,.) - a\right\|_{L^2(S^{N-1})} \right.$$
$$\left. + \frac{s}{\tau}\left(1 + \frac{1}{\ln(\tau/s)}\right)^{(N-1)/2} \left\|v(\tau,.) - b\right\|_{L^2(S^{N-1})}\right). \tag{1.1.25}$$

If we use (1.1.13) we obtain

$$\left\|v(s,.) - \overline{v}(s)\right\|_{L^{\infty}(S^{N-1})} \le C''\left(s + \left(1 + \frac{1}{\ln(s/\rho)}\right)^{(N-1)/2} \rho^{(N-1)/(N-2)} \right.$$
$$\left. + \frac{s}{\tau}\left(1 + \frac{1}{\ln(\tau/s)}\right)^{(N-1)/2} \tau^{1/(N-2)}\right), \tag{1.1.26}$$

for some positive constant C''. If we take $s = 2\rho = \tau/2$, (1.1.5) follows.

Step 2. The case N = 2. The idea is essentially the same, that is to obtain a L^2-isotropy estimate and then to improve up to a uniform estimate. We consider the following change of variable and unknown specific to N = 2

$$t = \ln(1/r) , \; v(t,\sigma) = u(r,\sigma) , \; \varphi(t,\sigma) = f(r,\sigma) \tag{1.1.27}$$

which transform (1.1.6) into

$$\frac{\partial^2 v}{\partial t^2} + \frac{\partial^2 v}{\partial \sigma^2} = e^{-2t}\left(g(v) - \varphi\right). \tag{1.1.28}$$

If we define X as in Step 1, we obtain

$$\int_{S^1} \frac{\partial^2}{\partial t^2}(v - \overline{v})^2 d\sigma - \int_{S^1}(v - \overline{v})^2 d\sigma \ge e^{-2t}\int_{S^1}(v - \overline{v})(\varphi - \overline{\varphi})d\sigma \tag{1.1.29}$$

and

$$\frac{d^2 X}{dt^2} - X \ge -Me^{-2t} \tag{1.1.30}$$

on $[S,\infty)$, for some S and M > 0, which implies

$$X(t) = \left\|v(t,.) - \overline{v}(t)\right\|_{L^2(S^1)} \le Ce^{-t} \tag{1.1.31}$$

from (2.1.3) and the maximum principle. Let us define H by

$$H(t) = \frac{1}{2}X^2(t) + \int_s^t \int_s^s \int_{S^1} e^{-2\tau} (\varphi - \overline{\varphi})(v - \overline{v}) \, d\sigma \, d\tau \, ds. \tag{1.1.32}$$

Inequality (1.1.29) means that H is convex. Moreover $H(t)/t$ is bounded and $\lim_{t \to \infty} H'(t) = 0$. Since (1.1.29) can also be written as

$$\int_{S^1} \frac{\partial^2}{\partial t^2}(v - \overline{v})^2 \, d\sigma - \int_{S^1} \left(\frac{\partial}{\partial \sigma}(v - \overline{v}) \right)^2 d\sigma \geq e^{-2t} \int_{S^1} (v - \overline{v})(\varphi - \overline{\varphi}) \, d\sigma, \tag{1.1.33}$$

we deduce that

$$\frac{d^2 H}{dt^2} \geq \int_{S^1} \left(\frac{\partial}{\partial \sigma}(v - \overline{v}) \right)^2 d\sigma. \tag{1.1.34}$$

Using Sobolev imbedding inequality on S^1 yields

$$\frac{d^2 H}{dt^2} + \int_{S^1} (v - v)^2 \, d\sigma \geq \theta \|(v - v)\|^2_{L^\infty(S^1)} \tag{1.1.35}$$

for some $\theta > 0$. If we integrate (1.1.35) and use (1.1.31), we deduce that

$$\int_s^\infty \|v(t,.) - \overline{v}(t)\|^2_{L^\infty(S^1)} \, dt < \infty. \tag{1.1.36}$$

Then there exists an increasing sequence $\{t_n\}$ tending to infinity with n such that $\lim_{n \to \infty} \|v(t_n,.) - \overline{v}(t_n)\|_{L^\infty(S^1)} = 0$. If we go back to u , it means that

$$\lim_{n \to \infty} \|u(r_n,.) - \overline{u}(r_n)\|_{L^\infty(S^1)} = 0 \tag{1.1.37}$$

where $r_n = e^{-t_n}$. We now proceed as in Step 1 by introducing the solution of an ordinary differential equation

$$\begin{cases} \dfrac{d^2 y}{dr^2} + \dfrac{1}{r}\dfrac{dy}{dr} = g(y) \quad \text{in } (r_{n+1}, r_n), \\ y(r_{n+1}) = \overline{u}(r_{n+1}) \quad, \quad y(r_n) = \overline{u}(r_n). \end{cases} \tag{1.1.38}$$

Applying again the maximum principle in the spherical shell $(r_{n+1}, r_n) \times S^1$ yields

$$u(r,\sigma) - y(r) \le$$

$$\max\left\{\left\|\left(u(r_n,.) - \overline{u}(r_n)\right)^+\right\|_{L^\infty(S^1)}, \left\|\left(u(r_{n+1},.) - \overline{u}(r_{n+1})\right)^+\right\|_{L^\infty(S^1)}\right\} + \beta_n(r), \qquad (1.1.39)$$

for $(r,\sigma) \in (r_{n+1}, r_n) \times S^1$ where β_n is the solution of

$$\begin{cases} \dfrac{d^2\beta_n}{dr^2} + \dfrac{1}{r}\dfrac{d\beta_n}{dr} = -k \quad \text{in } (0, r_n), \\[2mm] \beta_n(r_n) = 0, \end{cases} \qquad (1.1.40)$$

with $k = \|f\|_{L^\infty(B(O,R))}$; a simple computation gives $\beta_n(r) = \dfrac{k}{4}(r_n^2 - r^2)$. Therefore

$$\limsup_{r \to 0} \left\|\left(u(r,.) - \overline{u}(r)\right)^+\right\|_{L^\infty(S^1)} = 0. \qquad (1.1.41)$$

In the same way

$$\limsup_{r \to 0} \left\|\left(u(r,.) - \overline{u}(r)\right)^-\right\|_{L^\infty(S^1)} = 0, \qquad (1.1.42)$$

which ends the proof of (2.1.5).

Proof of Theorem 2.1. Step 1. The case $N \ge 3$. We keep the notations of (1.1.7), (1.1.8).

Case 1. We assume that $\overline{v}(s)$ is uniformly bounded on $(0, S]$. Then there exist some constant c and some sequence $\{s_n\}$ going to 0 when n tends to infinity such that $\lim_{n\to\infty} \overline{v}(s_n) = c$.

We first assume that $c > 0$. From (2.1.5) there exists some integer n_0 such that

$$v(s_n, \sigma) \ge c/2, \qquad \left(\forall n \ge n_0, \ \forall \sigma \in S^{N-1}\right). \qquad (1.1.43)$$

By changing f we can assume that $g(0) = 0$. Let $V = V_n$ be the solution of

$$\begin{cases} s^2 \dfrac{d^2V}{ds^2} = A_N s^{N/(N-2)} g\left(\dfrac{V}{s(N-2)}\right) \quad \text{in } (s_n, s_{n_0}), \\[2mm] V(s_n) = V(s_{n_0}) = c/2, \end{cases} \qquad (1.1.44)$$

and let $w = w_n$ be the one of

$$\begin{cases} s^2 \dfrac{\partial^2 w}{\partial s^2} + \dfrac{1}{(N-2)^2} \Delta_{S^{N-1}} w + A_N s^{N/(N-2)} |\varphi| = 0 \quad \text{in } (s_n, s_{n_0}) \times S^{N-1}, \\[2mm] w(s_n,.) = w(s_{n_0},.) = 0 \qquad\qquad\qquad\qquad\quad \text{in } S^{N-1}. \end{cases} \qquad (1.1.45)$$

Clearly V and w are nonnegative and V-w is a subsolution of (1.1.8). Therefore $v \geq V - w$ in $(s_n, s_{n_0}) \times S^{N-1}$. Moreover, since φ is bounded, there exists some positive constant k independent of n such that $w(s, \sigma) \leq ks$ in $\left[s_n, s_{n_0}\right] \times S^{N-1}$. Denote now $v_k(s, \sigma) = v(s, \sigma) + ks$, $\bar{v}_k(s)$ its average on S^{N-1} and $g_k(r) = g(r - k/(N-2)) - g(-k/(N-2))$. Then

$$s^2 \frac{d^2 v_k}{ds^2} = A_N \left(\overline{g_k\left(\frac{v_k}{s(N-2)}\right)} + g\left(\frac{-k}{N-2}\right) - \overline{\varphi} \right). \tag{1.1.46}$$

As $g_k\left(\dfrac{v_k}{s(N-2)}\right)$ is nonnegative we deduce that the function H, which is defined by

$$H(s) = \bar{v}_k(s) + A_N \int_a^s \int_a^t \tau^{N/(N-2)}(\overline{\varphi}(\tau) - g(-k/(N-2)))d\tau dt, \tag{1.1.47}$$

is convex (here a is some real number in (0, S)) and it admits a limit as s goes to 0. Since

$$\lim_{s_n \to 0} H(s_n) = c + A_N \int_a^0 \int_a^t \tau^{N/(N-2)}(\overline{\varphi}(\tau) - g(-k/(N-2)))d\tau dt, \tag{1.1.48}$$

it implies that

$$\lim_{s \to 0} \bar{v}_k(s) = \lim_{s \to 0} \bar{v}(s) = \lim_{s \to 0} v(s, \sigma) = c \tag{1.1.49}$$

uniformly on S^{N-1}.

If $c < 0$ we proceed similarly. If $c = 0$, then $\lim\limits_{s \to 0} \bar{v}(s) = 0$, otherwise we would obtain a contradiction by the same process as above. Again this implies that $\lim\limits_{s \to 0} v(s, \sigma) = c$, uniformly on S^{N-1}.

Case 2. We assume that $\bar{v}(s)$ is not bounded near 0 and , for example, that there exists a sequence $\{s_n\}$ going to 0 such that $\lim\limits_{n \to \infty} \bar{v}(s_n) = \infty$. Then we can again define the convex function H by (2.1.47) which implies that

$$\lim_{s_n \to 0} H(s_n) = \lim_{s \to 0} H(s) = \lim_{s \to 0} \bar{v}_k(s) = \lim_{s \to 0} \bar{v}(s) = \lim_{s \to 0} v(s, \sigma) = \infty, \tag{1.1.50}$$

uniformly on S^{N-1}.

Step 2. The case N = 2. If u(x) remains bounded in B(O,R), then $\lim\limits_{x \to 0} u(x)/\mu(x) = 0$. Therefore we may assume that $\bar{u}(r)$ is not bounded and that there exists a monotone

53

sequence $\{r_n\}$ with limit 0 when n tends to infinity such that $\lim_{n\to\infty} \overline{u}(r_n) = \infty$. From Proposition 2.1 there exists a sequence $\{\alpha_n\}$ tending to infinity with n such that

$$u(r_n,\sigma) \ge \alpha_n, \qquad \left(\forall \sigma \in S^1\right). \tag{1.1.51}$$

Let ψ_n be the solution of

$$\begin{cases} -\Delta\psi_n + g(\psi_n) = -\|f\|_{L^\infty(B(O,R)} & \text{in} \quad B(O,r_n), \\ \quad\psi_n(x) = \alpha & \text{on} \quad \partial B(O,r_n), \end{cases} \tag{1.1.52}$$

where α is some positive number (we have also assumed that $g(0) = 0$ by changing f). The function ψ_n is rotationnaly symmetric and nondecreasing and $\dfrac{d\psi_n}{dr}(0) = 0$. Consequently

$$\begin{cases} \dfrac{d}{dr}\left(r\dfrac{d\psi_n}{dr}\right) \le r\left(g(\alpha) + \|f\|_{L^\infty(B(O,R))}\right) & \text{in} \quad B(0,r_n), \\ \quad\psi_n(r_n) = \alpha, \qquad \dfrac{d\psi_n}{dr}(0) = 0. \end{cases} \tag{1.1.53}$$

Integrating (1.1.53) twice gives

$$\psi_n(0) = \min_{x\in B(O,r_n)} \psi_n(x) \ge \alpha - \frac{r_n^2}{4}\left(g(\alpha) + \|f\|_{L^\infty(B(O,R))}\right). \tag{1.1.54}$$

For any α arbitrary fixed, we can find an integer n_0 such that for any $n \ge n_0$ we have $\alpha_n \ge \alpha$ and $\dfrac{r_n^2}{4}\left(g(\alpha) + \|f\|_{L^\infty(B(O,R))}\right) < \dfrac{\alpha}{2}$. Moreover, from the maximum principle, we have $u(x) \ge \psi_n(x)$ for $0 < |x| \le r_n$. This implies clearly $\liminf_{x\to 0} u(x) \ge \alpha/2$, and finally

$$\lim_{x\to 0} u(x) = \infty. \tag{1.1.55}$$

In order to end the proof of (1.1.4) we only have to prove that if (1.1.55) is satisfied then there exists $c \in [0,\infty) \cup \{\infty\}$ such that

$$\lim_{x\to 0} u(x)/\ln(1/|x|) = c. \tag{1.1.56}$$

But from (1.1.55), the boundedness of f and the monotonicity of g, either

$$\lim_{x\to 0}\left(g(u(x)) - f(x)\right) = \infty, \tag{1.1.57}$$

or there exists some constant M such that

$$|g(u(x)) - f(x)| \leq M \qquad \text{in } B(O,R). \qquad (1.1.58)$$

We use again the notations of Proposition 2.1, Step 1: if (1.1.58) is satisfied then $\dfrac{d^2\overline{v}}{dt^2}$ is integrable on (S,∞). Therefore $\dfrac{d\overline{v}}{dt}(t)$ admits a finite limit c when t goes to infinity and it is the same with $\overline{v}(t)/t$ since $\overline{v}(t)/t = \overline{v}(S)/t + t^{-1}\int_S^t \dfrac{d\overline{v}}{dt}(s)ds$. Finally (1.1.56) holds with c finite. If (1.1.57) holds, then $\overline{v}(t)$ is convex for t large enough which implies again that $\overline{v}(t)/t$ has a limit, finite or not, when t goes to infinity and again we have (1.1.56).

Step 3. We assume now that c = 0. Then for $\varepsilon > 0$ the function w_ε defined by

$$w_\varepsilon(x) = \varepsilon\mu(x) + \varPhi(x), \qquad (1.1.59)$$

where $\varPhi$ is the solution of

$$\begin{cases} -\Delta\varPhi = f^+ & \text{in } B(O,R), \\ \varPhi(x) = u^+(x) & \text{on } \partial B(O,R), \end{cases} \qquad (1.1.60)$$

is a supersolution of (1.1.1) larger than u near O. Therefore $u(x) \leq w_\varepsilon(x)$ in $B(O,R)$. Letting ε go to 0 implies $u(x) \leq \varPhi(x)$. In the same way u is minorized on $B(O,R)$ by a continuous function. Finally u can be extended as a C^1 solution of (1.1.1) in $B(O,R)$.

Remark 2.1. Estimate (1.1.3) will be obtained for any (resp. any nonnegative) solution of

$$-\Delta u + |u|^{q-1}u = 0 \qquad (1.1.61)$$

provided $1 < q < (N+1)/(N-1)$ (resp. $q > 1$).

Remark 2.2. A similar isotropy result holds (see [Ve3]) for any solution of

$$-\Delta u + g(u) = 0 \qquad (1.1.62)$$

in an exterior domain of $\mathbb{R}^N$: if we assume that g is nondecreasing and u satifies

$$\liminf_{r\to\infty} r^{-1}\left(\int_{S^{N-1}} |u(r,\sigma) - \overline{u}(r)|^2 d\sigma\right)^{1/2} = 0, \qquad (1.1.63)$$

there exists $c \in \mathbb{R} \cup \{\pm\infty\}$ such that

$$\lim_{|x|\to\infty} u(x)/\mu(x) = c. \qquad (1.1.64)$$

2-1-2 Strong singularities for solutions of $-\Delta u + |u|^{q-1}u = 0$

The so-called Emden-Fowler equations are of the following type

$$-\Delta u + \varepsilon |u|^{q-1} u = 0 \qquad\qquad (1.2.1)$$

with $\varepsilon = \pm 1$ and $q > 1$. Since they play an important role in many physical or geometrical problems they have been intensely studied since the end of the nineteenth century in assuming that the solutions are radial (see [Em], [Fo] and [Ch] as the initiators). Later on, very precised asymptotic expansions of a radial solution near a singularity were obtained in the framework of the Thomas-Fermi theory of atoms and molecules ($N = 3$, $q = 3/2$) by Sommerfeld [So] and Hille [Hi1], [Hi2]. In the case $\varepsilon = 1$, the structure of the problem can be understood if we look for specific solutions $u(r,\sigma) = r^{-2/(q-1)}\omega(\sigma)$, in spherical coordinates. For such a function u, ω satisfies

$$-\Delta_{S^{N-1}}\omega - \frac{2}{q-1}\left(\frac{2q}{q-1} - N\right)\omega + |\omega|^{q-1}\omega = 0 \qquad\qquad (1.2.2)$$

on S^{N-1}. The set $\mathcal{E}$ of the solutions of (1.2.2) is described by the following result:

PROPOSITION 2.2. *(i) If $q \geq N/(N-2)$ there exists no non-trivial solution of* (1.2.2).
(ii) If $q \geq (N+1)/(N-1)$, all the solutions of (2.2.2) are constant.
(iii) If $1 < q < (N+1)/(N-1)$, there exists non-constant solutions of (1.2.2).
(iv) If $1 < q < N/(N-2)$, there exists only one positive solution of (1.2.2).

Proof. Set $1_{q,N} = \dfrac{2}{q-1}\left(\dfrac{2q}{q-1} - N\right)$. If $q \geq N/(N-2)$, then $1_{q,N} \leq 0$; multiplying (1.2.2) by ω and integrating over S^{N-1} yields

$$\int_{S^{N-1}}\left(|\nabla\omega|^2 - 1_{q,N}\omega^2 + |\omega|^{q+1}\right)d\sigma = 0,$$

and $\omega = 0$. If $q \geq (N+1)/(N-1)$, it is worth noticing that $1_{q,N} \leq N-1 = \lambda_1$. Since

$$-\int_{S^{N-1}}(\omega - \overline{\omega})\Delta_{S^{N-1}}(\omega - \overline{\omega})d\sigma \geq \lambda_1\int_{S^{N-1}}(\omega - \overline{\omega})^2 d\sigma$$

and

$$\int_{S^{N-1}}(\omega - \overline{\omega})\left(\omega|\omega|^{q-1} - \overline{\omega|\omega|^{q-1}}\right)d\sigma = \int_{S^{N-1}}(\omega - \overline{\omega})\left(\omega|\omega|^{q-1} - \overline{\omega}|\overline{\omega}|^{q-1}\right)d\sigma,$$

$$\geq 2^{1-q}\int_{S^{N-1}}|\omega - \overline{\omega}|^{q+1}d\sigma, \qquad\qquad (1.2.3)$$

we obtain

$$\int_{S^{N-1}} \left((\lambda_1 - 1_{q,N})(\omega - \overline{\omega})^2 + 2^{1-q}|\omega - \overline{\omega}|^{q+1}\right)d\sigma \le 0, \tag{1.2.4}$$

which implies $\omega = \overline{\omega}$.

If we define the functional J on $W^{1,2}(S^{N-1}) \cap L^{q+1}(S^{N-1})$ by

$$J(\varphi) = \int_{S^{N-1}} \left(\frac{1}{2}|\nabla \varphi|^2 - \frac{1_{q,N}}{2}\varphi^2 + \frac{1}{q+1}|\varphi|^{q+1}\right)d\sigma, \tag{1.2.5}$$

then the minimum of J is achieved for $\varphi = 0$ if $q \ge N/(N-2)$ and for $\varphi = \pm 1_{q,N}^{1/(q-1)}$ if $1 < q < N/(N-2)$; in that case the minimal value of the functional J is $\dfrac{1-q}{2(q+1)}|S^{N-1}|1_{q,N}^{q+1}$.

From the maximum principle, any non-constant solution of (1.2.2) must change sign. If $1 < q < (N+1)/(N-1)$, we can construct such non-constant solutions as follows: let S_+^{N-1} be an hemisphere of S^{N-1} with equator S^{N-2} and J^* the restriction of J to $W_0^{1,2}(S_+^{N-1}) \cap L^{q+1}(S_+^{N-1})$. Since $\lambda_1 = N-1$ is the first eigenvalue of $-\Delta_{S^{N-1}}$ in $W_0^{1,2}(S_+^{N-1})$, J^* takes on negative values and its minimum is achieved by some solution ω of (1.2.2) in S_+^{N-1}, vanishing on S^{N-2}. If we reflect ω through S^{N-2}, we define a new function ω^* which is a solution of (1.2.2) on S^{N-1} and is non-constant .

For proving the uniqueness of positive solutions of (1.2.2) we follow a technique of [BBL]: Let ω and $\tilde{\omega}$ be two positive solutions of (1.2.2) on S^{N-1}, then

$$\int_{S^{N-1}} \left(\frac{\Delta_{S^{N-1}}\omega}{\omega} - \frac{\Delta_{S^{N-1}}\tilde{\omega}}{\tilde{\omega}} \right)(\omega^2 - \tilde{\omega}^2)d\sigma = \int_{S^{N-1}} \left(\omega^{q-1} - \tilde{\omega}^{q-1}\right)\left(\omega^2 - \tilde{\omega}^2\right)d\sigma. \tag{1.2.6}$$

By Green's formula

$$-\int_{S^{N-1}} \left(\left|\nabla\omega - \frac{\omega}{\tilde{\omega}}\nabla\tilde{\omega}\right|^2 + \left|\nabla\tilde{\omega} - \frac{\tilde{\omega}}{\omega}\nabla\omega\right|^2 \right)d\sigma = \int_{S^{N-1}} \left(\omega^{q-1} - \tilde{\omega}^{q-1}\right)\left(\omega^2 - \tilde{\omega}^2\right)d\sigma \tag{1.2.7}$$

which implies that $\omega = \tilde{\omega}$.

When $N = 2$, we can give a complete desription of $\mathcal{E}$ (see [CMV] for example).

PROPOSITION 2.3. *Assume* $N = 2$. *Then*

(i) If $q \ge 3$, $\mathcal{E}$ *is reduced to the three elements* $0, \pm 1_{q,2}^{1/(q-1)}$.

(ii) If $1 < q < 3$, *we call* k_q *the largest integer smaller than* $\sqrt{1_{q,2}} = 2/(q-1)$. $\mathcal{E}$ *is the union of the isolated elements* $0, \pm 1_{q,2}^{1/(q-1)}$ *and of the* $\mathcal{E}_j$ $(1 \le j \le k_q)$ *which are closed curves of solutions of the differential equation*

$$\frac{d^2\omega}{d\sigma^2} + \left(\frac{2}{q-1}\right)^2 - |\omega|^{q-1}\omega = 0 \tag{1.2.8}$$

generated by the S^1-action from a particular (π/j)-anti-periodic solution .

Proof. For simplicity we denote $\lambda = (2/(q-1))^2$ and we assume (ii), that is $\lambda > 1$. Any non-constant solution of (1.2.8) must vanish at some point, say at 0, where the derivative is not zero. Let ω be a non-constant solution of (1.2.8) with $\omega(0) = 0$, $\frac{d\omega}{d\sigma}(0) = \alpha > 0$. For some interval $(0,\varepsilon) \subset (0,\pi)$, $\varepsilon > 0$, ω is increasing. Since

$$\frac{1}{2}\frac{d}{d\sigma}\left(\frac{d\omega}{d\sigma}\right)^2 = \frac{d}{d\sigma}\left(\frac{|\omega|^{q+1}}{q+1} - \frac{\lambda}{2}\omega\right) \tag{1.2.9}$$

and $\frac{d\omega}{d\sigma} > 0$ on $(0,\varepsilon)$, we obtain

$$\sigma = \int_0^{\omega(\sigma)} \frac{ds}{\sqrt{\alpha^2 + 2|\omega|^{q+1}/(q+1) - \lambda\omega^2}}, \tag{1.2.10}$$

and this formula is valid as long as ω remains smaller than the first zero of $r \mapsto \psi(\alpha,r) = \alpha^2 + 2|r|^{q+1}/(q+1) - \lambda r^2$. Since this function decreases on $\left(0, \lambda^{2/(q-1)}\right)$, we have only one possibility: $\alpha^2 < \lambda\lambda^{2/(q-1)} - 2\lambda^{(q+1)/(q-1)}/(q+1) = \frac{q-1}{q+1}\lambda^{(q+1)/(q-1)}$. In that case the function $r \mapsto \psi(\alpha,r)$ admits a simple zero $s(\alpha)$ in $\left(0, \lambda^{2/(q-1)}\right)$ and $\frac{\partial\psi}{\partial r}(\alpha, s(\alpha)) \neq 0$; setting

$$\sigma(\alpha) = \int_0^{s(\alpha)} \frac{ds}{\sqrt{\psi(\alpha,s)}}, \tag{1.2.11}$$

then $\sigma(\alpha) < \infty$, $\omega(\sigma(\alpha)) = s(\alpha)$ and $\frac{d\omega}{d\sigma}(\sigma(\alpha)) = 0$. By performing a symmetry about $\sigma(\alpha)$, we see that ω is decreasing on $(\sigma(\alpha), 2\sigma(\alpha))$, vanishes at $2\sigma(\alpha)$ and $\frac{d\omega}{d\sigma}(2\sigma(\alpha)) = -\alpha^2$. Therefore ω is a non-constant solution of (1.2.8) if and only if $\pi/(2\sigma(\alpha))$ is a positive integer.

Step 1. We claim that the mapping $\alpha \mapsto s(\alpha)$ is increasing and convex on $\left[0, \dfrac{q-1}{q+1}\lambda^{(q+1)/(q-1)}\right]$. From the implicit function theorem this function is C^2. Since

$$\frac{d}{d\alpha}\big(\psi(\alpha,s(\alpha))\big) = \frac{\partial\psi}{\partial\alpha}(\alpha,s(\alpha)) + \frac{ds}{d\alpha}\frac{\partial\psi}{\partial s}(\alpha,s(\alpha)),$$

we have

$$\frac{ds}{d\alpha} = \frac{\alpha}{\lambda s(\alpha) - s^q(\alpha)}, \tag{1.2.12}$$

which positive on $\left(0, \lambda^{1/(q-1)}\right)$ and

$$\frac{d^2 s}{d\alpha^2} = \frac{\big(\lambda s(\alpha) - s^q(\alpha)\big)^2 - \alpha^2\big(\lambda - q s^{q-1}(\alpha)\big)}{\big(\lambda s(\alpha) - s^q(\alpha)\big)^3}. \tag{1.2.13}$$

Therefore, the sign of the second derivative of $s(.)$ is the same as the one of

$$\tau(\alpha) = \big(\lambda s(\alpha) - s^q(\alpha)\big)^2 - \alpha^2\big(\lambda - q s^{q-1}(\alpha)\big). \tag{1.2.14}$$

Since $\tau(0) = 0$ and $\dfrac{d\tau}{d\alpha}(s(\alpha)) = \alpha^2 q(q-1)s^{q-2}(\alpha)$, the function s is convex.

Step 2. The mapping $\alpha \mapsto \sigma(\alpha)$ is increasing and continuous on $\left[0, \dfrac{q-1}{q+1}\lambda^{(q+1)/(q-1)}\right)$.

If u belongs to $[0,\alpha]$, the mapping $s \mapsto \psi(u,s)$ admits a first zero at $s = s(u)$ and $\psi(\alpha,s(u)) = \alpha^2 - u^2$. Therefore

$$\sigma(\alpha) = \int_0^\alpha \frac{ds}{du}(u)\frac{du}{\sqrt{\alpha^2 - u^2}} = \int_0^1 \frac{ds}{du}(\alpha\theta)\frac{d\theta}{\sqrt{1-\theta^2}}. \tag{1.2.15}$$

Since $u \mapsto \dfrac{ds}{du}(u)$ is increasing and C^1, it is the same with $\alpha \mapsto \sigma(\alpha)$.

Step 3. End of the proof. Since $\lim\limits_{\alpha\to 0} s(\alpha) = 0$, we have

$$\lim_{\alpha\to 0} \frac{s(\alpha)}{\alpha} = \frac{1}{\sqrt{\lambda}} \tag{1.2.16}$$

and

$$\lim_{\alpha \to 0} \sigma(\alpha) = \frac{\pi}{\sqrt{\lambda}}. \tag{1.2.17}$$

Moreover

$$\lim_{\alpha \uparrow \frac{q-1}{q+1}\lambda^{(q+1)/(q-1)}} s(\alpha) = \infty. \tag{1.2.18}$$

Therefore the function $\alpha \mapsto \sigma(\alpha)$ is a continuous bijection from $\left[0, \dfrac{q-1}{q+1}\lambda^{(q+1)/(q-1)}\right)$ onto $\left[\dfrac{\pi}{2\sqrt{\lambda}}, \infty\right)$. Since $\sqrt{\lambda} = 2/(q-1)$, there exist exactely k_q non-zero-integers j smaller than $2/(q-1)$. To each of those integers is associated a value α_j of the derivative at 0 of a non-constant solution ω_j of (1.2.8) with minimal antiperiod π/j. In that case $\mathcal{E}_j = \left\{\omega_j(.+\delta): \delta \in (0, 2\pi/j]\right\}$.

Remark 2.3. It is easy to check that in the case $N = 2$, the functional J defined in (1.2.5) takes on different values on each connected component of $\mathcal{E}$. The range of values of J goes from $\dfrac{1-q}{q+1}\pi\lambda^{(q+1)/(q-1)}$ corresponding to $\pm l_{q,2}^{1/(q-1)}$, to 0 corresponding to the zero solution. Moreover we have the following exclusion inequalities $\dfrac{1-q}{q+1}\pi\lambda^{(q+1)/(q-1)} < J_{|\mathcal{E}_j} < J_{|\mathcal{E}_1} < 0$, if $j < 1$.

  With the knowledge of the set $\mathcal{E}$ of the solutions of (1.2.2) we can start the describtion of the isolated singularities of the solution of

$$-\Delta u + |u|^{q-1} u = 0. \tag{1.2.19}$$

The classification of isotropic singularities was given in [Vel], but for the non-isotropic ones new techniques were needed. We present here a more global proof based upon some dynamical systems techniques and Lyapounov type analysis (see [CMV] for example).

THEOREM 2.2. *Let Ω be some open subset of $\mathbb{R}^N$, $N > 1$, containing O, $\Omega^* = \Omega \setminus \{O\}$ and u a solution of (1.2.19) in Ω^* with $1 < q < N/(N-2)$. Then there exists a connected component $\mathcal{E}^*$ of the set $\mathcal{E}$ such that*

$$\lim_{r \to 0} \mathrm{dist}_{C^2}\left(r^{2/(q-1)}u(r,.), \mathcal{E}^*\right) = 0 \tag{1.2.20}$$

where

$$\mathrm{dist}_{C^2}\left(r^{2/(q-1)}u(r,.), \mathcal{E}^*\right) = \inf\left(\left\|r^{2/(q-1)}u(r,.) - \omega(.)\right\|_{C^2(S^{N-1})} : \omega \in \mathcal{E}^*\right). \tag{1.2.21}$$

The following *a priori* estimate essentially due to Osserman [Os] and, separately, Keller [Ke], plays a fundamental role into the singularity theory of nonlinear elliptic equations with absorption.

PROPOSITION 2.4. *Assume* A *and* B *are two given constants, respectively positive and nonnegative, and* $\psi \in L^1(B(O,R))$ *with* $\Delta\psi \in L^1(B(O,R))$ *satisfies the following inequality*

$$-\Delta\psi + A\psi^q \leq B,\tag{1.2.22}$$

almost everywhere in the set $G = \{x \in B(O,R) : \psi(x) > 0\}$. *Then*

$$\psi(O) \leq \left(\frac{c(q,N)}{AR^2}\right)^{1/(q-1)} + \left(\frac{B}{A}\right)^{1/(q-1)}.\tag{1.2.23}$$

Proof. The idea of the proof is simply to introduce a function $x \mapsto w(x) = \lambda\left(R^2 - |x^2|\right)^{-2/(q-1)} + \mu$ where the parameters λ and μ are chosen in such a way that

$$-\Delta w + Aw^q \geq B\tag{1.2.24}$$

in $B(O,R)$. A good choice is $\lambda = \left(\frac{c(q,N)}{A}R^2\right)^{1/(q-1)}$ with $c(q,N) = 16\max\left(\frac{2N}{q-1}, \frac{4(q+1)}{(q-1)^2}\right)$ and $\mu = \left(\frac{B}{A}\right)^{1/(q-1)}$.

Remark 2.4. If we assume that ψ satisfies the hypotheses of Proposition 2.4 in some open subset Ω of $\mathbb{R}^N$ instead of $B(O,R)$, then (12.23) has to be replaced by

$$\psi(x) \leq \left(\frac{c(q,N)}{A\left(\mathrm{dist}(x,\Omega^c)\right)^2}\right)^{1/(q-1)} + \left(\frac{B}{A}\right)^{1/(q-1)}.\tag{1.2.25}$$

Proof of Theorem 2.1. The idea is to transform (1.2.19) into an autonomous equation in an infinite cylinder. We can assume that $B(O,2) \subset \Omega$ and set

$$u(r,s) = r^{-2/(q-1)}v(t,s), \quad t = \ln(1/r).\tag{1.2.26}$$

Then v satisfies

$$v_{tt} + \left(2\frac{q+1}{q-1} - N\right)v_t + 1_{q,N}^{q-1}v + \Delta_{S^{N-1}}v - |v|^{q-1}v = 0 \quad \text{in } (-\ln 2, \infty) \times S^{N-1}. \quad (1.2.27)$$

Step 1. We claim that there exists a constant $M > 0$ such that

$$\|v(t,.)\|_{C^3(S^{N-1})} + \|v_t(t,.)\|_{C^2(S^{N-1})} + \|v_{tt}(t,.)\|_{C^1(S^{N-1})}$$
$$+ \|v_{ttt}(t,.)\|_{C(S^{N-1})} + \|\nabla v_t(t,.)\|_{C1(S^{N-1})} \leq M \quad (1.2.28)$$

for $t \geq 0$. In fact, from Proposition 2.4, v is uniformly bounded on $[-\ln 3/2, \infty) \times S^{N-1}$ and it is the same for the nonlinear term $|v|^{q-1}v$. If we apply Agmon-Douglis-Nirenberg local estimates to the elliptic operator

$$w \mapsto Lw = w_{tt} + \left(2\frac{q+1}{q-1} - N\right)w_t + \Delta_{S^{N-1}}w \quad (1.2.29)$$

in the domain $(T - 5/4, T + 5/4) \times S^{N-1}$, $T \geq 0$, we deduce that for any $p \in (1, \infty)$, there exists a constant $C > 0$ such that

$$\|v\|_{W^{2,p}\left((T-8/7, T+8/7) \times S^{N-1}\right)} \leq C. \quad (1.2.30)$$

holds for any $T \geq 0$. But, for any $\gamma \in (0,1)$, we can take p large enough so that the space $W^{2,p}\left((T - 8/7, T + 6/7) \times S^{N-1}\right)$ is imbedded into $C^{1,\gamma}\left([T - 8/7, T + 6/7] \times S^{N-1}\right)$. Therefore v and $|v|^{q-1}v$ remain uniformly bounded into $C^{1,\gamma}\left([T - 8/7, T + 6/7] \times S^{N-1}\right)$ provided $0 < \gamma < \min(1, q-1)$. Applying now Schauder local estimates in $(T - 8/7, T + 8/7) \times S^{N-1}$ yields

$$\|v\|_{C^3\left([T-1, T+1] \times S^{N-1}\right)} \leq M \quad (1.2.31)$$

for some $M > 0$ and any $T \geq 0$, which implies the claim.

Step 2. We claim that

$$\lim_{t \to \infty} \|v_t\|_{L^2(S^{N-1})} = \lim_{t \to \infty} \|v_{tt}\|_{L^2(S^{N-1})} = 0. \quad (1.2.32)$$

By multiplying (1.2.27) by v_t the following energy variation is derived

$$\left(2\frac{q+1}{q-1} - N\right)\int_{S^{N-1}} v_t^2 d\sigma = \frac{d}{dt}E(v) \quad (1.2.33)$$

where $E(v) = \dfrac{1}{2}\displaystyle\int_{S^{N-1}}\left(|\nabla v|^2 - v_t^2 - 1_{q,N}v^2 + \dfrac{2}{q+1}|v|^{q+1}\right)d\sigma$. From Step 1 the E (v) is bounded. Therefore

$$\int_0^\infty\int_{S^{N-1}} v_t^2\,d\sigma dt < \infty, \tag{1.2.34}$$

since $\left(2\dfrac{q+1}{q-1} - N\right) \neq 0$, or, equivalently, $q \neq \dfrac{N+2}{N-2}$. From Step 1 v_t is uniformly continuous on $[0,\infty)\times S^{N-1}$, therefore

$$\lim_{t\to\infty}\|v_t\|_{L^2(S^{N-1})} = 0. \tag{1.2.35}$$

We differentiate (1.2.27) with respect to t and set $v_t = w$. Then

$$w_{tt} + \left(2\dfrac{q+1}{q-1} - N\right)w_t + 1_{q,N}^{q-1}w + \Delta_{S^{N-1}}w - q|v|^{q-1}w = 0 \tag{1.2.36}$$

holds in in $(-\ln 2,\infty)\times S^{N-1}$, which implies that

$$\left(2\dfrac{q+1}{q-1} - N\right)\int_{S^{N-1}} w_t^2\,d\sigma = \dfrac{d}{dt}\tilde{E}(w) + q\int_{S^{N-1}} w_t w|v|^{q-1}d\sigma, \tag{1.2.37}$$

where $\tilde{E}(w) = \dfrac{1}{2}\displaystyle\int_{S^{N-1}}\left(|\nabla w|^2 - w_t^2 - 1_{q,N}w^2\right)d\sigma$. Since $\tilde{E}(w)$ is uniformly bounded from (1.2.28), we deduce from (1.2.34), with the help of Young's inequality, that

$$\int_0^\infty\int_{S^{N-1}} w_t^2\,d\sigma dt < \infty, \tag{1.2.38}$$

which again implies

$$\lim_{t\to\infty}\|w_t\|_{L^2(S^{N-1})} = 0, \tag{1.2.39}$$

by using the uniform continuity of w_t.

Step 3. End of the proof. Let $\mathcal{T}$ be the (positive) trajectory of v defined by

$$\mathcal{T} = \bigcup_{t\geq 0}\{v(t,.)\}. \tag{1.2.40}$$

From Step 1, $\mathcal{T}$ is bounded in $C^3(S^{N-1})$, henceforth relatively compact in $C^2(S^{N-1})$. If we define the ω-limit set of the trajectory Γ^+ by

$$\Gamma^+ = \bigcap_{t \geq 0} \left(\overline{\bigcap_{\tau \geq t} \{v(t,.)\}}^{C^2(S^{N-1})} \right), \tag{1.2.41}$$

then Γ^+ is a decreasing intersection of non-empty, compact and connected subsets of $C^2(S^{N-1})$, therefore it is non-empty, compact and connected. Assume ω is some element of Γ^+, then there exists a sequence $\{t_n\}$ going to infinity with n such that

$$\lim_{n \to \infty} \|v(t_n,.) - \omega(.)\|_{C^2(S^{N-1})} = 0 \tag{1.2.42}$$

and from Step 2, $\lim_{n \to \infty} \|v_t(t_n,.)\|_{L^2(S^{N-1})} = \lim_{n \to \infty} \|v_{tt}(t_n,.)\|_{L^2(S^{N-1})} = 0$, which implies that ω satisfies (1.2.2).

In some particular cases the limit set $\mathcal{E}^*$ is reduced to a single element:

COROLLARY 2.1. *Assume* $1 < q < N/(N-2)$ *and* u *is a nonnegative solution of* (1.2.19) *in* $\Omega^* = \Omega \setminus \{O\}$. *Then*
(i) either

$$\lim_{x \to 0} |x|^{2/(q-1)} u(x) = 1_{q,N}^{1/(q-1)}, \tag{1.2.43}$$

(ii) or

$$\lim_{x \to 0} |x|^{2/(q-1)} u(x) = 0. \tag{1.2.44}$$

Assume $(N+1)/(N-1) \leq q < N/(N-2)$ *and* u *is a solution of* (1.2.19) *in* $\Omega^* = \Omega \setminus \{O\}$ *Then* $\lim_{x \to 0} |x|^{2/(q-1)} u(x)$ *exists and it can only take the three values* $\pm 1_{q,N}^{1/(q-1)}, 0$.

In the above theorem it may happen that the limit set of v is reduced to the zero element. We shall see in the next section that u has a singularity of linear type in that case. Another natural question deals with the uniqueness of the limit element ω in $\mathcal{E}^*$; when N = 2, Chen, Matano and Véron [CMV] proved that it always holds.

THEOREM 2.3. *Assume* N = 2, q > 1 *and* u *is a solution of* (2.2.19) *in* $\Omega^* = \Omega \setminus \{O\}$. *Then there exists a* 2π-*periodic solution* ω *of the differential equation* (1.2.8) *such that*

$$\lim_{r \to 0} r^{2/(q-1)} u(r,.) = \omega(.), \tag{1.2.45}$$

in the $C^2(S^1)$-*topology.*

Proof. From the description of $\mathcal{E}$ and Theorem 2.2, the notations of which are kept, we may assume that the limit set Γ^+ of the trajectory of v is included into a set $\mathcal{E}_j$ generated by a non-constant solution ω_j of (1.2.8) with minimal anti-period π / j. Assume now that Γ^+ contains two different elements of $\mathcal{E}_j$, say ω and ω' . Then there exists two sequences $\{t_n\}$ and $\{t'_n\}$ tending to infinity with n such that $\lim_{n\to\infty} v(t_n,.) = \omega(.)$ and $\lim_{n\to\infty} v(t'_n,..) = \omega'(.)$ in the $C^2(S^1)$-topology.

The key ingredient for obtaining the contradiction is the following: *If $\sigma_0 \in [0, 2\pi]$ is such that $\dfrac{d\omega}{d\sigma}(\sigma_0) > 0$ (resp. $\dfrac{d\omega}{d\sigma}(\sigma_0) < 0$ in the same way), then there exists $T > 0$ such that $\dfrac{\partial v}{\partial \sigma}(t, \sigma_0)$ is positive (resp. negative) for any $t \geq T$.* In fact, if we assume that $\dfrac{\partial v}{\partial \sigma}(t, \sigma_0)$ does not keep a constant sign at infinity, there exists a sequence $\{\tau_n\}$ tending to infinity with n such that

$$(-1)^n \frac{\partial v}{\partial \sigma}(\tau_n, \sigma_0) > 0. \tag{1.2.46}$$

We perform a reflection about σ_0 by defining

$$\begin{cases} \tilde{v}(t, \sigma) = v(t, \sigma) - v(t, 2\sigma_0 - \sigma), \\ \tilde{u}(t, \sigma) = u(t, \sigma) - u(t, 2\sigma_0 - \sigma), \\ \tilde{\omega}(\sigma) = \omega(\sigma) - \omega(2\sigma_0 - \sigma). \end{cases} \tag{1.2.47}$$

Then $\tilde{v}(t, \sigma_0) = \tilde{v}(t, \sigma_0 + \pi) = \tilde{\omega}(\sigma_0) = \tilde{\omega}(\sigma_0 + \pi) = 0$, $\dfrac{\partial \tilde{v}}{\partial \sigma}(\tau_n, \sigma_0) = 2\dfrac{\partial v}{\partial \sigma}(\tau_n, \sigma_0)$, as for $\tilde{u}$, it satisfies

$$- \Delta \tilde{u} + h\tilde{u} = 0, \tag{1.2.48}$$

where h is a nonnegative function, in the angular domain $G = (0, 2) \times (\sigma_0, \sigma_0 + \pi)$. Moreover $\tilde{u}$ vanishes on the lateral boundary of G and $(-1)^n \dfrac{\partial \tilde{u}}{\partial \sigma}(\rho_n, \sigma_0) > 0$, with $\rho_n = e^{-\tau_n}$. By continuity, every point (ρ_n, σ_0) belongs to a connected component G_n of G where $(-1)^n \tilde{u} > 0$. From the maximum principle, any of the G_n must intersect $\{x : |x| = 1\}$ or contains O in its closure. Since the G_n are arc-connected, we have two possibilities:

(i) either all the G_n intersect $\{x : |x| = 1\}$,

(ii) or there exists some n_0 such that G_{n_0} contains O in its closure and, in that case, any G_n such that $n \geq n_0$ contains O in its closure.

We define $\mathbf{Z}(\tilde{u}(r,.))$ to be the number of change sign of the function $\sigma \mapsto \tilde{u}(r,\sigma)$ when σ ranges from σ_0 to $\sigma_0 + \pi$. In case (i), we have $\mathbf{Z}(\tilde{u}(r,.)) = \infty$ for any $0 < r < 1$, and in case (ii) $\lim_{r \to 0} \mathbf{Z}(\tilde{u}(r,.)) = \infty$. Consequently, either $\mathbf{Z}(\tilde{v}(t_n,.)) = \infty$ for any $n > 0$, or $\lim_{n \to \infty} \mathbf{Z}(\tilde{v}(t_n,.)) = \infty$. Since the convergence of $\tilde{v}(t_n,.)$ toward $\tilde{\omega}$ occurs in the $C^2([\sigma_0,\sigma_0 + \pi])$-topology, the function $\tilde{\omega}$ must have a double zero in the interval $[\sigma_0,\sigma_0 + \pi]$. But $\tilde{\omega}$ satisfies a second order linear differential equation

$$\frac{d^2 \tilde{\omega}}{d\sigma^2} + 1_{q,2}\tilde{\omega} - \tilde{h}\tilde{\omega} = 0 \tag{1.2.49}$$

where $\tilde{h}$ is bounded. Therefore $\tilde{\omega} = 0$, which contradicts the fact that $\frac{d\tilde{\omega}}{d\sigma}(\sigma_0) > 0$.

If Γ^+ contains two different elements ω and ω' in $\mathcal{E}_j$, there exists some $\sigma_0 \in [0,2\pi]$ such that $\frac{d\omega}{d\sigma}(\sigma_0) > 0$ and $\frac{d\omega'}{d\sigma}(\sigma_0) < 0$, which yields a contradiction.

2-1-3 Weak singularities for solutions of $-\Delta u + |u|^{q-1}u = 0$

The following results complete Theorems 2-2 and 2-3

THEOREM 2.4. *Let Ω be some open subset of $\mathbb{R}^N$, $N > 1$, containing O, $\Omega^* = \Omega \setminus \{O\}$, and u be a solution of (1.2.19) in Ω^* with $1 < q < N/(N-2)$, such that*

$$\lim_{x \to 0} |x|^{2/(q-1)} u(x) = 0. \tag{1.3.1}$$

Then, if $2/(q-1)$ is not an integer, we have the following
(i) either there exist an integer $k \in [N-1, 2/(q-1))$ and a non-zero spherical harmonics Φ of degree $k + 2 - N$ such that

$$\lim_{r \to 0} r^k u(r,.) = \Phi(.), \tag{1.3.2}$$

in the $C^2(S^{N-1})$-topology,
(ii) either there exists a non-zero real number c such that

$$\lim_{r \to 0} u(r,.))/\mu(r) = c \tag{1.3.3}$$

in the $C^2(S^{N-1})$-topology,
(iii) or u can be extended to Ω as a C^3 solution of (1.2.19) in Ω.

This theorem is settled upon the following construction (see [CMV]-Appendix).

LEMMA 2.1. *Let ρ be a continuous positive function defined on $[0,\infty)$ satisfying*

$$\lim_{t\to\infty} \rho(t) = 0, \tag{1.3.4}$$

$$\limsup_{t\to\infty} e^{t\varepsilon}\rho(t) = \infty, \tag{1.3.5}$$

for any $\varepsilon > 0$. Then there exists a function η of class C^∞ on $[0,\infty)$ such that

$$\eta > 0, \quad \eta' < 0, \quad \lim_{t\to\infty} \eta(t) = 0, \tag{1.3.6}$$

$$0 < \limsup_{t\to\infty} \rho(t)/\eta(t) < \infty, \tag{1.3.7}$$

$$\lim_{t\to\infty} e^{t\varepsilon}\eta(t) = \infty, \tag{1.3.8}$$

for any $\varepsilon > 0$ and

$$(\eta'/\eta)', \quad (\eta'/\eta)'' \in L^1(0,\infty), \tag{1.3.9}$$

$$\lim_{t\to\infty} (\eta'/\eta)(t) = \lim_{t\to\infty} (\eta''/\eta)(t) = 0. \tag{1.3.10}$$

Proof of Theorem 2.4. We follow again the notations (1.2.26) of Theorem 2.2. First we use the function η to prove that there is some exponential decay for v ; therefore, thanks to classical linear analysis techniques, we prove the exact behaviour.

Step 1. We claim that there exists some $\gamma > 0$ such that $e^{-t\gamma}v(t,\sigma)$ is uniformly bounded on $[0,\infty)\times S^{N-1}$. Proceeding by contradiction we set $\rho(t) = \|v(t,.)\|_{L^\infty(S^{N-1})}$ and $w(t,\sigma) = v(t,\sigma)/\eta(t)$. The function w satisfies

$$w_{tt} + \left(2\frac{q+1}{q-1} - N + 2\frac{\eta'}{\eta}\right)w_t + \left(1_{q,N} + \frac{\eta''}{\eta} + \left(2\frac{q+1}{q-1} - N\right)\frac{\eta'}{\eta}\right)w \tag{1.3.11}$$

$$+ \Delta_{S^{N-1}}w - \eta^{q-1}|w|^{q-1}w = 0,$$

and is uniformly bounded. Henceforth, as in Theorem 2.2, we use Agmon-Douglis-Nirenberg and Schauder estimates and obtain

$$\|w(t,.)\|_{C^3(S^{N-1})} + \|w_t(t,.)\|_{C^2(S^{N-1})} + \|w_{tt}(t,.)\|_{C^1(S^{N-1})}$$

$$+ \|w_{ttt}(t,.)\|_{C(S^{N-1})} + \|\nabla w_t(t,.)\|_{C1(S^{N-1})} \leq M, \tag{1.3.12}$$

for some $M > 0$ and any $t \geq 0$. Multiplying (1.3.12) by w_t and integrating over S^{N-1} implies

$$\left(2\frac{q+1}{q-1} - N + 2\frac{\eta'}{\eta}\right)\int_{S^{N-1}} w_t^2 d\sigma =$$

$$\frac{1}{2}\frac{d}{dt}\int_{S^{N-1}}\left(|\nabla w|^2 - w_t^2 - \left(1_{q,N} + \frac{\eta''}{\eta} + \left(2\frac{q+1}{q-1} - N\right)\frac{\eta'}{\eta}\right)w^2\right)d\sigma \qquad (1.3.13)$$

$$\frac{1}{2}\int_{S^{N-1}}\left(\frac{\eta''}{\eta} + \left(2\frac{q+1}{q-1} - N\right)\frac{\eta'}{\eta}\right)' w^2 d\sigma + \int_{S^{N-1}} \eta^{q-1}|w|^{q-1} w w_t d\sigma.$$

But

$$\int_{S^{N-1}} \eta^{q-1}|w|^{q-1} w w_t d\sigma = \frac{d}{dt}\left(\frac{1}{q+1}\int_{S^{N-1}} \eta^{q-1}|w|^{q+1} d\sigma\right)$$

$$-\frac{q-1}{q+1}\eta'\,\eta^{q-2}\int_{S^{N-1}}|w|^{q+1}d\sigma, \qquad (1.3.14)$$

and moreover, for any $0 \leq t_1 \leq t_2$, there exists $\tau \in [t_1, t_2]$ such that

$$\frac{1-q}{q+1}\int_{t_1}^{t_2}\eta'\,\eta^{q-2}\int_{S^{N-1}}|w|^{q+1}d\sigma = \frac{1}{q+1}\left(\eta^{q-1}(t_1) - \eta^{q-1}(t_2)\right)\int_{S^{N-1}}|w(\sigma,\tau)|^{q+1}d\sigma. \quad (1.3.15)$$

Consequently, we deduce from (1.3.9), (1.3.10) and (1.3.12)- (1.3.15) that

$$\int_0^\infty\int_{S^{N-1}} w_t^2 d\sigma dt < \infty. \qquad (1.3.16)$$

Multiplying (1.3.11) by w_{tt} implies

68

$$\int_{S^{N-1}} w_{tt}^2 \, d\sigma =$$

$$-\frac{d}{dt}\int_{S^{N-1}}\left[\left(\frac{q+1}{q-1}-\frac{N}{2}+\frac{\eta'}{\eta}\right)w_t^2 -\left(1_{q,N}+\frac{\eta''}{\eta}+\left(2\frac{q+1}{q-1}-N\right)\frac{\eta'}{\eta}\right)ww_t\right.$$

$$\left.+\nabla w_t \cdot \nabla w -|\nabla w_t|^2 +\eta^{q-1}|w|^{q-1}ww\right]d\sigma$$

$$+\int_{S^{N-1}}\left[\left(\frac{\eta'}{\eta}\right)' w_t^2 +\left(\frac{\eta''}{\eta}+\left(2\frac{q+1}{q-1}-N\right)\frac{\eta'}{\eta}\right)' ww_t\right]d\sigma$$

$$\tag{1.3.17}$$

$$+\int_{S^{N-1}}\left[\left(1_{q,N}+\frac{\eta''}{\eta}+\left(2\frac{q+1}{q-1}-N\right)\frac{\eta'}{\eta}\right)w_t^2\right]d\sigma$$

$$-\int_{S^{N-1}}\left[q\eta^{q-1}|w|^{q-1}w_t^2 -(q-1)\eta'\,\eta^{q-2}|w|^{q-1}ww_t\right]d\sigma.$$

Because $(q-1)\int_{t_1}^{t_2}\eta'\,\eta^{q-2}\int_{S^{N-1}}w^q w_t \, d\sigma dt$ is bounded independently of t_1 and $t_2 > 0$, as in (1.3.15), we deduce from (1.3.9), (1.3.10) and (1.3.12)

$$\int_0^{\infty}\int_{S^{N-1}} w_{tt}^2 \, d\sigma dt < \infty, \tag{1.3.18}$$

which, in turn, implies

$$\lim_{t\to\infty}\left\|w_t(t,.)\right\|_{L^2(S^{N-1})} = \lim_{t\to\infty}\left\|w_{tt}(t,.)\right\|_{L^2(S^{N-1})} = 0 \tag{1.3.19}$$

as in Theorem 2.2. From (1.3.12) the trajectory $\mathcal{T}_w$ of w is relatively compact in $C^2(S^{N-1})$ and its limit set at infinity Γ_w^+ is non-empty, compact and connected. From (1.3.7) there exists a sequence $\{t_n\}$ tending to infinity with n and a positive real number θ such that $\lim_{n\to\infty}\left\|w(t_n,.)\right\|_{L^{\infty}(S^{N-1})} = \theta$. Therefore we can extract a subsequence $\{t_{n_k}\}$ such that $w(t_{n_k},.)$ converges in the $C^2(S^{N-1})$-topology to some non-zero ω. But this element ω, like any element of Γ_w^+, is a solution of

$$\Delta_{S^{N-1}}\omega + 1_{q,N}\omega = 0. \tag{1.3.20}$$

Since $\dfrac{2}{q-1}+2-N$ is not an integer, $1_{q,N} = \left(\dfrac{2}{q-1}\right)\left(\dfrac{2}{q-1}+2-N\right)$ is not an eigenvalue of $-\Delta_{S^{N-1}}$ in $W^{1,2}(S^{N-1})$, contradiction.

Step 2. Let k_q be the largest integer smaller than $2/(q-1)$. We claim that

$$\|u(r,.)\|_{W^{1,2}(S^{N-1})} \le Mr^{-k_q} \tag{1.3.21}$$

for some $M > 0$. We recall that the eigenvalues of $-\Delta_{S^{N-1}}$ are the $\lambda_k = k(k+N-2)$ with the corresponding eigenspaces H_k of dimension d_k (see Theorem 1.2). If $\{e_j\}$, $1 \le j \le k$, is an orthonormal basis of H_k, we write

$$v(t,.) = \sum_0^\infty \sum_1^{d_k} a_k^j(t)e_k^j \tag{1.3.22}$$

and

$$v|v|^{q-1}(t,.) = \sum_0^\infty \sum_1^{d_k} A_k^j(t)e_k^j. \tag{1.3.23}$$

With the above notations, (1.3.21) reads as follows

$$\sum_0^\infty \sum_1^{d_k} (1+\lambda_k)\left(a_k^j(t)\right)^2 \le Me^{-2t(2/(q-1)-k_q)} = Me^{-2t\delta_q}, \tag{1.3.23}$$

where $\delta_q = 2/(q-1) - k_q$. If we project (1.2.27) orthogonally onto e_k^j, we obtain

$$\frac{d^2 a_k^j}{dt^2} + \left(2\frac{q+1}{q-1} - N\right)\frac{da_k^j}{dt} + \left(1_{q,N} - \lambda_k\right)a_k^j = A_k^j. \tag{1.3.25}$$

Let r_1 and r_2 be the two roots of $X \mapsto X^2 + \left(2\frac{q+1}{q-1} - N\right)X + 1_{q,N} - \lambda_k$, that is

$$\begin{cases} \text{(i)} \quad r_1 = -\dfrac{2}{q-1} - k, \\[4mm] \text{(ii)} \quad r_2 = -\dfrac{2}{q-1} + k + N - 2, \end{cases} \tag{1.3.26}$$

then (up to a small modification when $k = 0$ and $N = 2$)

$$a_k^j(t) = (r_1 - r_2)^{-1}(\frac{da_k^j}{dt}(0) - r_2 a_k^j(0))e^{r_1 t} + (r_2 - r_1)^{-1}(\frac{da_k^j}{dt}(0) - r_1 a_k^j(0))e^{r_2 t}$$
$$+ (r_1 - r_2)^{-1}\int_0^t \left(e^{r_1(t-s)} - e^{r_2(t-s)}\right)A_k^j(s)ds, \tag{1.3.27}$$

for $0 \le k \le k_q + 2 - N$ and $1 \le j \le d_k$, and

$$a_k^j(t) = a_k^j(0)e^{r_1 t} + (r_1 - r_2)^{-1}e^{r_1 t}\int_0^t \left(e^{-r_1 s} - e^{-r_2 s}\right)A_k^j(s)ds$$

$$+(r_1 - r_2)^{-1}\left(e^{r_1 t} - e^{r_2 t}\right)\int_t^\infty e^{-r_2 s}A_k^j(s)ds, \tag{1.3.28}$$

for $k \geq k_q + 3 - N$ and $1 \leq j \leq d_k$.

- **Estimate of** A_k^j. From Step 1 we know that $z(t,\sigma) = e^{t\gamma}v(t,\sigma)$ remains uniformly bounded and z satisfies

$$z_{tt} + \left(2\frac{q+1}{q-1} - N - 2\gamma\right)z_t + \left(1_{q,N} + \gamma^2 - \left(2\frac{q+1}{q-1} - N\right)\gamma\right)z$$

$$+ \Delta_{S^{N-1}}z - e^{-(q-1)t\gamma}|z|^{q-1}z = 0. \tag{1.3.29}$$

From Agmon-Douglis-Nirenberg and Schauder estimates we deduce that z is bounded in $C^3\left([0,\infty)\times S^{N-1}\right)$, and in particular,

$$\left\|v^q(t,.)\right\|^2_{W^{1,2}(S^{N-1})} = \sum_{k=0}^\infty \sum_{j=1}^{d_k}(1 + \lambda_k)\left(A_k^j(t)\right)^2 \leq C\|v(t,.)\|^{2q}_{C^1(S^{N-1})} \leq C' e^{-2tq\gamma}. \tag{1.3.30}$$

- **Estimate of** a_k^j **for** $0 \leq k \leq k_q + 2 - N$ **and** $1 \leq j \leq d_k$. From (1.3.26) we have

$$\left(a_k^j(t)\right)^2 \leq 8(r_1 - r_2)^{-2}\left[\left(\frac{da_k^j}{dt}(0)\right)^2\left(e^{2r_1 t} + e^{2r_2 t}\right) + \left(a_k^j(0)\right)^2\left(r_2^2 e^{2r_1 t} + r_1^2 e^{2r_2 t}\right)\right.$$

$$\left. + \left(\int_0^t\left(e^{r_1(t-s)} - e^{r_2(t-s)}\right)A_k^j(s)ds\right)^2\right], \tag{1.3.31}$$

and, from (1.3.30),

$$\left(\int_0^t\left(e^{r_1(t-s)} - e^{r_2(t-s)}\right)A_k^j(s)ds\right)^2 \leq C\left(e^{-qt\gamma} + e^{-(1+\gamma)t} + e^{r_1 t} + e^{r_2 t}\right)^2, \tag{1.3.32}$$

since we can always assume that $(r_i - q\gamma)(r_i - 1 - \gamma) \neq 0$ for $i = 1, 2$. But $r_1 < r_2 \leq -\delta_q$ in this range of k. Therefore

$$\sum_{k=0}^{k_q + 2 - N}(1 + \lambda_k)\sum_{j=1}^{d_k}\left(a_k^j(t)\right)^2 \leq M\left(e^{-2t\delta_q} + e^{-2(1+\gamma)t} + e^{-tq\gamma}\right), \tag{1.3.33}$$

for some constant $M > 0$.

- Estimate of a_k^j **for** $k \geq k_q + 3 - N$ **and** $1 \leq j \leq d_k$. From (1.3.28), we have

$$\left(a_k^j(t)\right)^2 = 6\left(a_k^j(0)\right)^2 e^{2r_1 t} + 6(r_1 - r_2)^{-2} e^{2r_1 t}\left(\int_0^t \left(e^{-r_1 s} - e^{-r_2 s}\right)A_k^j(s)ds\right)^2$$

$$+ 6(r_1 - r_2)^{-2}\left(e^{r_1 t} - e^{r_2 t}\right)^2\left(\int_t^\infty e^{-r_2 s}A_k^j(s)ds\right)^2. \qquad (1.3.34)$$

But from (1.3.30)

$$\left(\int_0^t \left(e^{-r_1 s} - e^{-r_2 s}\right)A_k^j(s)ds\right)^2 \leq \frac{e^{2t\gamma} - 1}{\gamma}\int_0^t e^{-2(r_1 + \gamma)t}\left(A_k^j(s)\right)^2 ds, \qquad (1.3.35)$$

therefore

$$\sum_{k=k_q+3-N}^\infty (1 + \lambda_k)\sum_{j=1}^{d_k}\left(a_k^j(t)\right)^2 \leq Ce^{-4qt/(q-1)}\sum_{k=k_q+3-N}^\infty (1 + \lambda_k)\sum_{j=1}^{d_k}\left(a_k^j(0)\right)^2$$

$$+ C'\left(e^{-2qt\gamma} + e^{-2(1+\gamma)t}\right). \qquad (1.3.36)$$

Combining (1.3.33) and (1.3.36) yields

$$\sum_{k=0}^\infty (1 + \lambda_k)\sum_{j=1}^{d_k}\left(a_k^j(t)\right)^2 \leq M\left(e^{-4qt/(q-1)} + e^{-2qt\gamma} + e^{-2(1+\gamma)t} + e^{-2t\delta_q}\right), \qquad (1.3.37)$$

and finally

$$\|v\|_{W^{1,2}(S^{N-1})} \leq M'\left(e^{-t\delta_q} + e^{-tq\gamma}\right) \qquad (1.3.38)$$

for some constant $M' > 0$ and any $t \geq 0$.

When $\delta_q \leq q\gamma$, we have (1.3.21). If $\delta_q > q\gamma$, we set $\tilde{\gamma} = q\gamma$ and, from Sobolev's inequality,

$$\|v\|_{L^{2(N-1)/(N-3)}(S^{N-1})} \leq M''e^{-t\tilde{\gamma}}, \qquad (1.3.39)$$

if $N > 3$, or

$$\|v\|_{L^p(S^{N-1})} \leq M_p''e^{-t\tilde{\gamma}}, \qquad (1.3.40)$$

for any $1 < p < \infty$ if $N \leq 3$.

In the case where $N > 3$, we deduce from (1.3.39) that

$$\left\|v^{q}\right\|_{L^{\alpha}(S^{N-1})} \leq M'' e^{-tq\tilde{\gamma}} \tag{1.3.41}$$

holds with $\alpha = \dfrac{2(N-1)}{q(N-3)}$ (which is larger than 2 since $q < N/(N-2)$). Writing the elliptic equation satisfied by $\tilde{z}(t,\sigma) = e^{t\tilde{\gamma}}v(t,\sigma)$ and using the Agmon-Douglis-Nirenberg estimates, yields

$$\left\|v\right\|_{W^{2,2}(S^{N-1})} \leq C e^{-t\tilde{\gamma}}, \tag{1.3.42}$$

which implies

$$\left\|v^{q}\right\|_{L^{p}(S^{N-1})} \leq C e^{-qt\tilde{\gamma}} \tag{1.3.43}$$

if $N = 4$ or 5 and

$$\left\|v^{q}\right\|_{L^{\tilde{\alpha}}(S^{N-1})} \leq C e^{-qt\tilde{\gamma}} \tag{1.3.44}$$

where $\tilde{\alpha} = \dfrac{2(N-1)}{q(N-5)}$ if $N > 5$. By iterating this process the following uniform estimate is derived

$$\left\|v\right\|_{L^{\infty}(S^{N-1})} \leq M e^{-q\lambda t}. \tag{1.3.45}$$

If we continue this type of computation, we reach the final estimate:

$$\left\|v\right\|_{W^{1,2}(S^{N-1})} \leq M'\left(e^{-t\delta_{q}} + e^{-tq\gamma^{n}}\right), \tag{1.3.46}$$

for any integer n, which is (1.3.21).

Step 3. We claim that there exists $M > 0$ such that

$$\left\|u(r,.)\right\|_{L^{\infty}(S^{N-1})} \leq M r^{-k_{q}}. \tag{1.3.47}$$

In the variable v, this last inequality means

$$\left\|v\right\|_{L^{\infty}(S^{N-1})} \leq M e^{-t\delta_{q}}. \tag{1.3.48}$$

We write now the equation satisfied by $y(t,\sigma) = e^{t\delta_{q}}v(t,\sigma)$

$$y_{tt} + \left(2\frac{q+1}{q-1} - N - 2\delta_{q}\right)y_{t} + \left(1_{q,N} + \delta_{q}{}^{2} - \left(2\frac{q+1}{q-1} - N\right)\delta_{q}\right)y$$
$$+ \Delta_{S^{N-1}}y - e^{-(q-1)t\delta_{q}}|y|^{q-1}y = 0. \tag{1.3.49}$$

and from (1.3.21), $|y|^{q-1}y$ is uniformly bounded in $L^{\alpha}(S^{N-1})$ where $\tilde{\alpha} = \dfrac{2(N-1)}{q(N-5)}$ if $N > 5$. Therefore, by the same elliptic estimates technique as the ones used in Step 2, (1.3.48) is achieved.

Step 4. We claim that there exists some $\Phi \in H_{k_q+2-N}$ such that

$$\lim_{r \to 0} r^{k_q} u(r,.) = \Phi(.) \tag{1.3.50}$$

in the $C^2(S^{N-1})$-topology. We keep the notations of Step 3: from (1.3.48) we have

$$\|y(t,.)\|_{C^3(S^{N-1})} + \|y_t(t,.)\|_{C^2(S^{N-1})} + \|y_{tt}(t,.)\|_{C^1(S^{N-1})}$$
$$+ \|y_{ttt}(t,.)\|_{C(S^{N-1})} + \|\nabla y_t(t,.)\|_{C1(S^{N-1})} \le M, \tag{1.3.51}$$

which implies, as in Step 1,

$$\lim_{t \to \infty} \|y_t(t,.)\|_{L^2(S^{N-1})} = \lim_{t \to \infty} \|y_{tt}(t,.)\|_{L^2(S^{N-1})} = 0. \tag{1.3.52}$$

Consequently the trajectory of y is relatively compact in $C^2(S^{N-1})$. Therefore any element of the limit set at infinity Γ_y^+ of this trajectory is included into the set of solutions of

$$\Delta_{S^{N-1}}\omega + \left(1_{q,N} + \delta_q^2 + \left(2\frac{q+1}{q-1} - N\right)\delta_q\right)\omega = 0 \tag{1.3.53}$$

since $1_{q,N} + \delta_q^2 - \left(2\dfrac{q+1}{q-1} - N\right)\delta_q = k_q(k_q + 2 - N) = \lambda_{k_q+2-N}$. Actually, if we denote P_k the orthogonal projector onto H_{k_q+2-N} and $y_k = P_k(y)$, then

$$y_{tt}^k + \left(2\frac{q+1}{q-1} - N - 2\delta_q\right)y_t^k - e^{-(q-1)t\delta_q}P_k\left(|y|^{q-1}y\right) = 0. \tag{1.3.54}$$

By a direct integration, we get

$$y_t^k(t) = e^{(N+2-k_q)t}y_t^k(0) + e^{(N+2-k_q)t}\int_0^t e^{-((q-1)\delta_q + N+2-k_q)s}P_k\left(y|y|^{q-1}(s)\right)ds. \tag{1.3.55}$$

Since $k_q > N - 2$, we deduce that $y_t^k \in L^1(0,\infty)$, and this implies that $y^k(t)$ admits a limit when t goes to infinity.

Step 5. The relation (1.3.50) remains valid if we replace k_q by any integer $k > N - 2$. In that case, estimate (1.3.47) has to be replaced by

$$\|u(r,.)\|_{L^\infty(S^{N-1})} \leq M r^{-k}. \tag{1.3.56}$$

The proof is similar to the one of Step 4.

Step 6. Assume that

$$\lim_{r \to 0} |x|^k u(x) = 0 \tag{1.3.57}$$

for some $k > N - 2$, then $|x|^{k-1} u(x)$ (resp. $u(x)/\ln(1/|x|)$) remains bounded near O if $k - 1 > 0$ (resp. $N = 2$ and $k = 1$). We give the proof when $k - 1 > 0$. In fact, we follow the method of Step 2 where γ is replaced by $v = 2/(q-1) - k$, and by using the Fourier decomposition of v, we finally obtain

$$\|v\|_{W^{1,2}(S^{N-1})} \leq M' \left(e^{-t(v+1)} + e^{-tqv^n} \right) \tag{1.3.58}$$

and

$$\lim_{r \to 0} r^{k-1} u(r,.) = \Phi(.), \tag{1.3.59}$$

for some $\Phi \in H_{k+1-N}$.

Step 7. End of the proof. We know from Theorem 2.1 that if

$$\lim_{x \to O} u(x)/\mu(x) = 0, \tag{1.3.60}$$

u can be extended as a C^2 solution in whole Ω; this corresponds to (iii) and the proof of Theorem 2.4 is completed.

In some cases the assumption that $2/(q-1)$ be not a integer in Theorem 2.4 can be withdrawn, since the following result is valid

THEOREM 2.5. *If in Theorem 2.4 we replace the assumption that $2/(q-1)$ is not an integer by one of the following :*
(H1) u is non negative,
(H2) $(N+1)/(N-1) \leq q < N/(N-2)$,
(H3) $N = 2$,
then the assertions (i)-(iii) remain valid.

Proof. 1- u is nonnegative. Following the proof of Theorem 2.4-Step 1, we deduce that if v does not decay exponentially, there exists a non-zero element ω in the limit set at infinity of $w(t,\sigma) = v(t,\sigma)/\eta(t)$, and ω is a solution of

$$\Delta_{S^{N-1}}\omega + l_{q,N}\omega = 0. \tag{1.3.61}$$

As $l_{q,N}$ is a non-zero eigenvalue of $-\Delta_{S^{N-1}}$, ω must change sign, contradiction.

2 - $(N+1)/(N-1) \le q < N/(N-2)$. If $u(x) = o(|x|^{1-N})$, we know from Theorem 2.1 that there exists $c \in \mathbb{R} \cup \{\pm\infty\}$ such that

$$\lim_{x \to O} u(x)/\mu(x) = c. \tag{1.3.62}$$

If $c = 0$, u is regular. If c is non-zero, u keeps a constant sign near O and we are under the scope of assumption H-1 for u or $-u$.

3- $N = 2$. The proof heavily relies on a very technical adaptation of the argument of Theorem 2.3 which asserts that there exists some $\gamma > 0$ such that

$$\|v(t,.)\|_{L^{\infty}(S^{N-1})} \le Me^{-t\gamma}. \tag{1.3.63}$$

With the notations of Theorem 2.4-Step 1, we know that

$$\Gamma_w^+ \subset \left\{ \sigma \mapsto A\sin(k_q\sigma + \varphi):\ A \in \mathbb{R},\ \varphi \in S^1 \right\} \tag{1.3.64}$$

and there exist $A^* \ne 0$ and $\varphi \in S^1$ such that the function $\sigma \mapsto A^*\sin(k_q\sigma + \varphi)$ belongs to Γ_w^+. By using the same argument of the one of Theorem 2.3, we can assume that

$$\Gamma_w^+ \subset \left\{ \sigma \mapsto A\sin(k_q\sigma):\ A \in \mathbb{R}^+ \right\}. \tag{1.3.65}$$

Denote

$$\ell(t) = \int_{S^1} w(t,\sigma)\sin(k_q\sigma)d\sigma \tag{1.3.66}$$

$$L(t) = \int_{S^1} |w|^{q-1} w(t,\sigma)\sin(k_q\sigma)d\sigma. \tag{1.3.67}$$

We write $\psi(\sigma) = \sin(k_q\sigma)$, and for each $\varepsilon > 0$, we define

$$I_\varepsilon = \left\{ \sigma \in S^1:\ \mathrm{dist}(\sigma, \psi^{-1}(0)) < \varepsilon \right\}. \tag{1.3.68}$$

We fix ε such that $\dfrac{d\psi}{d\sigma}$ does not vanish on $I_{3\varepsilon}$. Applying a simple modification of Theorem 2.3, it follows that there exists $T > 0$ such that

$$w(t,\sigma)\psi(t,\sigma) > 0 \tag{1.3.69}$$

for $t > T$, $\sigma \in S^1 \setminus I_\varepsilon$, and

$$w_\sigma(t,\sigma)\frac{d\psi}{d\sigma}(t,\sigma) > 0$$

for $t > T$, $\sigma \in I_{3\varepsilon}$, which implies that

$$\int_{I_{3\varepsilon}-I_\varepsilon} v(t,\sigma)\psi(\sigma)d\sigma > \left|\int_{I_\varepsilon} v(t,\sigma)\psi(\sigma)d\sigma\right| \tag{1.3.70}$$

and $\ell(t) > 0$ for $t > T$. In the same way $L(t) > 0$ and

$$\frac{d^2\ell}{dt^2}(t) + \left(\frac{4}{q-1}\right)\frac{d\ell}{dt}(t) = F(t) > 0. \tag{1.3.71}$$

Because $\displaystyle\lim_{t\to\infty}\frac{d\ell}{dt}(t) = \lim_{t\to\infty}\ell(t) = 0$, we deduce from (1.3.71) that

$$\frac{d\ell}{dt}(t) + \left(\frac{4}{q-1}\right)\ell(t) \le 0, \tag{1.3.72}$$

for $t > T$. Therefore

$$0 < \ell(t) \le e^{-(4/(q-1))(t-T)}\ell(T) \tag{1.3.73}$$

and

$$\limsup_{t\to\infty} f(t)/\eta(t) \le \limsup_{t\to\infty} e^{t\varepsilon}f(t)\Big/\liminf_{t\to\infty} e^{t\varepsilon}\eta(t) = 0,$$

for $\varepsilon > 0$ small enough. From (1.3.6) we find that $\Gamma_w^+ = \{0\}$, contradiction. Consequently (1.3.63) holds, and this ends the proof.

2-1-4 Existence of singular solutions

If Ω is some N-dimensional domain ($N \ge 2$) containing O, g a continuous nondecreasing real valued function defined on $\mathbb{R}$ and u is a singular solution of

$$-\Delta u + g(u) = 0, \tag{1.4.1}$$

in $\Omega^* = \Omega \setminus \{O\}$, we say that u admits a weak positive singularity at O if there exists a positive c such that

$$\lim_{x\to O} u(x)/\mu(x) = c, \tag{1.4.2}$$

and a strong positive singularity if

$$\lim_{x \to 0} u(x)/\mu(x) = \infty. \tag{1.4.3}$$

One natural way to construct singular of solutions of (1.4.1) is to construct radial solutions, that means solutions of

$$\frac{d^2 u}{dr^2} + \frac{N-1}{r}\frac{du}{dr} = g(u). \tag{1.4.4}$$

The following results [Ve9] give conditions for the existence of such types of singularities

THEOREM 2.6. *Assume that* $N > 2$. *Then* (1.4.1) *admits solutions with weak positive singularities if and only if* g *satisfies*

$$\int_1^\infty g(s)s^{-2(N-1)/(N-2)}ds < \infty. \tag{1.4.5}$$

If we assume moreover that g *satisfies* (2.4.4) *and*

$$\int_\alpha^\infty \frac{ds}{\sqrt{j(s)}} < \infty \tag{1.4.6}$$

for some $\alpha > 0$, *where* $j(s) = \int_0^s g(\sigma)d\sigma$, *there exist solutions of* (1.4.1) *with strong positive singularities* .

Proof. Step 1. We assume that (1.4.5) holds, and for $\varepsilon > 0$ and $c > 0$, we denote y_ε the solution of the following Dirichlet problem

$$\begin{cases} \dfrac{d^2 y_\varepsilon}{ds^2} = A_N(s+\varepsilon)^{N/(N-2)-2}g\left(\dfrac{y_\varepsilon}{(N-2)(s+\varepsilon)}\right) & \text{on } (0,1), \\ y_\varepsilon(0) = c, \quad y_\varepsilon(1) = 0. \end{cases} \tag{1.4.7}$$

where $A_N = (N-2)^{(4-N)/(N-2)}$ (by a scaling transformation we can assume that $B(O,2) \subset \Omega$). For simplicity we shall assume $g(0) = 0 = g^{-1}(0)$ which implies that y_ε is nonnegative, nonincreasing and bounded above by c. By integration of (1.4.7)

$$\frac{dy_\varepsilon}{ds}(s) = \frac{dy_\varepsilon}{ds}(1) - A_N\int_s^1 (\sigma+\varepsilon)^{N/(N-2)-2}g\left(\frac{y_\varepsilon}{(s+\varepsilon)(N-2)}\right)d\sigma. \tag{1.4.8}$$

Since y_ε and $\dfrac{d^2 y_\varepsilon}{ds^2}$ remain bounded on $[a,1]$, $\dfrac{dy_\varepsilon}{ds}(1)$ also remains bounded independently of ε and

$$\left| y_\varepsilon(t) - y_\varepsilon(t') \right| \leq C(t'-t) + A_N \int_t^{t'} \int_s^1 (\sigma + \varepsilon)^{N/(N-2)-2} g\left(\frac{c}{(N-2)(\sigma+\varepsilon)} \right) d\sigma ds \quad (1.4.9)$$

holds for $0 < t \leq t' \leq 1$. Moreover

$$\int_t^{t'} \int_s^1 (\sigma + \varepsilon)^{N/(N-2)-2} g\left(\frac{c}{(N-2)(\sigma+\varepsilon)} \right) d\sigma ds$$

$$\leq \int_{t+\varepsilon}^{t'+\varepsilon} \int_s^2 \sigma^{N/(N-2)-2} g\left(\frac{c}{(N-2)\sigma} \right) d\sigma ds \qquad (1.4.10)$$

If we define Φ on $(0,2]$ by

$$\Phi(x) = \int_x^2 \int_s^2 \sigma^{N/(N-2)-2} g\left(\frac{c}{(N-2)\sigma} \right) d\sigma, \qquad (1.4.11)$$

then

$$\lim_{x \to 0} \Phi(x) = \int_0^2 \sigma^{N/(N-2)-2} g\left(\frac{c}{(N-2)\sigma} \right) d\sigma, \qquad (1.4.12)$$

and this expression is finite from assumption (1.4.5). Therefore Φ can be extended by continuity at 0 as being a uniformly continuous fonction $\tilde{\Phi}$ and (1.4.9) reads as

$$\left| y_\varepsilon(t) - y_\varepsilon(t') \right| \leq C|t'-t| + \left| \tilde{\Phi}(t') - \tilde{\Phi}(t) \right|, \qquad (1.4.13)$$

which implies that the family of functions $\{y_\varepsilon\}_{0 < \varepsilon < 1}$ is equicontinuous on $[0,1]$. From the Arzela-Ascoli theorem there exist a sequence $\{\varepsilon_n\}$ and $y \in C([0,1])$ such that $\{y_{\varepsilon_n}\}$ converges to y uniformly on $[0,1]$. By construction $x \mapsto u(x) = |x|^{2-N} y\left(|x|^{N-2} / (N-2)\right)$ is a solution of (1.4.1) in $B(O,1)$ which satisfies $\lim_{x \to O} u(x)/\mu(x) = c$.

Step 2. We assume that (2.4.5) does not hold an that there exists a solution u of (2.4.1) such that

$$\lim_{x \to O} u(x)/\mu(x) = c \qquad (1.4.14)$$

for some $c > 0$. If $\overline{u}(r)$ denotes the average of $u(r, \sigma)$ on S^{N-1}, then $\lim\limits_{r \to 0} r^{N-2}\overline{u}(r) = c$ and

$$\frac{d}{dr}\left(r^{N-1}\frac{d\overline{u}}{dr}\right) \geq r^{N-1}g\left(\frac{c}{2r^{N-2}}\right) \tag{1.4.15}$$

holds on some interval $(0, a]$. By integration

$$-r^{N-1}\frac{d\overline{u}}{dr}(r) \geq -a^{N-1}\frac{d\overline{u}}{dr}(a) + \int_r^a s^{N-1}g\left(\frac{c}{2s^{N-2}}\right)ds \tag{1.4.16}$$

for $0 < r \leq a$. But this implies

$$\lim_{r \to 0}\left(r^{N-1}\frac{d\overline{u}}{dr}(r)\right) = -\infty, \qquad \lim_{r \to 0}\left(r^{N-2}\overline{u}(r)\right) = \infty, \tag{1.4.17}$$

contradiction.

Step 3. Let y be the function defined in Step 1 as being the limit of y_ε, and let β be defined by

$$t = \int_{\beta(t)}^{\infty} \frac{ds}{\sqrt{2\,j(s)}}, \tag{1.4.18}$$

where $j(s) = \int_0^s g(\sigma)d\sigma$. The function β is positive and satisfies

$$\begin{cases} \dfrac{d^2\beta}{dr^2} = g(\beta) & \text{on } (0, \infty) , \\[2mm] \beta(0) = 0 \ , \quad \beta(\infty) = 0. \end{cases} \tag{1.4.19}$$

For $\tau > 0$ we set $\beta^\tau(r) = \beta(r - \tau)$ and we have

$$-\frac{d^2\beta^\tau}{dr^2} - \frac{N-1}{r}\frac{d\beta^\tau}{dr} + g(\beta^\tau) \geq 0 \quad \text{on } (\tau, \infty), \tag{1.4.20}$$

with blow-up at $r = \tau$. If we set

$$s = r^{N-2} / (N-2) \qquad (\forall r > \tau), \tag{1.4.21}$$

we deduce from the maximum principle that

$$y(s) \leq (N-2)s\beta\left(((N-2)s)^{1/(N-2)} - \tau\right) \qquad \left(\forall s \geq \tau^{N-2}/(N-2)\right). \tag{1.4.22}$$

Therefore, for any compact interval $[a,1]$ with $a > 0$, there exists a constant $C(a) > 0$ such that

$$y(s) \leq C(a), \quad \text{for } s \in [a,1] \text{ and } \varepsilon \in (0, a/2]. \tag{1.4.23}$$

Since the family of functions $u_c(x) = |x|^{2-N} y\left(|x|^{N-2} / (N-2)\right)$, $c > 0$, defined through Step 1 and satisfying (1.4.14), is increasing with respect to c, it converges in the locally uniform topology of $\overline{B}(O,1) \setminus \{O\}$ to some u_∞ which satisfies (2.4.1) in $\overline{B}(O,1) \setminus \{O\}$ and

$$\lim_{x \to 0} u_\infty(x)/\mu(x) = \infty. \tag{1.4.24}$$

Remark 2.5. The condition (1.4.5) was first introduced by Bénilan and Brezis [BB] for proving by a functional analysis method that for any $c > 0$, there is a solution of

$$-\Delta u + g(u) = c_N c \delta_0 \tag{1.4.25}$$

with $c_N = (N-2)|S^{N-1}|$. In fact it is easy to check from the singular behaviour at O that there is an equivalence between the u which solve (1.4.25) in the sense of distributions in Ω and the u wich satisfy (1.4.2). The condition (1.4.6) has been introduced by Keller [Ke] and Osserman [Os] and also by Vazquez [Va1] in the more general context of p-Laplace operator. It implies the following property: for any compact subset $K \subset \Omega^*$ there exists $C(K) > 0$ such that any solution u of (1.4.1) in Ω^* is bounded above by $C(R)$ on K.

Remark 2.6. There exist nondecreasing functions g such that

$$\int_\alpha^\infty s^{-2(N-1)/(N-2)} g(s)ds = \int_\alpha^\infty \frac{ds}{\sqrt{j(s)}} ds = \infty, \tag{1.4.26}$$

for any $\alpha > 0$. As a consequence [VV3], with such a g, there exist solutions of (1.4.1) with strong singularities at O, but no with weak singularity.

In the 2-dimensional case the method is due to Vazquez [Va2] who introduced the exponential order of growth of g as the natural hypothesis to replace (1.4.4):

$$a_g^+ = \inf\left\{a \geq 0 : \int_0^\infty g(s)e^{-as}ds < \infty\right\} \tag{1.4.27}$$

The 2-dimensional analogue of theorem 2.6 is

THEOREM 2.7. *Assume that* $N = 2$. *Then* (1.4.1)) *admits solutions with weak nonnegative singularities of order* $c \geq 0$, *that is such that* (1.4.2) *holds, if and only if*

$$0 \leq c \leq 2/a_g^+. \tag{1.4.28}$$

Proof. The idea for the existence of a nonnegative solution with a weak singularity at O is an adaptation of Steps 1 and 2 of Theorem 2.6: for $\varepsilon > 0$ we introduce the function y_ε which satisfies

$$\begin{cases} \dfrac{d^2 y_\varepsilon}{dt^2} = (t+\varepsilon)^{-3} e^{-2/(t+\varepsilon)} g\left(\dfrac{y_\varepsilon}{t+\varepsilon}\right) & \text{in } (0,1] \\ y_\varepsilon(0) = c \quad , \quad y_\varepsilon(1) = 0. \end{cases} \tag{1.4.29}$$

If c satisfies $0 \leq c < 2/a_g^+$, an integration shows that the family of function $\{y_\varepsilon\}_\varepsilon$ is equicontinuous on $[0,1]$. If we set $u_c(x) = \ln(1/|x|) y(-1/\ln(|x|))$, then u_c satisfies

$$\begin{cases} \Delta u_c = g(u_c) & \text{in } B(O, e^{-1}), \\ u_c(x) = 0 & \text{for } |x| = e^{-1}, \\ \lim_{x \to 0} u_c(x)/\ln(1/|x|) = c. \end{cases} \tag{1.4.30}$$

If $c = 2/a_g^+$, we consider an increasing sequence $\{c_n\}$ with $\lim_{n \to \infty} c_n = c$ and the corresponding solution $y(t) = y(c_n, t)$. Since $\{y(c_n, .)\}$ is an increasing sequence of convex functions bounded above by c and below by 0, by Dini s'theorem it converges to some solution of (1.4.29) (unique in fact). If $c > 2/a_g^+$, a simple integration implies that no positive solution of (1.4.1)-(1.4.2) exists. If we assume that $a_g^+ = 0$ and that (1.4.6) holds, then there exist solutions with weak positive singularities of any order and a least a solution with strong positive singularity [Ve9] but if $a_g^+ = 0$

and $\displaystyle \int_\alpha^\infty \frac{ds}{\sqrt{j(s)}} ds = \infty$ for any $\alpha > 0$, Remark 2.6 is valid.

Remark 2.7. When $N = 2$ and $0 \leq c \leq 2/a_g^+$, any solution of (1.4.1)-(1.4.2) satisfies

$$-\Delta u + g(u) = 2\pi c \delta_0 \tag{1.4.31}$$

in the sense of distributions in Ω.

Remark 2.8. It is important to notice that the positive singularities have some very strong stability or approximation properties. The construction of changing sign singularities is more difficult it and has only been done in the context of the equation

$$-\Delta u + |u|^{q-1} u = 0 \qquad (1.4.32)$$

when $1 < q < N/(N-2)$, in a ball (see [CMV]). In the 2-dimensional case for simplicity and if $2/(q-1) > k$ for some positive integer k, the idea is to construct a positive singular solution u_k in the angular domain $B_k = \{(r,\sigma) : 0 < r \leq 1, \ \sigma \in (0, \pi/k)\}$ which vanishes on the boundary $\partial B_k \setminus \{O\}$ and then to perform a reflection about the sides of the domain.

2-1-5 Prescribed singular submanifolds

Let Ω be an open subset of $\mathbb{R}^N$ and Σ a d-dimensional submanifold of Ω. The singular submanifold problem we shall be interested deals with the construction of solutions of

$$-\Delta u + g(u) = 0, \qquad (1.5.1)$$

in the case where $g(u) = |u|^{q-1} u$ or $g(u) = e^u$, in $\Omega \setminus \Sigma$ which are singular on Σ. In fact it is possible to prescribed the rate of blow-up of u near Σ. The following existence and uniqueness result is proved by Grillot ([Gr1], [Gr2]).

THEOREM 2.8. *Let* $N > 2$, Ω *be a bounded domain of* $\mathbb{R}^N$ *with a smooth boundary and* Σ *a d-dimensional* C^2, *compact submanifold of* Ω *without boundary. If* h *is a given continuous function on* $\partial\Omega$ *and q satisfies*

$$1 < q < (N-d)/(N-d-2) \qquad (1.5.2)$$

then for any $\alpha \in [-\infty, \infty]$, *there exists a unique function* u *belonging to* $C(\overline{\Omega} \setminus \Sigma) \cap C^2(\Omega \setminus \Sigma)$, *satisfying* $u = h$ *on* $\partial\Omega$,

$$-\Delta u + |u|^{q-1} u = 0 \qquad (1.5.3)$$

in $\Omega \setminus \Sigma$ *and such that*

$$\lim_{x \to O} u(x)/\mu_{N-d}(x) = \alpha \qquad (1.5.4)$$

where μ_{N-d} *is the fundamental solution of the Laplacian in* $\mathbb{R}^{N-d}$. *Moreover if* $\alpha = \infty$, *there holds*

$$\lim_{x \to O} |x|^{2/(q-1)} u(x) = l(q, N-d) = \left(\left(\frac{2}{q-1} \right) \left(\frac{2q}{q-1} - N + d \right) \right)^{1/(q-1)}. \qquad (1.5.5)$$

LEMMA 2.2. *Let Σ be a d-dimensional C^2, compact submanifold of $\mathbb{R}^N$ without boundary and $\delta(x)$ the distance from x to Σ, then there exists a neighborhood $\mathbf{V}$ of Σ and a positive constant K such that*

$$(N-d-1)-K\delta(x) \le \delta(x)\Delta\delta(x) \le (N-d-1)+K\delta(x) \tag{1.5.6}$$

for any x in $\mathbf{V}\setminus\Sigma$.

Proof. The proof is due to Loewner and Nirenberg [LN] and its idea is to represent locally the function $x \mapsto \delta(x)$ in the following way

$$\delta^2(x) = \sum_{i=1}^{N-d} \Phi_i^2(x) \tag{1.5.7}$$

where the Φ_i are defined locally, have linearly independent gradients and are bounded in the C^2-norm. Therefore

$$\delta(x)\frac{\partial\delta}{\partial x_i}(x) = \sum_{j=1}^{N-d} \Phi_j(x)\frac{\partial\Phi_j}{\partial x_i}(x) \tag{1.5.8}$$

and

$$\delta^2(x)\left|\nabla\delta(x)\right|^2 = \sum_{i=1}^{N}\left(\sum_{j=1}^{N-d}\Phi_j(x)\frac{\partial\Phi_j}{\partial x_i}(x)\right)^2. \tag{1.5.9}$$

Because $\left|\nabla\delta(x)\right|^2 = 1$, we have

$$\sum_{i=1}^{N}\left(\frac{\partial\Phi_j}{\partial x_i}(x)\right)^2 = 1 \qquad \left(\forall j = 1,...,N-d, \ \forall x \in \Gamma_j\right), \tag{1.5.10}$$

where $\Gamma_j = \left\{x: \ \Phi_1(x) = \Phi_2(x) =...\Phi_{j-1}(x) = \Phi_{j+1}(x) = \Phi_{N-d}(x) = 0\right\}$. Differentiating again (1.5.8) with respect to x_i and summing from $i = 1$ to N yields

$$1 + \delta(x)\Delta\delta(x) = \sum_{i=1}^{N}\sum_{j=1}^{N-d}\left(\frac{\partial\Phi_j}{\partial x_i}(x)\right)^2 + \sum_{j=1}^{N-d}\Phi_j(x)\Delta\Phi_j(x). \tag{1.5.11}$$

Since $\left|\Phi_j(x)\right| \le \delta(x)$, we deduce from the C^2 bound on the Φ_i, (1.5.10) and (1.5.11) that

$$\delta(x)\Delta\delta(x) - (N-d-1) = O(\delta(x)) \tag{1.5.12}$$

near Σ.

By an easy adaptation of the types of results that have been proved before when $K = 0$, we have the following classification and existence result.

LEMMA 2.3. *Let* K *be any constant,* q *a real number satisfying* (1.5.2) *and* φ *a positive continuous function on* $(0,R]$, *solution of*

$$\varphi_{rr} + \frac{N-1-d}{r}\varphi_r + K\varphi_r - \varphi^q = 0 \tag{1.5.13}$$

on $(0,R)$. *Then the following alternative holds near* 0:
(i) either

$$\lim_{r\to 0} r^{2/(q-1)}\varphi(x) = l(q,N-d), \tag{1.5.14}$$

(ii) or there exists some $\gamma \geq 0$ *such that*

$$\lim_{r\to 0} \varphi(r)/\mu(r) = \gamma. \tag{1.5.15}$$

Moreover, for any $\sigma \geq 0$, *there exists a solution* φ_γ *of* (1.5.13) *on* $(0,R)$ *such that* $\varphi_\gamma(R) = \sigma$ *and*

$$\lim_{r\to 0} \varphi_\gamma(r)/\mu_{N-d}(r) = \gamma. \tag{1.5.16}$$

When γ *increases,* φ_γ *increases and converges to some* φ_∞ *which satisfies* (1.5.14) *on* $(0,R)$, $\varphi_\infty(R) = \sigma$ *and*

$$\lim_{r\to 0} r^{2/(q-1)}\varphi_\infty(x) = l(q,N-d). \tag{1.5.17}$$

Proof of Theorem 2.8. The blow-up coefficient α is supposed to be positive, the negative case being similar.
Step 1. Construction of local super and subsolutions. From the maximum principle, we can suppose that $\alpha \in (0,\infty)$. Let $\varepsilon > 0$ be fixed, we denote $\mathrm{TUB}_\varepsilon(\Sigma) = \{x \in \Omega: \delta(x) \leq \varepsilon\}$ and $\mathrm{TUB}_\varepsilon^*(\Sigma) = \mathrm{TUB}_\varepsilon(\Sigma)\backslash \Sigma$. We look for a supersolution v of (1.5.3) in $\mathrm{TUB}_\varepsilon^*(\Sigma)$ under the form $v(x) = \varphi(\delta(x))$. Then φ should satisfy

$$-(\varphi''\circ\delta) - \Delta\delta(\varphi'\circ\delta) + (\varphi\circ\delta)|(\varphi\circ\delta)|^{q-1} \geq 0, \tag{1.5.18}$$

for $0 < \delta(x) \leq \varepsilon$ and

$$\lim_{\delta(x)\to 0} \varphi(\delta(x))/\mu_{N-d}(\delta(x)) = \alpha. \tag{1.5.19}$$

If K is the positive constant defined in Lemma 2.2, it is sufficient to construct a decreasing solution of

$$\varphi_{rr} + \frac{N-1-d}{r}\varphi_r - K\varphi_r - \varphi^q = 0 \quad \text{on} \quad (0,\varepsilon], \tag{1.5.20}$$

subject to the singular condition (1.5.19) and vanishing at $r = \varepsilon$. Such a solution exists from Lemma 2.3. We proceed similarly for the construction of the subsolution ψ, solution of

$$\psi_{rr} + \frac{N-1-d}{r}\psi_r + K\psi_r - \psi^q = 0 \quad \text{on} \quad (0,\varepsilon], \tag{1.5.21}$$

satisfying the same condition (1.5.19) and vanishing at $r=\varepsilon$. Moreover φ and ψ can be constructed in such a way that $0 \leq \psi \leq \varphi$ in $(0,\varepsilon)$. Finally, if we let α go to infinity, the corresponding solutions φ and ψ increase and converge to some solutions of (1.5.20) and (1.5.21) respectively satisfying (1.5.17).

Step 2. Construction of global super and subsolutions. Let ψ_α be the solution of (1.5.22) with blow-up coefficient α at 0 (see (1.5.21)). If M is the minimum of h on $\partial\Omega$, we denote U the solution of the regular Dirichlet problem

$$\begin{cases} -\Delta U + |U|^{q-1}U = 0 & \text{in} \quad \Omega, \\ \qquad\qquad U = M & \text{on} \quad \partial\Omega. \end{cases} \tag{1.5.22}$$

We take $\sigma > 0$ large enough, $0 < \eta < \varepsilon$ small enough and define χ_η and χ_Ω as the characteristic functions of $\mathrm{TUB}_\eta^*(\Sigma)$ and Ω respectively. The function

$$x \mapsto \Psi(x) = \left[\sup\left(\psi_\alpha(\delta(x)) - \sigma, U(x)\right)\right]\chi_\eta(x) + \left(\chi_\Omega(x) - \chi_\eta(x)\right)U(x) \tag{1.5.23}$$

is continuous in $\overline{\Omega} \setminus \Sigma$, belongs to $W^{1,2}_{loc}(\Omega \setminus \Sigma)$ and is a subsolution of (1.5.2) in $\Omega \setminus \Sigma$. If N is the maximum of $|h|$ on $\partial\Omega$, we call V_η the solution of

$$\begin{cases} -\Delta V_\eta + |V_\eta|^{q-1}V_\eta = 0 & \text{in} \quad \Omega_\eta, \\ \displaystyle\lim_{\mathrm{dist}(x,\partial\mathrm{TUB}_\eta(\Sigma))} V_\eta(x) = \infty, \\ \qquad V_\eta = N & \text{on} \quad \partial\Omega, \end{cases} \tag{1.5.24}$$

where $\Omega_\eta = \Omega \setminus \mathrm{TUB}_\eta(\Sigma)$, such solutions will be constructed in Section 4-2. Taking σ and ε as above implies that the function

$$\begin{aligned} x \mapsto \Phi(x) = {}& \left(\varphi_\alpha(\delta(x)) + \sigma\right)\chi_\eta(x) \\ & + \left[\inf\left(\varphi_\alpha(\delta(x)) + \sigma, V_\eta(x)\right)\right]\left(\chi_\varepsilon(x) - \chi_\eta(x)\right) + \left(\chi_\Omega(x) - \chi_\varepsilon(x)\right)V_\eta(x) \end{aligned} \tag{1.5.25}$$

86

has the same regularity properties as the function defined in (1.5.24) and is a supersolution of (1.5.3).

Step 3. Existence. From Step 2, for any $\alpha > 0$ there exist two functions Ψ and Φ which are continuous in $\overline{\Omega} \setminus \Sigma$ and belong to $W^{1,2}_{loc}(\Omega \setminus \Sigma)$ and are respectively sub and supersolutions of (1.5.3) in $\Omega \setminus \Sigma$ and satisfies

$$\lim_{x \to 0} \Psi(x) / \mu_{N-d}(x) = \lim_{x \to 0} \Phi(x) / \mu_{N-d}(x) = \alpha \qquad (1.5.26)$$

and $\Psi(x) \leq \Phi(x)$ in $\Omega \setminus \Sigma$, and $\Psi(x) \leq h(x) \leq \Phi(x)$ on $\partial\Omega$. Consequently there exists a solution u of (1.5.3) in $\Omega \setminus \Sigma$ which takes the value h on $\partial\Omega$ and such that $\Psi \leq u \leq \Phi$. The blow-up estimate (1.5.4) follows. Letting α go to infinity yields (1.5.5) from Lemma 2.2.

Step 4. Uniqueness in the case where α is finite. Suppose that u and $\tilde{u}$ are two solutions satisfying (1.5.4) and, for $\varepsilon > 0$, set $u_\varepsilon = (1 + \varepsilon)(u + N)$, N being the maximum of $|h|$ on $\partial\Omega$. It is clear that the function u_ε is a supersolution which dominates $\tilde{u}$ both near Σ and on $\partial\Omega$. Consequently $u_\varepsilon \geq \tilde{u}$. Letting ε go to 0 and reversing the roles of u and $\tilde{u}$ infers $\|u - \tilde{u}\|_{L^\infty(\Sigma)} \leq 2N$. If we set $w = (u - \tilde{u})^+$, then w is continuous bounded and subharmonic in $\Omega \setminus \Sigma$; it vanishes on $\partial\Omega$, therefore it is identically 0.

Step 5. Uniqueness in the case where $\alpha = \infty$. If $u = u_\infty$ is a solution of (1.5.3) in $\Omega \setminus \Sigma$ satisfying

$$\lim_{x \to 0} u_\infty(x) / \mu_{N-d}(x) = \infty, \qquad (1.5.27)$$

then for any $\alpha > 0$, it is bounded from below in $TUB^*_\varepsilon(\Sigma)$ by the function $x \mapsto \psi_\alpha(\delta(x))$ constructed in Step 1 from Lemma 2.3. Letting α go to infinity yields

$$\liminf_{x \to 0} |x|^{2/(q-1)} u_\infty(x) \geq l(q, N - d). \qquad (1.5.28)$$

The upper bound is constructed following the same principle by considering solution β_η of

$$\begin{cases} \beta_{\eta_{rr}} + \dfrac{N - 1 - d}{r} \beta_{\eta_r} - K\beta_\eta - \beta_\eta^q = 0 \quad \text{on } (\eta, \varepsilon), \\[2mm] \beta_\eta(\varepsilon) = \max_{\delta(x) = \varepsilon} u_\infty(x) \quad \text{and} \quad \lim_{r \to \eta} \beta_\eta(r) = \infty. \end{cases} \qquad (1.5.29)$$

From the maximum principle and Step 1,

$$\beta_\eta(\delta(x)) \geq u_\infty(x) \qquad \left(\forall x \in TUB_\varepsilon(\Sigma) \setminus TUB_\eta(\Sigma) \right). \qquad (1.5.30)$$

When η go to zero, β_η decreases and converges to a solution φ of (1.5.20) on $(0,\varepsilon]$ which is larger near O than all the φ_α constructed in Step 1. From Lemma 2.3,

$$\lim_{r\to 0} r^{2/(q-1)}\varphi(r) = l(q, N-d).\tag{1.5.31}$$

Since $\varphi \circ \delta$ dominates u_∞ in $\mathrm{TUB}_\varepsilon^*(\Sigma)$, we deduce from (1.5.28) and (1.5.31) that

$$\lim_{x\to O} |x|^{2/(q-1)} u_\infty(x) = l(q, N-d).\tag{1.5.32}$$

The completion of the proof is as in Step 4.

Remark 2.9. It is worth noticing that the solution u_∞ is a complete solution of (1.5.3) in $\Omega \backslash \Sigma$ in the sense defined in Chapter 4, Corollary 4.3.

Following the same techniques the following result is proved in [Gr1], [Gr2]

THEOREM 2.9. *Let* $N > 2$, Ω *be a bounded domain of* $\mathbb{R}^N$ *with a smooth boundary and* Σ *a 2-codimensional* C^2, *compact submanifold of* Ω *without boundary. If h is a given continuous function on* $\partial\Omega$, *then, for any* $\alpha \in [0,2]$, *there exists a unique function* u *belonging to* $C(\overline{\Omega}\backslash \Sigma) \cap C^2(\Omega\backslash \Sigma)$, *satisfying* $u = h$ *on* $\partial\Omega$,

$$-\Delta u + e^u = 0\tag{1.5.33}$$

in $\Omega\backslash \Sigma$ *and*

$$\lim_{\delta(x)\to 0} u(x)/\ln(1/\delta(x)) = \alpha.\tag{1.5.34}$$

2-2 Removable singularities

In this section we shall first present Brezis and Véron's removability result concerning isolated singularities of solutions of

$$-\Delta u + g(u) = 0\tag{2.1}$$

and then Baras and Pierre's extension to a more general type of singular set.

2-2-1 Removable isolated singularities

Let Ω be an open subset of $\mathbb{R}^N$, $N \geq 3$, such that $O \in \Omega$ and $\Omega^* = \Omega\backslash \{O\}$. The first result is *a priori* local upper bound for any function u which satisfies

$$-\Delta u + |u|^{q-1} u \leq C\tag{2.1.1}$$

in Ω^*, provided $q \geq N/(N-2)$.

THEOREM 2.10. *Assume that* $u \in L^\infty_{loc}(\Omega^*)$ *satisfies the condition* $\Delta u \in L^1_{loc}(\Omega^*)$ *(in the sense of distributions) and that*

$$-\Delta u + a u^q \leq C \qquad a.e. \quad on \quad \{x \in \Omega: u(x) \geq 0\}, \tag{2.1.2}$$

for some constants positive constants a *and* C. *If* $q \geq N/(N-2)$, *then* $u^+ \in L^\infty_{loc}(\Omega)$.

Proof. By a change of scale we can assume that $B(O,1) \subset \Omega$. From Proposition 2.4 there exists a positive constant A depending on q, N, a and C such that

$$u(x) \leq A|x|^{2-N} \qquad for \quad 0 < |x| \leq 1/2. \tag{2.1.3}$$

Step 1. We claim that $u^+ \in L^{N/(N-2)}_{loc}(\Omega)$. We first recall Kato's inequality:

$$\Delta|u| \geq \mathrm{sign}(u)\Delta u, \tag{2.1.4}$$

holds in the sense of distributions in Ω^* since u and Δu are locally integrable in Ω^*. This implies

$$\Delta u^+ \geq \mathrm{sign}^+(u)\Delta u, \tag{2.1.5}$$

and finally

$$-\Delta u^+ + a(u^+)^q \leq C, \tag{2.1.6}$$

in the sense of distributions in Ω^*. Moreover we can assume that $q = N/(N-2)$ by changing a and C. Let $\{\zeta_n\}_{n>0}$ be a sequence of C^∞ functions defined in $\mathbb{R}^N$ such that

$$\zeta_n(x) = \begin{cases} 0 & \text{if } |x| \leq 1/(2n), \\ 1 & \text{if } |x| \geq 1/n, \end{cases} \tag{2.1.7}$$

with $0 \leq \zeta_n \leq 1$, $|\nabla \zeta_n| \leq Kn$ and $|\Delta \zeta_n| \leq Kn^2$. If $\varphi \in C^\infty_0(\Omega)$, $\varphi \geq 0$, we deduce from (2.1.6) that

$$-\int_\Omega u^+ \Delta(\zeta_n \varphi)dx + a\int_\Omega (u^+)^q \zeta_n \varphi\, dx \leq C\int_\Omega \zeta_n \varphi\, dx. \tag{2.1.8}$$

In view of (2.1.3), we have

$$\left|\int_\Omega u^+ \Delta(\zeta_n \varphi)dx\right| \leq \int_\Omega u^+ |\Delta\varphi|\, dx + C_1 n\int_{|x|<1/n} |x|^{2-N}dx + C_1 n^2 \int_{|x|<1/n} |x|^{2-N}dx \tag{2.1.9}$$

and the right-hand side of (2.1.9) stays bounded independently of n. By letting n go to infinity, it implies that $(u^+)^q \varphi \in L^1(\Omega)$.

Step 2. We fix $\mu \ge (C/a)^{1/q}$ and there holds

$$\Delta(u-\mu)^+ \ge 0 \qquad\qquad (2.1.10)$$

in the sense of distributions in Ω^* by using (2.1.5). If $\varphi \in C_0^\infty(\Omega)$ and $\{\zeta_n\}_{n>0}$ is the sequence defined in Step 1, we have

$$\int_\Omega (u-\mu)^+ \Delta(\zeta_n \varphi) dx \ge 0. \qquad\qquad (2.1.11)$$

But

$$\int_\Omega (u-\mu)^+ \Delta(\zeta_n \varphi) dx$$
$$= \int_\Omega (u-\mu)^+ \zeta_n \Delta\varphi dx + \int_\Omega (u-\mu)^+ \varphi \Delta\zeta_n dx + 2\int_\Omega (u-\mu)^+ \nabla\varphi . \nabla\zeta_n dx. \qquad (2.1.12)$$

Since

$$\left| \int_\Omega (u-\mu)^+ \varphi \Delta\zeta_n dx \right| \le C_2 \left(\int_{|x|\le 1/n} \left((u-\mu)^+\right)^{N/(N-2)} dx \right)^{1-2/N} \qquad (2.1.13)$$

and

$$\left| \int_\Omega (u-\mu)^+ \nabla\varphi . \nabla\zeta_n dx \right| \le \frac{C_3}{n} \left(\int_{|x|\le 1/n} \left((u-\mu)^+\right)^{N/(N-2)} dx \right)^{1-2/N} , \qquad (2.1.14)$$

we deduce from (2.1.12) that

$$\int_\Omega (u-\mu)^+ \Delta\varphi \, dx \ge 0, \qquad\qquad (2.1.15)$$

and (2.1.10) holds in $\mathcal{D}'(\Omega)$. If we assume also that $\mu \ge \sup_{1/2 \le |x| \le 3/4} u(x)$, we deduce from the maximum principle that $u(x) \le \mu$ for $|x| \le 1/2$.

As a consequence of this result, we have a removability result for any solution of (2.1) provided g satisfies

$$\begin{cases} \liminf_{t \to \infty} g(t)/t^{N/(N-2)} > 0, \\ \limsup_{t \to -\infty} g(t)/(-t)^{N/(N-2)} < 0. \end{cases} \tag{2.1.16}$$

THEOREM 2.11. *Assume that g is a continuous real valued function which satisfies (2.1.16) and u belongs to* $L^\infty_{loc}(\Omega^*)$ *and satisfies (2.1) in* $\mathcal{D}'(\Omega^*)$. *Then there exists a* $C^1(\Omega)$*-function which coincides with u on* Ω^* *and satisfies (2.1) in* $\mathcal{D}'(\Omega)$.

Proof. From (2.1.16) there exist two positive constants a and C such that (2.1.6) holds in $\mathcal{D}'(\Omega^*)$ with $q = N/(N-2)$. Therefore u^+ remains locally bounded in Ω. Similarly $u^- \in L^\infty_{loc}(\Omega)$ and (2.1) holds in $\mathcal{D}'(\Omega)$. From the elliptic equations regularity theory u is $C^{1,\gamma}$, for any $\gamma \in (0,1)$.

Remark 2.10. This result can also be proved by using Theorem 2.1: from (2.1.16) the function u which satisfies (2.1) is majorized in $B(O,1)\backslash \{O\}$ by the solution v of

$$-\Delta v + av^q = C, \tag{2.1.17}$$

in the same punctured ball (with $q = N/(N-2)$ and $v(x) = \max_{|x|=1} u^+(x)$ on $\partial B(O,1)$). From Theorem 2.1 and Keller Osserman estimate there exists a nonnegative real k such that

$$\lim_{x \to O} |x|^{N-2} v(x) = k. \tag{2.1.18}$$

If k were not 0 there would exist a neighborhood of O in which

$$0 < C_2 \le -|x|^{-N} \Delta v \le C_1, \tag{2.1.19}$$

and a double integration would imply

$$\lim_{r \to 0} r^{N-2}\overline{v}(r) = \infty \tag{2.1.20}$$

where $\overline{v}(r)$ is the average of v on the sphere of center O and radius r, contradiction.

Remark 2.11. Theorem 2.11 admits some improvements [VV3] by replacing the growth assumption (2.1.16) by the weaker one

$$\begin{cases} \liminf_{t \to \infty} g(t)\ln t/t^{N/(N-2)} > 0, \\ \limsup_{t \to -\infty} g(t)\ln(-t)/(-t)^{N/(N-2)} < 0, \end{cases} \tag{2.1.21}$$

or its different generalizations such as

$$\begin{cases} \liminf\limits_{t \to \infty} g(t) \ln t \ \ln(\ln t)/t^{N/(N-2)} > 0, \\ \limsup\limits_{t \to -\infty} g(t) \ln(-t) \ \ln(\ln(-t)/(-t)^{N/(N-2)} < 0. \end{cases} \qquad (2.1.22)$$

From this growth assumptions the same *a priori* estimate (2.1.3) holds. The remaining of the proof is as in Remark 2.10.

The technique developped in Theorems 2.10-2.11 can be adapted to prove the removability of k-dimensional singular sets [Ve10]. For that we consider a k-dimensional compact C^2-submanifold Σ imbedded into the open subset Ω of $\mathbb{R}^N$, with $N > k+2$, and a continuous real valued function g wich satisfies

$$\begin{cases} \liminf\limits_{t \to \infty} g(t)/t^{(N-k)/(N-2-k)} > 0, \\ \limsup\limits_{t \to -\infty} g(t)/(-t)^{(N-k)/(N-2-k))} < 0. \end{cases} \qquad (2.1.23)$$

THEOREM 2.12. *Assume that g satisfies (2.1.23), and u is locally bounded in $\Omega \backslash \Sigma$ and satisfies (2.1) in $\mathcal{D}'(\Omega \backslash \Sigma)$. Then there exists a $C^1(\Omega)$-function which coincides with u on $\Omega \backslash \Sigma$ and satisfies (2.1) in $\mathcal{D}'(\Omega)$.*

We shall see in the next section a much more general result where the size of the singular set to be removed is expressed in term of capacity. For codimension-2 singular sets the critical growth of the nonlinear term g has to be of super-exponential order which means that g satisfies

$$\begin{cases} \liminf\limits_{t \to \infty} e^{-\alpha t} g(t) > 0, \\ \limsup\limits_{t \to -\infty} e^{\alpha t} g(t) < 0, \end{cases} \qquad (2.1.24)$$

for any $\alpha > 0$. For the sake of simplicity we restrict ourself to a single point although an easy adaptation of Theorem 2.12 yields the removability of $(N-2)$-dimensional singular sets under the assumption of the super-exponential growth [VV2].

THEOREM 2.13. *Assume that $N = 2$, g is a continuous real valued function such that (2.1.24) holds for any $\alpha > 0$ and u belongs to $L^{\infty}_{loc}(\Omega^*)$ and satisfies (2.1) in $\mathcal{D}'(\Omega^*)$. Then there exists a $C^1(\Omega)$-function which coincides with u on Ω^* and is a solution of (2.1) in $\mathcal{D}'(\Omega)$.*

Proof. By a change of scale we may assume that $B(O,2) \subset \Omega$. The key point of the proof is an *a priori* estimate of Keller-Osserman type. Under the assumption (2.1.24) for any $\alpha > 0$ there exist a and $C > 0$ such that

$$-\Delta u + a e^{\alpha u} \leq C \qquad \text{a.e. on } \left\{ x \in \Omega^* : u(x) \geq 0 \right\}. \qquad (2.1.25)$$

Step 1. We claim that

$$u(x) \le \frac{2}{\alpha} \ln(1/|x|) + B, \qquad \text{for } 0 < |x| \le 1,$$
(2.1.26)

with $B = B(\alpha, a, C)$. We set $\psi(x) = \lambda \ln\left(1/\left(R^2 - |x - x_0|^2\right)\right) + \mu = \lambda \ln\left(1/\left(R^2 - r^2\right)\right) + \mu$ where x_0 is such that $0 < |x_0| \le 1$ and $0 < R < |x_0|$. Then

$$-\Delta\psi + ae^{\alpha\psi} = -\frac{4\lambda}{\left(R^2 - r^2\right)^2} + a\frac{e^{\alpha\mu}}{\left(R^2 - r^2\right)^{2\lambda}}.$$
(2.1.27)

If we take

$$\lambda = \frac{2}{\alpha}, \ \mu = \frac{1}{\alpha}\ln((\alpha c + 8)/a\alpha),$$
(2.1.28)

we have

$$-\Delta\psi + ae^{\alpha\psi} \le C.$$
(2.1.29)

From the maximum principle $u < \psi$ and in particular $u(x_0) = \psi(x_0)$. Letting x_0 go to R infers the estimate.

Step 2. Since α is arbitrary, (2.1.26) implies that $u^+(x) = o\left(\ln(1/|x|)\right)$ near O. If β is the solution of

$$\begin{cases} -\Delta\beta = C^+ & \text{in } B(O,1), \\ \beta(x) = \max_{|y|=1} u^+(y) & \text{on } \partial B(O,1), \end{cases}$$
(2.1.30)

then, for any $\varepsilon > 0$

$$u(x) \le \varepsilon \ln(1/|x|) + \beta(x) \text{ in } B(O,1)\backslash \{O\}.$$
(2.1.31)

Letting ε go to 0 implies the boundedness of u^+. The completion of the proof is similar to the one of Theorem 2.11.

2-2-2 Removable singularities for positive solutions

The duality methods introduced by Baras and Pierre [BP] can be applied to general differential operators of order m such as

$$L_m u = \sum_{0 \leq |\alpha| \leq m} D^\alpha (a_\alpha u) \tag{2.2.1}$$

where $D^\alpha = \dfrac{\partial^{\alpha_1 + \alpha_2 + \ldots + \alpha_N}}{\partial x_1^{\alpha_1} \partial x_2^{\alpha_2} \ldots \partial x_N^{\alpha_N}}$ and where the a_α are bounded and measurable in Ω (an open subset of $\mathbb{R}^N$). If T is a distribution in Ω, the relation

$$L_m u = T \qquad (\text{resp. } L_m u \leq T) \tag{2.2.2}$$

in the sense of distributions in Ω means that

$$\int_\Omega u L_m^* \zeta \, dx = \langle T, \zeta \rangle \qquad (\text{resp. } \int_\Omega u L_m^* \zeta \, dx \leq \langle T, \zeta \rangle) \tag{2.2.3}$$

for any $\zeta \in C_0^\infty(\Omega)$ (resp. $\zeta \in C_0^\infty(\Omega)$, $\zeta \geq 0$) where L_m^* is the dual operator defined by

$$L_m^* \zeta = \sum_{0 \leq |\alpha| \leq m} (-1)^{|\alpha|} a_\alpha D^\alpha \zeta. \tag{2.2.4}$$

The definition of the capacity of a subset E relatively closed in Ω has been given in 1-1-3.

Definition. A property is said to be true $c_{m,q'}$-quasi-everywhere if it is true excepted for a set of $c_{m,q'}$-capacity zero. Given a Radon measure μ in an open subset Ω of $\mathbb{R}^N$, we say that μ does not load the sets with $c_{m,q'}$-capacity zero if

$$\forall E \subset \Omega, \; c_{m,q'}(E) = 0 \implies |\mu|(E) = 0. \tag{2.2.5}$$

THEOREM 2.14. *Assume that* $q > 1$, $q' = q/(q-1)$, *E is a relatively closed subset in* Ω, μ *a Radon measure on* Ω *which does not load the sets with* $c_{m,q'}$-*capacity zero and* g *is a continuous real valued function such that*

$$\liminf_{t \to \infty} g(t)/t^q > 0. \tag{2.2.6}$$

If $c_{m,q'}(E) = 0$ *and u is a nonnegative function such that* u *and* $g(u) \in L^1_{loc}(\Omega \setminus E)$ *which satisfies*

$$L_m u + g(u) \leq \mu \tag{2.2.7}$$

in $\mathcal{D}'(\Omega \setminus E)$, *then* u *and* $g(u) \in L^1_{loc}(\Omega)$; *moreover (2.2.7) holds in* $\mathcal{D}'(\Omega)$.

Proof. Since that $c_{m,q'}(E) = 0$, $|E| = 0$ and u is defined almost everywhere.

Step 1. We claim that $g(u) \in L^1_{loc}(\Omega)$. Take $\zeta \in C^\infty_0(\Omega)$, $0 \le \zeta \le 1$ with support S and let $\{\eta_n\}_{n \ge 0}$ be a sequence such that $0 \le \eta_n \le 1$, $\eta_n \equiv 1$ in some neighborhood of $E \cap S$, $\eta_n \to 0$ in $W^{m,q'}(\Omega)$ and a.e. in Ω (see Prop.1.1) and in fact $\eta_n \to 0$, $c_{m,q'}$-quasi-everywhere as n goes to infinity. We set $\zeta_n = (1 - \eta_n)\zeta$, then $\zeta_n \in C^\infty_0(\Omega \backslash E)$, $0 \le \zeta_n \le \zeta$, $\zeta_n \to \zeta$ in $W^{m,q'}(\mathbb{R}^N)$ and $c_{m,q'}$-quasi-everywhere as n goes to infinity. Taking $p > mq'$ yields

$$\int_\Omega \left(u L^*_m \zeta^p_n + g(u)\zeta^p_n \right) dx \le \int_\Omega \zeta^p_n d\mu . \tag{2.2.8}$$

Since $0 \le \zeta^p_n \le \zeta^p$, the right-hand side of (2.2.9) is bounded from above independently of n and

$$\int_\Omega g(u)\zeta^p_n dx \le C + \int_\Omega u \left| L^*_m \zeta^p_n \right| dx \tag{2.2.9}$$

(in the sequel of the proof, C will be any constant independent on n). Moreover

$$\left| L^*_m \zeta^p_n \right| \le C \sum_{0 \le |\alpha| \le m} \left| D^\alpha \zeta^p_n \right| . \tag{2.2.10}$$

The estimate of the different terms can be done in the following way. For $\alpha = 0$

$$\int_\Omega u \zeta^p_n dx \le \left(\int_\Omega u^q \zeta^p_n dx \right)^{1/q} \left(\int_\Omega \zeta^p_n dx \right)^{1/q'} \le C \left(\int_\Omega u^q \zeta^p_n dx \right)^{1/q} \left\| \zeta_n \right\|^m_{m,q'} . \tag{2.2.11}$$

For $|\alpha| \ge 1$

$$D^\alpha \zeta^p_n = \sum_{j=1}^{|\alpha|} c_j \zeta^j_n \sum_{\substack{\beta_1 + \beta_2 + ... + \beta_j \\ |\beta_k| \ge 1}} c_{\beta_1 \beta_2 ... \beta_j} D^{\beta_1}\zeta_n ... D^{\beta_j}\zeta_n \tag{2.2.12}$$

where the c_j and $c_{\beta_1 \beta_2 ... \beta_j}$ depend on the quantities in indices. Finally we have to estimate a finite numbers of terms in the form

$$A = \int_\Omega u \zeta^{p-j}_n \left| D^{\beta_1}\zeta_n ... D^{\beta_j}\zeta_n \right| dx . \tag{2.2.13}$$

From Hölder's inequality

$$A \le \left(\int_\Omega u^q \zeta^p_n dx \right) \left(\int_\Omega \zeta^{p-jq'}_n \left| D^{\beta_1}\zeta_n ... D^{\beta_j}\zeta_n \right|^{q'} dx \right)^{1/q'} . \tag{2.2.14}$$

Since $p \geq mq' \geq jq'$, we have $0 \leq \zeta_n^{p-jq'} \leq 1$. By applying again Hölder's inequality,

$$A \leq \left(\int_\Omega u^q \zeta_n^p dx \right)^{1/q} \prod_{i=1}^{j} \left(\int_\Omega |D^{\beta_i} \zeta_n|^{|\alpha|q'/|\beta_i|} dx \right)^{|\beta_i|/|\alpha|q'}, \tag{2.2.15}$$

since $\sum_{1 \leq i \leq j} |\beta_i| = |\alpha|$. From the Gagliardo-Nirenberg inequality

$$\left(\int_\Omega |D^{\beta_i} \zeta_n|^{q'|\alpha|/|\beta_i|} dx \right)^{|\beta_i|/q'|\alpha|} \leq \|\zeta_n\|_{\beta_i, q'|\alpha|/\beta_i} \tag{2.2.16}$$
$$\leq C\|\zeta_n\|_{|\alpha|,q'}^{|\beta_i|/|\alpha|} \|\zeta_n\|_\infty^{1-|\beta_i|/|\alpha|} \leq C\|\zeta_n\|_{m,q'}^{|\beta_i|/|\alpha|},$$

and (2.2.15) becomes

$$A \leq \left(\int_\Omega u^q \zeta_n^p dx \right)^{1/q} \|\zeta_n\|_{m,q'}. \tag{2.2.17}$$

Since ζ_n is bounded in $W^{m,q'}(\mathbb{R}^N)$ we deduce that

$$\int_\Omega g(u) \zeta_n^p dx \leq C + C\left(\int_\Omega u^q \zeta_n^p dx \right)^{1/q}. \tag{2.2.18}$$

From (2.2.6) there exists a and b positive such that $g(t) \geq at^q - b$ for $t \geq 0$. Therefore

$$\int_\Omega (g(u) + b) \zeta_n^p dx \leq C + C\left(\int_\Omega (g(u) + b) \zeta_n^p dx \right)^{1/q}. \tag{2.2.19}$$

Letting n go to infinity implies that $(g(u) + b)\zeta^p$ is integrable in Ω, which is the claim.

Step 2. End of the proof. From Step 1 and the growth assumption on g, $u \in L^q_{loc}(\Omega)$. In order to prove that u satisfies (2.2.7) in Ω, we consider $\zeta \in C_0^\infty(\Omega)$, $\zeta \geq 0$, and the $\zeta_n = (1 - \eta_n)\zeta$ where η_n is defined in Step 1. From (2.2.7) we have

$$\int_\Omega \left(uL_m^* \zeta_n + g(u)\zeta_n \right) dx \leq \int_\Omega \zeta_n d\mu. \tag{2.2.20}$$

Since μ does not load the sets with $c_{m,q'}$-capacity zero, ζ_n converges to ζ, μ-almost everywhere. By Lebesgue's theorem

$$\lim_{n \to \infty} \int_\Omega \zeta_n d\mu = \int_\Omega \zeta \, d\mu \quad \text{and} \quad \lim_{n \to \infty} \int_\Omega g(u)\zeta_n dx = \int_\Omega g(u)\zeta \, dx. \tag{2.2.21}$$

Finally $L_m^* \zeta_n$ converges to $L_m^* \zeta$ in $L^{q'}(\Omega)$ as n goes to infinity and , because $u \in L^q(\Omega)$, we deduce from (2.2.20) and (2.2.21) that

$$\int_\Omega \left(u L_m^* \zeta + g(u)\zeta \right) dx \le \int_\Omega \zeta \, d\mu , \tag{2.2.22}$$

which ends the proof.

Remark 2.12. If we take $L_m = \Delta$ and replace (2.2.7) by

$$\Delta u + u^q = 0 \quad \text{in} \quad \mathcal{D}'(\Omega \setminus E) \tag{2.2.23}$$

where $E = \Sigma$ is a k-dimensional compact submanifold of $\mathbb{R}^N$, then E has $c_{m,q'}$-capacity zero if and only if $N - 2q' \ge k$. Therefore u can be extended as a solution of (2.2.3) in $\mathcal{D}'(\Omega)$ if we assume that $q \ge (N-k)/(N-2-k)$. However, we shall see in Section 3 that u still can have a singularity near Σ, but this singularity is not detectable in the distributions level.

2-2-3 General removable singularities

In this paragraph L is the second order elliptic operator defined by

$$Lu = -\sum_{1 \le i,j \le N} \frac{\partial}{\partial x_j} \left(a_{ij} \frac{\partial u}{\partial x_i} \right) + \sum_{1 \le i \le N} \frac{\partial}{\partial x_i} (b_i u) + au \tag{2.3.1}$$

where the a_{ij} and b_i belong to $C^1(\overline{\Omega})$ and $a \in L^\infty(\Omega)$ where Ω is a N-dimensional domain with a regular boundary. We assume the uniform ellipticity condition

$$\sum_{i,j} a_{ij} \xi_i \xi_j \ge \alpha |\xi|^2 \qquad \left(\forall \xi = (\xi_1, \ldots, \xi_N) \in \mathbb{R}^N \right), \tag{2.3.2}$$

where $\alpha > 0$. We assume also

$$\min \left(a, \, a + \sum_i \frac{\partial b_i}{\partial x_i} \right) \ge 0, \quad \text{a.e. in } \Omega. \tag{2.3.3}$$

The adjoint operator is

$$L^* \varphi = -\sum_{i,j} a_{ij} \frac{\partial^2 \varphi}{\partial x_i \partial x_j} - \sum_{i,j} \frac{\partial a_{ij}}{\partial x_i} \frac{\partial \varphi}{\partial x_j} - \sum_i b_i \frac{\partial \varphi}{\partial x_i} + a\varphi \tag{2.3.4}$$

Assume E is a compact subset of Ω, then the main result is

THEOREM 2.15. *Let* $q > 1$ *and* u *belong to* $L^q_{loc}(\Omega \backslash E)$ *and satisfy*

$$Lu + |u|^{q-1} u = 0 \qquad\qquad\qquad (2.3.5)$$

in $\mathcal{D}'(\Omega \backslash E)$. *Then* u *belongs in fact to* $L^q_{loc}(\Omega)$ *and satisfies* (2.3.6) *in* $\mathcal{D}'(\Omega)$ *if and only if* $c_{2,q'}(E) = 0$. *Moreover such a* u *is an element of* $W^{2,p}_{loc}(\Omega)$ *for any* $p \in [1, \infty)$ *and* (2.3.5) *holds almost everywhere in* Ω.

Remark 2.13. In fact there are three aspects in the theorem.

(i) - Our condition $c_{2,q'}(E) = 0$ implies E is a removable singularity contains a stronger result: if $c_{2,q'}(E) > 0$ there exists a nonnegative function u belonging to $L^q(\Omega) \cap W^{1,1}_0(\Omega)$ and a nonnegative measure μ with support E such that

$$Lu + u^q = \mu \quad \text{in } \mathcal{D}'(\Omega). \qquad\qquad\qquad (2.3.6)$$

(ii) - In the previous removability result, we need not assume that u is nonnegative. In fact we prove that if u is a solution, $|u|$ is a subsolution.

(iii) - It is important to notice that in Theorem 2.15 the singular set E becomes removable not only for the equation but also for the solution u itself. This is in sharp contrast with the case of Theorem 2.14 and Remark 2.11.

LEMMA 2.4. *Assume* u *is locally integrable in* Ω *and* Lu *is a Radon measure in* Ω. *Then* $u \in W^{1,\gamma}_{loc}(\Omega)$ *for any* $\gamma \in [1, N/(N-1))$. *Moreover if* Ω_1 *and* Ω_2 *are two open subsets such that* $\Omega_1 \subset \overline{\Omega_1} \subset \Omega_2 \subset \overline{\Omega_2} \subset \Omega$, *there exists a constant* $C = C(\gamma, N, L, \Omega_1, \Omega_2)$ *such that*

$$\|u\|_{W^{1,\gamma}(\Omega_1)} \le C \int_{\Omega_2} (|Lu| + |u|) dx. \qquad\qquad (2.3.7)$$

Proof. We may assume $a = 0$ and $\partial\Omega_2$ is smooth. We fix $\zeta \in C^\infty_0(\Omega_2)$ such that $\zeta = 1$ on Ω_1. For any $\psi \in C^\infty(\Omega_2)$ $\int_\Omega \zeta\psi Lu \, dx = \int_\Omega uL^*(\zeta\psi) dx$, which means

$$\int_\Omega \zeta\psi Lu \, dx = \int_\Omega u \left(\zeta L^*\psi + \psi L^*\zeta - \sum_{i,j} (a_{ij} + a_{ji}) \frac{\partial\zeta}{\partial x_j} \frac{\partial\psi}{\partial x_i} \right) dx. \qquad (2.3.8)$$

If $\theta \in C^\infty_0(\Omega_2)$ we can take for the function ψ the solution in $W^{1,p}_0(\Omega_2) \cap W^{2,p}(\Omega_2)$ of $L^*\psi = \theta$ and p is arbitrary in $[1, \infty)$; actually, since the coefficients of L^* are not smooth enough, an approximation of the coefficients of L^* has to be performed. Therefore (2.3.8) yields

$$\int_\Omega u\zeta\theta \, dx \le C\|\psi\|_{L^\infty(\Omega_2)}\int_{\Omega_2}(|Lu|+|u|)dx + C\|\nabla\psi\|_{L^\infty(\Omega_2)}\int_{\Omega_2}|u|\,dx, \qquad (2.3.9)$$

and C depend on the structural constants, Ω_1, Ω_2 and ζ. The L^p-regularity theory gives

$$\|\nabla\psi\|_{L^\infty(\Omega_2)}+\|\psi\|_{L^\infty(\Omega_2)} \le C(p)\|\theta\|_{L^p(\Omega_2)}, \qquad (2.3.10)$$

for $p > N$. Therefore

$$\int_\Omega u\zeta\theta \, dx \le C\|\theta\|_{L^p(\Omega_1)}\int_{\Omega_2}(|Lu|+|u|)dx, \qquad (2.3.11)$$

which proves that $u \in L^\gamma(\Omega_1)$, by duality, and

$$\|u\|_{L^\gamma(\Omega_1)} \le C(\gamma)\int_{\Omega_2}(|Lu|+|u|)dx. \qquad (2.3.12)$$

We assume now that $\theta = \operatorname{div}(f) = \sum_i \dfrac{\partial f_j}{\partial x_j}$ where $f \in C_0^\infty(\Omega_2)^N$ and (2.3.10) becomes (by approximation [BP], [GT])

$$\|\nabla\psi\|_{L^p(\operatorname{supp}(\zeta))}+\|\psi\|_{L^\infty(\Omega_2)} \le C(p)\|f\|_{(L^p(\Omega_2))^N} \qquad (2.3.13)$$

and (2.3.9) reads as

$$\int_\Omega u\zeta\operatorname{div}(f) \, dx \le C\|\psi\|_{L^\infty(\Omega_2)}\int_{\Omega_2}(|Lu|+|u|)dx + C\left(\int_{\operatorname{supp}(\zeta)}|u|^\gamma\right)^{1/\gamma}\|\nabla\psi\|_{L^p(\operatorname{supp}(\zeta))} \qquad (2.3.14)$$

where $\gamma = p/(p-1)$. By using again (2.3.13) $u \in W^{1,\gamma}(\Omega_1)$ and

$$\|u\|_{W^{1,\gamma}(\Omega_1)} \le C\left(\int_{\Omega_2}(|Lu|+|u|)dx + \int_{\operatorname{supp}(\zeta)}|u|^\gamma\right)^{1/\gamma}, \qquad (2.3.15)$$

which completes the proof.

The following existence result is due to Brezis and Strauss [BS]

LEMMA 2.5. *Assume that* f *is integrable in* Ω *and* g *is a continuous nondecreasing real valued function vanishing at* 0. *Then there exists a unique* $u \in W_0^{1,1}(\Omega)$ *such that* g(u) *is integrable in* Ω *solution of*

$$Lu + g(u) = f \qquad (2.3.16)$$

in $\mathcal{D}'(\Omega)$. Moreover $u \in W^{1,\gamma}(\Omega)$ for any $\gamma \in [1, N/(N-1))$ with

$$\|u\|_{W^{1,\gamma}} + \|g(u)\|_{L^1} \le C(\gamma)\|f\|_{L^1}, \tag{2.3.17}$$

and the mapping $f \mapsto u$ is nondecreasing .

Proof. The crucial point is the following inequality which is proved by approximations in [BS]: if $v \in W_0^{1,1}(\Omega)$ satisfies $Lv \in L^1(\Omega)$ and ℓ a bounded measurable function such that $\ell(x) \in \left[h^-(v(x)), h^+(v(x))\right]$ almost everywhere in Ω where h is a nondecreasing real valued function, then

$$\int_\Omega hLv \, dx \ge 0. \tag{2.3.18}$$

This inequality implies the following *a priori* estimates

$$\int_\Omega |g(u)| dx \le \int_\Omega |f| dx \tag{2.3.19}$$

and

$$\int_\Omega |g(u) - g(\hat{u})| dx \le \int_\Omega |f - \hat{f}| dx \tag{2.3.20}$$

if u and $\hat{u}$ are respectively associated to f and $\hat{f}$. Moreover $|\ |$ can be replaced by $(\ )^+$ in (2.3.19)-(2.3.20). This implies the uniqueness and (2.3.17), by using Lemma 2.1. Existence is proved by approximating f by a sequence of bounded measurable functions and solving the approximate problems.

LEMMA 2.6. *Let μ be a Radon measure in Ω and $u \in L_{loc}^1(\Omega)$ such that*

$$Lu = \mu \qquad in \ \mathcal{D}'(\Omega). \tag{2.3.21}$$

Then for any bounded open subset ω with $\overline{\omega} \subset \Omega$, there exists $u_n \in W^{2,p}(\omega)$, $1 \le p < \infty$, satisfying:

(i) $\lim\limits_{n \to \infty} u_n = u$ in $L^1(\omega)$ and weakly in $W^{1,\gamma}(\omega)$ for $1 \le \gamma < N/(N-1)$;

(ii) $Lu_n = \mu * \rho_n$ in $\mathcal{D}'(\Omega)$ where $\{\rho_n\}$ is a smoothing sequence in $\mathbb{R}^N$.

Proof. From Lemma 2.4, we know that u belongs to $W_{loc}^{1,\gamma}(\Omega)$. Let $\zeta \in C_0^\infty(\Omega)$, $0 \le \zeta \le 1$, such that $\zeta = 1$ in a neighborhood of $\overline{\omega}$. Then

$$L(u\zeta) = \zeta Lu - \sum_{i,j}\left(a_{ij}\frac{\partial u}{\partial x_i}\frac{\partial \zeta}{\partial x_j} - \frac{\partial}{\partial x_j}\left(a_{ij}u\frac{\partial \zeta}{\partial x_i}\right)\right) + \sum_i b_i u \frac{\partial \zeta}{\partial x_i}, \tag{2.3.22}$$

100

and the right-hand side of (2.3.22) can be written as $\zeta\mu + f$ where f is an integrable function which is identically O in a neighborhood of $\overline{\omega}$. If $\{\rho_n\}$ is a smoothing sequence we call u_n the unique solution in $W_0^{1,1}(\Omega)$ of

$$Lu_n = (\zeta\mu + f) * \rho_n \quad \text{in } \mathcal{D}'(\Omega). \tag{2.3.23}$$

Then

$$\|u_n\|_{W^{1,\gamma}} \le C\|(\zeta\mu + f)*\rho_n\|_{L^1} \le C\left(\|f\|_{L^1} + \int_\Omega d|\mu|\right), \tag{2.3.24}$$

and in particular $\{u_n\}$ is relatively compact in $L^1(\Omega)$ and in $W_0^{1,\gamma}(\Omega)$-weak. There exist a subsequence still quoted $\{u_n\}$ and v belonging to $W_0^{1,\gamma}(\Omega)$ such that $\lim_{n\to\infty} u_n = v$ in $L^1(\omega)$ and weakly in $W^{1,\gamma}(\omega)$. Since v solves

$$Lv = \zeta\mu + f \quad \text{in } \mathcal{D}'(\Omega), \tag{2.3.25}$$

$v = \zeta u$. Since $u_n \in W^{2,p}(\Omega)$, we have $(\zeta\mu + f)*\rho_n = \mu*\rho_n$ on ω, for n large enough, which ends the proof.

LEMMA 2.7. *Let* u *be an element of* $L^q_{loc}(\Omega)$ *which solves*

$$Lu + |u|^{q-1} u = 0 \tag{2.3.26}$$

in $\mathcal{D}'(\Omega)$. *Then* $u \in W^{2,p}_{loc}(\Omega)$ *for any* $1 \le p < \infty$ *and* (2.3.26) *holds a.e. in* Ω.

Proof. As in the proof of Theorem 2.10, we have the following Kato's inequality:

$$L|u| \le -|u|^{q-1} u \le 0 \tag{2.3.27}$$

in the sense of distributions in Ω. In particular, $L|u|$ is a non-positive Radon measure in Ω and, for any bounded open subset ω with $\overline{\omega} \subset \Omega$, there exists $\{v_n\} \in W^{2,p}(\omega)$ $(1 \le p < \infty)$, which converges to $|u|$ in the topologies of Lemma 2.3 and the v_n satisfy

$$Lv_n \le 0 \qquad \text{in } \mathcal{D}'(\Omega). \tag{2.3.28}$$

Moreover $\|v_n\|_{W^{1,\gamma}}$ remains bounded independently of n for $1 \le \gamma < N/(N-1)$. From the weak maximum principle (see [GT]), for any ball $B(m,R) \subset \omega$ there holds

$$\max_{B(m,R/2)} v_n \le C(R)\|v_n\|_{L^\gamma(B(m,R))}. \tag{2.3.29}$$

Since this inequality is stable when n goes to infinity u remains locally bounded in Ω. We conclude by using the local regularity theory for elliptic equations.

LEMMA 2.8. *Let μ be a Radon measure with compact support in Ω such that both μ^+ and μ^- belong to $W^{-2,q}(\Omega)$. Then there exists a unique $u \in L^q(\Omega) \cap W_0^{1,1}(\Omega)$ solution of*

$$Lu + |u|^{q-1}u = \mu \tag{2.3.30}$$

in $\mathcal{D}'(\Omega)$. Moreover the mapping $\mu \mapsto u$ is nondecreasing.

Proof. We shall call μ the extension of μ to $\mathbb{R}^N$ by 0 outside Ω.
Step 1. Assume μ is nonnegative and therefore belongs to $W^{-2,q}(\Omega)$. Let $\mu_n = \mu * \rho_n$ where $\{\rho_n\}$ is a smoothing sequence and let $u_n \in L^q(\Omega) \cap W_0^{1,1}(\Omega)$ be the solution of

$$Lu_n + u_n^q = \mu_n \tag{2.3.31}$$

in the sense of distributions in Ω. Then u_n is nonnegative and

$$\|u_n\|_{W^{1,\gamma}} + \int_\Omega u_n^q dx \leq C(\gamma) \int_\Omega \mu_n dx \leq C(\gamma) \int_\Omega d\mu \tag{2.3.32}$$

for $1 \leq \gamma < N/(N-1)$. There exists a subsequence still quoted $\{u_n\}$ which converges in $L^\gamma(\Omega)$ and a.e. in Ω to some u belonging to $L^q(\Omega) \cap W_0^{1,1}(\Omega)$. Since Lu_n converges to Lu in $\mathcal{D}'(\Omega)$, the main point is to prove that u_n^q converges to u^q in $\mathcal{D}'(\Omega)$. This will be acheived provided we can find a sequence $\{v_n\}$ which converges in $L^q(\Omega)$ and satisfies $0 \leq u_n \leq v_n$. It is clear that the sequence $\{v_n\}$ defined in $W_0^{1,1}(\Omega)$ by

$$Lv_n = \mu_n \tag{2.3.33}$$

in $\mathcal{D}'(\Omega)$ satisfies $0 \leq u_n \leq v_n$. We first have to study the convergence properties of $\{\mu_n\}$. Since μ also belongs to $W^{-2,q}(\mathbb{R}^N)$, we set $\sigma = (I - \Delta)^{-1}\mu$ in $\mathbb{R}^N$ and $\sigma \in L^q(\mathbb{R}^N)$. Then $\mu_n = \mu * \rho_n = (I - \Delta)(\sigma * \rho_n)$. The fact that $\sigma * \rho_n$ converges to σ in $L^q(\Omega)$ implies that μ_n converges to in $W^{-2,q}(\Omega)$. For $\theta \in L^{q'}(\Omega)$ we denote by $w \in W^{2,q'}(\Omega) \cap W_0^{1,1}(\Omega)$ the solution of

$$L^* w = \theta \qquad \text{in } \mathcal{D}'(\Omega). \tag{2.3.34}$$

Clearly $\|w\|_{W^{2,q'}} \leq C\|\theta\|_{L^{q'}}$; moreover, if $\varphi \in C_0^\infty(\Omega)$ with $\varphi = 1$ in a neighborhood of the support of μ, there holds

$$\int_\Omega v_n \theta dx = \int_\Omega \mu_n w dx = \int_\Omega \mu_n \varphi w dx. \tag{2.3.35}$$

Therefore

$$\int_{\Omega}(v_p - v_q)\theta = \langle w\varphi, \mu_p - \mu_q\rangle \leq$$
$$\|w\varphi\|_{W_0^{2,q'}}\|\mu_p - \mu_q\|_{W^{-2,q}} \leq C\|\theta\|_{L^q}\|\mu_p - \mu_q\|_{W^{-2,q}}, \tag{2.3.36}$$

and

$$\|v_p - v_q\|_{L^q} \leq C\|\mu_p - \mu_q\|_{W^{-2,q}}, \tag{2.3.37}$$

which implies the convergence of the $\{v_n\}$ and of the $\{u_n\}$ in $L^q(\Omega)$.

Step 2. The general case. We just have to notice that the solution $u_n \in L^q(\Omega) \cap W_0^{1,1}(\Omega)$ of the approximate problem

$$Lu_n + |u|_n^{q-1}u = \mu_n = \mu * \rho_n, \tag{2.3.38}$$

satisfies

$$-w_n \leq u_n \leq \tilde{w}_n, \tag{2.3.39}$$

where w_n and $\tilde{w}_n$ belong to $L^q(\Omega) \cap W_0^{1,1}(\Omega)$ and solve

$$Lw_n + w_n^q = \mu^- * \rho_n \quad \text{and} \quad L\tilde{w}_n + \tilde{w}_n^q = \mu^+ * \rho_n. \tag{2.3.40}$$

From the construction of Step 1, w_n and $\tilde{w}_n$ converge in $L^q(\Omega)$. The remainining of the proof is straightforward.

Step 3. Monotonicity. Assume μ_n and $\tilde{\mu}_n$ are given Radon measures with compact support in Ω and u and $\tilde{u}$ are the elements of $L^q(\Omega) \cap W_0^{1,1}(\Omega)$ which satisfy

$$Lu + |u|^{q-1}u = \mu \quad \text{and} \quad L\tilde{u} + |\tilde{u}|^{q-1}\tilde{u} = \tilde{\mu} \tag{2.3.41}$$

in the sense of distributions in Ω. We denote by θ_n the solution, belonging to $W_0^{1,1}(\Omega)$, of the following equation

$$L\theta_n = (\mu - \tilde{\mu}) * \rho_n + |u|^{q-1}u - |\tilde{u}|^{q-1}\tilde{u}. \tag{2.3.42}$$

From the above estimates in $W^{1,\gamma}$, θ_n converges in $L^1(\Omega)$. Its limit is necessarily $\tilde{u} - u$ and $\int_{\Omega} \text{sgn}^+ \theta_n L\theta_n dx \geq 0$ by Lemma 2.2. Therefore

$$\tilde{\mu} \leq \mu \Rightarrow \int_{\Omega} \text{sgn}^+ \theta_n (\tilde{u}|\tilde{u}|^{q-1} - u|u|^{q-1})dx \leq 0. \tag{2.3.43}$$

Letting n go to infinity implies $\tilde{u} \le u$.

Proof of Theorem 2.15. Step 1. We assume that $c_{2,q'}(E) = 0$. From kato's inequality

$$L|u| \le |u|^q \tag{2.3.44}$$

in $\mathcal{D}'(\Omega \backslash E)$. From Theorem 2.14, $u \in L^q_{loc}(\Omega)$. If $\zeta \in C_0^\infty(\Omega)$ and $\{\eta_n\}$ is a sequence of $C_0^\infty(\Omega)$ as in Proposition 1.1, then

$$\int_\Omega \left(uL^*\left(\zeta(1-\eta_n)\right) + |u|^{q-1}u\zeta(1-\eta_n)\right)dx = 0. \tag{2.3.45}$$

Since $u \in L^q_{loc}(\Omega)$, $\lim_{n\to\infty}\eta_n = 0$ a.e. and $\lim_{n\to\infty}L^*\left(\zeta(1-\eta_n)\right) = L^*\zeta$ in $L^{q'}(\Omega)$, we obtain

$$\int_\Omega \left(uL^*\zeta + |u|^{q-1}u\zeta\right)dx = 0, \tag{2.3.46}$$

which means that (2.3.5) holds in $\mathcal{D}(\Omega)$. We complete the proof with Lemma 2.4.

Step 2. Let us assume that $c_{2,q'}(E) > 0$. Then by a classical result ([Me]) on capacity theory, there exists a nonnegative Radon measure μ in Ω such that

$$\mu \ne 0, \quad \sup p(\mu) \subset E, \quad \mu \in W^{-2,q'}(\Omega). \tag{2.3.47}$$

From Lemma 2.5 there exists a non-zero u with the appropriate regularity, solution of (2.3.30) in $\mathcal{D}'(\Omega)$. Therefore we have (2.3.5) in $\mathcal{D}'(\Omega \backslash E)$ but not in $\mathcal{D}'(\Omega)$.

Remark 2.14. In Lemma 2.5 we have seen that the equation (2.3.30) is solvable in the class $L^q(\Omega) \cap W_0^{1,1}(\Omega)$ provided μ has compact support and μ^+ and μ^- both belong to $W^{-2,q}(\Omega)$. In fact [BP], the same equation is solvable in the same class of u if μ is a bounded Radon in Ω which does not load the sets with $c_{2,q'}$-capacity zero. The space of such measures can be identified with $W^{-2,q}(\Omega) + L^1(\Omega)$.

2-2-4 Elliptic operators with discontinuous coefficients

A natural question is to know whether the removability results of sections 2.2.3 for equation

$$Lu + |u|^{q-1}u = 0 \tag{2.4.1}$$

still holds if the coefficients of L are discontinuous. A counter-example due to Aviles [Av2] proves that it not the case at least for operators which are not in divergence form. If we consider the operator

$$\varphi \mapsto L\varphi = \sum_{i,j} \left(\delta_{ij} + \frac{x_i x_j}{|x|^2} \right) \varphi, \tag{2.4.2}$$

then the function

$$x \mapsto u(x) = ((N-1)(N-2))^{(N-2)/2} |x|^{2-N} \tag{2.4.3}$$

satisfies

$$-Lu + u^{N/(N-2)} = 0 \tag{2.4.4}$$

in $\mathbb{R}^N \setminus \{O\}$. For an operator in divergence form as the L defined by (2.3.1), it is proved in [Av2] that Theorem 2.9 is still valid if the Laplacian is replaced by L, provided the coefficients a_{ij} and b_i are Hölder continuous. The results of Kondratiev and Landis [KL] are valid for strongly elliptic operators with bounded measurable coefficients in some domain Ω of $\mathbb{R}^N$ such as

$$Lu = -\sum_{i,j} \frac{\partial}{\partial x_j} \left(a_{ij} \frac{\partial u}{\partial x_i} \right) \tag{2.4.5}$$

and

$$\mathcal{L}u = -\sum_{i,j} a_{ij} \frac{\partial^2 u}{\partial x_i \partial x_j} \tag{2.4.6}$$

where the matrix $(a_{ij}(x))$ is supposed to symmetric. The following results extends Theorem 2.9 of Section 2.2.2. Let assume that Ω contains O and Ω^* denote $\Omega \setminus \{O\}$.

THEOREM 2.16. *Assume that the coefficients of* L *given by* (2.4.6) *are bounded and measurable in* Ω, $N > 2$, *and* $u \in C(\Omega^*) \cap W^{1,2}_{loc}(\Omega^*)$ *is a weak solution of* (2.4.1) *in* Ω^*. *If* $q > N / (N-2)$, u *can be extended by continuity as a continuous weak solution of* (2.4.1) *in* Ω.

In this result it is clear that the assumption of continuity is unnecessary since any weak solution is locally Hölder continuous.

THEOREM 2.17. *Assume that the coefficients of* $\mathcal{L}$ *given by* (2.4.7) *are Dini continuous at* O *and* $u \in C^2(\Omega^*)$ *satisfies*

$$\mathcal{L}u + |u|^{q-1} u = 0 \tag{2.4.7}$$

in Ω^*. *If* $q > N/(N-2)$, u *can be extended by continuity as a* $C^2(\Omega)$*solution of* (2.4.8) *in* Ω.

The proof of the two theorems is settled upon an *a priori* estimate near the singularity O and in the case of Theorem 2.14, it also heavily relies on integral inequalities (see [KL]).

LEMMA 2.9. *Assume that the coefficients of* L *are as in Theorem* 2.14, *and* u $\in C(B(m,1)) \cap W^{1,2}_{loc}(B(m,1))$ *is a weak solution of* (2.4.1) *in* B(m,1). *Then there exists a positive constant* $\alpha = \alpha(q,N)$*such that*

$$|u(m)| \leq \alpha. \tag{2.4.8}$$

The same estimate holds for the equation (2.4.7) without the continuity assumption on the coefficients of $\mathcal{L}$.

LEMMA 2.10. *Assume the coefficients of* $\mathcal{L}$ *are bounded and measurable . Then there exists some positive constant* $\alpha = \alpha(q,N)$ *such that such that any solution* $u \in C^2(B(m,1))$ *of* (2.4.8) *in* B(m,1) *satisfies* (2.4.8).

The proof of this Lemma is obtained by computing directly a supersolution in some subdomain G of B(m,1) containing m and goes to infinity on its boundary. From Lemmas 2.6-2.7 we deduce by a simple scaling argument that under the assumptions of Theorems 2.16-2.17 there exists some $C > 0$ such that

$$|u(x)| \leq \alpha|x|^{-2/(q-1)} \qquad \text{for } 0 < |x| \leq C. \tag{2.4.9}$$

Proof of Theorem 2.16. Under the assumptions on the a_{ij} there exists a fundamental solution Γ of L in B(O,C) such that

$$a^{-1}|x|^{2-N} \leq \Gamma(x) \leq a|x|^{2-N} \tag{2.4.10}$$

for some $a < 0$. Since Lemma 2.9 means that $|u(x)| = o(\Gamma(x))$ we conclude by the maximum principle.

Proof of Theorem 2.17. Under the assumption of the Dini continuity of the a_{ij} at O we apply the technique of the proof Theorem 1.3 which relies on the existence of a nonnegative function h defined in a neighborhood of O and which satisfies

$$\mathcal{L}h \geq 0 \text{ in } B^*(O,C) = B(O,C)\backslash \{O\}, \quad h(x) \approx |x|^{2-N} \text{ near O}. \tag{2.4.11}$$

Since Lemma 2.10 means that $|u(x)| = o(h(x))$, we conclude again by the maximum principle.

3 Singularities of solutions of elliptic equations with source term

3-1 Isolated singularities in differential inequalities

3-1-1 The Brezis-Lions Lemma

Definition. Assume that Ω is an open subset of $\mathbb{R}^N$ ($N > 1$) and μ is a nonnegative Borel Measure on Ω. For $p > 1$ and $u \in L^1_{loc}(\Omega, d\mu)$, we define

$$\|u\|_{M^p(\Omega, d\mu)} = \inf\left\{ c \in [0, \infty] : \int_K |u| d\mu \leq c\left(\int_K d\mu\right)^{1-1/p}, \right.$$
$$\left. (\forall K \subset \Omega),\ K\ \mu\text{-meas.} \right\} \tag{1.1.1}$$

and

$$M^p(\Omega, d\mu) = \left\{ u \in L^1_{loc}(\Omega, d\mu) : \|u\|_{M^p(\Omega, d\mu)} < \infty \right\} \tag{1.1.2}$$

The above space is called a Marcinkiewicz space. When μ is the Lebesgue measure in Ω, we denote $M^p(\Omega, dx) = M^p(\Omega)$. The following properties of the Marcinkiewicz spaces can be found in [BBC].

PROPOSITION 3.1. *Set* $1 \leq q < p < \infty$ *and* $u \in L^1_{loc}(\Omega, d\mu)$; *then*

$$C(p)\|u\|^p_{M^p(\Omega, d\mu)} \leq \sup_{\lambda > 0}\left\{ \lambda^p \int_{\{|u| > \lambda\}} d\mu \right\} \leq \|u\|^p_{M^p(\Omega, d\mu)}. \tag{1.1.3}$$

Moreover

$$\int_K |u|^q d\mu \leq C(p, q)\|u\|^p_{M^p(\Omega, d\mu)}\left(\int_K d\mu\right)^{1-q/p}. \tag{1.1.4}$$

The importance of the Marcinkiewicz class comes from the fact that for $N > \alpha > 0$ the function $x \mapsto |x|^{-\alpha}$ belongs to $M^{N/\alpha}(\mathbb{R}^N)$ but not to $L^{N/\alpha}(\mathbb{R}^N)$, and it has the following stability property ([BBC]):

PROPOSITION 3.2. *If* E *belongs to* $M^p(\mathbb{R}^N)$, $1 < p < \infty$, *and* f *is integrable in* $\mathbb{R}^N$, *then the convolution* $E * f$ *is well defined, belongs to* $M^p(\mathbb{R}^N)$ *and satisfies*

$$\|E * f\|_{M^p} \leq \|E\|_{M^p}\|f\|_{L^1}. \tag{1.1.5}$$

Remark 3.1. In particular, if we take

$$
E_N(x) = \begin{cases} \dfrac{1}{(N-2)\left|S^{N-1}\right|}|x|^{2-N} = \dfrac{\mu(x)}{(N-2)\left|S^{N-1}\right|} & \text{if } N \ge 3, \\[2ex] \dfrac{1}{2\pi}\ln(1/|x|) = \dfrac{\mu(x)}{2\pi} & \text{if } N = 2. \end{cases}
\tag{1.1.6}
$$

then $-\Delta E_N = \delta_O$ in $\mathcal{D}'(\mathbb{R}^N)$. Since $E_N \in M^{N/(N-2)}(\mathbb{R}^N)$ when $N > 2$ and $E_2 \in M^p(\mathbb{R}^2)$ for any $1 < p < \infty$, we can write a solution of $-\Delta v = f$ in $\mathbb{R}^N$ under the form $v = E_N * f$, provided $f \in L^1$. Moreover uniqueness is ensured by some decay at infinity. If we replace $\mathbb{R}^N$ by a ball (or a bounded subset Ω), then any integrable solution v of

$$
-\Delta v = f
\tag{1.1.7}
$$

in Ω belongs to $M^{N/(N-2)}(\Omega)$ if $N > 2$, and $M^p(\Omega)$ for any $1 < p < \infty$ if $N = 2$. If we assume some prescribed boundary data, for example $v \in W_0^{1,1}(\Omega)$, then v is uniquely determined and

$$
\|v\|_{M^{N/(N-2)}(\Omega)} + \|\nabla v\|_{M^{N/(N-1)}(\Omega)} \le C\|f\|_{L^1(\Omega)}.
\tag{1.1.8}
$$

The following result due to Brezis and Lions provides a natural way for studying isolated singularities of positive solutions of

$$
-\Delta u = u^q
\tag{1.1.9}
$$

for $q > 1$. We set $B^*(O,R) = B(O,R) \setminus \{O\}$.

THEOREM 3.1. *Assume* $u \in L^1_{loc}\big(B^*(O,R)\big)$ *is nonnegative and satisfies*

$$
\Delta u \in L^1_{loc}\big(B^*(O,R)\big)
\tag{1.1.10}
$$

where the Laplacian is taken is the sense of distributions and

$$
\Delta u(x) \le au(x) + f(x) \quad \text{a.e. in } B(O,R),
\tag{1.1.11}
$$

where a is positive and f *is locally integrable in* $B(O,R)$. *Then* $u \in L^1_{loc}(B(O,R))$ *and there exist* $\varphi \in L^1_{loc}(B(O,R))$ *and* $\alpha \ge 0$ *such that*

$$
-\Delta u = \varphi + \alpha \delta_O \quad \text{in } \mathcal{D}'(B(O,R)).
\tag{1.1.12}
$$

In particular, it follows that $u \in M_{loc}^p(B(O,R))$ *where* $p = N/(N-2)$ *if* $N > 2$ *and* $p \in (1, \infty)$ *is arbitrary when* $N = 2$.

Proof. Step 1. We claim that $u \in L_{loc}^1(B(O,R))$. We recall that $\bar{u}(r)$ is the spherical average of $u(x)$ on the sphere $|x| = r$. From hypotheses we have

$$\frac{1}{r^{N-1}} \frac{d}{dr}\left(\frac{d\bar{u}}{dr}\right) \in L_{loc}^1(0,R) \tag{1.1.13}$$

and

$$\frac{1}{r^{N-1}} \frac{d}{dr}\left(r^{N-1}\frac{d\bar{u}}{dr}\right) \leq a\bar{u} + \bar{f} \quad \text{on} \quad (0,R). \tag{1.1.14}$$

In particular $\bar{u} \in C^1(0,R)$. Let $R' < R$ be fixed. Integrating (1.1.14) over (r,R') yields

$$-r^{N-1}\frac{d\bar{u}}{dr}(r) \leq a\int_r^{R'} \bar{u}(s)s^{N-1}ds + C = a\psi(r) + C, \tag{1.1.15}$$

where C is a constant. Therefore

$$u(r) \leq a\int_r^{R'} \psi(s)s^{1-N}ds + Cr^{2-N} + C, \tag{1.1.16}$$

if $N > 2$, and

$$\bar{u}(r) \leq a\int_r^{R'} \psi(s)s^{-1}ds + C|\ln r| + C, \tag{1.1.17}$$

if $N = 2$. In particular

$$r^{N-1}\bar{u}(r) \leq a\int_r^{R'} \psi(s)ds + Cr \leq aR'\,\psi(r) + Cr, \tag{1.1.18}$$

with an obvious modification if $N = 2$. Integrating (1.1.17) and using again the fact that ψ is nonincreasing implies

$$\psi(r) \leq a(R')^2\,\psi(r) + C \quad \text{for } 0 < r < R'. \tag{1.1.19}$$

If we choose R' small enough we conclude that ψ remains bounded when r goes to 0 and $u \in L_{loc}^1(B(O,R))$. Going back to (1.1.13) we also find that

$$\bar{u}(r) \leq \begin{cases} Cr^{2-N} + C & \text{if} \quad N \geq 3, \\ C|\ln r| + C & \text{if} \quad N = 2. \end{cases} \tag{1.1.20}$$

Step 2. We claim that the function φ defined by

$$\varphi(x) = \Delta u(x) \quad \text{a.e. in } B(O,R), \tag{1.1.21}$$

is integrable in $B(O,R)$. By assumption $\varphi \in L^1_{loc}(B^*(BO,R))$. We introduce a sequence of functions $\zeta_\varepsilon \in C^\infty(B(O,R))$ with $0 \le \zeta_\varepsilon \le 1$ on $B(O,R)$ and

$$\begin{cases} \lim_{\varepsilon \to 0} \zeta_\varepsilon(x) = 1 \quad (\forall x \ne O), \\[2mm] \lim_{\varepsilon \to 0} \nabla \zeta_\varepsilon(x) = 0, \text{ uniformly on any compact subset of } B^*(O,R), \\[2mm] \zeta_\varepsilon(x) = 0 \quad \text{on } B(O,\varepsilon), \\[2mm] \Delta \zeta_\varepsilon \ge 0 \quad \text{on } B(O,R). \end{cases} \tag{1.1.22}$$

In fact, such a function can be constructed from a nonnegative C^∞ convex function Φ defined on $[0,\infty)$ and satisfying $\Phi(0) = 1$ and $\Phi(1) = 0$ by setting

$$\zeta_\varepsilon(x) = \begin{cases} \Phi\left(\varepsilon |x|^{2-N}\right) & \text{when } N \ge 3, \\[2mm] \Phi\left(\ln(|x|)/\ln\varepsilon\right) & \text{when } N = 3. \end{cases} \tag{1.1.23}$$

Let $\eta \in C^\infty_0(B(O,R))$, $0 \le \eta \le 1$ and $\eta = 1$ in a neighborhood of O. We have

$$\int_{B(O,R)} \varphi \zeta_\varepsilon \eta \, dx = -\int_{B(O,R)} u \Delta(\zeta_\varepsilon \eta) dx,$$

$$= \int_{B(O,R)} u\eta \Delta \zeta_\varepsilon dx - 2\int_{B(O,R)} u \nabla \eta . \nabla \zeta_\varepsilon dx - \int_{B(O,R)} u \zeta_\varepsilon \Delta \eta dx, \tag{1.1.24}$$

$$\le -2\int_{B(O,R)} u \nabla \eta . \nabla \zeta_\varepsilon dx - \int_{B(O,R)} u \zeta_\varepsilon \Delta \eta dx,$$

by the positivity of u and $\Delta \zeta_\varepsilon$. Since $\varphi \ge -au - f$ which is locally integrable in $B(O,R)$, we can let ε go to 0 and deduce from Fatou's lemma that $\varphi \in L^1_{loc}(B(O,R))$ and

$$\int_{B(O,R)} \varphi \eta \, dx \le -\int_{B(O,R)} u \Delta \eta \, dx. \tag{1.1.25}$$

Step 3. End of the proof. We consider the distribution $T = -\Delta u - \varphi$ whose support is reduced to the point O. From a classical result, T is a linear combination of the Dirac measure and derivatives of the Dirac measure at O, that is

$$T = \sum_{|p| \le m} c_p D^p \delta_O. \tag{1.1.26}$$

The only thing we have to prove is $c_p = 0$ for $|p| \geq 1$ and $c_0 \geq 0$. Indeed let $\zeta \in C^\infty(B(O,R))$ be any fixed function such that

$$(-1)^{|p|}(D^p \zeta)(O) = c_p, \quad \text{for every } |p| \leq m. \tag{1.1.27}$$

We set $\zeta_\varepsilon(x) = \zeta(x / \varepsilon)$ and deduce from (1.1.27) that

$$-\int_{B(O,R)} u \Delta \zeta_\varepsilon \, dx = \int_{B(O,R)} \varphi \zeta_\varepsilon \, dx + \sum_{|p| \leq m} \frac{c_p^2}{\varepsilon^{|p|}}. \tag{1.1.28}$$

On the other hand

$$\left| \int_{B(O,R)} u \Delta \zeta_\varepsilon \, dx \right| = \frac{1}{\varepsilon^2} \left| \int_{B(O,R)} u(x) \Delta \zeta\left(\frac{x}{\varepsilon}\right) dx \right| \leq \frac{C}{\varepsilon^2} \int_{B(O,R\varepsilon)} u \, dx \leq \frac{C}{\varepsilon^2} \int_0^{R\varepsilon} \overline{u}(s) s^{N-1} ds. \tag{1.1.29}$$

Using (1.1.16), we obtain

$$\left| \int_{B(O,R)} u \Delta \zeta_\varepsilon \, dx \right| \leq \begin{cases} C & \text{if } N \geq 3, \\ C \ln(1/\varepsilon) & \text{if } N = 2. \end{cases} \tag{1.1.30}$$

Comparing (1.1.28) and (1.1.26) implies that $c_p = 0$ for $|p| \geq 1$ and

$$-\Delta u = \varphi + c_0 \delta_O \quad \text{in } \mathcal{D}'(B(O,R)). \tag{1.1.31}$$

If we choose $\eta \in C^\infty(B(O,R))$ with $0 \leq \eta \leq 1$ and $\eta = 1$ in a neighborhood of O, we have, from inequality (1.1.25),

$$\int_{B(O,R)} \varphi \eta \, dx = -\int_{B(O,R)} u \Delta \eta \, dx - c_0 \leq -\int_{B(O,R)} u \Delta \eta \, dx. \tag{1.1.32}$$

Therefore c_0 is nonnegative. As for the estimate in the Marcinkiewicz class, it comes from Remark 3.1.

Remark 3.2. The assumption of nonnegativity upon u in Theorem 3.1 can be replaced by

$$u \geq g \quad \text{a.e. in } B(O,R), \qquad \text{where } g \in M^{N/(N-2)}(B(O,R)). \tag{1.1.33}$$

In that case the sign of α is not kept excepted if g satisfies a stronger assumption like $g \in L_{\text{loc}}^{N/(N-2)}(B(O,R))$.

A typical application of Theorem 3.1 is the following

COROLLARY 3.1. *Assume that* g *is a continuous real valued function defined on* $\mathbb{R}^+$ *such that*

111

$$\liminf_{t \to \infty} g(t)/t > -\infty \tag{1.1.34}$$

and assume that u *is nonnegative, belongs to* $L^1_{loc}\left(B^*(O,R)\right)$ *with* $g(u) \in L^1_{loc}\left(B^*(O,R)\right)$ *and satisfies*

$$-\Delta u = g(u) \quad in \quad \mathcal{D}'(B^*(O,R)). \tag{1.1.35}$$

Then $u \in L^1_{loc}(B(O,R))$, $g(u) \in L^1_{loc}(B(O,R))$ *and there exists* $\alpha \geq 0$ *such that*

$$-\Delta u = g(u) + \alpha \delta_o \quad in \quad \mathcal{D}'(B(O,R)). \tag{1.1.36}$$

3-1-2 Isotropic singularities of differential inequalities

In this paragraph we present Richard-Véron's results [RV] on the isolated singularities of nonnegative solutions of

$$\Delta u \leq g(u) + f \tag{1.2.1}$$

in $\Omega^* = \Omega \setminus \{O\}$ where Ω is an open subset of $\mathbb{R}^N$ ($N \geq 2$) and g a nondecreasing real valued function. We recall that μ is the fundamental solution of the Laplace equation in $\mathbb{R}^N$.

PROPOSITION 3.3. *Assume that* $B(O,R) \subset \overline{B}(O,R) \subset \Omega$, g *vanishes at* 0 *and let* $f \in L^1_{loc}(\Omega)$ *and* $u \in C^2(\Omega^*)$ *be both nonnegative and satisfy* (1.2.1) *a.e. in* Ω. *If* $v \in C^2(B^*(O,R))$ *is a radial nonnegative solution of*

$$\Delta v = g(v) \tag{1.2.2}$$

in $B^*(O,R)$ *such that* $g(v + \tilde{\delta}) \in L^1(B(O,R))$ *for some* $\tilde{\delta} > 0$, *then there exists* $\alpha \geq 0$ *such that,*

$$\lim_{r \to 0} \left\| \alpha - \inf(u,v)(r,.)/\mu(r) \right\|_{L^q(S^{N-1})} = 0, \tag{1.2.3}$$

for any $q \in [1, \infty)$.

Before proving this result which will provide a basis for Richard-Véron's technique, we need

LEMMA 3.1. *Assume that* $h \in L^1(B(O,R))$ *is radial and* φ *is a nonnegative radial solution of*

$$-\Delta \varphi = h \tag{1.2.4}$$

in $\mathcal{D}'(B^*(O,R))$ *(resp.* $\mathcal{D}'(B(O,R))$*). Then there exists* $v \in [0,\infty)$ *such that*

$$\lim_{x \to O} \varphi(x)/\mu(x) = v \quad (resp. \lim_{x \to O} \varphi(x)/\mu(x) = 0). \tag{1.2.5}$$

Proof. From Theorem 3.1, there exists $v \in [0,\infty)$ such that

$$-\Delta\varphi = h + C(N)v\delta_O \quad \text{in} \quad \mathcal{D}'(B^*(O,R)) \tag{1.2.6}$$

where $C(N)$ is the constant defined in (1.1.6). If we set $\tilde{\varphi} = \varphi - v\mu$, then $\tilde{\varphi}$ satisfies (1.2.4) in $\mathcal{D}'(B(O,R))$. Since

$$-\frac{d}{dr}\left(r^{N-1}\frac{d\tilde{\varphi}}{dr}\right) = r^{N-1}h, \tag{1.2.7}$$

we have from the integrability of h

$$\left[s^{N-1}\frac{d\tilde{\varphi}}{dr}(s)\right]_{r'}^{r} = \int_{r'}^{r} s^{N-1}h(s)ds \to 0 \quad \text{as } r, r' \to 0, \tag{1.2.8}$$

and $\lim\limits_{r \to 0} r^{N-1}\dfrac{d\tilde{\varphi}}{dr}(r) = \gamma$, for some constant γ, which implies that

$$\lim_{r \to 0} \tilde{\varphi}(r)/\mu(r) = \gamma/(N-2), \qquad (resp. \ \gamma) \tag{1.2.9}$$

if $N > 2$ (resp. $N = 2$). Since (1.2.4) and (1.2.9) imply

$$-\Delta\tilde{\varphi} = h + C(N)\gamma\delta_O, \tag{1.2.10}$$

we would have a contradiction with (1.2.6) if γ were not 0.

Proof of Proposition 3.3. Let us define a $C^{1,1}$ even convex function p by

$$p(t) = \begin{cases} |t| - \delta/2 & \text{for } |t| \geq \delta > 0, \\ t^2/2\delta & \text{for } |t| \leq \delta, \end{cases} \tag{1.2.11}$$

and set $\omega_\delta = \dfrac{1}{2}(u + v - p(u-v))$. Then

$$\Delta\omega_\delta = \frac{1}{2}\Delta(u+v) - \frac{1}{2}p'(u-v)\Delta(u-v) - \frac{1}{2}p''(u-v)|\nabla(u-v)|^2$$

$$\leq \frac{1}{2}\Delta(u+v) - \frac{1}{2}p'(u-v)\Delta(u-v) = F. \tag{1.2.12}$$

Moreover $\Delta\omega_\delta \in L^1_{loc}\left(B^*(O,R)\right)$. We decompose $B^*(O,R)$ as $G_1 \cup G_2 \cup G_3$ where

$$\begin{cases} G_1 = \left\{ x \in B^*(O,R) : (u(x)-v(x)) > \delta \right\}, \\ G_2 = \left\{ x \in B^*(O,R) : (u(x)-v(x)) < -\delta \right\}, \\ G_3 = \left\{ x \in B^*(O,R) : |u(x)-v(x)| \le \delta \right\}. \end{cases} \qquad (1.2.13)$$

On G_1, $p'(u-v) = 1$ and $F = \Delta v = g(v) = g(\omega_\delta - \delta/4)$. On G_2, $p'(u-v) = -1$ and $F \le g(v) + f$. On G_3, $p'(u-v) = (u-v)/\delta$, henceforth

$$\begin{aligned} F &= \frac{1}{2}\left(1 - \frac{u-v}{\delta}\right)\Delta u + \frac{1}{2}\left(1 + \frac{u-v}{\delta}\right)\Delta v \\ &\le \frac{1}{2}\left(1 - \frac{u-v}{\delta}\right)g(u) + \frac{1}{2}\left(1 + \frac{u-v}{\delta}\right)g(v) + f. \end{aligned} \qquad (1.2.14)$$

In any case, ω_δ satisfies

$$\Delta\omega_\delta = F \le g(\omega_\delta + 3\delta/4) + f \le g(v+\delta) + f. \qquad (1.2.15)$$

For $0 < \delta \le \tilde{\delta}$, the right-hand side of (1.2.15) is locally integrable in $B(O,R)$. Consequently there exist $\Phi \in L^1_{loc}(B(O,R)$ and $\alpha \ge 0$ such that

$$-\Delta\omega_\delta = \Phi + \alpha\, C(N)\delta_O, \qquad (1.2.16)$$

in the sense of distributions in $B(O,R)$. Setting $\varphi_\delta = \omega_\delta - \alpha\mu$ and $\overline{\varphi}_\delta$ (resp. $\overline{\Phi}$) the spherical average of φ (resp. Φ), then

$$-\Delta\overline{\varphi}_\delta = \overline{\Phi}. \qquad (1.2.17)$$

Since $\overline{\Phi}$ is locally integrable $\lim_{r\to 0}\int_{S^{N-1}}|\alpha - \omega_\delta(r,.)/\mu(r)|d\sigma = 0$, which implies, thanks to the uniform boundedness,

$$\lim_{r\to 0}\left(\int_{S^{N-1}}|\alpha - \omega_\delta(r,.)/\mu(r)|^q d\sigma\right)^{1/q} = 0, \qquad (1.2.18)$$

for any $q \in [1,\infty)$. Moreover $0 \le \omega \le \omega_\delta \le \omega + \delta/4$, therefore

$$\lim_{r\to 0}\left(\int_{S^{N-1}}|\alpha - \omega_\delta(r,.)/\mu(r)|^q d\sigma\right)^{1/q} = 0, \qquad (1.2.19)$$

114

which is (1.2.3).

Remark 3.3. Since Φ is integrable in B(O,R), it is the same with $p''(u-v)|\nabla(u-v)|^2$.

We recall now the following

Definition. Let (E,Σ,v) be an abstract measured space where Σ is a σ-algebra of subsets of E and v a nonnegative σ-additive and complete measure such that $v(E)<\infty$. If $\{\psi_r\}_{r\in(0,R]}$ is a subset of v-measurable real valued functions, we say that ψ_r converges in measure to some v-measurable function ψ, if for any $\varepsilon>0$ there holds

$$\lim_{r\to 0} v\left(\left\{x\in E:\ |\psi(x)-\psi_r(x)|>\varepsilon\right\}\right)=0. \tag{1.2.20}$$

It is equivalent to say that from any sequence $\{r_n\}$ going to 0, we can extract a subsequence $\{r_{n_k}\}$ such that $\psi_{r_{n_k}}$ converges to ψ v-a.e. on E.

THEOREM 3.2. *Suppose that* $N>2$, g *is nondecreasing and*

$$\int_0^1 g(r^{2-N})r^{N-1}dr<\infty, \tag{1.2.21}$$

f *is locally integrable in* Ω *and radial near* O, *and* $u\in C^2(\Omega^*)$ *is nonnegative and satisfies*

$$\Delta u\le g(u)+f \tag{1.2.22}$$

in Ω^*. *Then we have the following alternative. Either*
i) $r^{N-2}u(r,.)$ *converges in measure on* S^{N-1} *to some nonnegative real number* γ *as* r *goes to* 0, *or*
(ii)

$$\lim_{x\to O} |x|^{N-2}u(x)=\infty.$$

Proof. For $\lambda>0$, let v_λ be the solution of

$$\begin{cases} \Delta v_\lambda = g(v_\lambda)+|f| & \text{in}\quad B^*(O,R),\\[4pt] v_\lambda = 0 & \text{on}\quad \partial B^*(O,R),\\[4pt] \lim_{x\to O} |x|^{N-2}v_\lambda(x)=\lambda. \end{cases} \tag{1.2.23}$$

The function v_λ is radial and positive near O since Lemma 3.1 implies that the presence of f does not affect the behaviour of v_λ near O. From Proposition 3.3 there exists $v(\lambda)\in[0,\lambda]$ such that

$$\lim_{r \to 0} r^{N-2} \inf\left(u(r,.), v_\lambda(r)\right) = v(\lambda) \tag{1.2.24}$$

in the $L^q(S^{N-1})$-norm, for any $q \in [1, \infty)$. Moreover the function $\lambda \mapsto v(\lambda)$ is nondecreasing. We have to distinguish two cases according the limit of $v(\lambda)$, when λ goes to infinity, is finite or not.

Case 1. Assume that $\lim_{\lambda \to \infty} v(\lambda) = \gamma < \infty$. For $\lambda > \gamma$, we have (1.2.24). Suppose now that $\{r_n\}$ is some sequence converging to 0, then there exists a subsequence $\{r_{n_k}\}$ such that

$$\lim_{n_k \to \infty} r_{n_k}^{N-2} \inf\left(u(r_{n_k}, \sigma), v_\lambda(r_{n_k})\right) = v(\lambda), \tag{1.2.25}$$

a.e. on S^{N-1}. Since $v(\lambda) \le \gamma < \lambda$ and $\lim_{n_k \to \infty} r_{n_k}^{N-2} v_\lambda(r_{n_k}) = \lambda$, we deduce that, for almost all σ on S^{N-1}, there exists n_{k_0} such that for $k \ge k_0$, $\inf\left(u(r_{n_k}, \sigma), v_\lambda(r_{n_k})\right) = u(r_{n_k}, \sigma)$ and

$$\lim_{n_k \to \infty} r_{n_k}^{N-2} u(r_{n_k}, ..) = v(\lambda), \tag{1.2.26}$$

a.e. on S^{N-1}. If we take $\lambda' > \lambda$ and replace $\{r_n\}$ by $\{r_{n_k}\}$, there would exist a subsequence $\{r_{n_{k_j}}\}$ from $\{r_{n_k}\}$ such that

$$\lim_{n_{k_j} \to \infty} r_{n_{k_j}}^{N-2} u(r_{n_{k_j}}, ..) = v(\lambda') \tag{1.2.27}$$

a.e. on S^{N-1}. Therefore $v(\lambda) = v(\lambda') = \gamma$ and we have (i).

Case 2. Assume that $\lim_{\lambda \to \infty} v(\lambda) = \infty$. For $\lambda, \delta > 0$, we set
$\tilde{\omega}_\delta = \frac{1}{2}\left(u + v_\lambda - p(u - v_\lambda)\right) + \frac{3}{4}$ where p is the function introduced in (1.2.11). From (1.2.15), we have

$$\Delta \tilde{\omega}_\delta \le g(\tilde{\omega}_\delta) + |f| \tag{1.2.28}$$

and $\lim_{r \to 0} r^{N-2} \left\| \tilde{\omega}_\delta(r,.) - v(\lambda) \right\|_{L^q(S^{N-1})}$, for any $1 \le q < \infty$. We denote now $w = v_{v(\lambda)}$ and set

$$s = \frac{r^{N-2}}{N-2}, \quad w^*(s) = r^{N-2} w(r), \quad f^*(s) = f(r), \quad \tilde{\omega}_\delta^*(s, \sigma) = r^{N-2} \tilde{\omega}_\delta(s, \sigma). \tag{1.2.29}$$

Then

$$s^2\left(\tilde{\omega}_\delta^*\right)_{ss} + \frac{1}{(N-2)^2} \Delta_{S^{N-1}} \tilde{\omega}_\delta^* \le ks^{N/(N-2)}\left(g\left(\frac{\tilde{\omega}_\delta^*}{s(N-2)}\right) + f^*\right), \tag{1.2.30}$$

and

$$s^2 w_{ss}^* = ks^{N/(N-2)}\left(g\left(\frac{w^*}{s(N-2)}\right) + |f^*|\right), \tag{1.2.31}$$

where $k = k(N) = (N-2)^{(4-N)/((N-2)}$. Denote ρ a C^∞ bounded function, identically 0 on $(-\infty, 0]$, with $\rho' > 0$ on $(0, \infty)$ and $j(r) = \int_0^r \rho(t)dt$. By a convexity argument

$$s^2 \frac{d^2}{ds^2} \int_{S^{N-1}} j(w^* - \tilde{\omega}_\delta^*)d\sigma \ge 0. \tag{1.2.32}$$

Since $\int_{S^{N-1}} j(w^* - \tilde{\omega}_\delta^*)d\sigma \le C\int_{S^{N-1}} |w^* - \tilde{\omega}_\delta^*|^+ d\sigma$ and $w^*(s)$ and $\tilde{\omega}_\delta^*(s,.)$ both converge to $v(\lambda)$ in $L^1(S^{N-1})$, we deduce that $\int_{S^{N-1}} j(w^* - \tilde{\omega}_\delta^*)d\sigma = 0$ on $\left(0, R^{N-2}/(N-2)\right]$. Therefore $w^* \le \tilde{\omega}_\delta^*$ or

$$v_{v(\lambda)}(r) \le \omega_\delta(r,\sigma) \le \inf\left(u(r,\sigma), v_\lambda(r)\right) + \delta/4, \tag{1.2.33}$$

which implies

$$v_{v(\lambda)} \le \liminf_{x \to O}\left(|x|^{N-2}\inf(u(x), v(x))\right) \le \liminf_{x \to O} |x|^{N-2}u(x) \tag{1.2.34}$$

and finally (ii) is derived.

The 2-dimensional version of Theorem 3.2 is the following.

THEOREM 3.3. *Let* $N = 2$, g *be nondecreasing,* f *locally integrable in* Ω *and radial near* O, *and* $u \in C^2(\Omega^*)$ *be nonnegative and satisfy*

$$\Delta u \le g(u) + f \tag{1.2.35}$$

in Ω^*. *Then, if we define* a_g^+ *by*

$$a_g^+ = \inf\left\{a \ge 0 : \int_0^\infty g(s)e^{-as}ds < \infty\right\}, \tag{1.2.36}$$

we have

-I- $a_g^+ = 0$ *the alternative of Theorem 3.2 holds with* $\ln(1/|x|)$ *instead of* $|x|^{2-N}$.

-II- $a_g^+ > 0$ *we have the following alternative : either*

(i) there exists $\gamma \in [0, 2/a_g^+)$ *such that* $u(r,.)/\ln(1/r)$ *converges in measure to on as* r *goes to* 0*, or*

(ii)

$$\liminf_{x \to O} u(x)/\ln(1/|x|) \geq 2/a_g^+. \tag{1.2.37}$$

Proof. Case 1. Assume that $a_g^+ = 0$. We define v_λ by

$$\begin{cases} \Delta v_\lambda = g(v_\lambda) + |f| & \text{in} \quad B^*(O,R), \\ v_\lambda = 0 & \text{on} \quad \partial B^*(O,R), \\ \lim_{x \to O} v_\lambda(x)/\ln(1/|x|) = \lambda. \end{cases} \tag{1.2.38}$$

and $v(\lambda)$ by

$$\lim_{r \to 0} \left(\ln(1/r)\right)^{-1} \inf\left(u(r,.), v_\lambda(r)\right) = v(\lambda). \tag{1.2.39}$$

Since v_λ exists for all $\lambda > 0$ and $\lambda \mapsto v(\lambda)$ is nondecreasing, we can proceed as in Theorem 3.2 if $\lim_{\lambda \to \infty} v(\lambda) < \infty$. If $\lim_{\lambda \to \infty} v(\lambda) = \infty$, we introduce again $\tilde{\omega}_\delta$ and $v_{v(\lambda)} = w$ and set

$$t = \ln(1/t), \quad w^*(t) = w(r), \quad \tilde{\omega}_\delta^*(t,\sigma) = \tilde{\omega}_\delta(r,\sigma), \quad f^*(t) = f(r). \tag{1.2.40}$$

Therefore $\tilde{\omega}_\delta$ and $v_{v(\lambda)}$ satisfy

$$\left(\tilde{\omega}_\delta^*\right)_{tt} + \left(\tilde{\omega}_\delta^*\right)_{\sigma\sigma} \leq e^{-2t}\left(g(\tilde{\omega}_\delta^*) + f^*\right), \tag{1.2.41}$$

and

$$\left(w^*\right)_{tt} = e^{-2t}\left(g(w^*) + |f^*|\right) \tag{1.2.42}$$

on $[T,\infty) \times S^1$. Consequently $\dfrac{d^2}{dt^2} \displaystyle\int_{S^{N-1}} j(w^* - \tilde{\omega}_\delta^*) d\sigma \geq 0$, which implies $j(w^* - \tilde{\omega}_\delta^*) = 0$ since $\left\| (w^* - \tilde{\omega}_\delta^*)(t,.) \right\|_{L^1(S^1)} = o(t)$ at infinity.

Case 2 . Assume that $a_g^+ > 0$ and set $\lim_{\lambda \uparrow 2/a_g^+} v(\lambda) = \gamma$. If $\gamma \in [0, 2/a_g^+)$, we proceed as in Theorem 3.2. If $\gamma = 2/a_g^+$, we obtain, as in case 1,

$$\inf\left(u(x),v_\lambda(x)\right)\geq v_{\nu(\lambda)}(x)-\delta/4, \tag{1.2.43}$$

for any $x\in B^*(O,R)$ any $\delta>0$ and any $\lambda\in\left[0,2/a_g^+\right]$, since in this range there always exists a solution of (1.2.38). If we take in particular $\lambda=2/a_g^+$, we deduce (ii).

Remark 3.4. In the case (i) of Theorems 3.2-3.3, there holds

$$u(x)\geq v_\gamma(x) \tag{1.2.44}$$

in $B^*(O,R)$.

3-2 Isolated singularities of solutions of $\Delta u\pm g(u)=0$

The isotropy result of the previous section plays an important role in the description of isolated singularities of positive solutions of the two types of equations

$$\Delta u = g(u) \tag{2.1}$$

and

$$-\Delta u = g(u) \tag{2.2}$$

where g is essentially a nondecreasing function. Although the phenomena corresponding to (2.1) have yet been studied in Chapter 2, we shall first give a brief idea of how the techniques of differential inequalities can be applied in this case. Throughout this section Ω is a open subset of $\mathbb{R}^N$ $(N>2)$ containing O and $\Omega^*=\Omega\setminus\{O\}$. We assume that $B(O,R)\subset\Omega$. We shall first expose the singular version[NS] of Gidas-Ni-Nirenberg classical result on symmetry of positive solutions of nonlinear elliptic equations in balls[GNN]. Such types of results are obtained by an adaptation of of Alexandrov-Serrin's moving plane method.

3-2-1 Symmetry properties of positive singular solutions

We assume here that g is a locally Lipchitz continuous function.

THEOREM 3.4. *Let* $N>1$ *and* $u\in C^2\left(B^*(O,R)\right)$ *be a positive solution of*

$$\Delta u + g(u) = 0 \tag{2.1.1}$$

in $B^*(O,R)$ *which vanishes on* $\partial B(O,R)$ *and satisfies*

$$\lim_{x\to O}u(x)=\infty. \tag{2.1.2}$$

Then u *is radially symmetric*

We first recall without proof the strong maximum principle and Hopf's boundary lemma.

LEMMA 3.2. *Suppose* $c \in L^{\infty}(\Omega)$, $z \in C^2(\Omega)$ *is nonnegative in* $\overline{\Omega}$ *and satisfies*

$$\Delta z + c(x)z \geq 0 \qquad\qquad (2.1.3)$$

in Ω. *If* z *vanishes at some point in* Ω, *then it is identically* 0.

LEMMA 3.3. *Assume the same conditions on* c *and* z *as in Lemma 3.2. In addition, suppose that*
  (i) Ω *satisfies the interior sphere condition at* $x_0 \in \partial\Omega$, *that is there exists an ball* $B \subset \Omega$ *with* $x_0 \in \partial B$,
  (ii) $z \in C(\Omega \cup \{x_0\})$, $z(x_0) = 0$ *and the outer normal derivative of* z *exists at* x_0.

If z *is not identically* 0, *then* $\dfrac{\partial z}{\partial \nu}(x_0) > 0$.

Proof of Theorem 3.4. We pick an arbitrary direction, say x_1, call T_0 the hyperplane $x_1 = 0$ and $\tilde{x}$ the reflection of a point x about T_0. We will prove that $u(x) = u(\tilde{x})$. Since the x_1-direction is chosen arbitrarily, it will imply the radial symmetry of u. Let T_λ be the hyperplane $x_1 = \lambda$, with $\lambda > 0$ for example. We set $\tilde{\Sigma}_\lambda = B(O,R) \cap \{x : x_1 > \lambda\}$ and Σ_λ the reflection of $\tilde{\Sigma}_\lambda$ about T_λ. For any $x \in \tilde{\Sigma}_\lambda$ we call x^λ the reflection of x about T_λ. Thanks to (2.1.2) we shall infer that for any $\lambda > 0$, there holds

$$u(x) > u(x^\lambda), \qquad\qquad (2.1.4)$$

which will imply

$$u(x) \geq u(\tilde{x}). \qquad\qquad (2.1.5)$$

Moving the plane T_λ from the left ($\lambda < 0$) will imply that the reverse inequality (2.1.5) also holds. As in any moving plane technique the first problem is to start the movement.

Step 1. We claim that there exists $\varepsilon > 0$ such that

$$\frac{\partial u}{\partial r}(r,\sigma) < 0, \ \text{for} \ R - \varepsilon < r < R, \ \sigma \in S^{N-1}. \qquad\qquad (2.1.6)$$

Case 1. $g(0) \geq 0$. Then

$$\Delta u + g(u) - g(0) = \Delta u + c(x)u = -g(0) \leq 0, \qquad\qquad (2.1.7)$$

where c is some locally bounded function. Since u is not identically 0, we have $\frac{\partial u}{\partial v}(x) < 0$ on the boundary and therefore for $R - \varepsilon < |x| < R$.

Case 2. $g(0) < 0$. If $x_0 \in \partial B(O,R)$, then

$$\Delta u(x_0) = -g(0) > 0 \tag{2.1.8}$$

and, always, $\frac{\partial u}{\partial v}(x_0) \leq 0$. If $\frac{\partial u}{\partial v}(x_0) < 0$, we are finished, otherwhise $\frac{\partial u}{\partial v}(x_0) = 0$. Since all the tangential derivatives of u on the boundary are 0, we deduce from (2.1.9) that

$$\Delta u(x_0) = \frac{\partial^2 u}{\partial r^2}(x_0) = -f(0) > 0. \tag{2.1.9}$$

If (2.1.6) were not true there would exist a sequence $\{x^n\}$ in $B(O,R)$ such that $\lim_{n \to \infty} x^n = x_0$ and $\frac{\partial u}{\partial r}(x^n) \geq 0$. Since $\frac{\partial u}{\partial r}(x_0) = 0$, we would have $\frac{\partial^2 u}{\partial r^2}(x_0) \leq 0$, contradiction.

Step 2. Suppose that for some $\lambda \in (0,R)$ we have $u(x) \geq u(x^\lambda)$, $x \in \Sigma_\lambda$, with at least one point with strict inequality in Σ_λ, then

$$u(x) > u(x^\lambda) \text{ for } x \in \Sigma_\lambda \quad \text{and} \quad \frac{\partial u}{\partial x_1}(x) < 0, \text{ for } x \in T_\lambda \cap B(O,R) . \tag{2.1.10}$$

Case 1 : $\lambda \in (0,R/2)$. We define $v(x) = u(x^\lambda)$ and $w(x) = v(x) - u(x)$ for x in Σ_λ. By the mean value theorem there exists a bounded function c such that

$$\Delta w + g(u) - g(v) = \Delta w + c(x)w = 0 \tag{2.1.11}$$

in Σ_λ. Since w is nonnegative and not identically 0 we deduce from Lemmas 3.2-3.3 that w is negative in Σ_λ and $\frac{\partial w}{\partial x_1} > 0$ on $T_\lambda \cap B(O,R)$ which implies (2.1.10).

Case 2 : $\lambda \in [R/2,R)$. For $\varepsilon > 0$, we set $\Sigma_\lambda^\varepsilon = \Sigma_\lambda \setminus B(O,\varepsilon)$. The function w satisfies (2.1.11) in $\Sigma_\lambda^\varepsilon$ where c remains bounded, therefore w is negative and $\frac{\partial w}{\partial x_1} > 0$ on $T_\lambda \cap B(O,R)$. Since $\lim_{\varepsilon \to 0} \sup_{|x|=\varepsilon} w(x) = -\infty$, we obtain (2.1.10).

Step 3. The set of positive λ for which (2.1.10) holds is open.

If the above claim were not true, there would exist a sequence $\{\lambda_n\}$ converging to some $\lambda \in (0,R)$ for which (2.1.10) is valid and a sequence $\{x^n\}$ in Σ^{λ_n}, such that

$$u(x^n) \leq u((x^n)^{\lambda_n}). \tag{2.1.12}$$

We call also assume that $\{x^n\}$ converges to some z belonging to $\overline{\Sigma}_\lambda$. Moreover there exists some $\varepsilon > 0$ such that $|x^n| \geq \varepsilon$ for any $n > 0$, and

$$u(z) \leq u(z^\lambda). \tag{2.1.13}$$

Since $u(x) > u(x^\lambda)$ for $x \in \Sigma_\lambda$, $z \in \partial\Sigma_\lambda$ and more precisely to T_λ because u vanishes on $\partial B(O,R)$. Therefore $\dfrac{\partial u}{\partial x_1}(z) \geq 0$, which contradicts (2.1.10).

Step 4. End of the proof. From Steps 1 and 3 the set of positive λ for which (2.1.10) holds is open and non-empty. Therefore this property holds on a maximal interval (λ_1, R) with $\lambda_1 \geq 0$. If λ_1 were positive, we would have $u(x) \geq u(x^{\lambda_1})$ in Σ_{λ_1} with at least one point with strict inequality. From Step 2, (2.1.10) would be valid for $\lambda = \lambda_1$, which is a contradiction. Therefore $\lambda_1 = 0$ and $u(x) \geq u(\tilde{x})$. Finally $u(x) = u(\tilde{x})$. Since the direction of reflection is arbitrary we have the radiality of u.

Remark 3.5 If u is a regular solution of (2.1.1), the radiality property also holds since the above proof is still valid. Moreover Theorem 3.4 is also valid if f depends on x in the following way:

$$g(x,u) = g(|x|, u) \quad \text{and} \quad x \mapsto g(|x|, u) \text{ is is nonincreasing.} \tag{2.1.14}$$

Remark 3.6. If we assume that f is nonincreasing the technique of Proposition 2.1 in Chapter 2 applies; consequently any solution u of (2.1.1) in $B^*(O,R)$ which is radial on $\partial B(O,R)$ and satisfies

$$\liminf_{r \to 0} |x|^{N-1} \|u(r,.) - \overline{u}(r)\|_{L^2(S^{N-1})} = 0, \tag{2.1.15}$$

is radial: by a simple convexity argument, it is easy to prove that $\|u(r,.) - \overline{u}(r)\|_{L^2(S^{N-1})} = 0$ for $0 < r \leq R$.

3-2-2 Isolated singularities of solutions of $\Delta u = g(u)$

THEOREM 3.5. *Let* $N > 2$ *and g be a nondecreasing locally Lipschitz continuous function vanishing at* 0 *which satisfies*

$$\int_0^1 g(r^{2-N}) r^{N-1} dr < \infty. \tag{2.2.1}$$

If $u \in C^2(\Omega^*)$ is a nonnegative solution of (2.1) in Ω^*, $|x|^{N-2} u(x)$ converges to some $\gamma \in [0,\infty]$ as x goes to O.

Proof. From Theorem 3.2 we can assume that there exists $\gamma \in [0,\infty)$ such that $r^{N-2} u(r,.)$ converges in measure to γ on S^{N-1}. Therefore there exists a sequence $\{r_n\}$ converging to 0 such that

$$\lim_{n \to \infty} r_n^{N-2} u(r_n,.) = \gamma \quad \text{a.e. on } S^{N-1}. \tag{2.2.2}$$

Case 1. Assume that $\gamma > 0$ and for $\varepsilon > 0$ denote $\Gamma_{\varepsilon,R} = B(O,R) \setminus B(O,\varepsilon)$ and w_ε the solution of

$$\begin{cases} \Delta w_\varepsilon = g(w_\varepsilon) & \text{in} \quad \Gamma_{\varepsilon,R}, \\ w_\varepsilon = u & \text{on} \quad \partial B(O,\varepsilon), \setminus B(O,\varepsilon), \\ w_\varepsilon = \max_{x \in \partial B(O,R)} u(x) & \text{on} \quad \partial B(O,R). \end{cases} \tag{2.2.3}$$

From the maximum pinciple and the monotonicity of g,

$$u \le w_\varepsilon \le u + w_\varepsilon(R), \tag{2.2.4}$$

and there exist a sequence $\{\varepsilon_n\}$ and a function $w \in C^2\big(B^*(O,R)\big)$ solution of (2.1) in $B^*(O,R)$ such that $\{w_{\varepsilon_n}\}$ converges to w in the C^1_{loc}-topology of $B^*(O,R)$. Moreover (2.1.4) holds with w instead of w_ε. From Remark 3.4, $\liminf_{x \to 0} |x|^{N-2} w(x) = \gamma$. Since w is nonnegative in $B^*(O,R)$, radial on $\partial B(O,R)$ and tends to infinity as x goes to O, it is radially symmetric by Theorem 3.4. From (2.2.2) and $u \le w \le u + w_\varepsilon(R)$, we obtain

$$\lim_{r_n \to 0} r_n^{N-2} w(r_n) = \lim_{r_n \to 0} r_n^{N-2} u(r_n,.) = \gamma, \tag{2.2.5}$$

and from Theorem 2.1

$$\lim_{r \to 0} r^{N-2} w(r) = \lim_{r \to 0} r^{N-2} u(r,.) = \gamma, \tag{2.2.6}$$

uniformly on S^{N-1}.

Case 2. Assume that $\gamma = 0$. For $\varepsilon, \nu > 0$, we denote $w_{\varepsilon,\nu}$ as being the solution of

$$
\begin{cases}
\Delta w_{\varepsilon,v} = g(w_{\varepsilon,v}) & \text{in } \Gamma_{\varepsilon,R}, \\
w_{\varepsilon,v} = u + v\varepsilon^{2-N} & \text{on } \partial B(O,\varepsilon), \\
w_\varepsilon = vR^{2-N} + \max_{x\in\partial B(O,R)} u(x) & \text{on } \partial B(O,R).
\end{cases}
\tag{2.2.7}
$$

As in case 1 we have

$$
u(x) \le w_{\varepsilon,v}(x) \le u(x) + v|x|^{2-N} + w_{\varepsilon,v}(R)
\tag{2.2.8}
$$

in $\Gamma_{\varepsilon,R}$. There exists some subsequence $\{\varepsilon_n\}$ converging to 0 such that $\{w_{\varepsilon_n}\}$ converges to some w in the C^1_{loc}-topology of $B^*(O,R)$ and w is a radial solution of (2.1) in $B^*(O,R)$. Since (2.2.2) holds as well as

$$
u(x) \le w(x) \le u(x) + v|x|^{2-N} + vR^{2-N} + \max_{x\in\partial B(O,R)} w(x),
\tag{2.2.9}
$$

we deduce that

$$
\limsup_{r_n\to 0} \, r_n^{N-2}\left\|u(r_n,.)\right\|_{L^\infty(S^{N-1})} \le v.
\tag{2.2.10}
$$

Using again Theorem 2.1, we deduce that $|x|^{N-2}u(x)$ admits a limit as x goes to O and this limit is necessarily 0.

The 2-dimensional version of this result is the following

THEOREM 3.6. *Let* $N = 2$ *and g be a nondecreasing locally Lipschitz continuous function vanishing at* 0. *If* $u \in C^2(\Omega^*)$ *is a nonnegative solution of* (2.1) *in* Ω^*, *we have the following*
-I *If* $a_g^+ = 0$, $u(x)/\ln(1/|x|)$ *admits a limit* $\gamma \in [0,\infty]$ *as* x *goes to* O.
-II *If* $a_g^+ > 0$, $u(x)/\ln(1/|x|)$ *admits a limit* $\gamma \in \left[0, 2/a_g^+\right]$ *as* x *goes to* O, *provided for any* $a \ge 0$, $\lim_{r\to\infty} e^{-ar}g(r)$ *exists in* $[0,\infty]$.

Proof. If $a_g^+ = 0$, we proceed as in Theorem 3.5. If $a_g^+ = \infty$, then the singularity of u at O is removable (Theorem 2.11). If $0 < a_g^+ < \infty$, we have two possibilities:

(i) either there exist $\gamma \in \left[0, 2/a_g^+\right)$ and a sequence $\{r_n\}$ converging to 0 such that

$$
\lim_{n\to\infty} u(r_n,.)/\ln(1/r_n) = \gamma \quad \text{a.e. on } S^1,
\tag{2.2.11}
$$

(ii) or

$$
\liminf_{x\to O} u(x)/\ln(1/|x|) \ge 2/a_g^+.
\tag{2.2.12}
$$

In case (i) we have $\lim_{x\to 0} u(x)/\ln(1/|x|) = \gamma$ as in Theorem 3.5. In case (ii) the growth estimate on g implies that for any $\varepsilon > 0$, there exist $B(\varepsilon)$ such that

$$u(x) \leq \left(\frac{2}{a_g^+} + \varepsilon \right) \ln(1/|x|) + B(\varepsilon) \tag{2.2.13}$$

near O. Therefore

$$\limsup_{x\to 0} u(x)/\ln(1/|x|) \leq 2/a_g^+. \tag{2.2.14}$$

Combining (2.2.12) and (2.2.14) completes the proof.

3-2-3 Isolated singularities of solutions of $-\Delta u = g(u)$

The first non-radial results concerning this type of nonlinearity are due to Lions [Ls] who gave a fairly complete description of isolated singularities of positive solutions of

$$-\Delta u = u^q \tag{2.3.1}$$

in the case $1 < q < N/(N-2)$. We shall present a slightly more general version, valid for (2.2), with the techniques developped in Theorems 3.2-3.3. The class of functions g for which this method applies is essentially the class of functions which satisfy the inequality (2.2.1). We recall that $\overline{B}(O,R) \subset \Omega$ for some $R > 0$.

THEOREM 3.7. *Let* $N > 2$ *and* g *be a continuous function defined on* $[0,\infty)$ *such that* $\liminf_{r\to\infty} g(r)/r > -\infty$. *If* $u \in C^2(\Omega^*)$ *is a nonnegative solution of* (2.2) *in* Ω^*, *there exists* $\gamma \in [0,\infty)$ *such that*

$$\lim_{x\to 0} |x|^{1-N} \int_{\{y:|y|=|x|\}} \left| \gamma - |x|^{N-2} u(y) \right| dS(y) = 0. \tag{2.3.2}$$

Moreover $g(u) \in L^1_{loc}(\Omega)$ *and* u *solves*

$$-\Delta u = g(u) + C(N)\gamma\delta_0 \tag{2.3.3}$$

in $\mathcal{D}'(\Omega)$. *Finally, if we suppose that*

$$\int_0^1 \inf_{\alpha \leq \theta \leq \beta} g(\theta r^{2-N}) \, r^{N-1} dr = \infty, \tag{2.3.4}$$

for any $\alpha, \beta > 0$, *then* $\gamma = 0$.

Proof. From Corollary 3.1, $g(u) \in L^1_{loc}(\Omega)$ and u solves (2.3.3) for some $\gamma \in [0, \infty)$. Let $\overline{u}(r)$ and $\overline{g(u)}(r)$ be the spherical averages of u and g(u) respectively, then

$$-\Delta\overline{u} = \overline{g(u)} \tag{2.3.5}$$

in $B^*(O,R) \subset \Omega^*$ and from Lemma 3.1 we have (2.3.2). Let us suppose now that g satisfies (2.3.4) for any α, $\beta > 0$ and that $\gamma > 0$. Since $r^{N-2}u(r,.)$ converges in measure to γ on S^{N-1}, for any $\eta \in \left(0, \left|S^{N-1}\right|\right)$ there exists $r_0 \in (0,R)$ with the property that for any $r \in (0, r_0)$ there exists a measurable subset $\omega(r) \subset S^{N-1}$ with $\text{meas}(\omega(r)) > \eta$ and $\left|r^{N-2}u(r,\sigma) - \gamma\right| < \gamma/2$ for $\sigma \in \omega(r)$. Because $\liminf_{r \to \infty} g(r)/r > -\infty$ and $u \in L^1_{loc}(B(O,R))$ there is no loss of generality to assume that g takes only nonnegative values, henceforth

$$\int_{B(O,r_0)} g(u)dx = \int_0^{r_0}\int_{S^{N-1}} g(u)d\sigma dr \geq \int_0^{r_0}\int_{\omega(r)} g(u)d\sigma dr. \tag{2.3.6}$$

Since $(\gamma/2)r^{2-N} \leq u(r,\sigma) \leq 2\gamma r^{2-N}$ for $\sigma \in \omega(r)$ and g is continuous, $g(u(r,\sigma))$ takes its values in $I_r = \left\{g(\theta r^{2-N}): \gamma/2 \leq \theta \leq 2\gamma\right\}$. Therefore $g(u(r,\sigma)) \geq \inf_{\gamma/2 \leq \theta \leq 2\gamma} g(\theta r^{2-N})$, which implies

$$\int_{B(O,r_0)} g(u(x))dx \geq \eta\int_0^{r_0} \inf_{\gamma/2 \leq \theta \leq 2\gamma} g(\theta r^{2-N})\, r^{N-1}dr = \infty, \tag{2.3.7}$$

which is a contradiction.

Under an assumption of monotonicity upon g and an upper domination of u by some supersolution of (2.2), we obtain a much more precise behaviour.

PROPOSITION 3.4. *Let* $N > 2$*, g be a nondecreasing locally Lipschitz continuous function defined on* $[0,\infty)$ *and* $u \in C^2(\Omega^*)$ *a nonnegative solution of (2.2) in* Ω^**. We also suppose that there exists some continuous function* Φ *in* $\overline{B}(O,R) \setminus \{O\}$ *which satisfies*

$$-\Delta\Phi \geq g(\Phi) \tag{2.3.8}$$

in the sense of distributions in $B^*(O,R)$ *and* $\Phi \geq u$*. Then, if one of the two following additional assumptions is satisfied*

(i) Φ *is radial,*

(ii) $\lim_{x \to O} u(x) = \infty$*,*

$|x|^{N-2}u(x)$ *converges to some nonnegative real number* γ *when x goes to O.*

Proof. From Theorem 3.7, (2.3.2) holds for some $\gamma \in [0, \infty)$. In case (i), Theorem 3.2 implies that $|x|^{N-2} \Phi(x)$ converges to some nonnegative γ' at O and $0 \le \gamma \le \gamma' < \infty$. If $\gamma' = 0$, then $\lim_{x \to O} |x|^{N-2} u(x) = 0$. If $\gamma' > 0$, Theorem 3.1 implies that

$$-\Delta \Phi = -\{\Delta \Phi\} + C(N)\gamma' \delta_O, \tag{2.3.9}$$

with $\{\Delta \Phi\} \in L^1_{loc}(B(O,R))$, which implies that the same holds for $g(\Phi)$. We consider the following iterative scheme $\{u^n\}$ with $u^0 = \Phi$, and

$$\begin{cases} -\Delta u^n = g(u^{n-1}) + C(N)\gamma \delta_O & \text{in } B(O,R), \\ \qquad\qquad u^n = \Phi & \text{on } \partial B(O,R). \end{cases} \tag{2.3.10}$$

for $n > 0$. The function u^n is radial and $u \le u^n \le u^{n-1} \le \Phi$. When n goes to infinity, $\{u^n\}$ converges locally uniformly in $\overline{B}(O,R) \backslash \{O\}$ to some radial function u^* which dominates u and satisfies

$$-\Delta u^* = g(u^*) + C(N)\gamma \delta_O \quad \text{in } \mathcal{D}'(B(O,R)). \tag{2.3.11}$$

Clearly $\lim_{x \to O} |x|^{N-2} u^*(x) = \gamma$ and the same holds with u, from Remark 3.4. In case (ii) we consider the following iterative scheme $\{\Phi^n\}$, with $\Phi^0 = \Phi$ and

$$\begin{cases} -\Delta \Phi^n = g(\Phi^{n-1}) + C(N)\gamma' \delta_O & \text{in } B(O,R), \\ \Phi^n(x) = \sup_{|y|=|x|} u(y) & \text{on } \partial B(O,R). \end{cases} \tag{2.3.12}$$

Then $u \le \Phi^n \le \Phi^{n-1} \le \Phi$ and $\{\Phi^n\}$ converges to some Φ^* which dominates u and satisfies

$$\begin{cases} -\Delta \Phi^* = g(\Phi^*) + C(N)\gamma' \delta_O & \text{in } B(O,R), \\ \Phi^*(x) = \sup_{|y|=|x|} u(y) & \text{on } \partial B(O,R). \end{cases} \tag{2.3.13}$$

From (ii) and Theorem 3.4, Φ^* is radial and we are back to case (i).

PROPOSITION 3.5. *Let* $N > 2$ *and* g *be a nondecreasing locally Lipschitz continuous function defined on* $[0, \infty)$ *which satisfies*

$$\sup\big(g'(\varphi), g'(\psi)\big) \in L^q_{loc}(\Omega) \tag{2.3.14}$$

for some $q > N/2$, *whenever* φ *and* ψ *are two continuous nonnegative defined in* Ω^* *and such that* $g(\varphi)$ *and* $g(\psi) \in L^1_{loc}(\Omega)$. *If* $u \in C^2(\Omega^*)$ *is a nonnegative solution of* (2.2) *in* Ω^*, *then* $|x|^{N-2} u(x)$ *converges to some nonnegative real number* γ *when* x *goes to* O.

Proof. From Theorem 3.7, (2.3.2) holds for some nonnegative γ.
Case 1. $\gamma = 0$. Without any loss of generality we can suppose that u is bounded from below near 0 by some positive constant and we can write (2.2) in the following form

$$\Delta u + d(x)u + g(0) = 0 \tag{2.3.15}$$

with $d(x) = (g(u) - g(0))/u$. Since $g(u) \in L^1_{loc}(\Omega)$, (2.3.14) implies that $d \in L^q_{loc}(\Omega)$ for some $q > N/2$. From Theorem 1.20, either u is regular at O or

$$0 < \liminf_{x \to O} |x|^{N-2} u(x) < \limsup_{x \to O} |x|^{N-2} u(x) < \infty, \tag{2.3.16}$$

which is a contradiction with the assumption that $\gamma = 0$.

Case 2 . $\gamma > 0$. Let v_γ be the solution of

$$\begin{cases} -\Delta v_\gamma = g(v_\gamma) + C(N)\gamma \delta_O & \text{in } B(O,R), \\ v_\gamma(x) = 0 & \text{on } \partial B(O,R). \end{cases} \tag{2.3.17}$$

The existence of such a solution is not obvious but can be easily derived by considering an increasing scheme of functions $\{v_\gamma^n\}$ dominated by u, with $v_\gamma^0 = 0$ and

$$\begin{cases} -\Delta v_\gamma^n = g(v_\gamma^{n-1}) + C(N)\gamma \delta_O & \text{in } B(O,R), \\ v_\gamma^n(x) = 0 & \text{on } \partial B(O,R). \end{cases} \tag{2.3.18}$$

for $n > 0$. Moreover $v_\gamma \leq u$ and v_γ is radially symmetric. If $w = u - v_\gamma$, then

$$\Delta w + d(x)w = 0, \tag{2.3.19}$$

with $d(x) = (g(u) - g(v_\gamma))/(u - v_\gamma)$ and $d \in L^q_{loc}(\Omega)$ from (2.3.14). If w were not regular at O we would have

$$0 < \liminf_{x \to O} |x|^{N-2} w(x) < \limsup_{x \to O} |x|^{N-2} w(x) < \infty, \tag{2.3.20}$$

which is impossible since

$$\gamma = \lim_{r \to 0} r^{N-2} v_\gamma(r) = \lim_{r \to 0} r^{N-2} u(r,.) \tag{2.3.21}$$

in $L^1(S^{N-1})$. Therefore the proof is complete.

Remark 3.7. Under the assumptions of Proposition 3.5, two solutions u_i of (2.3.3) ($i = 1, 2$) are such that $u_1 - u_2 \in L^\infty_{loc}(\Omega)$. Concerning the existence of singular solutions, we have the following.

PROPOSITION 3.6. *Let Ω be bounded with a C^1 boundary, $N > 2$, and g be a nondecreasing real valued function defined on $[0, \infty)$ satisfying (2.2.1) and also $\lim_{r \to 0} g(r) / r = 0$. Then there exists $\gamma^* \in (0, \infty]$ with the following properties:*

(i) for any $\gamma \in [0, \gamma^*)$ there exists at least one nonnegative function $u \in C^1(\overline{\Omega} \setminus \{O\})$ vanishing on $\partial\Omega$ solution of (2.3.3);
(ii) for $\gamma > \gamma^*$ no such u exists.

Proof. Step 1. We first assume that Ω is the ball $B(O, R)$, and we look for a solution u under the form $u(r) = v(t)$, $t = r^{2-N}$. Then v satisfies

$$
\begin{cases}
\dfrac{d^2 v}{dt^2} + \dfrac{t^{-2(N-1)/(N-2)}}{(N-2)^2} g(v) = 0 \quad \text{on} \quad (R^{2-N}, \infty), \\[2mm]
v(R^{2-N}) = 0, \\[2mm]
\lim_{t \to \infty} v(t) / t = \gamma.
\end{cases}
\tag{2.3.22}
$$

Since v is concave, the last condition of (2.3.22) is equivalent to

$$
\lim_{t \to \infty} \frac{dv}{dt}(t) = \gamma.
\tag{2.3.23}
$$

For $\alpha > 0$, let v^α be the maximal solution of

$$
\begin{cases}
\dfrac{d^2 v^\alpha}{dt^2} + \dfrac{t^{-2(N-1)/(N-2)}}{(N-2)^2} g(v^\alpha) = 0 \quad \text{on} \quad (R^{2-N}, T^*), \\[2mm]
v^\alpha(R^{2-N}) = 0, \\[2mm]
\dfrac{dv^\alpha}{dt}(R^{2-N}) = \alpha.
\end{cases}
\tag{2.3.24}
$$

If $T^* < \infty$, $\lim_{t \uparrow T^*} v^\alpha(t) = 0$ and there exists $T \in (R^{2-N}, T^*)$ such that $\dfrac{dv^\alpha}{dt}(T) = 0$. If $T^* = \infty$, the same holds with $T = \infty$. Therefore, if no solution of (2.40) exists for any $\gamma > 0$, we have

$$(N-2)^2\alpha = \int_{R^{2-N}}^{T} t^{-2(N-1)/(N-2)}g(v^\alpha(t))dt, \tag{2.3.25}$$

and the concavity implies

$$(N-2)^2\alpha R^{-N} < \int_0^\infty (1+t)^{-2(N-1)/(N-2)}g(\alpha R^{2-N}t)dt, \tag{2.3.26}$$

or equivalently,

$$(N-2)^2 R^{-2} < \int_0^\infty t(1+t)^{-2(N-1)/(N-2)}\frac{g(\alpha R^{2-N}t)}{\alpha R^{2-N}t}dt. \tag{2.3.27}$$

Since $g(r) = o(r)$ near 0, for any $\varepsilon > 0$ there exists $\eta > 0$ such that $\alpha R^{2-N}t < \eta$ implies $g(\alpha R^{2-N}t) < \alpha R^{2-N}t$. Consequently the right-hand side of (2.3.27) is majorized by

$$\frac{R^{N-2}}{\alpha}\int_{R^{2-N}\eta/\alpha}^\infty (1+t)^{-2(N-1)/(N-2)}g(\alpha R^{2-N}t)dt + \varepsilon\int_0^{R^{2-N}\eta/\alpha} t(1+t)^{-2(N-1)/(N-2)}dt, \tag{2.3.28}$$

which, in turn, is dominated by

$$\alpha^{2(N-1)/(N-2)}\int_\eta^\infty (R^{N-2}s+\alpha)^{-2(N-1)/(N-2)}g(s)ds + \varepsilon\int_0^\infty t(1+t)^{-2(N-1)/(N-2)}dt. \tag{2.3.29}$$

Therefore

$$\lim_{\alpha\to 0}\int_0^\infty t(1+t)^{-2(N-1)/(N-2)}\frac{g(\alpha R^{2-N}t)}{\alpha R^{2-N}t}dt = 0, \tag{2.3.30}$$

which contradicts (2.3.27). As a consequence, there exists $\alpha^* > 0$ such that for any $\alpha \in (0,\alpha^*)$ the solution v^α of (2.3.24) is defined on $\left[R^{2-N},\infty\right)$ and satisfies (2.3.23) for some $\gamma > 0$.

Step 2. The general case. We may assume that $\Omega \subset B(O,R)$ for some $R > 0$. Let $\tilde\gamma > 0$ be such that for any $\gamma \in [0,\tilde\gamma]$ there exists a solution v of (2.3.22). Then the following sequence $\{u_n\}$, defined by $u_0 = 0$ and

$$\begin{cases} -\Delta u_n = g(u_{n-1}) + C(N)\gamma\delta_O & \text{in } \mathcal{D}'(\Omega), \\ \quad u_n = 0 & \text{on } \partial\Omega, \end{cases} \tag{2.3.31}$$

for $n \geq 1$, is increasing and majorized by v. It converges to some u which vanishes on $\partial\Omega$ and satisfies (3.3.3). From the same construction, the set of $\gamma > 0$ such that there exists a nonnegative solution u of (3.3.3) which vanishes on $\partial\Omega$ is an interval.

The 2-dimensional version of Theorem 3.7 is the following.

THEOREM 3.8. *Let* $N = 2$, *g be a continuous real valued function defined on* $\mathbb{R}^+$ *such that* $\lim_{r \to \infty} g(r)/r > -\infty$ *and* $u \in C^2(\Omega^*)$ *is a nonnegative solution of* (2.2) *in* Ω^*. *Then there exists* $\gamma \in [0, \infty)$ *such that*

$$\lim_{x \to O} |x|^{-1} \int_{\{|y|=|x|\}} |\gamma - u(y)/\ln(1/|x|)| dS(y) = 0, \tag{2.3.32}$$

$g(u) \in L^1_{loc}(\Omega)$ *and u solves*

$$-\Delta u = g(u) + 2\pi\gamma\delta_O \tag{2.3.33}$$

in $\mathcal{D}'(\Omega)$. *If we suppose moreover that*

$$\int_0^1 \inf_{\alpha \leq \theta \leq \beta} g(\theta \ln(1/r)) r dr = \infty \tag{2.3.34}$$

for any $0 < \alpha < \beta$, *then* $\gamma = 0$.

The proof of Theorem 3.8 is similar to the one of Theorem 3.7 up to minor modifications. Moreover, if $a_g^+ = 0$, Proposition 3.4 is also valid with $|x|^{2-N}$ replaced by $\ln(1/|x|)$. It gives a criterion for proving that

$$\lim_{x \to O} u(x)/\ln(1/|x|) = \gamma. \tag{2.3.35}$$

In order to have more precise convergence estimates we introduce the following complementary assumption on g:

$$\forall \sigma > 0, \quad \lim_{r \to \infty} e^{-r\sigma} g(r) = l(\sigma) \text{ exists in } [0, \infty]. \tag{2.3.36}$$

It is clear that

$$a_g^+ = \sup\{\sigma > 0 : l(\sigma) = \infty\} = \inf\{\sigma > 0 : l(\sigma) = 0\}. \tag{2.3.37}$$

PROPOSITION 3.7. *Let* $N = 2$, *g be a continuous real valued function defined on* $\mathbb{R}^+$ *satisfying* $\lim_{r \to \infty} g(r)/r > -\infty$ *and* (2.3.36) *with* $a_g^+ < \infty$ *and* $u \in C^2(\Omega^*)$ *a nonnegative solution of* (2.2) *in* Ω^*. *Suppose also*

(i) either $a_g^+ = 0$,

(ii) or $a_g^+ > 0$ *and* $\displaystyle\int_0^1 g\left(\frac{2}{a_g^+}\ln(1/r)\right)r\,dr = \infty$.

Then there exists $\gamma \in \left[0, \dfrac{2}{a_g^+}\right)$ *such that* $u - \gamma\ln(1/r)$ *remains locally bounded in* Ω.

One important tool for proving this result is John and Nirenberg's theorem (see [GT]) that we recall.

LEMMA 3.4. *Let G be a convex domain in* $\mathbb{R}^2$ *and* $u \in W^{1,1}(G)$ *satisfy*

$$\int_{G \cap B_r} |\nabla u|\,dx \leq Kr \tag{2.3.38}$$

for any ball B_r *of radius* r, *for some constant* $K > 0$. *Then there exists two positive constants* μ_0 *and* C *such that*

$$\int_G \exp\left(\frac{\mu}{K}|u - u_G|\right)dx \leq C(\mathrm{diam}(G))^2, \tag{2.3.39}$$

where $\mu = \mu_0|G|$ *and* $u_G = \dfrac{1}{|G|}\displaystyle\int_G u(x)\,dx$.

Proof of Proposition 3.7. From Theorem 3.8, $g(u)$ is locally integrable in Ω and there exists $\gamma \geq 0$ such that $u(r,.)/\ln(1/r)$ converges to γ in $L^1(S^1)$ as r goes to 0. We set $w = u - \gamma\ln(1/r)$ and w solves

$$-\Delta w = g(u) \tag{2.3.40}$$

in $\mathcal{D}'(\Omega)$. From the L^1-elliptic equation theory (see [BBC]) ∇w belongs to the Marcinkiewicz space $M^2_{\mathrm{loc}}(\Omega)$. Taking $G = B(O,R) \subset \Omega$, it means that ∇w satisfies (2.3.38) for some $K > 0$ and therefore, there exist α and $C(\rho) > 0$ such that

$$\int_{B(O,\rho)} e^{\alpha w}\,dx \leq C(\rho) \tag{2.3.41}$$

for any $0 < \rho \leq R$.

Case 1. Suppose that $a_g^+ = 0$. Then for any $\varepsilon > 0$, we have

$$|g(r)| \leq K_\varepsilon e^{\varepsilon r} \tag{2.3.42}$$

for some positive constant K_ε and any $r \geq 0$. Since (2.3.41) implies

$$\int_{B(O,\rho)} e^{\alpha u}|x|^{\alpha\gamma}\,dx \leq C(\rho),\tag{2.3.43}$$

we have, if $\gamma > 0$, p, $\sigma > 1$ and $\lambda > 0$,

$$\int_{B(O,\rho)} e^{p\varepsilon u}\,dx \leq \left(\int_{B(O,\rho)} e^{p\varepsilon\sigma u}|x|^{\sigma\lambda}\,dx\right)^{1/\sigma}\left(\int_{B(O,\rho)} |x|^{-\sigma'\lambda}\,dx\right)^{1/\sigma'}\tag{2.3.44}$$

where $\sigma' = \sigma/(\sigma-1)$. We set $\sigma p\varepsilon = \alpha$, $\sigma\lambda = \alpha\gamma$, then $\lambda = \gamma p\varepsilon$, $\sigma = \alpha/p\varepsilon$ and $\sigma'\lambda = \alpha\gamma p\varepsilon/(\alpha-p\varepsilon)$. Consequently, for any $p > 1$, we can take ε small enough so that $\sigma'\lambda < 2$ and $\sigma > 1$. It implies that $g(u) \in L^p(B(O,\rho))$ and $w \in L^\infty(B(O,\rho))$. If $\gamma = 0$ the same holds for $g(u)$ and $u \in L^\infty(B(O,\rho))$.

Case 2. Assume now that $a_g^+ > 0$ and $\displaystyle\int_0^1 g\left(\frac{2}{a_g^+}\ln(1/r)\right)r\,dr = \infty$.

Step 1. We claim that $0 \leq \gamma < 2/a_g^+$. We proceed by contradiction in assuming $\gamma \geq 2/a_g^+$. Since $\lim_{r\to\infty} g(r) = \infty$, $g(r) \geq -L$ for some $L > 0$ and any $r > 0$. We can minorize u in $B^*(O,R)$ by the solution z_γ of

$$-\Delta z_\gamma + L = 2\pi\gamma\delta_O \quad\text{in}\quad \mathscr{D}'(B(O,R))\tag{2.3.45}$$

which vanishes on $\partial B(O,R))$. Therefore there exists $m > 0$ such that $u(x) \geq \gamma\ln(1/|x|) - m$; this implies $\int_{B(O,R)} g(u)\,dx = \infty$, a contradiction.

Step 2. We claim that for any α there exist $\rho \in (0,R]$ and $C(\rho) > 0$ such that (2.3.41) holds. If we consider $0 < R' < R$, where R' is to be fixed later on, we can decompose w into $w_1 + w_2$ where w_1 is harmonic in $B(O,R')$ and takes the value w on $\partial B(O,R')$ and w_2 vanishes on $\partial B(O,R')$ and satisfies

$$-\Delta w_2 = g(u)\tag{2.3.46}$$

in $B(O,R')$. Since ∇w_1 belongs to $L^2(B(O,R'))$ we have

$$\lim_{\rho\to 0} \left\|\nabla w_1\right\|_{L^2(B(O,\rho))} = 0.\tag{2.3.47}$$

As for w_2, we have

$$\left\|\nabla w_2\right\|_{M^2(B(O,R'))} \leq C\|g(u)\|_{L^1(B(O,R'))}\tag{2.3.48}$$

and the constant C is independent of R' as it can be easily checked by a change of scale. For $\varepsilon > 0$ we first fix R' so that

$$\int_{B(O,R')\cap B_r} |\nabla w_2| dx \leq \varepsilon r / 2 \tag{2.3.49}$$

for any ball $B_r \subset B(O,R') = G'$. We now use (2.3.47) and take $0 < R'' < R'$ in such a way that

$$\int_{B(O,R'')\cap B_r} |\nabla w| dx \leq \varepsilon r \tag{2.3.50}$$

whenever $B_r \subset B(O,R'') = G''$. Consequently the coefficient K of John and Nirenberg's theorem can be taken as small as we need and we have the claim.

Step 3. End of the proof. From the definition of a_g^+, for any $\varepsilon > 0$ there exists $K_\varepsilon > 0$ such that

$$|g(r)| \leq K_\varepsilon e^{(a_g^+ + \varepsilon)r} \tag{2.3.51}$$

for $r \geq 0$. Since

$$\int_{B(O,\rho)} e^{p(a_g^+ + \varepsilon)u} dx \leq \left(\int_{B(O,\rho)} e^{p\sigma(a_g^+ + \varepsilon)u} |x|^{\lambda\sigma} dx \right)^{1/\sigma} \left(\int_{B(O,\rho)} |x|^{-\lambda\sigma'} dx \right)^{1/\sigma'}, \tag{2.3.52}$$

we take $\sigma p(a_g^+ + \varepsilon) = \alpha$, $\sigma\lambda = \alpha\gamma$ (here we assume that $\gamma > 0$ otherwise $g(u) \in L^p_{loc}(\Omega)$ and w remains locally bounded), $\lambda = \gamma p(a_g^+ + \varepsilon)$, $\sigma = \alpha/p(a_g^+ + \varepsilon)$ and finally $\lambda\sigma' = \alpha\gamma p(a_g^+ + \varepsilon)/(\alpha - p(a_g^+ + \varepsilon))$. Since $\gamma a_g^+ < 2$, we can take $p > 1$, $\varepsilon > 0$, $\alpha > 0$ and have $\sigma'\lambda < 2$. Therefore $g(u) \in L^p_{loc}(\Omega)$ and we complete the proof as in Case 1.

For the existence of solutions of (2.2), the following result can be proved by O.D.E. technique as in Proposition 3.6.

PROPOSITION 3.8. *Let $N = 2$, Ω be bounded with a C^2 boundary, $\overline{\Omega}^* = \overline{\Omega}\setminus\{O\}$ and g a continuous nondecreasing function defined on $\mathbb{R}^+$ such that $a_g^+ \in (0,\infty]$ and $g(r) = o(r)$ near 0. Then there exists $\gamma^* \in \left(0, a_g^+\right]$ with the following properties.*

(i) If $\gamma \in [0, \gamma^*)$ there exists at least one nonnegative solution $u \in C^1(\overline{\Omega}^*)$ of (2.3.33) in $\mathcal{D}'(\Omega)$ vanishing on $\partial\Omega$.

(ii) If $\gamma > \gamma^*$ no such solution exists.

Remark 3.8. If $a_g^+ = \infty$, then $\gamma = 0$ from Theorem 3.8; we do not believe that Proposition 3.7 is still valid, however we conjecture that $\lim_{x \to 0} u(x)/\ln(1/|x|) = 0$.

3-3 Singularities of solutions of $-\Delta u = u^q$

3-3-1 The case $1 < q < N/(N-2)$

We apply here the results of the previous Section in supposing that Ω is a bounded open subset of $\mathbb{R}^N$ ($N > 1$) containing O, with a C^1 boundary $\partial\Omega$ and $\Omega^* = \Omega \setminus \{O\}$. The equation under consideration is the following

$$-\Delta u = u^q \tag{3.1.1}$$

in Ω^*, and u is always supposed to be nonnegative, continuous in Ω^* and therefore C^2.

THEOREM 3.9. *Let $N > 2$, $1 < q < N/(N-2)$ and $u \in C^2(\Omega^*)$ be a nonnegative solution of (3.1.1) in Ω^*. Then there exists $\gamma \geq 0$ such that*

$$\lim_{x \to O} |x|^{N-2} u(x) = \gamma, \tag{3.1.2}$$

and u *solves*

$$-\Delta u = u^q + C(N)\gamma\delta_O \tag{3.1.3}$$

in $\mathcal{D}'(\Omega)$; moreover $u \in C^2(\Omega)$ if $\gamma = 0$. Conversely, there exists $\gamma^* > 0$ such that for any $\gamma \in [0, \gamma^*)$, there exists a nonnegative solution of (3.1.2)-(3.1.3) vanishing on $\partial\Omega$ and no such solution exists whenever $\gamma > \gamma^*$.

The proof of this result is a direct application of Propositions 3.5-3.6 and we believe that the solution still exists when $\gamma = \gamma^*$. When $N = 2$, we have $a_g^+ = 0$ if $g(r) = r^q$, $q > 1$. Therefore Theorem 3.9 extends to

THEOREM 3.10. *Let $N = 2$, $q > 1$ and $u \in C^2(\Omega^*)$ be a nonnegative solution of (3.1.1) in Ω^*. Then there exists $\gamma \geq 0$ such that*

$$u(x) - \gamma \ln(1/|x|) \in L^\infty_{loc}(\Omega), \tag{3.1.4}$$

and u *solves*

$$-\Delta u = u^q + 2\pi\gamma\delta_O \tag{3.1.5}$$

in $\mathcal{D}'(\Omega)$; moreover $u \in C^2(\Omega)$ if $\gamma = 0$. Conversely, there is a $\gamma^* > 0$ such that for any $\gamma \in [0, \gamma^*)$, there exists a nonnegative solution of (3.1.4)-(3.1.5) vanishing on $\partial\Omega$, but no such solution exists whenever $\gamma > \gamma^*$.

The remaining of this section is devoted to the case $q \geq N/(N-2)$ and in fact we shall see that we have to treat separately the cases $q = N/(N-2)$, $N/(N-2) < q < (N+2)/(N-2)$, $q = (N+2)/(N-2)$ and $q > (N+2)/(N-2)$.

3-3-2 The *a priori* estimate

The first step consists in looking for solutions of (3.1.1) under the form

$$u(r,\sigma) = r^{-2/(q-1)}\omega(\sigma) \tag{3.2.1}$$

where (r,σ) are the radial coordinates in $\mathbb{R}^N \setminus \{O\}$, and we see that ω satisfies

$$-\Delta_{S^{N-1}}\omega - \frac{2}{q-1}\left(\frac{2q}{q-1} - N\right)\omega - \omega^q = 0 \tag{3.2.2}$$

on S^{N-1}. If we set $l_{q,N} = \dfrac{2}{q-1}\left(\dfrac{2q}{q-1} - N\right)$, then we have seen in Section 2.1 that $l_{q,N}$ is negative (resp. positive) if and only if $q > N/(N-2)$ (resp. $1 < q < N/(N-2)$) and it vanishes only when $q = N/(N-2)$. Therefore there exist singular radial positive solutions of (3.1.1) under the form (3.2.1) only if $q > N/(N-2)$. As in the absorption case developped in Section 2.1 , an important key-stone element is an *a priori* estimate near the singular point. This universal *a priori* estimate is known only in the case $1 < q \leq (N+2)/(N-2)$. In the case $1 < q < (N+2)/(N-2)$, the original proof is due to Gidas and Spruck in their celebrated paper [GSk1]. Later this estimate was improved and simplified in [B.VV] and [B.VR]. The *a priori* estimate when $q = (N+2)/(N-2)$ has been obtained very recentlyby a completely different technique (see Section 3.3.6) by Chen and Li [CL], in a much more general framework than the one of isolated singularities.

THEOREM 3.11. *Let* $N > 2$, $1 < q < (N+2)/(N-2)$ *and* $u \in C^2(B^*(O,R))$ *be a nonnegative solution of* (3.1.1) *in* $B^*(O,R)$. *Then there exists a constant* $C = C(N,q) > 0$ *such that*

$$u(x) \leq C|x|^{-2/(q-1)} , \tag{3.2.3}$$

for $0 < |x| \leq R/4$.

LEMMA 3.5. *Let   be an open subset of* $\mathbb{R}^N (N \geq 1)$, *then for any* $w \in C^2(G)$, $w \geq 0$ *and* $\eta \in C_0^\infty(G)$, $\eta \geq 0$, *and any real numbers* (d,m) *with* $d \neq m+2$, *the following inequality holds*

$$\frac{1}{4N}\left(2(N-m)d-(N-1)(m^2+d^2)\right)\int_G \eta w^{m-2}|\nabla w|^4 dx$$

$$-\frac{N-1}{N}\int_G \eta w^m(\Delta w)^2 dx$$

$$-\frac{1}{2N}\left(2(N-1)m+(N+2)d\right)\int_G \eta w^{m-1}|\nabla w|^2\Delta w dx$$

$$\leq \frac{m+d}{2}\int_G w^{m-1}\nabla w.\nabla\eta dx+\int_G w^m\Delta w\nabla w.\nabla\eta dx+\frac{1}{2}\int_G w^m|\nabla w|^2\Delta\eta dx$$

(3.2.4)

Proof. We apply the Bochner-Weitzenböck formula (Theorem 1.1) to $f=w^t$, and get from Schwarz' inequality

$$\frac{1}{2}\Delta|\nabla f|^2=|\mathrm{Hess}(f)|^2+\nabla(\Delta f).\nabla f\geq\frac{1}{N}(\Delta f)^2+\nabla(\Delta f).\nabla f.$$

(3.2.5)

Since $\nabla f=tw^{t-1}\nabla w$ and $\Delta f=t(t-1)w^{t-2}|\nabla w|^2+tw^{t-1}\Delta w$, if we multiply (3.2.5) by $t^{-2}\eta w^d$ and integrate over G, we obtain

$$\left(\frac{(t-1)^2}{N}+(t-1)(t-2)\right)\int_G \eta w^{d+2t-4}|\nabla w|^4 dx$$

$$+\frac{N+2}{N}(t-1)\int_G \eta w^{d+2t-3}|\nabla w|^2\Delta w dx$$

$$+\int_G \eta w^{d+2t-2}\nabla w.\nabla(\Delta w)dx+\frac{1}{N}\int_G \eta w^{d+2t-2}(\Delta w)^2 dx$$

$$+(t-1)\int_G \eta w^{d+2t-3}\nabla w.\nabla(|w|^2)dx-\frac{1}{2}\int_G \eta w^d\Delta(w^{2t-2}|\nabla w|^2)dx\leq 0.$$

(3.2.6)

Integrating by parts yields

$$\int_G \eta w^{d+2t-2}\nabla w.\nabla(\Delta w)dx=-\int_G w^{d+2t-2}\Delta w\nabla w.\nabla\eta dx$$

$$-(d+2t-2)\int_G w^{d+2t-3}|\nabla w|^2\Delta w dx-\int_G \eta w^{d+2t-2}(\Delta w)^2 dx,$$

(3.2.7)

$$\int_G \eta w^{d+2t-3}\nabla w.\nabla(|\nabla w|^2)dx=-\int_G w^{d+2t-3}|\nabla w|^2\nabla w.\nabla\eta dx$$

$$-(d+2t-3)\int_G \eta w^{d+2t-4}|\nabla w|^4 dx-\int_G \eta w^{d+2t-3}|\nabla w|^2\Delta w dx,$$

(3.2.8)

and

$$\int_G \eta w^d \Delta(w^{2t-2}|\nabla w|^2)dx = \int_G w^{d+2t-2}|\nabla w|^2 \Delta\eta dx$$

$$+2d\int_G w^{d+2t-3}|\nabla w|^2\nabla w.\nabla\eta dx \tag{3.2.9}$$

$$+d(d-1)\int_G \eta w^{d+2t-4}|\nabla w|^4 dx + d\int_G \eta w^{d+2t-3}|\nabla w|^2\Delta w dx.$$

Plugging this estimate into (3.2.6) and setting $m = d + 2t - 2$, allows us to conclude that (3.2.4) holds.

LEMMA 3.6. *Let* $q > 1$ *and* $u \in C^2(B^*(O,1))$ *be a nonnegative solution of* (3.1.1) *in* $B^*(O,1)$. *Then for any open subset* G *such that* $\overline{G} \subset B^*(O,1)$, *any nonnegative* $\zeta \in C_0^\infty(G)$ *and any triple of real numbers* (d,m,r) *such that* $r > 0$, $d \neq m + 2$, *the following inequality holds*

$$\frac{N-1}{4N}\left(\frac{2d(N-m)}{N-1} - m^2 - d^2\right)\int_G \zeta^r u^{m-2}|\nabla u|^4 dx$$

$$+\left(\frac{(N+2)d}{2N} - \frac{q(N-1)}{N}\right)\int_G \zeta^r u^{q+m-1}|\nabla u|^2 dx$$

$$\leq \frac{1}{2}\int_G u^m |\nabla u|^2 \Delta(\zeta^r)dx + \frac{r(m+d)}{2}\int_G \zeta^{r} u^{m-1}|\nabla u|^2 \nabla u.\nabla\zeta dx \tag{3.2.10}$$

$$-\frac{r}{N}\int_G \zeta^r u^{q+m}\nabla u.\nabla\zeta dx.$$

Proof. We take (3.2.4) with $\eta = \zeta^r$ and use the fact that

$$\int_G \zeta^r u^m(\Delta u)^2 dx = -\int_G \zeta^r u^{m+q}\Delta u dx$$

$$= (q+m)\int_G \zeta^r u^{m+q-1}|\nabla u|^2 dx + r\int_G \zeta^{r-1}u^{m+q}\nabla u.\nabla\zeta dx. \tag{3.2.11}$$

The following result is also immediate from the equation (3.1.1).

LEMMA 3.7. *Under the assumptions of Lemma 3.6, the following identity holds*

$$\int_G \zeta^r u^{2q+m}dx = (q+m)\int_G \zeta^r u^{q+m-1}|\nabla u|^2 dx + r\int_G \zeta^{r-1}u^{q+m}\nabla u.\nabla\zeta dx. \tag{3.2.12}$$

LEMMA 3.8. *Whenever* $1 < q < (N+2)/(N-2)$, *there exist two real numbers* d *and* m *such that* $d \neq m + 2$, $d > 2(N-1)q/(N+2)$, $m > \max(-2,(N-4)q-N)/2)$ *and* $d^2 - 2d(N-m)/(N-1) + m^2 < 0$.

Proof. We first choose d such that

$$2q\frac{N-1}{N+2} < d < \min\left(2\frac{N-1}{N-2}, (N-1)q\right). \tag{3.2.13}$$

Then the polynomial $m \mapsto m^2 - 2d(N-m)/(N-1) + d^2$ has two roots $m_1 < m_2$, and m_2 satisfies $m_2 > c = \max(-2, -q, (N-4)q - N/2)$. We take now $m \in \left(\max(c, m_1), m_2\right)$ with $m \neq d - 2$.

In the sequel $\varepsilon \in (0,1)$ is a parameter, C and C_ε are different positive constants such that $C = C(N, q, m, d, r)$ and $C_\varepsilon = C_\varepsilon(N, q, m, d, r, \varepsilon)$.

LEMMA 3.9. *Let* $1 < q < (N+2)/(N-2)$ *and the assumptions of Lemma 3.6 be fulfilled with* d *and* m *as in Lemma 3.8 and* $r > 4$. *Then there exists a constant* C *such that*

$$\int_G \zeta^r u^{q+m-1}|\nabla u|^2\, dx \leq C\int_G \zeta^{r-4} u^{m+2}\left(|\nabla\zeta|^4 + \zeta^2(\Delta\zeta)^2 + |x|^{-4}\zeta^4\right)dx. \tag{3.2.14}$$

Proof. We start from (3.2.10) and set

$$C_1 = \frac{N-1}{4N}\left(\frac{2d(N-m)}{N-1} - m^2 - d^2\right),$$
$$\tag{3.2.15}$$
$$C_2 = \frac{N+2}{2N}d - \frac{N-1}{N}q.$$

The C_j are positive from Lemma 3.7. With the help of Hölder's inequality the following is derived to estimate the right-hand side of (3.2.10)

$$\int_G u^m |\nabla u|^2 |\Delta(\zeta^r)|\,dx$$
$$\leq \varepsilon\int_G \zeta^r u^{m-2}|\nabla u|^4\,dx + C_\varepsilon \int_G \zeta^{r-4}u^{m+2}\left(|\nabla\zeta|^4 + \zeta^2(\Delta\zeta)^2\right)dx, \tag{3.2.16}$$

$$\int_G \zeta^{r-1}u^{m-1}|\nabla u|^2|\nabla u.\nabla\zeta|\,dx \leq \varepsilon\int_G \zeta^r u^{m-2}|\nabla u|^4\,dx + C_\varepsilon\int_G \zeta^{r-4}u^{m+2}|\nabla\zeta|^4\,dx, \tag{3.2.17}$$

and

$$\int_G \zeta^{r-1}u^{m+q}|\nabla u.\nabla\zeta|\,dx$$
$$\leq \varepsilon\int_G \zeta^r u^{q+m-1}|\nabla u|^2\,dx + C_\varepsilon\int_G \zeta^{r-2}u^{q+m-1}|\nabla\zeta|^2\,dx + C_\varepsilon\int_G \zeta^r |x|^{-2}u^{q+m+1}\,dx, \tag{3.2.18}$$
$$\leq \varepsilon\int_G \zeta^r u^{q+m-1}|\nabla u|^2\,dx + \varepsilon\int_G \zeta^r u^{2q+m}\,dx + C_\varepsilon\int_G \zeta^{r-4}u^{m+2}\left(|\nabla\zeta|^4 + \zeta^4|x|^{-4}\right)dx.$$

Using (3.2.12) and (3.2.18) yields

$$\int_G \zeta^r u^{2q+m}\,dx \leq C\left(\int_G \zeta^r u^{q+m-1}|u|^2\,dx + \int_G \zeta^{r-4}u^{+m+2}\left(|\nabla\zeta|^4 + \zeta^4|x|^{-4}\right)dx\right). \tag{3.2.19}$$

and, by using again (3.2.18), we obtain

$$\int_G \zeta^{r-1} u^{q+m} |\nabla u . \nabla \zeta| dx$$

$$\leq \varepsilon \int_G \zeta^r u^{q+m-1} |\nabla u|^2 dx + C_\varepsilon \int_G \zeta^{r-4} u^{m+2} \left(|\nabla \zeta|^4 + \zeta^4 |x|^{-4} \right) dx. \tag{3.2.20}$$

By changing ε, we deduce from (3.2.20), (3.2.10), (3.2.16), (3.2.17) that

$$(C_1 - \varepsilon) \int_G \zeta^r u^{m-2} |\nabla u|^4 dx + (C_2 - \varepsilon) \int_G \zeta^r u^{q+m-1} |\nabla u|^2 dx$$

$$\leq C_\varepsilon \int_G \zeta^{r-1} u^{m+2} \left(|\nabla \zeta|^4 + \zeta^2 (\Delta \zeta)^2 + \zeta^4 |x|^{-4} \right) dx, \tag{3.2.21}$$

holds, which finally gives (3.2.14).

LEMMA 3.10. *Under the assumptions of Lemma 3.9, there holds*

$$\int_G \zeta^r u^{2q+m} dx \leq C \int_G \zeta^{r-2s} u^{m+2} \left(|\nabla \zeta|^2 + \zeta \Delta \zeta + |x|^{-2} \zeta^2 \right)^s dx. \tag{3.2.22}$$

where $s = (2q+m)/(q-1)$, *and* $s > N/2$.

Proof. From (3.2.14) and (3.2.19) the following is derived

$$\int_G \zeta^r u^{2q+m} dx \leq C \int_G \zeta^{r-4} u^{m+2} \left(|\nabla \zeta|^4 + \zeta^2 (\Delta \zeta)^2 + \zeta^4 |x|^{-4} \right) dx \tag{3.2.23}$$

with $m + 2 > 0$, and from Lemma 3.8 and Hölder's inequality

$$\int_G \zeta^{r-4} u^{m+2} \left(|\nabla \zeta|^4 + \zeta^2 (\Delta \zeta)^2 + \zeta^4 |x|^{-4} \right) dx$$

$$\leq \varepsilon \int_G \zeta^r u^{2q+m} dx + C_\varepsilon \left(\int_G \zeta^{r-2s} \left(|\nabla \zeta|^2 + \zeta |\Delta \zeta| + \zeta^2 |x|^{-2} \right) dx \right)^s. \tag{3.2.24}$$

Therefore (3.2.22) follows from (3.2.23) and (3.2.24), and $s > N/2$ from Lemma 3.8.

Proof of Theorem 3.11. Since the equation (3.1.1) and the estimate (3.2.3) are both invariant under the transformation $u \mapsto u_k$ where $u_k(x) = k^{2/(q-1)} u(kx)$ $(k > 0)$, we can suppose that $R = 1$. We write (3.1.1) under the form

$$\Delta u + wu = 0 \tag{3.2.25}$$

with $w = u^{q-1}$. We choose d and m satisfying the assumptions of Lemma 3.8 and $r > \max(4, 2s)$ with $s = (2q+m)/(q-1)$. Let $x_0 \in B^*(O, 1/4)$ and $B_{x_0} = B(x_0, |x_0|/2)$. We set $G = G_{x_0} = B(x_0, 3|x_0|/4)$ and $\zeta \in C_0^\infty(G_{x_0})$ with $0 \leq \zeta \leq 1$ and $\zeta \equiv 1$ on B_{x_0} and $|x_0| |\nabla \zeta| + |x_0|^2 |\Delta \zeta| \leq C = C(N, q)$ in G_{x_0}. Applying (3.2.22) we get

140

$$\int_{B_{x_0}} w^s dx = \int_{B_{x_0}} u^{2q+m} dx \le C|x_0|^{N-2s}. \tag{3.2.26}$$

From $[GT]$ u satisfies uniformly the Harnack inequality

$$\sup_{x \in B(x_0,r)} u(x) \le C \inf_{x \in B(x_0,r)} u(x), \tag{3.2.27}$$

for any $r < |x_0|/2$; therefore, from (3.2.26), (3.2.27), we also have

$$|x_0|^N u^{(q-1)s}(x_0) \le C \int_{B_{x_0}} u^{2q+m} dx \le C^2 |x_0|^{N-2s} \tag{3.2.28}$$

for any $x_0 \in B^*(O,1/4)$. This proves Theorem 3.11.

Remark 3.9. In the previous Theorem, it is important to notice that the constant C does not depend upon the particular solution u. We conjecture that for any q > 1 and any nonnegative solution u of (3.1.1) in $B^*(O,R)$, there exists an *a priori* bound of type (3.2.3) in $B^*(O,R/4)$ with a constant C depending also on u. In fact it is possible to obtain (see [BV]) an estimate of the following form

$$\|u(r,.)\|_{L^\alpha(S^{N-1})} \le C(N,q,u,\alpha) r^{-2/(q-1)} \tag{3.2.29}$$

in $B^*(O,R/4)$, for any $\alpha \in [1,N/(N-2))$. In the particular case N > 3, $q = (N+1)/(N-3)$, the set of positive solutions ω of (3.2.2) is not uniformly bounded since it corresponds to the solutions of Obata's equation (or extremal of Sobolev's inequality in $\mathbb{R}^N$ up to a stereographic projection, [Ob], [Au1]); therefore no estimate independent of u is possible in that case.

3-3-3 Equations on compact Riemannian manifolds

If we want to describe the possible singular behaviour at O of any nonnegative solutions u of (3.1.1) we first have to give a description of those written under the form (3.2.1), which in fact means a description of the set of positive solutions of (3.2.2). Such a description is actually very difficult except when the only positive solutions are constants (or in the case $q = (N+1)/(N-3)$). It is in fact more interesting to imbed this problem in a more general one. Let (M,g) be a n-dimensional Riemannian manifold without boundary and Δ_g the Laplacian on M. Under what condition on (M,g), q > 1 and λ is any positive solution of

$$\Delta_g u - \lambda u + u^q = 0 \qquad \text{on M,} \tag{3.3.1}$$

a constant (by a solution we always mean a $C^2(M)$-function and in fact u is C^∞)? The first general answer to this problem is found in [GSk1]. Later on, a simpler and more general proof is given in [B.VV]. Both proofs amphazise the role of the Ricci curvature. We present here an even more general result due to Licois and Véron [LV1], [LV2] in which appears also the first nonzero eigenvalue of $-\Delta_g$ that we denote by λ_1. It is clear that $\lambda > 0$ is needed is order to have a positive solution of (3.3.1). Moreover, if $(q-1)\lambda = \lambda_1$, the linearized operator

$$\varphi \mapsto \mathbf{L}\varphi = -\Delta_g \varphi - (q-1)\lambda\varphi \tag{3.3.2}$$

associated to the constant solutions of (3.3.1) is singular. Therefore $(u,\lambda) = \left((\lambda_1 / (q-1))^{1/(q-1)}, \lambda_1 / (q-1)\right)$ is generically a point of bifurcation for (3.3.1), and it also holds if $(M,g) = (S^n, g_0)$. Consequently, in a small enough neighbourhood of this point the exists a continuum of non constant positive solutions of (3.3.1). Let Ricc_g be the Ricci tensor of g and dv_g the volume element. Then we prove:

THEOREM 3.12.-*Suppose that*

$$\mathrm{Ricc}_g \geq Rg, \tag{3.3.3}$$

for some nonnegative R *, that* $\lambda \geq 0$ *and*

$$1 < q \leq (n+2)/(n-2) \tag{3.3.4}$$

and that

$$(q-1)\lambda \leq \lambda_1 + \frac{qn(n-1)}{q+n(n+2)}\left(R - \frac{n-1}{n}\lambda_1\right). \tag{3.3.5}$$

Assume also that one of the two inequalities (3.3.4)-(3.3.5) is strict if (M,g) is conformally diffeomorphic to $\left(S^n, g_0\right)$*, then any nonnegative solution* u *of (3.3.1) is a constant.*

Proof of Theorem 3.12. Step 1. It is essentially an algebraic computation based upon Bochner-Weitzenböck formula that we recall

$$\frac{1}{2}\Delta_g\left|\nabla_g v\right|^2 = \left|\mathrm{Hess}\ v\right|^2 + \langle \nabla_g(\Delta_g v), \nabla_g v\rangle + \mathrm{Ricc}\left(\nabla_g v, \nabla_g v\right). \tag{3.3.6}$$

Since u satisfies (3.3.1), if we set $u = v^{-\beta}$ where $\beta \in \mathbb{R}_*$, then v satisfies

$$-\Delta_g v + (\beta+1)\frac{\left|\nabla_g v\right|^2}{v} + \frac{1}{\beta}\left(v^{1+\beta-\beta q} - \lambda v\right) = 0 \tag{3.3.7}$$

on M. The following identities play an important role in the proof:

LEMMA 3.11. *For any* $\gamma \neq -2$ *and* $\beta \in \mathbb{R}_*$ *the following identity is verified*

$$A\int_M v^{\gamma-2}\left|\nabla_g v\right|^4 dv_g = \frac{\beta q}{\gamma}\int_M \left(v^\gamma J + v^\gamma \mathrm{Ricc}(\nabla_g v, \nabla_g v)\right)dv_g$$

$$+ \frac{n+2}{2n}\lambda(q-1)\int_M v^\gamma \left|\nabla_g v\right|^2 dv_g - B\int_M \left(\Delta_g(v^{(\gamma+2)/2})\right)^2 dv_g,$$

(3.3.8)

where

$$A = \frac{n+2}{2n}\left(\left(\beta+1+\frac{\gamma}{4}\right)(\beta q - \gamma) - (\beta+1)^2\right) + \frac{\beta q(\gamma-4)}{8},$$

(3.3.9)

$$B = \frac{2}{n(\gamma+2)^2}\left(n+2+2\frac{\beta q}{\gamma}(n-1)\right),$$

(3.3.10)

$$J = \left(\left|\mathrm{Hess}(v)\right|^2 - \frac{1}{n}\left(\Delta_g v\right)^2\right).$$

(3.3.11)

Moreover, in the case where $\gamma = -2$, the above relation becomes

$$A\int_M \left|\nabla_g(\ln\, v)\right|^4 dv_g = -\frac{\beta q}{2}\int_M \left(v^{-2}J + v^{-2}\mathrm{Ricc}(\nabla_g v, \nabla_g v)\right)dv_g$$

$$+ \frac{n+2}{2n}\lambda(q-1)\int_M \left|\nabla_g(\ln\, v)\right|^2 dv_g - B\int_M \left(\Delta_g(\ln\, v)\right)^2 dv_g,$$

(3.3.12)

where

$$A = \frac{n+2}{2n}\left(\left(\beta+\frac{1}{2}\right)(\beta q + 2) - (\beta+1)^2\right) - \frac{3\beta q}{4},$$

(3.3.13)

$$B = \frac{1}{2n}\left(n+2-\beta q(n-1)\right).$$

(3.3.14)

Proof of lemma 3.11. We multiply (3.3.7) by $v^{\gamma-1}\left|\nabla_g v\right|^2$ et $v^\gamma \Delta_g v$, integrate on M and get

$$\int_M v^{\gamma-1}\Delta_g v\left|\nabla_g v\right|^2 dv_g =$$

$$(\beta+1)\int_M v^{\gamma-2}\left|\nabla_g v\right|^4 dv_g + \frac{1}{\beta}\int_M \left(v^{\beta-\beta q+\gamma} - \lambda v^\gamma\right)\left|\nabla_g v\right|^2 dv_g,$$

(3.3.15)

$$\int_M v^\gamma (\Delta_g v)^2 dv_g = (\beta + 1)\int_M v^{\gamma-1}\Delta_g v |\nabla_g v|^2 dv_g$$
$$-\frac{1}{\beta}\int_M (1+\beta - \beta q + \gamma)\left(v^{\beta-\beta q+\gamma} - \lambda(\gamma+1)v^\gamma\right)|\nabla_g v|^2 dv_g. \tag{3.3.16}$$

By a linear combination we can eliminate the term $\int_M v^{\beta-\beta q+\gamma}|\nabla_g v|^2 dv_g$ between (3.3.15) and (3.3.16), which yields

$$(\gamma - \beta q)\int_M v^{\gamma-1}\Delta_g v |\nabla_g v|^2 dv_g + \int_M v^\gamma (\Delta_g v)^2 dv_g$$
$$+(\beta+1)(\beta q - \gamma - \beta - 1)\int_M v^{\gamma-2}|\nabla_g v|^4 dv_g = \lambda(q-1)\int_M v^\gamma |\nabla_g v|^2 dv_g. \tag{3.3.17}$$

We multiply the formula (3.3.6) by v^γ, integrate on M and replace the term $|\text{Hess } v|^2$ by $J+\frac{1}{n}(\Delta_g v)^2$ (where J defined by (3.3.11)) is nonnegative from the Schwarz inequality); then the following is derived:

$$\frac{3\gamma}{2}\int_M v^{\gamma-1}\Delta_g v |\nabla_g v|^2 dv_g + \frac{\gamma(\gamma-1)}{2}\int_M v^{\gamma-2}|\nabla_g v|^4 dv_g + \frac{n-1}{n}\int_M v^\gamma (\Delta_g v)^2 dv_g$$
$$= \int_M Jv^\gamma dv_g + \int_M v^\gamma \text{Ricc}(\nabla v, \nabla v)dv_g. \tag{3.3.18}$$

If $\gamma \neq -2$, we have

$$v^{\gamma-1}|\nabla_g v|^2 \Delta_g v = \frac{4}{\gamma(\gamma+2)^2}\left(\Delta_g\left(v^{(\gamma+2)/2}\right)\right)^2 - \frac{\gamma}{4}v^{\gamma-2}|\nabla_g v|^4 - \frac{1}{\gamma}v^\gamma(\Delta_g v)^2, \tag{3.3.19}$$

and if $\gamma = -2$, (3.3.19) reads

$$v^{-3}|\nabla_g v|^2 \Delta_g v = -\frac{1}{2}\left(\Delta_g(\ln v)\right)^2 + \frac{1}{2}v^{-4}|\nabla_g v|^4 + \frac{1}{2}v^{-2}(\Delta_g v)^2. \tag{3.3.20}$$

In (3.3.17)-(3.3.18), we replace $\int_M v^{\gamma-1}\Delta_g v |\nabla_g v|^2 dv_g$ by the right-hand side of (3.3.19) or (3.3.20) and obtain,

$$\frac{\beta q}{\gamma}\int_M v^\gamma(\Delta_g v)^2 dv_g + \left[\left(\beta+1+\frac{\gamma}{4}\right)(\beta q - \gamma) - (\beta+1)^2\right]\int_M |\nabla_g v|^4 v^{\gamma-2}dv_g$$
$$-\frac{4(\beta q - \gamma)}{\gamma(\gamma+2)^2}\int_M \left(\Delta_g(v)^{(\gamma+2)/2}\right)^2 dv_g = \lambda(q-1)\int_M v^\gamma |\nabla_g v|^2 dv_g \tag{3.3.21}$$

and

$$\frac{6}{(\gamma+2)^2}\int_M\left(\Delta_g(v)^{(\gamma+2)/2}\right)^2 dv_g + \frac{\gamma(\gamma-4)}{8}\int_M v^{\gamma-2}\left|\nabla_g v\right|^4 dv_g$$

$$-\frac{n+2}{2n}\int_M v^\gamma\left(\Delta_g v\right)^2 dv_g = \int_M v^\gamma J\, dv_g + \int_M v^\gamma Ricc_g\left(\nabla_g v, \nabla_g v\right) dv_g \tag{3.3.22}$$

if $\gamma \neq -2$, with an easy modification in the case $\gamma = -2$. In those two identities the terms $\int_M\left(\Delta_g(v)^{(\gamma+2)/2}\right)^2 dv_g$ and $\int_M v^\gamma\left(\Delta_g v\right)^2 dv_g$ are nonnegative but give no estimate; if we eliminate one of them between (3.3.22)) and (3.3.23)) , for example $\int_M v^\gamma\left(\Delta_g v\right)^2 dv_g$, we get (3.3.8). We obtain (3.3.12) in the same way.

Proof of Theorem 3.12-Step 2. From the nonnegativity of J, Lemma 3.11 and the classical Fourier analysis relation

$$\int_M\left(\Delta_g(v)^{(\gamma+2)/2}\right)^2 dv_g \geq \frac{(\gamma+2)^2}{4}\lambda_1\int_M v^\gamma\left|\nabla_g v\right|^2 dv_g, \tag{3.3.23}$$

if $\gamma \neq -2$, with an immediate modification if $\gamma = -2$, it is sufficient to exhibit a couple (β,γ) such that

$$A \geq 0\,,\ B \geq 0\ \text{et}\ \frac{\beta}{\gamma} \leq 0. \tag{3.3.24}$$

In fact, if such a couple exists, we deduce from the previous relations that

$$A\int_M v^{\gamma-2}\left|\nabla_g v\right|^4 dv_g \leq \frac{\beta q}{\gamma}\int_M v^\gamma J dv_g$$

$$+\left[\frac{n+2}{2n}(\lambda(q-1)-\lambda_1)+\frac{\beta q}{\gamma}\left(R-\lambda_1\frac{n-1}{n}\right)\right]\int_M v^\gamma\left|\nabla_g v\right|^2 dv_g. \tag{3.3.25}$$

We set

$$X = \frac{\beta}{\gamma}\,,\ \delta = \frac{1}{\gamma}+\frac{1}{2}\ \text{and}\ \tilde{A} = \frac{2n}{(n+2)\gamma^2}A \tag{3.3.26}$$

and the problem is reduced to maximize X in $\left[-(n+2)/\left(2q(n-1)\right),0\right]$ under the constraint

$$\tilde{A} = -\delta^2 + 2\frac{q-(n+2)}{n+2}\delta X + (q-1)X^2 + \frac{q(n-1)}{2(n+2)}X \geq 0. \tag{3.3.27}$$

If we compute the derivative of $\tilde{A}$ with respect to δ we find

$$\frac{d\tilde{A}}{d\delta} = -2\left[\delta - \frac{q-(n+2)}{n+2}\right]. \tag{3.3.28}$$

Therefore the maximum of $\tilde{A}$ is achieved for $\delta = \delta_0 = \dfrac{q-(n+2)}{n+2}$ which gives

$$\tilde{A}(\delta_0,X) = X^2\left[q-1+\left(\frac{q-(n+2)}{n+2}\right)^2\right] + \frac{q(n-1)}{2(n+2)}X. \tag{3.3.29}$$

If X_0 is the negative root of the above polynomial in X, then

$$X_0 = -\frac{(n+2)(n-1)}{2(q+n(n+2))}, \tag{3.3.30}$$

and the condition

$$X_0 \geq -\frac{n+2}{2q(n-1)} \tag{3.3.31}$$

is equivalent to

$$q \leq (n+2)/(n-2). \tag{3.3.32}$$

For this specific value of X_0, we get

$$\left[\frac{n+2}{2n}(\lambda(q-1)-\lambda_1)+\frac{\beta q}{\gamma}\left(R-\lambda_1\frac{n-1}{n}\right)\right] =$$
$$\frac{n+2}{2n}\left[\lambda(q-1)-\lambda_1-\frac{qn(n-1)}{q+n(n+2)}\left(R-\lambda_1\frac{n-1}{n}\right)\right]. \tag{3.3.33}$$

Therefore, if we assume that (3.3.32) is fulfilled and that

$$\lambda(q-1) \leq \lambda_1 + \frac{qn(n-1)}{q+n(n+2)}\left(R-\lambda_1\frac{n-1}{n}\right), \tag{3.3.34}$$

holds, we have two possibilities:

  i) either (M,g) is not conformally diffeomorphic to $\left(S^n,g_0\right)$ and there exists no nonconstant positive solution to the equation $J=0$, or

ii) (M,g) is conformally diffeomorphic to (S^n, g_0) and, unless $v^{(\gamma+2)/2}$ is an eigenfunction of the Laplacian, the relation (3.3.23) is strict and B is positive if $q < (n+2)/(n-2)$. In that case v has also to be constant if (3.3.34) is fulffiled.

Remark 3.10. It worth noticing that, in estimate (3.3.5), the term $R - \lambda_1(n-1)/n$ is always nonpositive from Lichnerowicz well known result [Li]. Moreover, it vanishes if and only if (M,g) is isometric to (S^n, g_0) [Ob]. If (M,g) is flat $(R = 0)$, in the flat torus case $(M,g) = (T^n, g_0)$ for example, formula (3.3.5) reads as

$$(q-1)\lambda \le \lambda_1 \frac{n(n+2-q(n-2))}{q+n(n+2)}. \tag{3.3.35}$$

Remark 3.11. In the case where (M,g) is isometric to (S^n, g_0) we have $\mathrm{Ricc}_g = (n-1)g$ and $\lambda_1 = n$. Therefore (3.3.4), (3.3.5) reads as

$$1 < q \le (n+2)/(n-2), \tag{3.3.36}$$

and

$$(q-1)\lambda \le \lambda_1, \tag{3.3.37}$$

with one of the two inequality being strict. This result is optimal and it has been obtained in [GSk1].

As a consequence of this result we obtain new estimates for the infimum of the following quotient

$$Q_{\lambda,q}(u) = \frac{\int_M \left(|\nabla_g u|^2 + \lambda u^2\right)dv_g}{\left(\int_M |u|^{q+1} dv_g\right)^{2/(q+1)}}. \tag{3.3.38}$$

COROLLARY 3.2. *Let the Ricci curvature of* g *satisfy* (3.3.3)*, and* (3.3.4) *and* (3.3.6) *hold. Then*

$$S_{\lambda,q} = \inf\left\{Q_{\lambda,q}(u) : u \in W^{1,2}(M) - \{0\}\right\} = \lambda(\mathrm{Vol}\, M)^{(q-1)/(q+1)}. \tag{3.3.39}$$

The proof is the same as the one of $\left[\mathrm{BVV, Cor}\ 6\text{-}2\right]$, by using directly the equation in the case $1 < q < (n+2)/(n-2)$ and the left upper semicontinuity of $q \mapsto S_{\lambda,q}$ at $q = (n+2)/(n-2)$.

Remark 3.12. As we have quoted it in Remark 3.11, the result of Theorem 3.12 is optimal if $(M,g) = (S^n, g_0)$. It has been noticed by H. Hamza [Ha] that, if

$q = (n+2)/(n-2)$, there exist non-constant positive solutions of (3.3.1) on (M,g) whenever $\lambda = \lambda_1 / (q-1) = (n-2)\lambda_1 / 4$ and

$$\lambda_1 > n\left(\frac{\operatorname{vol} S^n}{\operatorname{vol} M}\right)^{2/n}. \tag{3.3.40}$$

Actually, we know from Aubin's results $[Au2]$ that

$$v = 4\frac{n-1}{n-2}S_{\lambda,(n+2)/(n-2)} \leq n(n-1)\left(\operatorname{vol} S^n\right)^{2/n}, \tag{3.3.41}$$

for any λ. If the only positive solutions of (3.3.1) were constant, it would imply that

$$\lambda_1(n-1)(\operatorname{vol} M)^{2/n} \leq n(n-1)\left(\operatorname{vol} S^n\right)^{2/n} \tag{3.3.42}$$

and consequently

$$\lambda_1 \leq n\left(\frac{\operatorname{vol} S^n}{\operatorname{vol} M}\right)^{2/n}. \tag{3.3.43}$$

If we take $(M,g) = \left(\mathbf{P}_n(\mathbb{R}), g_0\right)$, the n-dimensional real projective space, then

$$\operatorname{vol} M = 1/2 \quad \text{and} \quad \lambda_1 = 2(n+1); \tag{3.3.44}$$

it is clear that (3.3.43) means $2(n+1) \leq n2^{2/n}$, which is never true for $n > 1$.

More generaly, if $q = (n+2)/(n-2)$, the fact that the only positive solutions of (3.3.1) are constant functions yields

$$\lambda \leq \frac{n(n-2)}{4}\left(\frac{\operatorname{vol} S^n}{\operatorname{vol} M}\right)^{2/n}. \tag{3.3.45}$$

which, in turn, implies that there exist positive non constant solutions of (3.3.1) whenever $\lambda > \dfrac{n(n-2)}{4}\left(\dfrac{\operatorname{vol} S^n}{\operatorname{vol} M}\right)^{2/n} = \lambda(M)$. Moreover, from the upper semicontinuity of $(\lambda,q) \mapsto S_{\lambda,q}$ on the left at $q = 2^* - 1 = (n+2)/(n-2)$, this result still holds in some neighborhood of the couple $\left(\lambda(M),(n+2)/(n-2)\right)$.

Remark 3.13. Another interesting application deals with the uniqueness of Einstein metric with constant scalar curvature that is a metric g where

$$\operatorname{Ricc}_g = kg \tag{3.3.46}$$

for some real number k, for such a metric the scalar curvature is constant and in fact

$$\text{Scal}_g = nk.$$ (3.3.47)

If g is some metric on M, the metric g' is said to be conformal to g if $g' = v(x)g$ for some $C^\infty(M)$, positive function v. If we write $v = u^{4/(n-2)}$ $(n \geq 3)$ and $g_u = g' = u^{4/(n-2)}g$, then (see [LP]) u satisfies

$$-4\frac{n-1}{n-2}\Delta_g u + \text{Scal}_g u - \text{Scal}_{g_u} u^{(n+2)/(n-2)} = 0.$$ (3.3.48)

Then, up to an homothety, g is the unique metric in its conformal class to have constant scalar curvature. Using (3.3.47) transforms (3.3.48) into

$$\Delta_g u - \frac{n(n-2)k}{4(n-1)}u + u^{(n+2)/(n-2)} = 0.$$ (3.3.49)

The case $k < 0$ being obvious from the maximum principle, we are left with the case $k > 0$. Lichnerowicz theorem implies that $\lambda_1 \geq nk / (n-1)$ and the condition (3.3.5) reads as

$$\frac{4n(n-2)k}{4(n-2)(n-1)} \leq \frac{n}{n-1}k,$$ (3.3.50)

which is obviously satisfied with equality. Therefore u is constant.

3-3-4 The convergence theorems

The classical transformation for studying the singularity problem is to use spherical coordinates and set

$$u(r,\sigma) = r^{-2/(q-1)}v(t,\sigma), \quad t = \ln(1/r)$$ (3.4.1)

as we have done it in Chapter 2. Since there is no restriction in assuming that $B(O,2) \subset \Omega$, the function v satisfies

$$v_{tt} + \left(2\frac{q+1}{q-1} - N\right)v_t + \frac{2}{q-1}\left(\frac{2q}{q-1} - N\right)v + \Delta_{S^{N-1}}v + v^q = 0$$ (3.4.2)

in $(-\ln 2, \infty) \times S^{N-1}$. We call $\mathcal{F}$ the set of nonnegative stationnary solutions of (3.4.2), that is the set of C^2 nonnegative solutions of

$$\frac{2}{q-1}\left(\frac{2q}{q-1} - N\right)\omega + \Delta_{S^{N-1}}\omega + \omega^q = 0$$ (3.4.3)

on S^{N-1}. Moreover it is important to notice that the coefficient of v_t (the damping term) in (3.4.2) vanishes only when $q = (N+2)/(N-2)$, in which case a specific treatment will be needed. If we know that v is bounded and $q \neq (N+2)/(N-2)$, we can apply the technique of Theorem 2.2 and obtain a weak type result

PROPOSITION 3.9. *Let* $q > 1$, $q \neq (N+2)/(N-2)$ *and* u *be a solution of* (3.1.1) *in* Ω^* *such that*

$$u(x) \leq C|x|^{-2/(q-1)} \tag{3.4.4}$$

near O, *for some* $C > 0$. *Then there exists a compact connected component* $\mathcal{F}^*$ *of* $\mathcal{F}$ *such that*

$$\lim_{r \to 0} \, \mathrm{dist}_{C^2}\left(r^{2/(q-1)} u(r,.), \mathcal{F}^* \right) = 0. \tag{3.4.5}$$

Remark 3.14. In the particular case where $1 < q < (N+2)/(N-2)$, estimate (3.4.4) is always satisfied and the set $\mathcal{F}$ is disconnected: reduced to the zero function if $1 < q \leq N/(N-2)$ or to the two constant functions 0 and $(-l_{q,N})^{1/(q-1)}$ if $N/(N-2) < q < (N+2)/(N-2)$. Since the case $1 < q < N/(N-2)$ has yet been treated, Proposition 3.9 implies

(i) *if* $q = N/(N-2)$, *then* $\displaystyle\lim_{x \to 0} |x|^{2/(q-1)} u(x) = \lim_{x \to 0} |x|^{N-2)} u(x) = 0$,

(ii) *if* $N/(N-2) < q < (N+2)/(N-2)$, *then* $\displaystyle\lim_{x \to 0} |x|^{2/(q-1)} u(x)$ *is either* 0 *or* $(-l_{q,N})^{1/(q-1)}$.

A more detailled analysis will be given later on. In fact the convergence of Proposition 3.9 is improved by Simon's results on asymptotics of analytic functionals ([Si1], [Si2]) .

THEOREM 3.13. *Let* (M,g) *be a* n-*dimensional Riemannian manifold without boundary,* h *a* C^2 *real valued function defined on* $\mathbb{R}$ *and* v *a uniformly bounded function defined in* $[0,\infty) \times M$ *where it satisfies*

$$v_{tt} + \alpha v_t + \Delta_g v + h(v) = 0 \tag{3.4.6}$$

with $\alpha \neq 0$; *assume also that* h *is real analytic in a neighborhoud of the closure of the range of* v. *Then there exists a* $C^\infty(M)$-*fonction* ω *solution of*

$$\Delta_g \omega + h(\omega) = 0 \tag{3.4.7}$$

in M *such that*

$$\lim_{t \to \infty} v(t,.) = \omega(.) \tag{3.4.8}$$

holds in the $C^k(M)$*-topology, for any* $k > 0$.

The proof of this result is a real monument of very deep ideas but also of difficulties and it is based upon the infinite dimensional generalization of Lojaciewicz inequalities defining semi-analytic sets ([Lo]). The assumption of analyticity is rather restrictive and in some cases it can be replaced by a notion of hyperbolicity:

Definition. Let h be a C^2 function and ω a $C^2(M)$-solution of (3.4.7). We say that the set $\mathcal{F}$ of solutions of (3.4.7) is hyperbolic at ω if there exists a C^2 neighborhood $\mathcal{N}_\omega$ of ω such that $\mathcal{F} \cap \mathcal{N}_\omega$ is a finite dimensional submanifold of $C^2(M)$ whose dimension satisfies

$$\dim \mathcal{F} \cap \mathcal{N}_\omega = \dim \, \mathrm{Ker}\Big(\Delta_g + h'(\omega)\mathrm{Id}\Big). \tag{3.4.9}$$

THEOREM 3.14. *Let* (M,g) *be a* n*-dimensional Riemannian manifold without boundary,* h *a* C^3 *real valued function defined on* $\mathbb{R}$ *and* v *a uniformly bounded function defined in* $[0,\infty) \times M$ *where it satisfies*

$$v_{tt} + \alpha v_t + \Delta_g v + h(v) = 0 \tag{3.4.10}$$

with $\alpha \neq 0$. *Assume also that the* $C^3(M)$*-limit set of the positive trajectory of* v *defined by* $\mathcal{T} = \underset{t \geq 0}{\cup}\{v(t,.)\}$ *contains an element* ω *of* $\mathcal{F}$ *with the property that* $\mathcal{F}$ *is hyperbolic at* ω. *Then*

$$\lim_{t \to \infty} v(t,.) = \omega(.) \tag{3.4.11}$$

in the $C^3(M)$*-topology.*

Remark 3.15. The hyperbolicity property is usually difficult to verify. However, if $q = (N+1)/(N-1) = (n+2)/(n-1)$ with $n = N-1$, if $\mathcal{F} = \mathcal{F}_n$ is the set of solutions of

$$\Delta_{S^n}\omega - \frac{n(n-2)}{4}\omega + \omega^{(n+2)/(n-2)} = 0, \tag{3.4.12}$$

and if ω is an element of $\mathcal{F}_n$, there exists a conformal transformation ϕ of S^n such that

$$\omega = \left(\frac{n(n-2)}{4}\right)^{(n-2)/4} \big(\det|d\phi|\big)^{(n-2)/4}. \tag{3.4.13}$$

Consequently, $\mathcal{F}_n$ has the structure of a n-dimensional non-compact manifold (see also Remark 3.9). If we define the conformal Laplacian on S^n by

$$L\psi = \Delta_{S^n}\psi + \frac{n(n-2)}{4}\psi \tag{3.4.14}$$

and set $\psi = \left(\det|d\phi|\right)^{(n+2)/(n-2)}\tilde{\psi}\circ\phi$, then

$$L\psi = \left(\det|d\phi|\right)^{(n+2)/(n-2)}(L\tilde{\psi})\circ\phi, \tag{3.4.15}$$

which yields

$$\psi \in \mathrm{Ker}\left(\Delta_{S^n} + (n-1)\mathrm{Id}\right)$$
$$\Leftrightarrow \psi\circ\phi \in \mathrm{Ker}\left(\Delta_{S^n} + \left(\frac{n+2}{n-2}\omega^{4/(n-2)} - \frac{n(n-2)}{4}\right)\mathrm{Id}\right). \tag{3.4.16}$$

Therefore the dimension of the kernel of the linearized operator at some $\omega \in \mathcal{F}_n$ is always n + 1, which proves the hyperbolicity property.

THEOREM 3.15. *Let* $q \in (N/(N-2),\infty)\setminus\{(N+2)/(N-2)\}$ *and* u *be a solution of* (3.1.1) *in* Ω^* *satisfying* (3.4.4) *near* O, *for some* $C > 0$. *Then,*
(i) either there exists a positive solution ω *of* (3.4.3) *such that*

$$\lim_{r\to 0} r^{2/(q-1)}u(r,.) = \omega(.) \tag{3.4.17}$$

in the $C^k(S^{N-1})$*-topology, for any* $k > 0$,
(ii) or u *is regular at* O.

Proof. Step 1. By using the transformation (3.4.1) and the classical analysis which in fact gives rise to Proposition 3.8, we know that $\dfrac{\partial^\alpha}{\partial t^\alpha}\nabla^\beta v$ is uniformly bounded on $[0,\infty)\times S^{N-1}$, up to the order $\alpha + |\beta| = 3$ (we denote by $\nabla^\beta v$ the covariant derivative of v to the order β). Consequently

$$\int_0^\infty\int_{S^{N-1}}\left(v_t^2 + v_{tt}^2 + |\nabla v_t|^2\right)d\sigma dt < \infty \tag{3.4.18}$$

holds and finally

$$\lim_{t\to\infty}\left(\left\|v_t\right\|_{L^2(S^{N-1})} + \left\|v_{tt}\right\|_{L^2(S^{N-1})} + \left\|\nabla v_t\right\|_{L^2(S^{N-1})}\right) = 0. \tag{3.4.19}$$

We define the limit set at infinity of the positive trajectory $\mathcal{T}=\bigcup_{t\geq 0}\{v(t,.)\}$ by

$$\Gamma^+ = \bigcap_{t\geq 0}\left(\overline{\bigcup_{\tau\geq t}\{v(t,.)\}}^{\,C^2(S^{N-1})}\right) \quad \text{and } \Gamma^+ \text{ is just a compact connected component of } \mathcal{F}.$$

Step 2. Suppose that $0 \notin \Gamma^+$. If we set

$$\delta = \inf_{\omega\in\Gamma^+}\ \inf_{\sigma\in S^{N-1}} \omega(\sigma) \tag{3.4.20}$$

then $\delta > 0$ from the compactness of Γ^+ and the strong maximum principle. Therefore there exists some $T > 0$ such that

$$v(t,\sigma) \geq \delta/2 \quad \text{if } (t,\sigma) \in [T,\infty)\times S^{N-1}. \tag{3.4.21}$$

Since the function $r \mapsto r^q$ is analytic on the intreval $(\delta/3,\infty)$ which contains the range of v for $t \geq T$, we deduce from Theorem 3.13 that there exists a positive element ω of $\mathcal{F}$ such that

$$\lim_{t\to\infty} v(t,.) = \omega(.) \tag{3.4.22}$$

in the $C^2(M)$-topology (and in fact C^k for any $k > 0$).

Step 3. Suppose that $0 \in \Gamma^+$. If we set

$$E(v)(t) = \int_{S^{N-1}}\left(\frac{1}{2}|\nabla v|^2 - \frac{1}{2}v_t^2 - \frac{l_{q,N}}{2}v^2 - \frac{1}{q+1}|v|^{q+1}\right)d\sigma, \tag{3.4.23}$$

then

$$\frac{d}{dt}E(v)(t) = \left(2\frac{q+1}{q-1} - N\right)\int_{S^{N-1}} v_t^2\,d\sigma, \tag{3.4.24}$$

and the function $t \mapsto E(v)(t)$ is monotonous. Therefore

$$\lim_{t\to\infty} E(v)(t) = E(0) = 0. \tag{3.4.25}$$

Since $E(\omega) = \dfrac{q-1}{2(q+1)}\displaystyle\int_{S^{N-1}}|\omega|^{q+1}d\sigma$ for any $\omega \in \mathcal{F}$, it implies that Γ^+ is reduced to the zero function and

$$\lim_{t\to\infty} \|v(t,.)\|_{C^2(S^{N-1})} = 0. \tag{3.4.26}$$

The completion of the proof is just an application of the perturbation theory of linear partial differential equations. The two linearly independent solutions of

$$v_{tt} + \left(2\frac{q+1}{q-1} - N\right)v_t + 1_{q,N}v = 0 \tag{3.4.27}$$

are the $\rho_j(t) = e^{v_j t}$, $j = 1, 2$, where v_1 (resp. v_2) is the positive (resp. negative) root of the polynomial

$$x \mapsto x^2 + \left(2\frac{q+1}{q-1} - N\right)x + 1_{q,N} \tag{3.4.28}$$

(remember that $1_{q,N}$ is negative in this range of values of q). For any $\varepsilon > 0$, there exists T > 0 such that v satisfies

$$v_{tt} + \left(2\frac{q+1}{q-1} - N\right)v_t + (1_{q,N} + \varepsilon)v + \Delta_{S^{N-1}}v \geq 0 \tag{3.4.29}$$

on $[T, \infty) \times S^{N-1}$,. Therefore, we deduce from the maximum principle and from the fact that we can always take $\varepsilon > 0$ small enough, that the following estimate holds:

$$\|v(t,.)\|_{L^\infty(S^{N-1})} \leq \|v(T,.)\|_{L^\infty(S^{N-1})}e^{v_{2,\varepsilon}t}, \tag{3.4.30}$$

where $v_{2,\varepsilon}$ is the negative root of $x \mapsto x^2 + \left(2\frac{q+1}{q-1} - N\right)x + 1_{q,N} + \varepsilon$. If we compute the roots of (3.4.28) we find $v_2 = -2/(q-1)$ and consequently $v_{2,\varepsilon} = -2/(q-1) + O(\varepsilon)$. If we go back to the function u, we have

$$u(x) = O\left(|x|^{-K\varepsilon}\right) \quad \text{and} \quad u^{q-1}(x) = d(x) = O\left(|x|^{-K(q-1)\varepsilon}\right) \tag{3.4.31}$$

near O, for some K > 0. If we write (3.1.1) as

$$\Delta u + d(x)u = 0, \tag{3.4.32}$$

then we notice that we can take ε so that $d \in L^{1+N/2}_{loc}(\Omega)$. We conclude from Serrin's classical result (see Theorem 1.19) that the singularity at O is removable.

In the case $q = N/(N-2)$ we have a new type of singular behaviour which was first obtained by Aviles [Av3] in the non radial case.

THEOREM 3.16. *Let* $q = N/(N-2)$ *and u be a solution of* (3.1.1) *in* Ω^*. *Then we have the following alternative :*
(i) either

$$\lim_{r \to 0} r^{N-2}(\ln(1/r))^{(N-2)/2}u(r,.) = \left(\frac{N-2}{\sqrt{2}}\right)^{N-2} \tag{3.4.33}$$

in the $C^2(S^{N-1})$-*topology,*
(ii) or u *is regular at* O.

Proof. Step 1. We claim that there exists $C > 0$ such that

$$u(x) \leq C|x|^{2-N}\left(\ln(1/|x|)\right)^{(2-N)/2},\tag{3.4.34}$$

for $0 < |x| \leq 1/2$. From Theorem 3.11

$$u(x) \leq C|x|^{2-N}\tag{3.4.35}$$

for $0 < |x| \leq 1$. Writing (3.1.1) under the form (3.4.32) we see that the coefficient $d(x)$ satisfies

$$\|d\|_{L^\infty(B(x,|x|/2))} \leq C'|x|^{-2},\tag{3.4.36}$$

which implies that the coefficient in the Harnack inequality is controlled ([GT]) and that there exists $C'' > 0$ such that

$$\max_{y \in B(x,|x|/2)} u(y) \leq C'' \min_{y \in B(x,|x|/2)} u(y).\tag{3.4.37}$$

Therefore, for some positive constant $C > 0$, we have

$$\max_{|x|=r} u(x) \leq C \min_{|x|=r} u(x),\tag{3.4.38}$$

for $0 < r \leq 1$. With the notations (3.4.1) the function v satisfies

$$v_{tt} + (N-2)v_t + \Delta_{S^{N-1}}v + v^{N/(N-2)} = 0\tag{3.4.39}$$

in $[-\ln 2, \infty) \times S^{N-1}$ where it remains bounded. From (3.4.39) we have

$$\bar{v}_{tt} + (N-2)\bar{v}_t + C^{-N/(N-2)}\bar{v}^{N/(N-2)} \leq 0.\tag{3.4.40}$$

where $\bar{v}$ is the average of v on S^{N-1}. Since $\bar{v}$ is nonnegative and goes to O at infinity, a mere integration implies that it decreases with t. If we differentiate (3.4.39) with respect to t, set $y = v_t$ and $\bar{y}$ its average, we have

$$y_{tt} + (N-2)y_t + \Delta_{S^{N-1}}y + \frac{N}{N-2}v^{2/(N-2)}y = 0\tag{3.4.41}$$

and

$$\bar{y}_{tt} + (N-2)\bar{y}_t + \frac{N}{N-2}C^{2/(N-2)}\bar{v}^{2/(N-2)}\bar{y} \le 0. \tag{3.4.42}$$

The function $\bar{y}$ is nonpositive and bounded and there exists a sequence $\{t_n\}$ going to infinity with n such that $\lim_{n\to\infty}\bar{y}(t_n) = 0$. The maximum principle implies that $\bar{y}$ is nondecreasing. Therefore $\bar{v}$ is convex and

$$(N-2)\bar{v}_t + C^{-N/(N-2)}\bar{v}^{N/(N-2)} \le 0 \tag{3.4.43}$$

from (3.4.38). The integration of (3.4.43) implies that $t^{(N-2)/2}\bar{v}(t)$ remains bounded and finally we deduce from (3.4.38) that (3.4.34) holds.

Step 2. For any $\varepsilon > 0$ there exists $C_\varepsilon > 0$ such that

$$\|v(t,.) - \bar{v}(t)\|_{L^\infty(S^{N-1})} \le C_\varepsilon e^{-(N-1-\varepsilon)t}. \tag{3.4.44}$$

Denote $w = v - \bar{v}$, then

$$w_{tt} + (N-2)w_t + \Delta_{S^{N-1}}w + v^{N/(N-2)} - \overline{v^{N/(N-2)}} = 0, \tag{3.4.45}$$

and $v^{N/(N-2)} - \overline{v^{N/(N-2)}} = v^{N/(N-2)} - \bar{v}^{N/(N-2)} + \bar{v}^{N/(N-2)} - \overline{v^{N/(N-2)}}$. Multiplying (3.4.45) by w, integrating over S^{N-1}, using (3.4.34) and the once used relation

$$-\int_{S^{N-1}}(v-\bar{v})\Delta_{S^{N-1}}(v-\bar{v})d\sigma \ge (N-1)\int_{S^{N-1}}(v-\bar{v})^2 d\sigma, \tag{3.4.46}$$

yields

$$\int_{S^{N-1}}ww_{tt}d\sigma + (N-2)\int_{S^{N-1}}ww_t d\sigma - (N-1-Ct^{-1})\int_{S^{N-1}}w^2 d\sigma \ge 0. \tag{3.4.47}$$

If we set $z(t) = \|w(t,.)\|_{L^2(S^{N-1})}$, we deduce that

$$z'' + (N-2)z' - (N-1-Ct^{-1})z \ge 0 \tag{3.4.48}$$

holds in $\mathcal{D}'(-\ln 2, \infty)$. Since the two roots of the polynomial $x \mapsto x^2 + (N-2)x - (N-1)$ are 1 and 1-N, we obtain an L^2 decay estimate, namely

$$\|w(t,.)\|_{L^2(S^{N-1})} \le C'_\varepsilon e^{-(N-1-\varepsilon)t}. \tag{3.4.49}$$

Applying the L^2-regularity theory for elliptic equations gives

$$\|w\|_{W^{2,2}\left([T-1,T+1]\times(S^{N-1})\right)}$$

$$\leq C\left(\|w\|_{L^2\left([T-2,T+2]\times(S^{N-1})\right)} + \left\|v^{N/(N-2)} - \overline{v^{N/(N-2)}}\right\|_{L^2\left([T-2,T+2]\times(S^{N-1})\right)}\right), \tag{3.4.50}$$

for any $T \geq 2$ and

$$\left\|v^{N/(N-2)} - \overline{v^{N/(N-2)}}\right\|_{L^p\left((\alpha,\beta)\times S^{N-1}\right)} \leq \frac{N}{N-2}\left(\max_{(\alpha,\beta)\times S^{N-1}} v^{2/(N-2)}\right)\|v - \overline{v}\|_{L^p\left((\alpha,\beta)\times S^{N-1}\right)}, \tag{3.4.51}$$

for any $p \geq 1$, α and β. Using Sobolev inequalities and (3.4.49)-(3.4.34) finally yields

$$\|w(t,.)\|_{L^{2^{**}}(S^{N-1})} \leq C_{\varepsilon,1}e^{-(N-1-\varepsilon)t} \tag{3.4.52}$$

with $2^{**} = 4N/(N-4)$ if $N > 4$ and the usual modifications if $N = 3,4$. Iterating this process gives (3.4.45).

Step 3. We claim that $\lim\limits_{t\to\infty} t^{(N-2)/2}v(t,.)$ exists and is either $\left((N-2)/\sqrt{2}\right)^{N-2}$ or 0. If we set $\rho(t,.) = t^{(n-2)/2}v(t,.)$, then ρ is bounded and satisfies

$$\rho_{tt} + (N-2)(1+t^{-1})\rho_t + \Delta_{s^{N-1}}\rho + t^{-1}\left(\rho^{N/(N-2)} - \frac{(N-2)^2}{2}\rho\right) + t^{-2}\rho = 0. \tag{3.4.53}$$

From the classical elliptic equation theory $\dfrac{\partial^\alpha}{\partial t^\alpha}\nabla^\beta\rho$ remains uniformly bounded for $\alpha + |\beta| \leq 3$. Moreover

$$(N-2)(1-t^{-1})\int_{S^{N-1}}\rho_t^2 d\sigma$$

$$= \frac{d}{dt}\int_{S^{N-1}}\left[\left(\frac{1}{2}|\nabla\rho|^2 - \frac{1}{2}\rho_t^2\right) + t^{-1}\left(\frac{(N-2)^2}{4}\rho^2 - \frac{N-2}{2(N-1)}\rho^{2(N-1)/(N-2)}\right)\right. \tag{3.4.54}$$

$$\left. + t^{-2}\frac{N(N-2)}{8}\rho^2\right]d\sigma + O(t^{-2}),$$

which implies that the classical Lyapounov type analysis is valid and this gives

$$\int_0^\infty\int_{S^{N-1}}\left(\rho_t^2 + \rho_{tt}^2 + |\nabla\rho_t|^2\right)d\sigma dt < \infty. \tag{3.4.55}$$

The $C^2(S^{N-1})$-limit set at infinity of the positive trajectory of ρ is therefore non-empty and reduced to harmonic functions on the sphere, which are in fact the constants (nonnegative as ρ is). If $\overline{\rho}$ is the spherical average of ρ it satisfies

$$\overline{\rho}_{tt} + (N-2)(1+t^{-1})\overline{\rho}_t + t^{-1}\left(\overline{\rho}^{N/(N-2)} - \frac{(N-2)^2}{2}\overline{\rho}\right) + t^{-2}\overline{\rho} = O(e^{-(N-1-\varepsilon)t}). \quad (3.4.56)$$

Denote $\ell_S = \limsup\limits_{t\to\infty}\overline{\rho}(t)$ and $\ell_I = \liminf\limits_{t\to\infty}\overline{\rho}(t)$ and suppose that $0 \le \ell_I < \ell_S$. Then there exist two sequences $\{t_n^1\}$ and $\{t_n^2\}$ going to infinity with n such that $t_n^1 < t_n^2 < t_{n+1}^1$ for any $n \in \mathbb{N}$ and

$$\lim_{n\to\infty}\rho(t_n^1) = \ell_S, \quad \lim_{n\to\infty}\rho(t_n^2) = \ell_I, \quad \overline{\rho}_t(t_n^j) = 0, \quad (-1)^j\overline{\rho}_{tt}(t_n^j) \ge 0 \quad \text{for } j = 1, 2. \quad (3.4.57)$$

Multiplying (3.4.56) by t and using (3.4.57) yields

$$\ell_I^{N/(N-2)} - \frac{(N-2)^2}{2}\ell_I \le 0 \le \ell_S^{N/(N-2)} - \frac{(N-2)^2}{2}\ell_S, \quad (3.4.58)$$

which implies

$$0 \le \ell_I \le \left((N-2)/\sqrt{2}\right)^{N-2} \le \ell_S. \quad (3.4.59)$$

If we suppose for example that $\left((N-2)/\sqrt{2}\right)^{N-2} < \ell_S$, there exist two sequences $\{\tau_n\}$, $\{\eta_n\}$ such that $t_n^1 < \tau_n < \eta_n < t_{n+1}^1$, $\overline{\rho}(\tau_n) = \overline{\rho}(\eta_n) = \frac{1}{2}\left(\ell_S + \left((N-2)/\sqrt{2}\right)^{N-2}\right)$ with $\overline{\rho}(t) \ge \left((N-2)/\sqrt{2}\right)^{N-2}$ on $\left[t_n^1, \tau_n\right] \cup \left[\eta_n, t_{n+1}^1\right]$. Integrating (3.4.56) on $\left[t_n^1, \tau_n\right] \cup \left[\eta_n, t_{n+1}^1\right]$ and letting n go to infinity gives

$$(N-2)\left[\ell_S - \left((N-2)/\sqrt{2}\right)^{N-2}\right]$$
$$+ \left(\int_{t_n^1}^{\tau_n} + \int_{\eta_n}^{t_{n+1}^1}\right) t^{-1}\left[\overline{\rho}^{N/(N-2)}(t) - (N-2)^2\overline{\rho}(t)/2\right]dt = o(1). \quad (3.4.60)$$

Since the second term in (3.4.60) is positive, we have a contradiction. Therefore $\left((N-2)/\sqrt{2}\right)^{N-2} = \ell_S$. In the same way $\left((N-2)^2/2\right)^{N-2} > \ell_I$ is impossible. Consequently $\ell_S = \ell_I = \ell \ge 0$. We see that $\lim\limits_{t\to\infty}\int_1^t s^{-1}\left(\overline{\rho}^{N/(N-2)}(s) - (N-2)^2\overline{\rho}(s)/2\right)ds$ exists. Therefore ℓ must satisfy $\ell^{N/(N-2)} - (N-2)^2\ell/2 = 0$, which is the claim since $\rho(t,.)$ and $\overline{\rho}(t)$ have the same limit..

Step 4. If $\lim\limits_{t\to\infty} t^{(N-2)/2}v(t,.) = 0$, then we claim that u is regular in whole Ω. For any $\varepsilon > 0$ there exists $T_\varepsilon > 0$ such that

$$\overline{\rho}_{tt} + (N-2)(1+t^{-1})\overline{\rho}_t - t^{-1}\left(\frac{(N-2)^2}{2} - \varepsilon\right)\overline{\rho} \geq 0 \tag{3.4.61}$$

on $[T_\varepsilon, \infty)$. For $0 < \beta < N-2$, the function $e_\beta : t \mapsto e_\beta(t) = e^{-\beta t}$, satisfies

$$\frac{d^2 e_\beta}{dt^2} + (N-2)(1+t^{-1})\frac{de_\beta}{dt} - t^{-1}\left(\frac{(N-2)^2}{2} - \varepsilon\right)e_\beta$$
$$= e_\beta\left(\beta(\beta - N + 2) + O(t^{-1})\right), \tag{3.4.62}$$

and the right-hand side f(3.4.62) is nonpositive for $t \geq T_\varepsilon' \geq T_\varepsilon$. By the maximum principle

$$\overline{\rho}(t) \leq \overline{\rho}(T_\varepsilon')e_\beta(t - T_\varepsilon'). \tag{3.4.63}$$

This implies

$$u(x) \leq C|x|^{2-N+\beta}\left(\ln(1/|x|)\right)^{(2-N)/2} \tag{3.4.64}$$

We conclude in the same way as in Theorem 3.15-Step 3, by putting the equation (3.1.1) under the form (3.4.32) and using Serrin's result, that the singularity at O is removable and finally u is a regular solution.

Remark 3.16. The techniques of Theorem 3.15 are immediately transferable to the study of the asymptotic behaviour of the nonnegative solutions of (3.1.1) in an exterior domain: *if $q \in (N/(N-2), \infty) \setminus \{(N+2)/(N-2)\}$ and if the estimate (3.4.4) is valid for $|x|$ large enough, then*
(i) either there exists a positive solution ω of (3.4.3) such that

$$\lim_{r \to \infty} r^{2/(q-1)}u(r,.) = \omega(.) \tag{3.4.65}$$

in the $C^k(S^{N-1})$-topology, for any $k > 0$,
(ii) or there exists $\gamma > 0$ such that

$$\lim_{r \to \infty} r^{N-2}u(r,.) = \gamma, \tag{3.4.66}$$

in the $C^2(S^{N-1})$-topology.

If $1 < q < N/(N-2)$, an easy concavity argument shows that there exists no positive solution of (3.1.1) in an exterior domain.

Remark 3.17. In the case $q = N/(N-2)$ it is possible to prove that there truly exist positive solutions of (3.1.1) such that (3.4.31) holds (see $[Av3]$ for an attempt of proof and $[B.VR]$).

We end this section with a global result.

THEOREM 3.17. *Let* $N > 2$, $q \in (N/(N-2),(N+1)/(N-3)] \setminus \{(N+2)/(N-2)\}$ *and* u *be a positive solution of* (3.1.1) *in* $\mathbb{R}^N \setminus \{O\}$ *such that*

$$|x|^{2/(q-1)} u(x) \in L^\infty(\mathbb{R}^N). \tag{3.4.67}$$

We have the following:

(I) *If* $N/(N-2) < q < (N+2)/(N-2)$, *then*
(i) *either*

$$u(x) \equiv \left(\frac{2}{q-1} \left(N - \frac{2q}{q-1} \right) \right)^{1/(q-1)} |x|^{-2/(q-1)}$$

$$= \left(-l_{q,N} \right)^{1/(q-1)} |x|^{-2/(q-1)} \quad (\forall x \in \mathbb{R}^N \setminus \{O\}), \tag{3.4.68}$$

(ii) *or* u *is singular at* O *and regular at infinity at in the sense that*

$$\lim_{r \to 0} r^{2/(q-1)} u(r,.) = \left(-l_{q,N} \right)^{1/(q-1)}, \quad \lim_{r \to \infty} r^{N-2} u(r,.) = \gamma, \tag{3.4.69}$$

for some $\gamma > 0$.

(II) *If* $(N+2)/(N-2) < q < (N+1)/(N-3)$, *then*
(i) *either* (3.4.69) *holds,*
(ii) *or* u *is regular at* O *and singular at infinity in the sense that* $u \in C^\infty(\mathbb{R}^N)$ *and*

$$\lim_{r \to \infty} r^{2/(q-1)} u(r,.) = \left(-l_{q,N} \right)^{1/(q-1)}. \tag{3.4.70}$$

(III) *If* $q = (N+1)/(N-3)$, $N > 3$, *then*
(i) *either there exists* $\omega \in \mathcal{F} = \mathcal{F}_n$, *the set of positive solutions of* (3.4.12), *such that*

$$u(r,\sigma) \equiv r^{-2/(q-1)} \omega(\sigma), \quad \forall (r,\sigma) \in (0,\infty) \times S^{N-1}, \tag{3.4.71}$$

(ii) *or* u *is regular at* O *and singular at infinity as in* (II-ii).

Proof. We define v on $(-\infty,\infty) \times S^{N-1}$ by (3.4.1)-(3.4.2) and denote ω_+ and ω_- the limit of v(t,.) at infinity and minus infinity respectively and both are element of $\mathcal{F}$. We recall that the energy function is defined by

$$E(\varphi) = \int_{S^{N-1}} \left(\frac{1}{2} |\nabla \varphi|^2 - \frac{l_{q,N}}{2} \varphi^2 - \frac{1}{q+1} |\varphi|^{q+1} \right) d\sigma \tag{3.4.72}$$

and if $\varphi \in \mathcal{F}$, $E(\varphi) = \dfrac{q-1}{2(q+1)} \int_{S^{N-1}} |\varphi|^{q+1} d\sigma$. Therefore

$$\left(2\frac{q+1}{q-1} - N\right) \int_{-\infty}^{\infty}\int\int_{S^{N-1}} v_t^2 d\sigma dt = E(\omega_+) - E(\omega_-). \tag{3.4.73}$$

If $q \in (N/(N-2), (N+1)/(N-3))$, $\mathcal{F}$ contains only one non-zero element (see Remark 3.11). If $q = (N+1)/(N-3)$ we define the Sobolev quotient in $W^{1,2}(S^{N-1}) \setminus \{0\}$ by

$$Q(\varphi) = \frac{\int_{S^{N-1}}\left(|\nabla\varphi|^2 + ((N-1)/(N-3)/4)\varphi^2\right)d\sigma}{\left(\int_{S^{N-1}} |\varphi|^{2(N-1)/(N-3)} d\sigma\right)^{(N-3)/(N-1)}}, \tag{3.4.74}$$

and its minimum, which is $S = ((N-1)(N-3))|S^{N-1}|^{2/(N-1)}\big/4$, is achieved by any element of $\mathcal{F}_{N-1}$. (see Corollary 3.2). This implies that E keeps the value $4S^{(N-1)/2}/(N-1)$ on the set $\mathcal{F}_{N-1} \setminus \{0\}$. Consequently, if ω_+ and ω_- are non-zero, v is constant and u takes the form (3.4.68) or (3.4.71). Since $(2(q+1)/(q-1) - N)$ changes sign for $q = (N+2)/(N-2)$, we deduce from Theorem 3.15 and Remark 3.16 that, if the assertions (i) of (I)-(II)-(III) do not hold, then the assertions (ii) hold.

Remark 3.18. The so-called isothermal gas sphere equation (see [Ch])

$$\Delta u + \lambda e^u = 0, \tag{3.4.75}$$

$\lambda > 0$, in $\mathbb{R}^3$ or $\mathbb{R}^3 \setminus \{O\}$, presents a structure which is in some way similar to the one of equation (3.1.1) when $q = (N+1)/(N-3)$. If we look for solutions under the form

$$u(r,\sigma) = \ln(1/r^2) + \ln(2/\lambda) + 2\omega(\sigma), \quad (r,\sigma) \in (0,\infty) \times S^2, \tag{3.4.76}$$

then ω satisfies

$$\Delta_{S^2}\omega + e^{2\omega} - 1 = 0 \tag{3.4.77}$$

on S^2. The set $\mathcal{F}_2$ of all the solutions of (3.4.77) is well known since any element ω can be written under the form $\omega = \ln\left(\sqrt{|\det(d\varphi)|}\right)$ where φ is a conformal transformation of S^2 (see [B.VV] for a detailed analysis).

3-3-5 Prescribed singularities

When $q \geq N/(N-2)$ we know that the isolated singularities of positive solutions of (3.1.1) are not visible at the distributions level, therefore the construction of solutions with a prescribed set of isolated singularities cannot be made in the framework of distributions theory. By using technique first introduced in differential geometry by Smale ([Sm1], [Sm2]), then developped by Paccard, Mazzeo (see their references) many non trivial examples of singular solutions of equations of type (3.1.1) or (3.4.75) have been constructed. The following result is proved in [Pc1]

THEOREM 3.18. *Let* $N > 10$, $\lambda > 0$ *and* $\Sigma = \{p_i\}_{i=1}^{k}$ *be a finite subset of* $\mathbb{R}^N$; *then there exists a connected open subset* Ω *of* $\mathbb{R}^N$ *containing* Σ *and a solution* u *of*

$$-\Delta u = \lambda e^{u} \tag{3.5.1}$$

in $\Omega \setminus \Sigma$ *such that*

$$u(x) = \ln\left(1/|x - p_i|^2\right) + 2\ln(2(N-2)/\lambda) + o(1) \tag{3.5.2}$$

when $x \to p_i$, *for any* $i = 1,\ldots, k$.

Abridged proof. First, it is always possible to suppose that $\lambda = 2(N-2)$ and therefore $x \mapsto \ln\left(1/|x|^2\right) = u_s(x)$ is a singular solution of (3.5.1) in $\mathbb{R}^N \setminus \{O\}$; for simplicity let us also suppose that $|p_i - p_j| > 2$ for $i \neq j$. Let $\varepsilon > 0$, we can join the k disjoint balls $B(p_i, 1)$ by small regular bridges with diameter less than ε so that the resulting open subset $\tilde{\Omega}^{\varepsilon}$ is C^4. Let $\Gamma_{p_i, p_j}(\varepsilon)$ be the bridge joining $B(p_i, 1)$ to $B(p_j, 1)$ and $\Gamma(\varepsilon) = \underset{1 \leq i < j \leq k}{\cup} \Gamma_{p_i, p_j}(\varepsilon)$. We can also constraint the diameter of $\Gamma_{p_i, p_j}(\varepsilon)$ to be contained in $\left[c_0^{-1}\varepsilon, c_0\varepsilon\right]$ for some fixed $c_0 > 1$ and the second fundamental form of $\partial\Gamma_{p_i, p_j}(\varepsilon)$ to be bounded by $c_0\varepsilon^{-1}$ (see [Sm1]). Set $\Omega^{\varepsilon} = \tilde{\Omega}^{\varepsilon} \setminus \underset{1 \leq j \leq k}{\cup} \{p_j\}$. On Ω^{ε} we construct a quasi-solution u_0 in the following way:

$$u_0(x) = \ln\left(1/|x - p_i|^2\right) \quad \text{on} \quad B(p_i, 1)\setminus \{p_i\}, \tag{3.5.3}$$

and $\|u_0\|_{C^{2,\alpha}(\overline{\Gamma}(\varepsilon))} \leq c_1$ for some positive constant c_1 and $0 < \alpha < 1$. The solution u of (3.5.1) is therefore looked for as a perturbation of u_0 under the form $u = v + u_0$ with

$$\begin{cases} v \in C^{2,\alpha}(\overline{\Omega}^{\varepsilon}), \\ -\Delta(v + u_0) = 2(N-2)e^{v+u_0} & \text{in} \quad \Omega^{\varepsilon}, \\ v = \Psi & \text{on} \quad \partial\Omega^{\varepsilon}, \\ |v(x)| \leq c|p_i - x|^{v_i} & \text{in} \quad B(p_i, 1), \end{cases} \tag{3.5.4}$$

where c is a positive constant, v_i is larger than 1 and Ψ is free. Denote $\xi_0 = -\Delta u_0 - 2(N-2)e^{u_0}$, then $\xi_0 \in C^{0,\alpha}(\overline{\Omega}_\varepsilon)$ with support in $\Gamma(\varepsilon)$. We define L and E by

$$Lv(x) = \Delta v(x) + 2(N-2)e^{u_0(x)}v(x), \tag{3.5.5}$$

$$E(v)(x) = e^{u_0(x)}\int_0^1 (1-t)v^2(x)e^{tv(x)}dt = e^{u_0(x)}\left(e^{v(x)} - v(x) - 1\right), \tag{3.5.6}$$

then the equation for v is

$$Lv = \xi_0 - 2(N-2)E(v). \tag{3.5.7}$$

In fact the theorem shows that for any given choice of $\{v_i\}_{i=1}^k$ with $v_i \geq 2$, there exists $\tilde{\varepsilon} > 0$ and a solution u of (3.5.1) in Ω_ε for any $0 < \varepsilon \leq \tilde{\varepsilon}$, under the form $u = v + u_0$ where v satisfies (3.5.4) for a suitably chosen Ψ. In the sequel we denote $\overline{B}^*(O,1) = \overline{B}(O,1)\setminus \{O\}$.

Step 1. Stability analysis.
For $0 < \delta < 1$ set $\Gamma_{\delta,1} = B(O,1)\setminus \overline{B}(O,\delta)$. If $U \in C^\infty\left(\overline{B}^*(O,1)\right)$ we define the operator $L_{U,\delta}$ by

$$L_{U,\delta}v(x) = -\Delta v(x) - 2(N-2)e^{U(x)}v(x). \tag{3.5.8}$$

The operator $|x|^2 L_{U,\delta}$ is uniformly elliptic and self-adjoint in $L^2(\Gamma_{\delta,1}, |x|^{-2}dx) \cap W_0^{1,2}(\Gamma_{\delta,1})$. Let $\lambda_1(U,\delta)$ be the infimum of its spectrum and $d_0(U) = \inf_{0<\delta<1} \lambda_1(U,\delta)$. The function U is stable (resp. strictly stable, unstable) if $d_0(U) \geq 0$ (resp. $d_0(U) > 0$, $d_0(U) < 0$). If $U(x) = u_s(x) = \ln\left(1/|x|^2\right)$, the operator $|x|^2 L_{U,\delta}$ is of Fuschian type. A straightforward computation shows that u_s is stable if $N \geq 10$, strictly stable if $N > 10$ and unstable if $2 < N < 10$.

Step 2. The linearized operator.
Let us suppose that $N > 10$ and for $\delta > 0$ set $\Omega_\delta^\varepsilon = \left\{x \in \Omega^\varepsilon : |x - p_i| > \delta,\ \forall 1 \leq i \leq k\right\}$ and define a positive C^4 function g by

$$g(x) = |x - p_i| \quad \text{on } B(p_i,1),\quad g(x) = 1 \text{ on } \Gamma(\varepsilon). \tag{3.5.9}$$

The operator $g^2 L$ is uniformly elliptic and self-adjoint in $L^2(\Omega_\delta^\varepsilon, g(x)^{-2}dx)$ with smallest eigenvalue $\lambda_1(g, \Omega_\delta^\varepsilon)$. From Step 1 there exist $\overline{d} > 0$ and $\overline{\varepsilon} > 0$ depending of the previous data $d_0(u_s)$, c_i and $|\Omega^{\overline{\varepsilon}}|$ such that $\lambda_1(g, \Omega_\delta^\varepsilon) > \overline{d}$ and

$$\int_{\Omega_\delta^\varepsilon}\left(|\nabla v|^2 + v^2 g^{-2}\right)dx \leq \overline{d}^{-2}\int_{\Omega_\delta^\varepsilon}(Lv)^2 g^2 dx, \tag{3.5.10}$$

for any $v \in C^2(\overline{\Omega}_\delta^\varepsilon) \cap C_0(\overline{\Omega}_\delta^\varepsilon)$. From this estimate it can be derived that, for any f $\in C^{0,\alpha}(\Omega^\varepsilon) \cap L^2(\Omega^\varepsilon, g^2 dx)$ and $\Psi \in C^{2,\alpha}(\overline{\Omega}^\varepsilon)$ vanishing in a neighborhood of the $p_i, 1 \leq i \leq k$, there exists a unique $v \in C^{2,\alpha}(\Omega^\varepsilon) \cap L^2(\Omega^\varepsilon, g^{-2}dx)$ satisfying

$$\begin{cases} Lv(x) = f(x) & \text{in } \Omega^\varepsilon, \\ v = \Psi & \text{on } \partial\Omega^\varepsilon. \end{cases} \tag{3.5.11}$$

Moreover

$$\int_{\Omega_\delta^\varepsilon}\left(|\nabla v|^2 + v^2 g^{-2}\right)dx \leq \overline{d}^{-2}\int_{\Omega_\delta^\varepsilon}\left[\left(f^2 + (L\Psi)^2\right)g^2 + \left(\Psi^2 + |\nabla\Psi|^2\right)g^{-2}\right]dx. \tag{3.5.12}$$

Step 3. Local behaviour near the singularities.
The technique of Smale is to introduce the space of h-time differentiable functions ψ in Ω^ε which decay like $|x - p_i|^{v_i}$ near p_i. Set $v = (v_1,\ldots,v_k)$ and $\Gamma_{r,2r}^i = \{x \in B(p_i, 1): r < |x - p_i| < 2r\}$, then

$$C_v^{h,\alpha}(\Omega^\varepsilon) = \left\{ v \in C^{h,\alpha}(\Omega^\varepsilon): \max_{0<r<1/2}\sum_{|\gamma|=0}^{h}\left\|D^\gamma v\right\|_{C(\overline{\Gamma}_{r,2r}^i)} r^{|\gamma|-v_i} \right.$$
$$\left. + \max_{0<r<1/2}\sum_{|\gamma|=0}^{h}\left\|D^\gamma v\right\|_{C^{0,\alpha}(\overline{\Gamma}_{r,2r}^i)} r^{|\gamma|+\alpha-v_i} < \infty, \ i = 1,\ldots k \right\}. \tag{3.5.13}$$

Moreover $C_v^{h,\alpha}(\Omega^\varepsilon)$ is a Banach space with the above norm. If λ^j is the j-th eigenvalue of the Laplace-Beltrami operator on S^{N-1} and

$$\gamma_\pm^j = \frac{1}{2}\left(2 - N \pm \sqrt{(N-2)^2 - 4\left(2(N-2) + \lambda - \lambda^j\right)}\right), \tag{3.5.14}$$

then the $\gamma_\pm^j$ are real since $N > 10$. Moreover, for any $i = 1,\ldots, k$, there exists $J_i \in \mathbb{N}$ such that $\gamma_+^{J_i} < v_i \leq \gamma_+^{J_i+1}$. Set $\overline{v} = v - (3/2,\ldots,3/2)$; if f belongs to $C_{\overline{v}}^{0,\alpha}(\Omega^\varepsilon)$ let $f_i^j(r,\sigma)$ denote the projection of the restriction of f to $B(p_i, 1)$ onto the j-th eigenspace of the Laplace Beltrami operator and define $F_i^j(f)$ by

$$F_i^j(f)(r,\sigma) = \begin{cases} r^{\gamma_+^j}\int_0^r \tau^{1-N-2\gamma_+^j}\int_0^\tau s^{N-1+\gamma_+^j}f_i^j(s,\sigma)ds d\tau & \text{if } j \leq J_j, \\ r^{\gamma_+^j}\int_1^r \tau^{1-N-2\gamma_+^j}\int_0^\tau s^{N-1+\gamma_+^j}f_i^j(s,\sigma)ds d\tau & \text{if } j > J_j. \end{cases} \tag{3.5.15}$$

The projection of the equation $Lv = f$ onto the eigenspaces of $-\Delta_{S^{N-1}}$ allows us to obtain a very precise asymptotic expansion of v near each p_i and more precisely the following series is convergent in $L^2(S^{N-1})$

$$v(p_i + r\sigma) = \sum_{j=0}^{\infty}\left(\alpha_i^j(\sigma)r^{\gamma_+^j} + F_i^j(f)(r,\sigma)\right), \tag{3.5.16}$$

with the α_i^j belonging to $\mathrm{Ker}\left(-\Delta_{S^{N-1}} - \lambda^j I\right)$. Moreover $\sum_{j=0}^{\infty}\left\|\alpha_i^j\right\|_{L^2(S^{N-1})}^2 < \infty$ and

$$\left\|\sum_{j>J_i}\left(\alpha_i^j(.)r^{\gamma_+^j} + F_i^j(f)(r,.)\right)\right\|_{L^2(S^{N-1})}^2$$

$$\leq c\left(\int_0^1 s^{3-2\nu_i}\int_{S^{N-1}} f^2(s,\sigma)\,d\sigma ds + \sum_{j>J_i}\int_{S^{N-1}}(\alpha_i^j(\sigma))^2\,d\sigma\right)r^{2\nu_i}, \tag{3.5.17}$$

where c is some positive constant.

Step 4. End of the proof. Since the problem is to find v as a fixed point of the mapping

$$v \mapsto L^{-1}\left(\xi_0 - 2(N-2)E(v)\right) \tag{3.5.18}$$

the first question deals with the solvability of

$$\begin{cases} Lw = \xi_0 - 2(N-2)E(v) & \text{in } \Omega^\varepsilon, \\ w = \Psi & \text{on } \partial\Omega^\varepsilon, \end{cases} \tag{3.5.19}$$

and, for that, we have to control the terms α_i^j for $j \leq J_i$ in the expansion of the solution of (3.5.11) and more precisely we need to prove that we can find conditions on Ψ in order the α_i^j be 0 for $j \leq J_i$. This choice is possible although extremely technical; it is proved in [Pc1] with the help of [Sm2] that there exists a closed subspace $K(\varepsilon,\alpha)$ of $C_v^{2,\alpha}(\Omega^\varepsilon)$ such that if Ψ satisfies the above mentioned conditions, and if v belongs to $K(\varepsilon,\alpha)$, then the solution w of (3.5.19) belongs also to $K(\varepsilon,\alpha)$. Moreover the mapping $v \mapsto w$ is compact in the $C_v^{2,\alpha}(\Omega^\varepsilon)$-topology. An application of Schauder Theorem implies the existence of a fixed point .

Remark 3.19. This method extends to the construction of positive solutions of

$$-\Delta u = u^p \tag{3.5.20}$$

which are singular on a finite subset $\Sigma = \{p_i\}_{i=1}^k$ of in $\mathbb{R}^N$, and satisfy

$$\lim_{x \to p_i} |x - p_i|^{2/(p-1)} = \left[\frac{2}{p-1} \left(N - \frac{2p}{p-1} \right) \right]^{1/(p-1)} \tag{3.5.21}$$

in the range of values $N > 2$ and

$$\frac{N}{N-2} < p < p^*, \tag{3.5.22}$$

where p^* is the root larger than 1 of

$$(N-2)^2 - \frac{8X}{X-1} \left(N - \frac{2X}{X-1} \right) = 0. \tag{3.5.23}$$

Under this condition on p, the singular solution of (3.5.20)

$$x \mapsto u_s(x) = \left[\frac{2}{p-1} \left(N - \frac{2p}{p-1} \right) \right]^{1/(p-1)} |x|^{-2/(p-1)} \tag{3.5.24}$$

is strictly stable in the sense of Theorem 3.18. Theorem 3.20 below is much more general.

When $p = N / (N-2)$ Pacard [Pc2] the existence of solutions with a more complicated singular set.

THEOREM 3.19. *Suppose* $N > 2$, Ω *is a connected bounded open subset of* $\mathbb{R}^N$ *and* $\Sigma = \{p_i\}_{i \in I}$ *is a set of points in* Ω *of one of the following types*
(I) *I is finite or*
(II) $I = \mathbb{N}^*$, $\lim_{i \to \infty} p_i = p_0 \in \Omega$ *and* $|p_i - p_0| \geq 3|p_{i+1} - p_0|$ *for any* $i \geq 1$.
Then there exist at least two weak solutions of

$$-\Delta u = u^{N/(N-2)} \quad in \quad \Omega \tag{3.5.25}$$

which are singular on the p_i *and vanish on* $\partial \Omega$.

The technique of the proof, much simpler than for Theorem 3.18, is a variational method. Very recently, Mazzeo and Pacard have produced a beautiful article [MP] to prove the existence of positive solutions of (3.5.20) with a finite or continuous singular set.

THEOREM 3.20. *Suppose* $N > 2$, Ω *is a connected bounded open subset of* $\mathbb{R}^N$ *with a smooth boundary and* $\Sigma = \{p_i\}_{i=1}^k$ *is a set of points in* Ω. *Suppose also that*

$$N / (N-2) \leq p < (N+2) / (N-2). \tag{3.5.26}$$

Then there exists a k-pamameters family of weak solutions of (3.5.20) *in* Ω *vanishing on* $\partial\Omega$ *with singular set* Σ. *In fact the solution space of this equation is locally a k-dimensional real analytic manifold.*

This result has been extended to cover the case where Σ is the union of disjoint, complete and compact submanifolds of Ω.

THEOREM 3.21. *Suppose* $N > 2$, Ω *is a connected bounded open subset of* $\mathbb{R}^N$ *with a smooth boundary and* $\Sigma = \left\{\Sigma_i\right\}_{i=1}^{k}$ *where* Σ_i *is a compact and complete* d_i*-dimensional submanifold of* Ω *with* $\Sigma_i \cap \Sigma_j = \varnothing$ *if* $i \neq j$ *and*

$$\frac{N-d_i}{N-2-d_i} \leq p < \frac{N+2-d_i}{N-2-d_i} \tag{3.5.27}$$

for $i = 1,...,k$. *Then there exists a weak solution* u *of* (3.5.20) *in* Ω *vanishing on* $\partial\Omega$ *with singular set* Σ. *In fact if one of the* d_i *is positive there is an infinite dimensional manifold of solutions.*

The method for proving these results is a very refined analysis of the linearized form of some operators of the type introduced in Theorem 3.18 near the singular set Σ. In the same article the authors treat the problem of finding complete metric on subdomains of S^N or of a N-dimensional compact complete manifold (M,g) with constant positive scalar Scal_g. In that case the equation for the conformal factor $u^{2/(N-2)}$ is just

$$-4\frac{N-1}{N-2}\Delta_g u + \mathrm{Scal}_g u - \mathrm{Scal}_{g_u} u^{(N+2)/(N-2)} = 0. \tag{3.5.28}$$

As a consequence of Theorem 3.21 they proved

THEOREM 3.22. *Suppose* $N > 2$, (M,g) *is a compact complete N-dimensional manifold with constant positive scalar curvature. Let* $\Sigma = \left\{\Sigma_i\right\}_{i=1}^{k}$ *be a finite disjoint union of* C^∞, d_i*-dimensional submanifolds of* M *with*

$$0 < d_i \leq (N-2)/2. \tag{3.5.29}$$

Then there is an infinite dimensional family of complete metrics $g_u = u^{4/(N-2)}g$ *on* $M \setminus \Sigma$ *with constant positive scalar curvature* Scal_{g_u}. *Equivalently there is an infinite dimensional family of positive solutions of* (3.5.29) *in* $M \setminus \Sigma$ *such that*

$$\int_0^1 u^{2/(N-2)}(\gamma(t))dt = \infty, \tag{3.5.30}$$

for any $\gamma \in C^\infty([0,1], M)$ *with* $|\gamma'(t)| = 1$ *and* $\gamma([0,1)) \subset M \setminus \Sigma$ *and* $\gamma(1) \in \Sigma_i$ *for some* i.

Previous article on the complete metrics problem have been published before by Mazzeo and Smale [MS], Schoen and Yau [SY] and Schoen [Sc]. In Schoen's article the case of the N-dimensional sphere is treated and this complete Theorem 3.22 since he proved the following

THEOREM 3.23. *Suppose* $N > 2$, $\Sigma = \{p_i\}_{i=1}^{k}$ *is a finite set of at least two points in the standard sphere* (S^N, g). *Then there exists a complete metric on* $M \backslash \Sigma$ *with constant positive scalar curvature.*

Remark 3.20. The result of Theorem (3.18) has been extended by Horsin-Molinaro [HM] to the case where the singular set of the solution of (3.5.1) is not a finite collection of points but a d-dimensional manifold Σ with some algebraic structure. In that case N must be larger than $10 + d$. In the opposite direction, if codim $(\Sigma) = 2$, local singular solutions of (3.5.20) in $\Omega \backslash \Sigma$ are constructed by Grillot [Gr1]. The codim $(\Sigma) = 3$-case remains completely open for this equation as well as the small codimension case for large value of p for equation (3.5.20); for example we believe that the case when codim $(\Sigma) = N$ and $p = (N+1)/(N-3)$ have some similarity with the codim $(\Sigma) = 3$-case of equation (3.5.1).

3-3-6 The asymptotic symmetry method

The method presented in this section has been introduced by Caffarelli, Gidas and Spruck ([CGS]) in particular to describe the isolated singularities of the positive solutions of

$$-\Delta u = u^{(N+2)/(N-2)} \tag{3.6.1}$$

in a punctured N-dimensional domain. In its full development, it is based on the moving planes technique. The following result has first been proved by Caffareli, Gidas and Spruck in [CGS], but their proof is made much simpler by a device due to C.M. Li.

THEOREM 3.24. *Suppose* $N > 2$, $1 < p \le (N+2)/(N-2)$ *and* u *is a nonnegative solution of* (3.5.20) *in* $\mathbb{R}^N \backslash \{O\}$.
(i) *if the origin is a non-removable singularity, then* u *is radially symmetric with about the origin* ;
(ii) *if the origin is a removable singularity, then* u *is radially symmetric about some point in* $\mathbb{R}^N$.

Proof. We shall suppose that O is a non-removable singularity, the other case being a simple adaptation of the proof below. Since u is superharmonic, nonzero and nonnegative in $B^*(O,1) = B(O,1) \backslash \{O\}$, we deduce from the Brezis-Lions Lemma that

$$u(x) \ge \min_{|y|=1} u(y) = \beta > 0 \qquad (\forall x \in B^*(O,1)). \tag{3.6.2}$$

Let $z \in \mathbb{R}^N \backslash \{O\}$, close to O, we perform a Kelvin transform about z and set

$$x = \frac{y-z}{|y-z|^2}, \qquad y = z + \frac{x}{|x|^2}, \qquad v(x) = |x|^{|2-N} u(y). \tag{3.6.3}$$

Then v satisfies

$$-\Delta v = |x|^{p(N-2)-(N+2)} v^p \tag{3.6.4}$$

in $\mathbb{R}^N \setminus \{O\} \cup \{-z/|z|^2\}$. The key argument is to prove that v admits the axis (O, z) as an axis of symmetry. For that we consider a direction of reflection τ orthogonal to this axis and for simplicity we suppose τ is the negative x_1 direction. For $\lambda \le 0$ we set

$$\Sigma_\lambda = \left\{ x = (x_1,...,x_N) : x_1 < \lambda \right\}, \qquad T_\lambda = \partial \Sigma_\lambda, \tag{3.6.5}$$

$x^\lambda = (2\lambda - x_1, x_2,...,x_N)$ is the symmetric point with respect to the plane T_λ,

$$v_\lambda(x) = v(x^\lambda) = v(2\lambda - x_1, x_2,...,x_N), \tag{3.6.6}$$

$$w_\lambda(x) = v_\lambda(x) - v(x). \tag{3.6.7}$$

The function w_λ is singular at O, at $x_\lambda = (2\lambda, 0,...,0)$, at $-z/|z|^2$ and $Z_\lambda = (z/|z|^2)^\lambda$. Denote $\tilde{\Sigma}_\lambda = \Sigma_\lambda \setminus \{x_\lambda, Z_\lambda\}$.

Step 1. We claim that there exists $\lambda_0 < 0$ such that $w_\lambda \ge 0$, $\forall x \in \tilde{\Sigma}_\lambda$, for any $\lambda \le \lambda_0$. We introduce a positive function g defined in Σ_0 such that for any $C > 0$ we have

$$C|x|^{-4} + \frac{\Delta g(x)}{g(x)} < 0 \tag{3.6.8}$$

for $|x|$ large enough, for example the function $g(x) = \ln(3 - x_1)$, and set $\overline{w}_\lambda(x) = w_\lambda(x) / g(x)$. We deduce also from the Brezis-Lions Lemma that

$$v(x) \ge \min_{|y|=1} v(y) = \alpha > 0 \qquad \left(\forall x \in B^*(O,1) \right). \tag{3.6.9}$$

Since $v(x)$ goes to 0 when $|x|$ goes to infinity, there exists $\lambda_1 < -1$ such that $v(x) \le \alpha$ for any $x \in B(x_\lambda, 1)$. For $\lambda \le \lambda_1$ we denote $m_\lambda = \inf_{x \in \Sigma_\lambda} \overline{w}_\lambda(x)$; m_λ is finite since v is nonnegative and regular at Z_λ.

Step 1-1. Let us suppose first that $m_\lambda < 0$. We claim that m_λ is achieved.

If x belongs to $B(x_\lambda,1)$, then x^λ belongs to $B(O,1)$ and $v(x^\lambda) \geq \alpha \geq v(x)$. Therefore $w_\lambda(x) \geq 0$ and it is the same with $\overline{w}_\lambda(x)$. Moreover $\liminf\limits_{|x|\to\infty} w_\lambda(x) \geq -\lim\limits_{|x|\to\infty} v(x) = 0$. Therefore there exists $A > 0$ such that $w_\lambda(x) > m_\lambda/2 > m_\lambda$ if $|x| \geq A$. Finally w_λ vanishes on $T_\lambda \cap \overline{B}(O,A)$ and there exists $b > 0$ such that $w_\lambda(x) > m_\lambda/2 > m_\lambda$ in $\overline{B}(O,A) \cap \{x:\ \lambda - b \leq x_1 \leq \lambda\}$. Denote $\tilde{\Sigma}_\lambda^* = \Sigma_\lambda \cap \overline{B}(O,A) \cap B^c(x_\lambda,1)$, then the infimum of $\overline{w}_\lambda$ cannot be obtained as the limit of $\overline{w}_\lambda(x_n)$ for some sequence $\{x_n\}$ converging to Z_λ since w_λ is singular and bounded below by 0 in $B(Z_\lambda,1)$, if we take z small enough. Therefore m_λ is achieved at some point $\tilde{x}$ in $\tilde{\Sigma}_\lambda^* = \Sigma_\lambda \cap \overline{B}(O,A) \cap B^c(x_\lambda,1) \cap B^c(Z_\lambda,1)$.

Step 1-2. We claim that there exists $R_0 > 0$, independent of λ, such that at any point $\tilde{x}$ with $m_\lambda = \inf\limits_{x \in \Sigma_\lambda} \overline{w}_\lambda(\tilde{x}) = \inf\limits_{x \in \tilde{\Sigma}_\lambda^*} \overline{w}_\lambda(\tilde{x}) < 0$, we have $|\tilde{x}| \leq R_0$.

For any $x \in \Sigma_\lambda$ and $\lambda \leq 0$ we have

$$
\begin{aligned}
-\Delta v_\lambda(x) = -\Delta v(x^\lambda) &= |x^\lambda|^{p(N-2)-(N+2)} v^p(x^\lambda), \\
&\geq |x|^{p(N-2)-(N+2)} v_\lambda^p(x),
\end{aligned}
\tag{3.6.10}
$$

since $|x| \geq |x^\lambda|$ and $1 < p \leq (N+2)/(N-2)$. Therefore

$$
\begin{aligned}
-\Delta w_\lambda(x) &\geq |x|^{p(N-2)-(N+2)}\left(v^p(x^\lambda) - v^p(x)\right), \\
&\geq c(x) w_\lambda(x),
\end{aligned}
\tag{3.6.11}
$$

where $c(x) = p|x|^{p(N-2)-(N+2)}\psi(x)$ and $\psi(x) = \theta v_\lambda(x) + (1-\theta)v(x)$ with $\theta = \theta(x) \in [0,1]$. Consequently

$$
-\Delta \overline{w}_\lambda - 2g^{-1}\nabla g.\nabla \overline{w}_\lambda \geq \left(c + g^{-1}\Delta g\right)\overline{w}_\lambda
\tag{3.6.12}
$$

in $\tilde{\Sigma}_\lambda$. From (3.6.3) $\lim\limits_{|x|\to\infty} |x|^{N-2} v(x) = u(z)$. Therefore, a straightforward verification shows that $v_\lambda(x) \leq C|x|^{2-N}$ for some $C > 0$, independent of λ, for any x in the set $\tilde{\Sigma}_\lambda^* \cap \{x:\ |x| \geq 1\}$ and there exists $C' > 0$ such that $c(x) \leq C'|x|^{-4}$ in the same set. Consequently

$$
C'|x|^{-4} + g^{-1}(x)\Delta g(x) < 0 \qquad \left(\forall x \in \tilde{\Sigma}_\lambda^* \cap \{x:\ |x| \geq R\}\right),
\tag{3.6.13}
$$

for some $R > 0$. From the strict maximum principle $\tilde{x}$ cannot be a point of minimum of $\overline{w}_\lambda$ if $|\tilde{x}| \geq R$.

Step 1-3. The claim of step 1 is fulfilled if we suppose that $m_\lambda \geq 0$.

Step 2. Denote Λ the set of the $\mu < 0$ such that Step 1 holds for any $\lambda \in (-\infty, \mu]$. Since v is superharmonic and smooth in Σ_0, the function w_λ is lower semicontinuous and the set Λ is closed. In order to see that it is open, we suppose that $w_{\lambda^*}(x) \geq 0$ for any x in $\tilde{\Sigma}_{\lambda^*}$. If Λ is not open there exist a sequence $\{\lambda_n\}$ converging to λ^* and a sequence of points $\{x^n\}$ in Σ_{λ_n} such that $v(x^{n \lambda_n}) < v(x^n)$. As in step 1 it is possible to prove that, up to a subsequence still quoted $\{x^n\}$, $\lim_{n \to \infty} x^n = x^*$, with $x^* \in \tilde{\Sigma}_{\lambda^*}$ and $v(x^*_{\lambda^*}) = v(x^*)$.

Case 1. $x_1^* = \lambda^*$ and $\dfrac{\partial v}{\partial x_1}(x_1^*) \geq 0$. In that case w_{λ^*} is a nonnegative, smooth, superharmonic function in a small half-ball $B^-(x^*, \varepsilon) = B(x^*, \varepsilon) \cap \{x: x_1 < \lambda^*\}$, which vanishes on $x_1 = \lambda^*$. From Hopf boundary lemma

$$\frac{\partial w_{\lambda^*}}{\partial x_1}(x^*) = -2 \frac{\partial v}{\partial x_1}(x^*) < 0, \tag{3.6.14}$$

which yields a contradiction.

Case 2. $x_1^* > \lambda^*$. Since w_{λ^*} is superharmonic and nonnegative in a neighborhood of x^* and vanishes at x^*, we also obtain a contradiction.

Consequently Λ is open and $\Lambda = (-\infty, 0)$.

Step 3. End of the proof. From Step 2 $v(x^\lambda) \geq v(x)$ for any $\lambda \leq 0$ and any x in Σ_λ. If we proceed in th same way with $\lambda \geq 0$ we deduce that $v(x^\lambda) = v(x)$ and therefore $\dfrac{\partial u}{\partial x_1} = 0$ on T_0. But the direction x_1 can be replaced by any direction orthogonal to (O, z) and therefore u is axisymmetric with axis (O, z). Since z is arbitrary, u is radially symmetric with respect to O.

A classical ordinary differential equations analysis implies the following consequence of Theorem 3.24.

COROLLARY 3.3. *Under the assumptions of Theorem 3.24, we have the following:*
 I- *if* u *is a nonnegative solution of* (3.5.20) *in* $\mathbb{R}^N \setminus \{O\}$ *with a non removable singularity at* O, *then*
 (i) if $1 < p \leq N / (N - 2)$, u *cannot exist,*

(ii) if $N / (N - 2) < p < (N + 2) / (N - 2)$, $u(x) = \left[\left(\dfrac{2}{p-1} \right) \left(N - \dfrac{2p}{p-1} \right) \right]^{1/(p-1)} |x|^{-2/(p-1)}$,

(iii) if $p = (N + 2) / (N - 2)$, $u(x) = |x|^{-(N-2)/2} \phi(|x|)$ *where* ϕ *is a periodic positive solution of*

$$\phi'' - (N(N-2)/4)\phi + \phi^{(N+2)/(N-2)} = 0. \tag{3.6.15}$$

II- *if* u *is a nonnegative solution of* (3.5.20) *in* $\mathbb{R}^N$, *then*

(iv) if $1 < p < (N+2)/(N-2)$, u *is identically* 0,

(v) if $p = (N+2)/(N-2)$, $u(x) = \left(\dfrac{\sigma\sqrt{N(N-2)}}{\sigma^2 + |x|^2} \right)^{(N-2)/2}$ *, for some* $\sigma > 0$.

Another application of the moving planes method due to Chen and Li is the following

THEOREM 3.25. *Suppose* $N = 2$ *and   is any solution of*

$$-\Delta u = e^u \tag{3.6.16}$$

in $\mathbb{R}^2$ *such that* $\int_{\mathbb{R}^2} e^u dx < \infty$, *then* u *is radial with respect to some point* a *in* $\mathbb{R}^2$ *and in fact there exists* $\lambda > 0$ *such that*

$$u(x) = \ln(32\lambda^2)\left(4 + \lambda^2|x - a|^2\right)^2, \qquad \left(\forall x \in \mathbb{R}^2\right). \tag{3.6.17}$$

Proof. In that case there is no Kelvin transform, but the estimate $\int_{\mathbb{R}^2} e^u dx < \infty$ implies that u is bounded from above by a result of Brezis and Merle [BM]. Moreover the use of the isoperimetric inequality (Ding) yields

$$\int_{\mathbb{R}^2} e^u dx \geq 8\pi. \tag{3.6.18}$$

Considering $x \mapsto w(x) = \dfrac{1}{2\pi} \int_{\mathbb{R}^2} \left(\ln(|x - y|) - \ln(|y|)\right)e^{u(y)}dy$, then $\Delta w(x) = e^{u(x)}$ and

$$\lim_{|x|\to\infty} \frac{w(x)}{\ln(|x|)} = \frac{1}{2\pi} \int_{\mathbb{R}^2} e^{u(y)}dy. \tag{3.6.19}$$

Setting $v(x) = w(x) + u(x)$, then v is harmonic in the plane and

$$v(x) \leq C + C' \ln(|x| + 1) \tag{3.6.20}$$

for some positive constants C and C', which implies that v is constant and

$$\lim_{|x|\to\infty} \frac{u(x)}{\ln(|x|)} = -\frac{1}{2\pi} \int_{\mathbb{R}^2} e^{u(y)}dy. \tag{3.6.21}$$

With this accurate estimate it is possible to start the moving plane method from infinity and to derive the radial symmetry of u as in Theorem 3.24.

The original result of Cafarelli, Gidas and Spruck is a local one and reads as follows

THEOREM 3.26. *Suppose* $N > 2$, g *is a nondecreasing, locally Lipschitz continuous function defined on* $\mathbb{R}^+$ *vanishing at* 0 *and such that*

(i) $t \mapsto t^{-(N+2)/(N-2)} g(t)$ *is nonincreasing,*

(ii) $g(t) \geq Ct^p$ *holds for any* $t \in \mathbb{R}^+$, *for some constants* $C > 0$ *and* $N/(N-2) \leq p$.

If u *is any nonnegative* $C^2(B^*(O,1))$-*function where* $B^*(O,1) = B(O,1) \setminus \{O\}$, *which satisfies*

$$-\Delta u = g(u) \tag{3.6.22}$$

in $B^*(O,1)$ *and has a non-removable singularity at* O, *then* u *is asymptotically radial near* O *in the sense that*

$$u(x) = \overline{u}(|x|)(1 + o(|x|)). \tag{3.6.23}$$

where $\overline{u}(|x|)$ *is the spherical average of* u. *More precisely there exists a positive radial solution* v *of* (3.6.15) *such that*

$$u(x) = v(|x|)(1 + o(|x|)). \tag{3.6.24}$$

The proof of this result is much more delicate and involves the kelvin transform about points in the neighborhood of O, a reflection theorem and a geometric measure argument on the set of the admissible directions in which the reflection is possible. Later on their proof was simplified by Li [Li1].

We end this paragraph with the recent result due to Chen and Lin [CL] dealing with the *a priori* estimate of any positive solution of (3.6.1) near a non-removable singular set with zero $c_{1,2}$-capacity.

THEOREM 3.27. *Suppose* $N > 2$, g *is a nondecreasing, locally Lipschitz continuous function defined on* $\mathbb{R}^+$ *vanishing at* 0 *and such that* $t \mapsto t^{-(N+2)/(N-2)} g(t)$ *is nonincreasing, with limit* 1 *at infinity. Let* Ω *be an open subset of* $\mathbb{R}^N$ *with* $\overline{B}(O,2R) \subset \Omega$ *and* $Z \subset B(O,R)$ *be a closed subset with* $\mathrm{cap}_{1,2}(Z) = 0$, *the capacity being relative to* B(O, 2R). *If* u *is any* $C^2(\Omega \setminus Z)$ *nonnegative function satisfying* (3.6.22) *in* $\Omega \setminus Z$ *and minorized by some* $m > 0$ *on* $\Omega \setminus Z$, *then*

$$u(x) \leq C(\mathrm{dist}(x,Z))^{-(N-2)/2} \qquad (\forall x \in B(O,R) \setminus Z), \tag{3.6.25}$$

for some positive constant C *depending on* R, m *and* N.

In the case where Z is a smooth k-dimensional compact submanifold $(0 \leq k \leq N-2)$ of Ω without boundary, the behaviour of a solution u of (3.6.22) near Z can be described precisely. Let N be a tubular neiborhood of Z such that any point x of N can be expressed uniquely as $x = y + v$ where $y \in Z$ and $v \in N_y Z = (T_y Z)^{\perp}$ which is the normal space to Z at y. Let Π be the orthogonal projection of N onto Z. For $r > 0$ small enough and $z \in Z$,

$$\Pi^{-1}(z) = \left\{ y \in N : \Pi(y) = z \text{ and } |y - z| = r \right\}. \tag{3.6.26}$$

The following result [CL] extends Theorem 3.26

THEOREM 3.28. *Under the assumptions of Theorem 3.27 and with the previous notations where* Z, N *and* Π *are as above, then for any positive solution* u *of* (3.6.21) *in* $\Omega \setminus Z$ *there holds*

$$u(x) = u(x')(1 + o(1)) \tag{3.6.27}$$

for x *and* x' *in* $\Pi^{-1}(z)$, *as* r *goes to* 0.

The deep meaning of this result is that the behaviour of u is essentially "radially symmetric with axis Z" near Z.

4 Boundary singularities of solutions of elliptic equations

4-1 Equations with absorption term

Let Ω be a open subset of $\mathbb{R}^N$ ($N > 1$), whose boundary $\partial\Omega$ contains O. The boundary singularity problem for semilinear elliptic equation is the following: let g be a continuous real valued function and $u \in C^2(\Omega) \cap C(\overline{\Omega} \setminus \{O\})$ be a solution of

$$-\Delta u + g(u) = 0 \qquad (1.1)$$

in Ω, which coincides on $\partial\Omega \setminus \{O\}$ with a continuous function φ defined on whole $\partial\Omega$.

Then is it possible to extend u as a continuous function defined in whole $\overline{\Omega}$? If it is not possible, can we described the behaviour of u near O ? In the absorption term case the function g is essentially nondecreasing, and a particular emphasis is made when $g(r) = r|r|^{q-1}$, $q > 1$. Much work on these problems has been done by Gmira and Véron [GV] in the regular case and Fabbri and Véron [FV] in the nonregular case.

4-1-1 The model example

Denote G the half space $\mathbb{R}^1_* \times \mathbb{R}^{N-1}$ and ∂G its boundary. We first look for solutions of

$$-\Delta u + |u|^{q-1} u = 0 \qquad (1.1.1)$$

($q > 1$) in G which vanishes on $\partial G \setminus \{O\}$, under the form

$$u(r, \sigma) = r^{-2/(q-1)} \omega(\sigma) \qquad (1.1.2)$$

where $r > 0$ and $\sigma \in S^{N-1}_+$ which is the hemisphere of S^{N-1} contained in G. Then ω satisfies

$$-\Delta_{S^{N-1}} \omega - \frac{2}{q-1} \left(\frac{2q}{q-1} - N \right) \omega + |\omega|^{q-1} \omega = 0 \qquad (1.1.3)$$

in S^{N-1}_+ and vanishes on the relative boundary $\partial S^{N-1}_+ = S^{N-2}$. Since N-1 is the first eigenvalue of the Laplacian in $W^{1,2}_0(S^{N-1}_+)$, we obtain

$$\int_{S^{N-1}_+} \left[\left((N-1) - \frac{2}{q-1} \left(\frac{2q}{q-1} - N \right) \right) \omega^2 + |\omega|^{q+1} \right] d\sigma \leq 0 \qquad (1.1.4)$$

"

from (1.1.3). Therefore a necessary condition for the existence of nontrivial solutions of (1.1.3) is

$$\frac{2}{q-1}\left(\frac{2q}{q-1}-N\right) \geq N-1 \Leftrightarrow 1 < q < (N+1)/(N-1).$$ (1.1.5)

Moreover, if this condition is fulfilled, the functional

$$J(\psi) = \int_{S_+^{N-1}} \left(\frac{1}{2}|\nabla \psi|^2 - \frac{1}{q-1}\left(\frac{2q}{q-1}-N\right)\psi^2 + \frac{1}{q+1}|\psi|^{q+1}\right) d\sigma$$ (1.1.6)

achieves its minimum in $W_0^{1,2}(S_+^{N-1}) \cap L^{q+1}(S_+^{N-1})$ at some ω which is a positive solution of (1.1.3) in S_+^{N-1} vanishing on ∂S_+^{N-1} (and in fact it is unique). If we write $x = (x_1, x')$ the coordinates in $G = \mathbb{R}_*^+ \times \mathbb{R}^{N-1}$, the Poisson kernel at O is defined by

$$P(x, O) = P(x_1, x', O) = K(N)\frac{x_1}{(x_1^2 + x'^2)^{N/2}}.$$ (1.1.7)

P is a harmonic function in G, vanishes on $\partial G \setminus \{O\}$ and, if ζ belongs to the space $C_0^{1,1}(\overline{G})$ of continuously differentiable functions with bounded second derivatives and compact support in $\overline{G}$, there holds

$$-\int_G P(x, O)\zeta(x)dx = \frac{\partial \zeta}{\partial x_1}(O) = -\frac{\partial \zeta}{\partial v}(O).$$ (1.1.8)

As we shall see it in the next sections, the example given above is the Ariane's Thread to understand the singularity problem for (1.1.1).

4-1-2 Measure boundary data

We assume here that Ω is a domain in $\mathbb{R}^N$ with a smooth boundary $\partial\Omega$. We call $\mathcal{M}(\partial\Omega)$ the space of Radon measures on $\partial\Omega$ and set $\rho(x)$ the distance from a point x in Ω to $\partial\Omega$; $C_0^{1,1}(\overline{G})$ is the space of C^1 functions vanishing on $\partial\Omega$ with Lipschitz continuous gradient.

Definition. Let $\mu \in \mathcal{M}(\partial\Omega)$; we say that u satisfies the following Dirichlet problem with measure boundary data

$$\begin{cases} -\Delta u + g(u) = 0 & \text{in } \Omega, \\ u = \mu & \text{on } \partial\Omega, \end{cases}$$ (1.2.1)

if $u \in L^1(\Omega)$, $\rho g(u) \in L^1(\Omega)$ and

$$\int_{\Omega}(-u\Delta\zeta + g(u)\zeta)dx = -\int_{\partial\Omega}\frac{\partial\zeta}{\partial v}d\mu \tag{1.2.2}$$

for any $\zeta \in C_0^{1,1}(\overline{\Omega})$.

The general existence result is the following.

THEOREM 4.1. *Let* g *be a nondecreasing function satisfying*

$$\int_1^{\infty}\left(g(s) + |g(-s)|\right)s^{-2N/(N-1)}ds < \infty. \tag{1.2.3}$$

Then for any $\mu \in \mathcal{M}(\partial\Omega)$, *there exists a unique solution* u *to* (1.2.1). *Moreover* $u \in M^{(N+1)/(N-1)}(\Omega,\rho dx)\cap M^{N/(N-1)}(\Omega,dx)$ *and the mapping* $\mu \mapsto u$ *is nondecreasing.*

We first present a construction and some estimates due to Brezis $[Br2]$

LEMMA 4.1. *Let* f *be a measurable function in* Ω *such that* $\rho f \in L^1(\Omega)$ *and* $\varphi \in L^1(\partial\Omega)$. *Then there exists a unique* $u \in L^1(G)$ *such that*

$$-\int_{\Omega}u\Delta\zeta dx = \int_{\Omega}f\zeta dx - \int_{\partial\Omega}\varphi\frac{\partial\zeta}{\partial v}dS \tag{1.2.4}$$

for any $\zeta \in C_0^{1,1}(\overline{G})$. *Moreover there exists* $C = C(\Omega) > 0$ *such that*

$$\|u\|_{L^1(\Omega)} \leq C\left(\|\rho f\|_{L^1(\Omega)} + \|\varphi\|_{L^1(\partial\Omega)}\right) \tag{1.2.5}$$

and u *satisfies*

$$-\int_{\Omega}|u|\Delta\zeta dx + \int_{\partial\Omega}|\varphi|\frac{\partial\zeta}{\partial v}dS \leq \int_{\Omega}f\,\zeta\,\mathrm{sgn}(u)dx \tag{1.2.6}$$

for any $\zeta \in C_0^{1,1}(\overline{\Omega})$, $\zeta \geq 0$.

Proof. The existence is obtained by the closure of the same equation in $C_0^2(\Omega)\times C^2(\partial\Omega)$ through the estimate (1.2.5). Denote $\gamma(r)$ a smooth, odd and increasing approximation of $\mathrm{sign}(r)$ and set η the solution of

$$\begin{cases} -\Delta\eta = \gamma(u) & \text{in } \Omega, \\ \quad \eta = 0 & \text{on } \partial\Omega, \end{cases} \tag{1.2.7}$$

then

$$\int_\Omega u\gamma(u)dx = \int_\Omega f\eta dx - \int_{\partial\Omega} \varphi \frac{\partial\eta}{\partial v}dS. \tag{1.2.8}$$

Let β be the solution of

$$\begin{cases} -\Delta\beta = 1 & \text{in } \Omega, \\ \quad \beta = 0 & \text{on } \partial\Omega, \end{cases} \tag{1.2.9}$$

then $-\beta \leq \eta \leq \beta$ and $0 \leq \beta(x) \leq C\rho(x)$ in $\overline{\Omega}$ where $C = C(\Omega) \geq 0$. Therefore

$$\int_\Omega u\gamma(u)dx \leq C\int_\Omega |f|\rho dx + \left\|\frac{\partial\beta}{\partial v}\right\|_{L^\infty(\partial\Omega)} \int_{\partial\Omega} |\varphi|dS, \tag{1.2.10}$$

which gives (1.2.5) when $\gamma(r)$ goes to $\text{sign}(r)$.

For proving (1.2.6) we take $\tilde{\zeta} = \gamma(u)\zeta$ as a test function where $\zeta \in C_0^{1,1}(\overline{\Omega})$, $\zeta \geq 0$. Then

$$-\int_\Omega u\Delta(\zeta\gamma(u))dx = \int_\Omega f\zeta\gamma(u)dx - \int_{\partial\Omega} \varphi \frac{\partial}{\partial v}(\zeta\gamma(u))dS \tag{1.2.11}$$

and

$$\begin{aligned} -\int_\Omega u\Delta(\zeta\gamma(u))dx &= \int_\Omega \gamma(u)\nabla u.\nabla\zeta dx + \int_\Omega \zeta\gamma'(u)|\nabla u|^2 dx \\ &\quad - \int_{\partial\Omega} u\gamma(u)\frac{\partial\zeta}{\partial v}dS - \int_{\partial\Omega} u\gamma'(u)\zeta\frac{\partial u}{\partial v}dS. \end{aligned} \tag{1.2.12}$$

Denote $j(r) = \int_0^r \gamma(s)ds$, then

$$\int_\Omega \gamma(u)\nabla u.\nabla\zeta dx = \int_\Omega \nabla j(u).\nabla\zeta dx = -\int_\Omega j(u)\Delta\zeta dx + \int_{\partial\Omega} j(u)\frac{\partial\zeta}{\partial v}dS. \tag{1.2.13}$$

Since $\int_\Omega \zeta\gamma'(u)|\nabla u|^2 dx \geq 0$ and $-\int_{\partial\Omega} \varphi \frac{\partial}{\partial v}(\gamma(u)\zeta)dS = -\int_{\partial\Omega} \varphi\gamma(u)\frac{\partial\zeta}{\partial v}dS$, we obtain

$$-\int_\Omega j(u)\Delta\zeta dx + \int_{\partial\Omega} j(\varphi)\varphi\frac{\partial\zeta}{\partial v}dS \leq \int_\Omega f\,\zeta\gamma(u)dx. \tag{1.2.14}$$

Letting $\gamma(r)$ go to $\text{sign}(r)$ implies (1.2.6).

LEMMA 4.2. *Let* g *be a nondecreasing real valued function,* f *be a measurable function in* Ω *such that* $\rho f \in L^1(\Omega)$ *and* $\varphi \in L^1(\partial\Omega)$*. Then there exists a unique* $u \in L^1(\Omega)$ *such that* $\rho g(u) \in L^1(\Omega)$ *satisfying*

$$-\int_\Omega u \Delta\zeta dx + \int_\Omega g(u)\zeta dx = \int_\Omega f\zeta dx - \int_{\partial\Omega} \varphi \frac{\partial\zeta}{\partial\nu} dS \tag{1.2.15}$$

for any $\zeta \in C_0^{1,1}(\overline{\Omega})$*. Moreover if* u *and* $\hat{u}$ *are the solutions associated respectively to the data* (f,φ) *and* $(\hat{f},\hat{\varphi})$ *in* $L^1(\Omega) \times L^1(\partial\Omega)$*, the following estimate holds*

$$\|u - \hat{u}\|_{L^1(\Omega)} + \|\rho(g(u) - g(\hat{u}))\|_{L^1(\Omega)} \leq C\Big(\|\varphi - \hat{\varphi}\|_{L^1(\partial\Omega)} + \|\rho(f - \hat{f})\|_{L^1(\Omega)}\Big), \tag{1.2.16}$$

where $C = C(\Omega)$*, and the correspondence* $(\varphi, f) \mapsto u$ *is nondecreasing.*

Proof. By minimization procedure and elliptic equations regularity theory, it is clear that for any $(f_n, \varphi_n) \in L^2(\Omega) \times W^{3/2,2}(\partial\Omega)$ there exists a unique $u_n \in W^{2,2}(\Omega)$ solution of

$$\begin{cases} -\Delta u_n + g(u_n) = f_n & \text{in } \Omega, \\ \qquad u_n = \varphi_n & \text{on } \partial\Omega, \end{cases} \tag{1.2.17}$$

and from (1.2.6), there holds

$$-\int_\Omega |u_n - u_m| \Delta\zeta dx + \int_\Omega (g(u_n) - g(u_m))\zeta \, \text{sgn}(u_n - u_m) dx$$

$$\leq -\int_{\partial\Omega} |\varphi_n - \varphi_m| \frac{\partial\zeta}{\partial\nu} dS + \int_\Omega |f_n - f_m| \zeta \, dx \tag{1.2.18}$$

for any $\zeta \in C_0^{1,1}(\overline{\Omega})$, $\zeta \geq 0$. If we take in particular ζ as the solution of

$$\begin{cases} -\Delta\zeta = 1 & \text{in } \Omega, \\ \quad \zeta = 0 & \text{on } \partial\Omega, \end{cases} \tag{1.2.19}$$

then $C^{-1}\rho \leq \zeta \leq C\rho$, for some positive constant $C = C(\Omega)$. Since

$$\int_\Omega \zeta(g(u_n) - g(u_m))\text{sign}(u_n - u_m) dx = \int_\Omega |g(u_n) - g(u_m)| dx, \tag{1.2.20}$$

with the convention $\text{sgn}(0) = 0$, we deduce that the inequality

$$\int_\Omega |u_n - u_m| dx + C^{-1}\int_\Omega |g(u_n) - g(u_m)|\rho dx$$

$$\leq -\int_{\partial\Omega} |\varphi_n - \varphi_m| \frac{\partial\zeta}{\partial\nu} dS + C\int_\Omega |f_n - f_m|\rho dx, \tag{1.2.21}$$

holds, and the problem is closable in $L^1(\partial\Omega)$. For proving the monotonicity of $(\varphi, f) \mapsto u$, a simple adaption of the proof of (1.2.6) yields

$$-\int_\Omega u^+ \Delta\zeta\, dx + \int_{\partial\Omega} \varphi^+ \frac{\partial\zeta}{\partial v}\, dS \le \int_\Omega f\, \zeta\, \mathrm{sgn}^+(u)\, dx \qquad (1.2.22)$$

for any $\zeta \in C_0^{1,1}(\overline{\Omega})$, $\zeta \ge 0$ which implies

$$\int_\Omega |u - \hat{u}|\, dx + C^{-1}\int_\Omega |g(u) - g(\hat{u})|\rho\, dx \le -\int_{\partial\Omega}|\varphi - \hat{\varphi}|\frac{\partial\zeta}{\partial v}\, dS + C\int_\Omega |f - \hat{f}|\rho\, dx \qquad (1.2.23)$$

for given (φ, f) and $(\hat{\varphi}, \hat{f})$ in $L^1(\Omega, \rho dx) \times L^1(\partial\Omega)$.

Remark 4.1. If, in lemma 4.2, we assume $f \in L^p_{\mathrm{loc}}(\Omega)$, $1 < p < \infty$, then $u \in W^{2,p}_{\mathrm{loc}}(\Omega)$.

Let $P(x, y)$ be the Poisson kernel in $\Omega \times \partial\Omega$. If $\varphi \in C(\overline{\Omega})$ is harmonic in Ω, the following representation ([GT], [St1])

$$\varphi(x) = \int_{\partial\Omega} P(x, y)\varphi(y)\, dS(y) \qquad (1.2.24)$$

holds for any $x \in \Omega$. Moreover there exists $K = K(\Omega) > 1$ such that

$$K^{-1}\rho(x)|x - y|^{-N} \le P(x, y) \le K\rho(x)|x - y|^{-N} \qquad (1.2.25)$$

for any $(x, y) \in \Omega \times \partial\Omega$.

LEMMA 4.3. *Let φ be defined by* (1.2.24), *then the following estimates are valid*

$$\|\varphi\|_{M^{(N+1)/(N-1)}(\Omega, \rho dx)} \le C\|\varphi\|_{L^1(\partial\Omega)}, \qquad (1.2.26)$$

$$\|\varphi\|_{M^{N/(N-1)}(\Omega)} \le C\|\varphi\|_{L^1(\partial\Omega)}, \qquad (1.2.27)$$

for some positive constant C depending on Ω.

Proof. We can assume that φ is nonnegative and, for $\lambda > 0$ and $y \in \partial\Omega$, we define

$$A_\lambda(y) = \{x \in \Omega : P(x, y) > \lambda\}, \qquad (1.2.28)$$

$$m_\lambda(y) = \int_{A_\lambda(y)} \rho(x)\, dx. \qquad (1.2.29)$$

Since $A_\lambda(y) \subset \tilde{A}_\lambda(y) = \{x \in \Omega : \rho(x)|x - y|^{-N} > \lambda / K\}$, we have

$$m_\lambda(y) \le \tilde{m}_\lambda(y) = \int_{\tilde{A}_\lambda(y)} \rho dx. \tag{1.2.30}$$

Moreover $\tilde{A}_\lambda(y) \subset B\!\left(y, (\lambda/K)^{1/(N-1)}\right)$, which implies

$$\tilde{m}_\lambda(y) \le \int_{B\left(y,(\lambda/K)^{1/(N-1)}\right)} |x-y| dx = K'(N,K)\lambda^{-(N+1)/(N-1)} \tag{1.2.31}$$

and

$$\lambda^{(N+1)/(N-1)} m_\lambda(y) \le K'(N,K). \tag{1.2.32}$$

We fix $y \in \partial\Omega$, $\lambda_0 > 0$ and take ω a measurable subset of Ω; then

$$\int_\omega P(x,y)\rho(x)dx \le \lambda_0 \int_\omega \rho(x)dx + \int_{A_{\lambda_0}(y)} P(x,y)\rho(x)dx. \tag{1.2.33}$$

Since

$$\int_{A_{\lambda_0}(y)} P(x,y)\rho(x)dx = -\int_{\lambda_0}^{\infty} \lambda dm_\lambda(y) = \lambda_0 m_{\lambda_0}(y) + \int_{\lambda_0}^{\infty} m_\lambda(y)d\lambda$$

$$\le \lambda_0 m_{\lambda_0}(y) + K'(N,K)\int_{\lambda_0}^{\infty} \lambda^{-(N+1)/(N-1)} d\lambda, \tag{1.2.34}$$

we obtain

$$\int_{A_{\lambda_0}(y)} P(x,y)\rho(x)dx \le \lambda_0 m_{\lambda_0(y)} + \frac{N-1}{2} K'(N,K)\lambda_0^{-2/(N-1)}. \tag{1.2.35}$$

Plugging this into (1.2.33) and using (1.2.32) yields

$$\int_\omega P(x,y)\rho(x)dx \le \lambda_0 \int_\omega \rho(x)dx + \frac{N+1}{2} K'(N,K)\lambda_0^{-2/(N-1)}. \tag{1.2.36}$$

If we choose $\lambda_0 = \left(\int_\omega \rho(x)dx\right)^{-(N-1)/(N+1)}$, we conclude that

$$\int_\omega P(x,y)\rho(x)dx \le K''(N,K)\left(\int_\omega \rho(x)dx\right)^{2/(N+1)} \tag{1.2.37}$$

holds, which is an estimate of $P(.,y)$ in $M^{(N+1)/(N-1)}(\Omega,\rho dx)$ independent of y. Consequently, the following is derived from (1.2.24):

$$\int_{\omega}\varphi(x)\rho(x)dx = \int_{\omega}\int_{\partial\Omega}P(x,y)\varphi(y)dS(y)\rho(x)dx,$$

$$= \int_{\partial\Omega}\int_{\omega}P(x,y)\varphi(y)dS(y)\rho(x)dx, \qquad (1.2.38)$$

$$\leq \left(\int_{\partial\Omega}\varphi(y)dS(y)\right)\max_{y\in\partial\Omega}\int_{\omega}P(x,y)\rho(x)dx.$$

Thus

$$\int_{\omega}\varphi(x)\rho(x)dx \leq K''(N,K)\left(\int_{\omega}\rho(x)dx\right)^{2/(N-1)}\|\varphi\|_{L^1(\partial\Omega)}, \qquad (1.2.39)$$

which is (1.2.26). For proving (1.2.27), we set

$$n_{\lambda}(y) = \int_{A_{\lambda}(y)} dx \qquad (1.2.40)$$

and

$$n_{\lambda}(y) \leq K_3(N,K)\lambda^{-N/(N-1)}. \qquad (1.2.41)$$

If ω is any measurable subset of Ω, we obtain by the same technique as above

$$\int_{\omega}P(x,y)dx \leq K_3(N,K)\left(\int_{\omega}dx\right)^{1/N}, \qquad (1.2.42)$$

which implies

$$\int_{\omega}\varphi(x)dx = \int_{\omega}\int_{\partial\Omega}P(x,y)\varphi(y)dS(y)dx,$$

$$\leq \left(\int_{\partial\Omega}\varphi(y)dS(y)\right)\max_{y\in\partial\Omega}\int_{\omega}P(x,y)dx, \qquad (1.2.43)$$

$$\leq K_3(N,K)\left(\int_{\omega}dx\right)^{1/N}\|\varphi\|_{L^1(\partial\Omega)},$$

and this is (1.2.27).

Proof of Theorem 4.1. Existence. Let $\{\mu_n\}$ be a sequence in $C(\partial\Omega)$ converging to μ in the topology of $\mathcal{M}(\partial\Omega)$ and $\{u_n\}$ the corresponding sequence of solutions of

$$\begin{cases} -\Delta u_n + g(u_n) = 0 & \text{in } \Omega, \\ \qquad u_n = \mu_n & \text{on } \partial\Omega. \end{cases} \qquad (1.2.44)$$

Step 1. We claim that $\{u_n\}$ remains uniformly bounded in the space $M^{(N+1)/(N-1)}(\Omega,\rho dx)$ $\cap M^{N/(N-1)}(\Omega)$. Denote $\tilde{g}(r) = g(r) - g(0)$ and h the solution of

$$\begin{cases} -\Delta h = |g(0)| & \text{in } \Omega, \\ \quad h = 0 & \text{on } \partial\Omega. \end{cases} \tag{1.2.45}$$

We define v_n by

$$v_n(x) = \int_{\partial\Omega} P(x,y)|\mu_n(y)|dS(y) \tag{1.2.46}$$

and set $w_n = v_n + h$. Since w_n satisfies

$$-\Delta w_n + \tilde{g}(w_n) \geq |g(0)|, \tag{1.2.47}$$

the comparison principle implies that $|u_n| \leq w_n$. We deduce from Lemmas 4.1-4.2-4.3 and the definition of v_n and w_n that

$$\begin{cases} \text{(i)} \quad \|u_n\|_{M^{(N+1)/(N-1)}(\Omega,\rho dx)} \leq C\left(1 + \|\mu_n\|_{L^1(\partial\Omega)}\right), \\ \text{(ii)} \quad \|u_n\|_{M^{N/(N-1)}(\Omega)} \leq C\left(1 + \|\mu_n\|_{L^1(\partial\Omega)}\right), \end{cases} \tag{1.2.48}$$

and also

$$\|\rho g(u_n)\|_{L^1(\Omega)} \leq C\left(1 + \|\mu_n\|_{L^1(\partial\Omega)}\right), \tag{1.2.49}$$

which is the claim.

Step 2. We claim that $\{g(u_n)\}$ is weakly relatively compact in $L^1(\Omega,\rho dx)$. The idea is to use Dunford-Pettis equi-integrability criterion for the measure ρdx. Set $\varepsilon > 0$, we have to find $\eta > 0$ such that for any measurable subset $\omega \subset \Omega$ and $n \in \mathbb{N}$, we have

$$\int_\omega \rho dx \leq \eta \Rightarrow \int_\omega |g(u_n)|\rho dx < \varepsilon. \tag{1.2.50}$$

We set $\hat{g}(r) = g(r) + |g(-r)|$, for $R > 0$, then

$$\int_\omega |g(u_n)|\rho dx = \int_{\omega \cap \{|u_n| \leq R\}} |g(u_n)|\rho dx + \int_{\omega \cap \{|u_n| > R\}} |g(u_n)|\rho dx,$$
$$\leq \hat{g}(R)\int_\omega \rho dx + \int_{\omega \cap \{|u_n| > R\}} \hat{g}(|u_n|)\rho dx. \tag{1.2.51}$$

We denote $B_n(\lambda) = \{x \in \Omega : |u_n(x)| > \lambda\}$ for $\lambda > 0$ and $\alpha_n(\lambda) = \int_{B_n(\lambda)} \rho dx$; since

$$\int_{\omega \cap \{|u_n| > R\}} \hat{g}(|u_n|)\rho dx \leq \int_{\{|u_n| > R\}} \hat{g}(|u_n|)\rho dx = -\int_R^\infty \hat{g}(\lambda)d\alpha_n(\lambda) \tag{1.2.52}$$

we deduce from Lemma 4.1 that

$$\alpha_n(\lambda) \leq C\lambda^{-(N+1)/(N-1)}.$$ (1.2.53)

Moreover, the following relation holds

$$-\int_R^\infty \hat{g}(\lambda)d\alpha_n(\lambda) = \hat{g}(R)\alpha_n(R) + \int_R^\infty \alpha_n(\lambda)d\hat{g}(\lambda);$$ (1.2.54)

therefore

$$\int_R^\infty \alpha_n(\lambda)d\hat{g}(\lambda) \leq C\int_R^\infty \lambda^{-(N+1)/(N-1)}d\hat{g}(\lambda),$$

$$\leq C\left(-\hat{g}(R)R^{-(N+1)/(N-1)} + \frac{N+1}{N-1}\int_R^\infty \lambda^{-2N/(N-1)}\hat{g}(\lambda)d\lambda\right).$$ (1.2.55)

If we first choose R such that

$$C\frac{N+1}{N-1}\int_R^\infty \lambda^{-2N/(N-1)}\hat{g}(\lambda)d\lambda \leq \varepsilon/2,$$ (1.2.56)

and set $\eta = \dfrac{\varepsilon}{2}(1+\hat{g}(r))^{-1}$, we obtain (1.2.50).

Step 3. We claim that $\{u_n\}$ is weakly relatively compact in $L^1(\Omega)$. From step 1 and Lemma 4.2, we have

$$\|u_n\|_{M^{N/(N-1)}(\Omega)} \leq C\left(1 + \|\mu_n\|_{L^1(\partial\Omega)}\right) \leq K,$$ (1.2.57)

which implies

$$\int_\omega |u_n|dx \leq K\left(\int_\omega dx\right)^{1/N},$$ (1.2.58)

and the equi-integrability condition follows.

Step 4. End of the proof. From Lemma 4.2, for any compact subset $K \subset \Omega$, there exists $C(K) > 0$ such that

$$\|\Delta u_n\|_{L^1(K)} + \|u_n\|_{L^1(K)} \leq C(K).$$ (1.2.59)

Therefore $\{u_n\}$ is relatively compact in $L^1_{loc}(\Omega)$ and there exist an integrable function u and a subsequence $\{n_k\}$ such that $u_{n_k} \to u$ a.e. in Ω and in $L^1(\Omega)$-weak. Moreover $g(u_{n_k}) \to g(u)$ in $L^1(\Omega, \rho dx)$-weak. If $\zeta \in C^{1,1}_0(\overline{\Omega})$, then ζ/ρ is bounded, and the following relations hold

$$\lim_{n_k \to \infty} \int_\Omega \left(-u_{n_k} \Delta\zeta + g(u_{n_k})\zeta\right)dx = \int_\Omega \left(-u\Delta\zeta + g(u)\zeta\right)dx, \tag{1.2.60}$$

$$\lim_{n_k \to \infty} \int_{\partial\Omega} \frac{\partial\zeta}{\partial v}\mu_n dS(x) = \int_{\partial\Omega} \frac{\partial\zeta}{\partial v}d\mu, \tag{1.2.61}$$

which means that u solves (1.2.1) in the sense defined by (1.2.2).

Uniqueness. Suppose that u and $\hat{u}$ are two solutions of the same problem and set $w = u - \hat{u}$. Then, we deduce from (1.2.7) that

$$\int_\Omega \left(-|w|\Delta\zeta + (g(u) - g(\hat{u}))\zeta\,\text{sgn}(w)\right)dx \leq 0, \tag{1.2.62}$$

for $\zeta \in C^{1,1}_0(\overline{\Omega})$, $\zeta \geq 0$. If we choose $\zeta \geq 0$ as being the solution of (1.2.19) as in Lemma 4.2, we deduce that $w = 0$ from the monotonicity of g.

Remark 4.2. As in Lemma 4.2, it is clear that the mapping $\mu \mapsto u$ is nondecreasing. Moreover the following continuity result holds

$$\|u - \hat{u}\|_{L^1(\Omega)} + C^{-1}\|\rho(g(u) - g(\hat{u}))\|_{L^1(\Omega)} \leq C\|\mu - \hat{\mu}\|_{M(\partial\Omega)}. \tag{1.2.63}$$

If the assumption (1.2.2) is not satisfied by g, it is not possible to solve (1.2.1) for any measure $\mu \in \mathcal{M}(\partial\Omega)$ since this measure has to be "not to much concentrated". For that we define the Poisson potentiel m_μ of μ by

$$m_\mu(x) = \int_{\partial\Omega} P(x,y)d\mu(y) \tag{1.2.64}$$

for $x \in \Omega$. The function m_μ is integrable and harmonic in Ω and solves

$$-\int_\Omega m_\mu \Delta\zeta dx = -\int_{\partial\Omega} \frac{\partial\zeta}{\partial v}d\mu \tag{1.2.65}$$

for any $\zeta \in C^{1,1}_0(\overline{\Omega})$.

Definition. Let g be a continuous real valued fonction defined on $\mathbb{R}$. If $\mu \in \mathcal{M}(\partial\Omega)$, with Lebesgue's decomposition $\mu = \mu_S + \mu_R$, where μ_S is the singular part and μ_R the absolutely continuous part with respect to the Lebesgue's measure, we say that μ is g-admissible if

$$\int_{\Omega} \left| g(m_{|\mu_s|}) \right| \rho \, dx < \infty \tag{1.2.66}$$

(it is clear that $|\mu_s| = |\mu|_s$).

The following result due to Marcus and Véron [MV3] extends Theorem 4.1 to the class of g-admissible measures.

THEOREM 4.2. *Suppose that* g *is nondecreasing, vanishes at 0 and satisfies the* Δ_2-*condition, namely that there exists* $\theta > 0$ *such that* $|g(r + r')| \leq \theta(|g(r)| + |g(r')|)$ *holds whenever* r *and* r' *are two real numbers with the same sign. If* $\mu \in \mathcal{M}(\partial\Omega)$ *is g-admissible, there exists a unique solution* u *to* (1.2.1).

Proof. For $k > 0$ we set

$$g_k(r) = \begin{cases} k & \text{if } g(r) > k, \\ g(r) & \text{if } -k \leq g(r) \leq k, \\ -k & \text{if } g(r) < -k. \end{cases} \tag{1.2.67}$$

It is clear that g_k satisfies the Δ_2-condition with the same θ.

Step 1. We suppose that $\mu = \mu_R + \mu_S$ is nonnegative. Let $\{\mu_R^n\}$ be a sequence of smooth nonnegative functions on $\partial\Omega$ converging to μ_R in $L^1(\partial\Omega)$ when n goes to infinity. We denote u_k^n the solution of

$$\begin{cases} -\Delta u_k^n + g_k(u_k^n) = 0 & \text{in } \Omega, \\ u_k^n = \mu_R^n + \mu_S & \text{on } \partial\Omega, \end{cases} \tag{1.2.68}$$

and v_k^n the one of

$$\begin{cases} -\Delta v_k^n + g_k(v_k^n) = 0 & \text{in } \Omega, \\ v_k^n = \mu_R^n & \text{on } \partial\Omega, \end{cases} \tag{1.2.69}$$

since Theorem 4.1 implies the existence of such solutions. But $0 \leq u_k^n \leq v_k^n + m_{\mu_s}$, from the maximum principle and

$$0 \leq g_k(u_k^n) \leq \theta\big(g_k(v_k^n) + g_k(m_{\mu_s})\big) \leq \theta\big(g_k(v_k^n) + g(m_{\mu_s})\big), \tag{1.2.70}$$

from the Δ_2-condition. When k goes to infinity, $\{v_k^n\}$ converges in the $C_{\text{loc}}^1(\Omega)$-topology and uniformly on $\overline{\Omega}$ to the solution v^n of

$$\begin{cases} -\Delta v^n + g(v^n) = 0 & \text{in } \Omega, \\ v^n = \mu_R^n & \text{on } \partial\Omega. \end{cases} \tag{1.2.71}$$

The function v^n is continuous in $\overline{\Omega}$ and $\{g_k(v_k^n)\}$ converges uniformly in $\overline{\Omega}$ to $g(v^n)$ from the regularity theory of elliptic equations and the uniqueness of the solution of (1.2.71). The set of functions $\{u_k^n\}$ remains locally bounded in Ω and therefore locally compact in the $C_{loc}^1(\Omega)$-topology. Therefore there exist a subsequence $\{u_{k_\ell}^n\}_{k_\ell}$ and a function u^n such that $\lim_{k_\ell \to \infty} u_{k_\ell}^n = u^n$ in the $C_{loc}^1(\Omega)$-topology, and $\lim_{k_\ell \to \infty} g_{k_\ell}(u_{k_\ell}^n) = g(u^n)$ a.e. in Ω. From Lebesgue's theorem, (1.2.70), the uniform convergence of $\{g_k(v_k^n)\}$ and the g-admissibility assumption on μ, we actually have $\lim_{k_\ell \to \infty} g_{k_\ell}(u_{k_\ell}^n) = g(u^n)$ in $L^1(\Omega, \rho dx)$. Since $0 \le u_k^n \le v_k^n + m_{\mu_s}$ and the integrability of v^n and m_{μ_s} we can also infer that $\lim_{k_\ell \to \infty} u_{k_\ell}^n = u^n$ in $L^1(\Omega)$. Therefore u^n is the unique solution of

$$\begin{cases} -\Delta u^n + g(u_k^n) = 0 & \text{in } \Omega, \\ u^n = \mu_R^n + \mu_s & \text{on } \partial\Omega. \end{cases} \tag{1.2.72}$$

Moreover $0 \le u^n \le v^n + m_{\mu_s}$ and

$$0 \le g(u^n) \le \theta\big(g(v^n) + g(m_{\mu_s})\big). \tag{1.2.73}$$

From estimate (1.2.16) in Lemma 4.2, there holds

$$\left\| v^n - v^m \right\|_{L^1(\Omega)} + \left\| \rho(g(v^n) - g(v^m)) \right\|_{L^1(\Omega)} \le C \left\| \mu_R^n - \mu_R^m \right\|_{L^1(\partial\Omega)} \tag{1.2.74}$$

for some positive constant C. When n goes to infinity the sequence $\{g(v^n)\}_n$ is a Cauchy one in $L^1(\Omega, \rho dx)$ and v^n converges in $L^1(\Omega)$ to the solution v of

$$\begin{cases} -\Delta v + g(v) = 0 & \text{in } \Omega, \\ v = \mu_R & \text{on } \partial\Omega. \end{cases} \tag{1.2.75}$$

Up to some subsequence $\{u^{n_\ell}\}$ converges to some u, in the $C_{loc}^1(\Omega)$-topology, a.e. in Ω and in $L^1(\Omega)$-weak (since v^n and m_{μ_s} are respectively equi-intgrable and integrable in Ω). Moreover $\lim_{n_\ell \to \infty} g(u^{n_\ell}) = g(u)$ a.e. in Ω. We deduce from (1.2.73), g-admissibility assumption and (1.2.74) that this last limit occurs also weakly in $L^1(\Omega, \rho dx)$ as a consequence of Dunford-Pettis criterion. If $\zeta \in C_0^{1,1}(\overline{\Omega})$, we can let n_ℓ go to infinity in

$$\int_{\Omega}\left(-u^{n_\ell}\Delta\zeta + g(u^{n_\ell})\zeta\right)dx = -\int_{\partial\Omega}\frac{\partial\zeta}{\partial\nu}d\mu \qquad (1.2.76)$$

and conclude that u is a solution of (1.2.1) in Ω, unique since u and g(u) belong respectively to $L^1(\Omega)$ and $L^1(\Omega,\rho dx)$.

Step 2. The case of a general g-admissible measure μ. By the above technique we successively construct the solutions u_k^n of (1.2.68) and $\overline{u}_k^n$ and $\underline{u}_k^n$ of

$$(i)\begin{cases} -\Delta\overline{u}_k^n + g_k(\overline{u}_k^n) = 0 & \text{in } \Omega, \\ \overline{u}_k^n = \left|\mu_R^n\right| + \left|\mu_S\right| & \text{on } \partial\Omega, \end{cases}$$

$$(ii)\begin{cases} -\Delta\underline{u}_k^n + g_k(\underline{u}_k^n) = 0 & \text{in } \Omega, \\ \underline{u}_k^n = -\left|\mu_R^n\right| + -\left|\mu_S\right| & \text{on } \partial\Omega, \end{cases} \qquad (1.2.77)$$

and the solutions $\overline{v}_k^n$ and $\underline{v}_k^n$ of

$$(i)\begin{cases} -\Delta\overline{v}_k^n + g_k(\overline{v}_k^n) = 0 & \text{in } \Omega, \\ \overline{v}_k^n = \left|\mu_R^n\right| & \text{on } \partial\Omega, \end{cases}$$

$$(ii)\begin{cases} -\Delta\underline{v}_k^n + g_k(\underline{v}_k^n) = 0 & \text{in } \Omega, \\ \underline{v}_k^n = -\left|\mu_R^n\right| & \text{on } \partial\Omega. \end{cases} \qquad (1.2.78)$$

Since $\underline{v}_k^n - m_{|\mu_S|} \le u_k^n \le \overline{v}_k^n + m_{|\mu_S|}$ and

$$\theta\left(g_k(\underline{v}_k^n) + g(-m_{|\mu_S|})\right) \le g_k(u_k^n) \le \theta\left(g_k(\overline{v}_k^n) + g(m_{|\mu_S|})\right), \qquad (1.2.79)$$

we conclude with the same limit procedures (first we let k go to infinity and then n) than the ones of Step 1 that there exists a solution u of (1.2.1) with u and g(u) in $L^1(\Omega)$ and $L^1(\Omega,\rho dx)$ respectively. The solution u is therefore unique.

Remark 4.3. The assumption that $g(0) = 0$ is unnecessary from the Δ_2-condition. It is clear that this condition is fulfilled for any function g with a power-like growth. It does not hold in the exponential case.

4-1-3 Removable singularities: the smooth case

In this Section Ω is a domain of $\mathbb{R}^N$ (N > 1) whose boundary $\partial\Omega$ is C^3 and contains O. We define $C_0^{1,1}(\overline{\Omega}\setminus\{O\})$ as the space of functions with Lipschitz gradient, vanishing on

$\partial\Omega$ and in a neighbourhood of O. We consider a continuous real valued function g satisfying the growth condition

$$\begin{cases} \liminf_{r\to\infty} g(r)r^{-(N+1)/(N-1)} > 0, \\[2ex] \limsup_{r\to-\infty} g(r)|r|^{-(N+1)/(N-1)} < 0, \end{cases} \tag{1.3.1}$$

and the main removability result for isolated singularities is the following

THEOREM 4.3. *Suppose* g *satisfies* (1.3.1) *and let* $u \in C^2(\Omega) \cap C(\overline{\Omega} \setminus \{O\})$ *and* $\varphi \in C(\partial\Omega)$ *be such that*

$$\int_\Omega (-u\Delta\zeta + g(u)\zeta)dx = -\int_{\partial\Omega} \varphi\frac{\partial\zeta}{\partial\nu}dS, \tag{1.3.2}$$

for any $\zeta \in C_0^{1,1}(\overline{\Omega} \setminus \{O\})$. *Then* u *can be extended to* $\overline{\Omega}$ *as a continuous function* $\hat{u}$ *which satisfies*

$$\int_\Omega (-\hat{u}\Delta\zeta + g(\hat{u})\zeta)dx = -\int_{\partial\Omega} \varphi\frac{\partial\zeta}{\partial\nu}dS \tag{1.3.3}$$

for any $\zeta \in C_0^{1,1}(\overline{\Omega})$.

The first estimate extends is an extension of Osserman-Keller *a priori* estimate to the boundary singularity situation. It is important to notice that no regularity assumption is made on the boundary of the domain

LEMMA 4.4. *Let* G *be a bounded domain in* $\mathbb{R}^N$, $N > 1$, *whose boundary contains* O *and suppose* $\varphi \in C(\partial G)$ *and* $u \in C(\overline{G} \setminus \{O\})$ *with* $\nabla u \in L^2_{loc}(G)$ *and* $\Delta u \in L^1_{loc}(G)$ *are such that* $u = \varphi$ *on* $\partial G \setminus \{O\})$ *and*

$$-\Delta u + au^q \leq b \tag{1.3.4}$$

a.e. on $\{x \in G: u(x) \geq 0\}$ *for some constants* $q > 1$, $a > 0$ *and* $b \geq 0$. *Then*

$$u(x) \leq C|x|^{-2/(q-1)} \qquad \left(\forall x \in \overline{G} \setminus \{O\}\right), \tag{1.3.5}$$

for some $C = C(N, q, \max_{x\in\partial G} \varphi^+) > 0$.

Proof. Let B be an open ball containing $\overline{G}$ and Φ the solution of

$$\begin{cases} -\Delta\Phi = b & \text{in } B, \\ \Phi = \max_{\partial G} \varphi^+ & \text{on } \partial B. \end{cases} \tag{1.3.6}$$

The function Φ is nonnegative and $\Phi_{|\partial G} \geq \varphi^+$. If $\varepsilon > 0$, we denote $u^* = u - \Phi - \varepsilon$; then

$$-\Delta u^* + a(u^* + \Phi + \varepsilon)^q \leq 0 \tag{1.3.7}$$

a.e. on $\{x \in G: \varepsilon + (u^* + \Phi)(x) \geq 0\}$. Therefore

$$-\Delta u^* + a(u^*)^q \leq 0$$

a.e. on $\{x \in G: u^*(x) \geq 0\}$. We define the function $j = j_\varepsilon$ by

$$j(r) = \begin{cases} 0 & \text{if } r \leq 0, \\ r^2/2\varepsilon & \text{if } 0 \leq r \leq \varepsilon, \\ r - \varepsilon/2 & \text{if } r \geq \varepsilon, \end{cases} \tag{1.3.8}$$

and set $v = j(u)$. Then $\Delta v = j'(u^*)\Delta u^* + J''|\nabla u^*|^2$ is locally integrable in G and $v^q = \left(j(u^*)\right)^q$ $\leq j'(u^*)(u^*)^q$. We obtain

$$-\Delta v + av^q \leq 0 \tag{1.3.9}$$

on the same subset of G as for u^*. Since u^* is continuous and negative on $\partial G \setminus \{O\}$, the function $\tilde{v}$ which is v in $\overline{G} \setminus \{O\}$ and 0 in $\mathbb{R}^N \setminus \overline{G}$ is continuous in $\mathbb{R}^N \setminus \{O\}$ and vanishes in an open neighbourhood of $\partial G \setminus \{O\}$. Therefore it satisfies

$$-\Delta \tilde{v} + a\tilde{v}^q \leq 0 \tag{1.3.10}$$

a.e. in $\mathbb{R}^N \setminus \{O\}$ and we conclude classicaly that it satisfies an Osserman-Keller type estimate which implies that (1.3.5) holds.

Remark 4.4. If we replace the point O by a closed subset F of ∂G, the previous construction is still valid. Therefore if $\varphi \in C(\partial G)$ and $u \in C(\overline{G} \setminus \{F\})$ with $\nabla u \in L^2_{\text{loc}}(G)$ and $\Delta u \in L^1_{\text{loc}}(G)$ are such that $u = \varphi$ on $\partial G \setminus \{F\}$) and (1.3.4) holds a.e. on $\{x \in G: u(x) \geq 0\}$, then

$$u(x) \leq C(\text{dist}(x,F))^{-2/(q-1)} \qquad \left(\forall x \in \overline{G} \setminus \{F\}\right). \tag{1.3.11}$$

LEMMA 4.5. *Let G be a Lipchitz domain in* $\mathbb{R}^N$, $N > 1$, *whose boundary contains O. We suppose that* $\partial G \setminus \{O\}$ *is* C^2 *with a uniformly bounded curvature and that* ∂G *admits a smooth tangent cone at* O. *If* $\varphi \in C(\partial G)$ *and* $u \in C(\overline{G} \setminus \{O\})$ *with* $\nabla u \in L^2_{loc}(G)$ *and* $\Delta u \in L^1_{loc}(G)$ *are such that* $u = \varphi$ *on* $\partial G \setminus \{O\})$ *and* (1.3.4) *holds a.e. on* $\{x \in G : u(x) \geq 0\}$ *for some constants* $q > 1$, $a > 0$ *and* $b \geq 0$, *then*

$$u(x) \leq C\rho(x)|x|^{-(q+1)/(q-1)} + D, \qquad \forall x \in \overline{G} \setminus \{O\}), \qquad (1.3.12)$$

for some $C, D > 0$ *depending on* $(N, q, \max_{x \in \partial G} \varphi^+, G)$.

Proof. Step 1. We first construct a nonnegative solution $U \in C^2(G) \cap C(\overline{G} \setminus \{O\})$ of

$$-\Delta U + aU^q = 0 \quad \text{in} \quad G, \qquad (1.3.13)$$

which vanishes on $\partial G \setminus \{O\}$ and such that $U + (b/a)^{1/q} + \max_{\partial G} \varphi^+ \geq u$. For $\varepsilon > 0$ small enough, we set $G_\varepsilon = G \setminus B(O, \varepsilon)$, $\partial G^1_\varepsilon = \partial G \setminus B(O, \varepsilon)$, $\partial G^2_\varepsilon = \partial B(O, \varepsilon) \cap \overline{G}$ and U_ε the solution of

$$\begin{cases} -\Delta U_\varepsilon + aU^q_\varepsilon = 0 & \text{in} \quad G_\varepsilon, \\ \quad U_\varepsilon = 0 & \text{on} \quad \partial G^1_\varepsilon, \\ \quad U_\varepsilon = \infty & \text{on} \quad \partial G^2_\varepsilon. \end{cases} \qquad (1.3.14)$$

Then $\tilde{U}_\varepsilon = U_\varepsilon + (b/a)^{1/q} + \max_{\partial G} \varphi^+$ is larger than u on ∂G_ε and

$$-\Delta \tilde{U}_\varepsilon + a\tilde{U}^q_\varepsilon \geq b \quad \text{in} \quad G_\varepsilon. \qquad (1.3.15)$$

Therefore it is larger than u in whole G_ε. Since U_ε is locally bounded in $\overline{G} \setminus \{O\}$, by use of the Osserman-Keller estimate: there exists a subsequence $\{\varepsilon_n\}$ such that U_{ε_n} converges in the $C_{loc}(\overline{G} \setminus \{O\})$-topology to some U which satisfies (1.3.13) in G. Moreover U dominates u up to the addition of $(b/a)^{1/q} + \max_{\partial G} \varphi^+$.

Step 2. The function U satisfies

$$U(x) \leq C\rho(x)|x|^{-(q+1)/(q-1)}, \qquad (1.3.16)$$

for some $C = C(N, q, a, G) > 0$. We proceed by contradiction in assuming that there exist sequences $\{x_n\}$ and $\{M_n\}$ with $\{x_n\} \subset G$, $\lim_{n \to \infty} M_n = \infty$ and

191

$$U(x_n) = M_n \rho(x_n) |x_n|^{-(q+1)/(q-1)}. \tag{1.3.17}$$

For any $\alpha > 0$ there holds

$$\max_{x \in G \cap B^c(O,\alpha)} U(x) \leq C(N,q)\alpha^{-2/(q-1)}, \tag{1.3.18}$$

and $|\nabla U|$ is uniformly bounded on $G \cap B^c(O,\alpha)$ by some constant depending on α. Therefore $x_n \to O$. We define now U_n by

$$U(x) = k_n^{2/(q-1)} U_n(k_n x) \tag{1.3.19}$$

with $k_n = |x_n|^{-1}$. If we set $y_n = x_n / |x_n|$, we obtain

$$U_n(y_n) = M_n \rho(x_n) / |x_n| = M_n \rho(|x_n| y_n) / |x_n|. \tag{1.3.20}$$

Therefore $\lim_{n \to \infty} \rho(x_n) / |x_n| = 0$, which means that x_n goes to O tangentially to ∂G; moreover $\rho(|x_n| y_n) / |x_n| = \rho(y_n)(1 + o(1))$. As for the function U_n, it is a solution of

$$-\Delta U_n + a U_n^q = 0 \tag{1.3.21}$$

in $G_n = \{y : |x_n| y \in G\}$, it vanishes on $\partial G_n \setminus \{O\}$ and satisfies

$$U_n(y) \leq C(N,q,a)|y|^{-2/(q-1)}. \tag{1.3.22}$$

As a consequence of (1.3.22) and elliptic equation regularity theory, $|\nabla U_n(y)|$ is uniformly bounded for $y \in G_n$, with $1/2 \leq |y| \leq 3/2$ which contradicts (1.3.20). Therefore U satisfies (1.3.16) with $C = C(N,q,a,G) > 0$ since U is the maximal solution of (1.3.13) in G, continuous in $\overline{G} \setminus \{O\}$ and vanishing on $\partial G \setminus \{O\}$. Finally we obtain (1.3.12) with $D = (b/a)^{1/q} + \max_{\partial G} \varphi^+$.

In the next lemmas we need more regularity on the domain.

LEMMA 4.6. *Suppose* $N > 2$ *and* Ω *is as in Theorem 4.3. If* $w \in L^1(\Omega)$, $f \in L^1(\Omega, \rho dx)$ *and* $\varphi \in \partial\Omega$ *are such that*

$$-\int_\Omega w \Delta \zeta dx = \int_\Omega f \zeta dx - \int_{\partial\Omega} \frac{\partial \zeta}{\partial v} \varphi dS, \tag{1.3.23}$$

for any $\zeta \in C_0^{1,1}(\overline{\Omega})$; *then, for any* $x_0 \in \Omega$, *there holds*

$$\lim_{\sigma \to 0} \sigma^{-2} \int_{\Omega \cap B(x_0,\sigma)} |w| dx = 0. \tag{1.3.24}$$

Proof. It is clear that w is the unique solution of

$$\begin{cases} -\Delta w = f & \text{in } \Omega, \\ w = \varphi & \text{on } \partial\Omega. \end{cases} \tag{1.3.25}$$

G (resp. P) being the Green kernel (resp. Poisson kernel) in Ω, we decompose w as $w = w_1 + w_2$ where $w_1(x) = \int_\Omega G(x,y)f(y)dy$ and $w_2(x) = \int_{\partial\Omega} P(x,y)\varphi(y)dS(y)$. For $R > 0$ we call χ_R the characteristic function of $\Omega \cap B(x_0,R)$ and we write

$$w_1(x) = \int_\Omega G(x,y)\chi_R(y)f(y)dy + \int_\Omega G(x,y)(1-\chi_R(y))f(y)dy. \tag{1.3.26}$$

By continuity

$$\int_\Omega \left(|w_2(x)| + \left| \int_\Omega G(x,y)f(y)(1-\chi_R(y))dy \right| \right) \rho(x)\chi_\sigma(x)dx = O(\sigma^{N+1}) \tag{1.3.27}$$

as $\sigma \to 0$. It remains to estimate

$$\int_\Omega \left| \int_\Omega G(x,y)f(y)\chi_R(y)dy \right| \rho(x)\chi_\sigma(x)dx = \ell_\sigma. \tag{1.3.28}$$

For $0 < \sigma < R$ we call ζ_σ the solution of

$$\begin{cases} -\Delta\zeta_\sigma = \rho\chi_\sigma & \text{in } \Omega, \\ \zeta_\sigma = 0 & \text{on } \partial\Omega. \end{cases} \tag{1.3.29}$$

Then

$$\begin{aligned}
\ell_\sigma &\le \int_\Omega |f|\chi_R\zeta_\sigma dx = \iint_{\Omega\times\Omega} G(x,y)\chi_R(x)\chi_\sigma(y)\rho(y)|f(x)|dxdy, \\
&\le j_\sigma = C_N \iint_{\Omega\times\Omega} |x-y|^{2-N}\chi_R(x)\chi_\sigma(y)\rho(y)|f(x)|dxdy,
\end{aligned} \tag{1.3.30}$$

where $C_N|x-y|^{2-N}$ is the Newton kernel in $\mathbb{R}^N$. Let Φ be a C^2 diffeomorphism from $B(x_0,2R)$ into $B(O,2R)$ such that $\Phi(x_0) = O$ and

$$\Phi(B(x_0,2R)\cap\Omega) \subset B^+(O,2R) = \left\{ x = (x_1,x') \in \mathbb{r}_*^+ \times \mathbb{r}^{N-1} \right\} \cap B(O,2R), \tag{1.3.31}$$

$$\Phi(B(x_0,2R)\cap\partial\Omega) \subset B(O,2R)\cap\{x = (0,x')\}, \tag{1.3.32}$$

$$\Phi(B(x_0,R)\cap\Omega) = B^+(O,R) = B(O,R)\cap\mathbb{R}_*^+\times\mathbb{R}^{N-1}, \tag{1.3.33}$$

$$\|D\Phi - I\|_{L^\infty(B^+(x_0,R))} \le 1/4, \tag{1.3.34}$$

193

$$D\Phi(x)(v_x) \in \mathbb{R}_*^+ \times \{0\} \quad \text{for any } x \in \overline{B}(x_0,R) \cap \partial\Omega. \tag{1.3.35}$$

Such a diffeomorphism exists for R small enough. Moreover there is some $\alpha > 0$ such that $\alpha^{-1} \le \rho(x)/\tilde{\rho}(\Phi(x)) \le \alpha$ for any $x \in B(x_0,R) \cap \Omega$ where $\tilde{\rho}((x_1,x')) = |x_1|$. There exists also a positive constant $C = C(N,R)$ such that the inequality

$$\begin{aligned}
j_\sigma &\le C\int_{B^+(O,R)}\int_{B^+(O,2\sigma)} |x-y|^{2-N}\tilde{\rho}(y)|f_\circ\Phi^{-1}(x)|\,dxdy, \\
&\le k_\sigma = \int_{B^+(O,R)} |f_\circ\Phi^{-1}(x)|\tilde{\zeta}_\sigma(x)\,dx,
\end{aligned} \tag{1.3.36}$$

holds for $\sigma < R/2$, in which formula $\tilde{\zeta}_\sigma$ is the solution of

$$\begin{cases}
-\Delta\tilde{\zeta}_\sigma(x) = \dfrac{C}{C_N}\tilde{\rho}(x)\chi_{B^+(O,2\sigma)}(x) & \text{in } \mathbb{R}^+ \times \mathbb{R}^{N-1}, \\[2mm]
\tilde{\zeta}_\sigma = 0 & \text{on } \{0\} \times \mathbb{R}^{N-1}.
\end{cases} \tag{1.3.37}$$

By uniqueness, $\tilde{\zeta}_\sigma(x) = \dfrac{\rho^3}{R^3}\tilde{\zeta}_R(\dfrac{R}{\rho}x)$. Moreover $\tilde{\zeta}_R$ satisfies the following estimate

$$\tilde{\zeta}_R(x) \le K\frac{x_1}{(1+x_1)(1+|x|^{N-2})} \tag{1.3.38}$$

where $x = (x_1,x')$. Consequently

$$k_\sigma \le K\frac{\sigma^2}{R^2}\int_{B^+(O,R)} |f_\circ\Phi(x)|\tilde{\rho}(x)\frac{\sigma}{\sigma+R\tilde{\rho}(x)}\frac{\sigma^{N-2}}{\sigma^{N-2}+|x|^{N-2}}\,dx, \tag{1.3.39}$$

which implies that $\lim\limits_{\sigma\to 0}\sigma^{-2}\ell_\sigma = 0$ and completes the proof.

LEMMA 4.7. *Let $N = 2$ and Ω be as in Theorem 4.3, $v \in L^3(\Omega,\rho dx)$, $f \in L^1(\Omega,\rho dx)$ and $\varphi \in L^1(\partial\Omega)$ are such that*

$$-\int_\Omega v\Delta\zeta\,dx \le \int_\Omega f\zeta\,dx - \int_{\partial\Omega}\frac{\partial\zeta}{\partial\nu}\varphi\,dS \tag{1.3.40}$$

for any $\zeta \in C_0^{1,1}(\overline{\Omega})$, $\zeta \ge 0$, vanishing in some neighbourhood of O. Then (1.3.38) remains valid for any $\zeta \in C_0^{1,1}(\overline{\Omega})$, $\zeta \ge 0$.

Proof. We take $\zeta \in C_0^{1,1}(\overline{\Omega})$, $\zeta \geq 0$ and $\eta_n \in C^\infty(\mathbb{R}^2)$ such that $0 \leq \eta_n \leq 1$ and $\eta_n(x) = 0$ (resp. 1) if $|x| \leq 1/2n$ (resp. $|x| \geq 1/n$). Assume also that $|\nabla \eta_n(x)| \leq Cn$, $|\Delta \eta_n(x)| \leq Cn^2$ for any $n \in \mathbb{N}_*$ and $x \in \mathbb{R}^2$. Thus

$$\int_\Omega -v(\zeta \Delta \eta_n + 2\nabla \zeta . \nabla \eta_n + \eta_n \Delta \zeta)dx \leq \int_\Omega f\zeta \eta_n dx - \int_{\partial\Omega} \varphi \frac{\partial}{\partial \nu}(\zeta \eta_n)dS. \tag{1.3.41}$$

But

$$\begin{cases} \text{(i)} \quad \lim_{n\to\infty} \int_{\partial\Omega} \varphi \frac{\partial}{\partial \nu}(\eta_n \zeta)dS = \lim_{n\to\infty} \int_{\partial\Omega} \varphi \, \eta_n \frac{\partial \zeta}{\partial \nu}dS = \int_{\partial\Omega} \varphi \frac{\partial \zeta}{\partial \nu}dS, \\ \text{(ii)} \quad \lim_{n\to\infty} \int_\Omega f\eta_n \zeta dx = \int_\Omega \zeta \, fdx. \end{cases} \tag{1.3.42}$$

Moreover, if $\Omega_n = \{x \in \Omega: 1/2n \leq |x| \leq 1/n\}$,

$$\left| \int_\Omega v\zeta \Delta \eta_n dx \right| \leq Kn^2 \int_{\Omega_n} |v|\rho dx \leq Kn^2 \left(\int_{\Omega_n} |v|^3 \rho dx \right)^{1/3} \left(\int_{\Omega_n} \rho dx \right)^{2/3},$$
$$\leq K(2\pi/3)^{2/3} \left(\int_{\Omega_n} |v|^3 \rho dx \right)^{1/3}, \tag{1.3.43}$$

which goes to 0 when n goes to infinity. We also have

$$\left| \int_\Omega v\nabla \eta_n . \nabla \zeta dx \right| \leq Cn \left(\int_{\Omega_n} |v|^3 \rho dx \right)^{1/3} \left(\int_{\Omega_n} \rho^{-1/2} dx \right)^{2/3} = o(1). \tag{1.3.44}$$

Finally we obtain (1.3.40) by letting n go to infinity in (1.3.41), .

When $N > 2$, the situation is more complicated since (1.3.44) has no N-dimensional counterpart, but we have a weaker removability result.

LEMMA 4.8. *Let $N > 2$ and Ω be as in Theorem 4.3, $v \in L^{(N+1)/(N-1)}(\Omega, \rho dx)$ $\cap M^{N/(N-1)}(\Omega)$, $v \geq 0$, $f \in L^1(\Omega, \rho dx)$ and $\varphi \in L^1(\partial\Omega)$ are such that*

$$-\int_\Omega v\Delta \zeta dx = \int_\Omega f\zeta dx - \int_{\partial\Omega} \frac{\partial \zeta}{\partial \nu} \varphi dS \tag{1.3.45}$$

for any $\zeta \in C_0^{1,1}(\overline{\Omega})$ vanishing in some neighbourhood of O. Then (1.3.45) remains valid for any $\zeta \in C_0^{1,1}(\overline{\Omega})$.

Proof. We take $\zeta \in C_0^{1,1}(\overline{\Omega})$ and η_n satisfying the same assumption as in Lemma 4.7. We also assume that η_n is radial and we have

$$\int_\Omega -v(\zeta\Delta\eta_n + 2\nabla\zeta.\nabla\eta_n + \eta_n\Delta\zeta)dx = \int_\Omega f\zeta\eta_n dx - \int_{\partial\Omega}\varphi\frac{\partial}{\partial v}(\zeta\eta_n)dS. \qquad (1.3.46)$$

The problem is to let n go to infinity in (1.3.46) and the only term which does not obviously converge to 0 is $\left|\int_\Omega v\nabla\eta_n.\nabla\zeta dx\right|$. Since $v \in M^{N/(N-1)}(\Omega)$, for any measurable subset ω, we have

$$\int_\omega vdx \le \|v\|_{M^{N/(N-1)}(\Omega)}\left(\int_\omega 1dx\right)^{1/N}. \qquad (1.3.47)$$

Thus there holds

$$0 \le n\int_{\Omega_n} vdx \le C_1. \qquad (1.3.48)$$

Let us assume that

$$\limsup_{n\to\infty} n\int_{G_n} vdx = \lim_{n_k\to\infty} n_k\int_{G_{n_k}} vdx = \gamma > 0. \qquad (1.3.49)$$

We may choose η_n such that

$$\nabla\eta_n = -2nx/|x| \qquad \text{for} \quad 1/2n+1/n^2 \le |x| \le 1/n - 1/n^2, \qquad (1.3.50)$$

and we deduce from (1.3.47)-(1.3.49) that

$$\lim_{n_k}\left(-2\int_\Omega v\nabla\eta_{n_k}.\nabla\zeta dx\right) = 4\gamma\frac{\partial\zeta}{\partial v}(0). \qquad (1.3.51)$$

If we let $n = n_k$ go to infinity in (1.3.46), we obtain

$$-\int_\Omega(v\Delta\zeta)dx = \int_\Omega f\,\zeta dx - \int_{\partial\Omega}\varphi\frac{\partial\zeta}{\partial v}dS - 4\gamma\frac{\partial\zeta}{\partial v}(0). \qquad (1.3.52)$$

Since there exists C(N) such that

$$-\int_\Omega P(x,O)\Delta\xi dx = C(N)\frac{\partial\xi}{\partial v}(0) \qquad (1.3.53)$$

for any $\xi \in C_0^{1,1}(\overline\Omega)$, we set $v(x) = w(x) + \dfrac{4\gamma}{C(N)}P(x,O)$ and w solves

$$-\int_\Omega(w\Delta\zeta)dx = \int_\Omega f\,\zeta dx - \int_{\partial\Omega}\varphi\frac{\partial\zeta}{\partial v}dS. \qquad (1.3.54)$$

Let W be the unique integrable function which solves

$$-\int_{\Omega} W\Delta\zeta dx = \int_{\Omega}|f|\zeta dx - \int_{\partial\Omega}\frac{\partial\zeta}{\partial\nu}|\varphi|dS \tag{1.3.55}$$

for any $\xi \in C_0^{1,1}(\overline{\Omega})$; by the maximum principle W is nonnegative, and $W \geq w \geq -W$ from Lemma 4.1. Since $v \in L^{(N+1)/(N-1)}(\Omega,\rho dx)$, we have

$$\lim_{\sigma\to 0} \sigma^{-2}\int_{\Omega\cap B(O,\sigma)} v\rho dx = 0. \tag{1.3.56}$$

If we apply Lemma 4.6 , we also have

$$\lim_{\sigma\to 0} \sigma^{-2}\int_{\Omega\cap B(O,\sigma)} W\rho dx = 0, \tag{1.3.57}$$

and finally

$$\lim_{\sigma\to 0} \sigma^{-2}\int_{\Omega\cap B(O,\sigma)} P(x,O)\rho(x)dx = 0. \tag{1.3.58}$$

But we have seen in (1.2.25) above that $P(x,O) \geq K^{-1}\rho(x)|x|^{-N}$. This implies that

$$\liminf_{\sigma\to 0} \sigma^{-2}\int_{\Omega\cap B(O,\sigma)} P(x,O)\rho(x)dx > 0, \tag{1.3.59}$$

which is a contradiction. Therefore $\gamma = 0$ in (1.3.50) and (1.3.53), which ends the proof.

Proof of Theorem 4.3. Since g satisfies (1.3.52) there exist a > 0 and b ≥ 0 such that

$$-\Delta u + au^{(N+1)/(N-1)} \leq b \tag{1.3.60}$$

a.e. on $\{x: u(x) \geq 0\}$. As in Lemma 4.5 we construct a maximal solution U of

$$-\Delta U + aU^{(N+1)/(N-1)} = 0, \quad U \geq 0, \tag{1.3.61}$$

in Ω which vanishes on $\partial\Omega\setminus\{O\}$ and is continuous in $\overline{\Omega}\setminus\{O\}$ where it satisfies $u(x) \leq U(x) + C$ for some positive $C = C(a,b,N,\max_{\partial G}\varphi^+)$.

Step 1. We claim the function U belongs to $L^{(N+1)/(N-1)}(\Omega,\rho dx)\cap M^{N/(N-1)}(\Omega)$. Because U satisfies

$$0 \leq U(x) \leq C\rho(x)|x|^{-N} \leq C'P(x,O) \tag{1.3.62}$$

from Lemma 4.5 and (1.2.25), and $x \mapsto P(x,O)$ belongs to $M^{N/(N-1)}(\Omega)$, from Lemma 4.3, it is the same with U. Let $\zeta \in C_0^{1,1}(\overline{\Omega})$, $\zeta \geq 0$ and $\eta_n \in C^{\infty}(\mathbb{R}^N)$ as in Lemma 4.7. Then

$$\int_{\Omega}-U(\zeta\Delta\eta_n + 2\nabla\eta_n\cdot\nabla\zeta + \eta_n\Delta\zeta)dx + a\int_{\Omega}U^{(N+1)/(N-1)}\zeta dx = 0.$$

But, as in Lemmas 4.7-4.8, we have

$$\left|\int_{\Omega} U\zeta\Delta\eta_n dx\right| \leq Cn^2\int_{\Omega_n} U\rho dx \leq C' \;, \quad \left|\int_{\Omega} U\nabla\zeta.\nabla\eta_n dx\right| \leq Cn^2\int_{\Omega_n} Udx \leq C'' \qquad (1.3.63)$$

where Ω_n denotes $\{x \in \Omega: 1/2n \leq |x| \leq 1/n\}$. Thus

$$\limsup_{n\to\infty} a\int_{\Omega} U^{(N+1)/(N-1)}\eta_n\zeta dx < \infty, \qquad (1.3.64)$$

which implies the claim (take for example a first positive eigenfunction of $-\Delta$ in $W_0^{1,2}(\Omega)$ for ζ).

Step 2. We claim that $U = 0$. From Step 1 and Lemmas 4.7-4.8, U satisfies

$$\int_{\Omega}\left(-U\Delta\zeta + aU^{(N+1)/(N-1)}\zeta\right)dx = 0 \qquad (1.3.65)$$

for any $\zeta \in C_0^{1,1}(\overline{\Omega})$. If we take again a first positive eigenfunction $-\Delta$ in $W_0^{1,2}(\Omega)$ for the function ζ, we conclude that U is zero.

Step 3. End of the proof. From Steps 1-2 the function u is bounded from above by some constant in Ω, and, in the same way, it is bounded from below by another constant; as a consequence, g(u) remains uniformly bounded. Taking $\zeta \in C_0^{1,1}(\overline{\Omega})$ and $\eta_n \in C^{\infty}(\mathbb{R}^N)$ as in Lemma 4.7, we have

$$\int_{\Omega}\left[-u\left(\zeta\Delta\eta_n + 2\nabla\eta_n.\nabla\zeta + \eta_n\Delta\zeta\right) + g(u)\zeta\eta_n\right]dx = -\int_{\partial\Omega}\varphi\frac{\partial}{\partial\nu}(\zeta\eta_n)dS. \qquad (1.3.66)$$

By letting n go to infinity, we conclude that the equation (1.1) is satisfied in the sense (1.2.2), with test functions in $C_0^{1,1}(\overline{\Omega})$.

4-1-4 More on removable singularities

The removability results of Section 4.1.4 admit extensions in two directions: the regularity of the boundary and the size of the singular set. In the first direction some results have been obtained by Fabbri and Véron ([FV]) and the problem can be formulated in the following way: we suppose that Ω is a domain of $\mathbb{R}^N$ whose boundary $\partial\Omega$ contains O; for the local geometry, we suppose that O is a conical point of $\partial\Omega$ and that the remaining of the boundary is either regular in the same way as in Lemma 4.5 or that it is locally a polyedra. If G is any subdomain of S^{N-1}, we call $\lambda_1(G)$ the first eigenvalue of $-\Delta_{S^{N-1}}$ in $W_0^{1,2}(G)$ and

$q_{G,N}$ the only zero of the function $x \mapsto \dfrac{2}{x-1}\left(\dfrac{2x}{x-1}-N\right) - \lambda_1(G)$ which is larger than 1.

As we have seen it in 4.1.1 (we have to replace S_+^{N-1} by G), the equation

$$-\Delta_{S^{N-1}}\omega - \frac{2}{q-1}\left(\frac{2q}{q-1}-N\right)\omega + |\omega|^{q-1}\omega = 0, \tag{1.4.1}$$

in G admits no non-zero solution vanishing on ∂G if

$$\frac{2}{q-1}\left(\frac{2q}{q-1}-N\right) \leq \lambda_1(G) \Leftrightarrow 1 < q \leq q_{G,N}. \tag{1.4.2}$$

Consequently, there exists no non-zero solution u of (1.1.1) in the cone $C_O(G)$ with vertex O and basis G which vanishes on $\partial C_O(G)\setminus\{O\}$ and has the form

$$u(r,\sigma) = r^{-2/(q-1)}\omega(\sigma), \quad (r,\sigma)\in\mathbb{R}_*^+\times G. \tag{1.4.3}$$

We introduce the two following hypotheses on Ω at O:

H1- $\bigcap\limits_{\delta>0}\left\{\bigcup\limits_{0<\varepsilon\leq\delta} H_{O,\varepsilon^{-1}}(S_\varepsilon(O)\cap\Omega)\right\}\subset G$, where $\mathcal{H}_{O,\varepsilon^{-1}}$ is the homothety with center O and dilatation factor ε^{-1} and G is a domain of S^{N-1}.

H2- Ω is locally a Lipschitz cone near O, which means that there exists a Lipschitz domain of S^{N-1} such that $\Omega\cap B(O,\rho) = \bigcup\limits_{0<\tau<\rho}\{\tau x: x/|x|\in G\}$ for some $\rho>0$.

THEOREM 4.4. *Let us suppose that* H1 *holds, g is a continuous real valued function defined on* $\mathbb{R}$ *satisfying*

$$\liminf_{r\to\infty} g(r)r^{-q} > 0, \quad \limsup_{r\to-\infty} g(r)|r|^{-q} < 0, \tag{1.4.4}$$

for some $q > q_{G,N}$, *and* $u\in C^2(\Omega)\cap C(\overline{\Omega}\setminus\{O\})$ *is a solution of (1.1.1) in* Ω *which coincides on* $\partial\Omega\setminus\{O\}$ *with a continuous function* φ *defined on whole* $\partial\Omega$. *Then* u *can be extended as a continuous function in* $\overline{\Omega}$.

Proof. Since $q > q_{G,N}$, we have $2/(q-1) < v_G$ where $-v_G$ is the negative root of the equation $y\mapsto y^2+(N-2)y-\lambda_1(G)=0$; moreover, if $\tilde{G}$ is any domain of S^{N-1} containing G and different from G, $\lambda_1(\tilde{G})<\lambda_1(G)$; this implies $v_{\tilde{G}}<v_G$. From H1 there exist a regular domain $\tilde{G}$ of S^{N-1} and $\delta_0>0$ such that

$$\bigcup\limits_{0<\varepsilon\leq\delta_0} H_{O,\varepsilon^{-1}}(S_\varepsilon(O)\cap\Omega)\subset\tilde{G}, \tag{1.4.5}$$

and $2/(q-1) < v_{\tilde{G}}$. From (1.4.4) u satisfies the inequality (1.3.4) for some a and b as in Lemma 4.4. Consequently the same *a priori* estimate (1.3.5) holds. We set

$$M = \max\left\{ \max_{x \in \Omega \cap S_{\delta_0}(O)} u(x), \ \max_{x \in \partial\Omega \cap B(O,\delta_0)} \varphi^+(x) \right\}, \tag{1.4.6}$$

and denote $\psi_{\tilde{G}}$ the first eigenfunction of $-\Delta_{S^{N-1}}$ in $W_0^{1,2}(\tilde{G})$ normalised by $0 \leq \psi_{\tilde{G}} \leq \max_{\tilde{G}} \psi_{\tilde{G}} = 1$.

For $\varepsilon > 0$, the function $x \mapsto \Psi_\varepsilon(x) = \varepsilon |x|^{-\nu_{\tilde{G}}} \psi_{\tilde{G}}(x/|x|) + M + (b/a)^{1/q}$ satisfies

$$-\Delta \Psi_\varepsilon + a\Psi_\varepsilon^q \geq b \tag{1.4.7}$$

in the truncated cone $C_O(\tilde{G}) \cap B(O,\delta_0)$ which contains $\Omega \cap B(O,\delta_0)$. From the choice of the boundary values of Ψ_ε and $2/(q-1) < \nu_{\tilde{G}}$, the function $x \mapsto (u(x) - \Psi_\varepsilon(x))^+$ has a compact support in this cone. It is therefore identically O; letting ε go to 0 yields

$$u(x) \leq M + (b/a)^{1/q} \quad \text{in} \quad C_O(\tilde{G}) \cap B(O,\delta_0). \tag{1.4.8}$$

In the same way u is bounded from below in this set. Since Ω satisfies the Wiener regularity criterion at O (see [GT]), u can be extended as a continuous function at O.

The next two results deal with the case where g at infinity has the critical growth of the function $r \mapsto r^{q_{G,N}}$. The technique developped here has been introduced by Licois [Li] in the study of conservative dynamical systems.

THEOREM 4.5. *Let us suppose that* H2 *holds*, g *is a continuous real valued function defined on* $\mathbb{R}$ *satisfying*

$$\liminf_{r \to \infty} g(r) r^{-q_{G,N}} > 0 \ , \quad \limsup_{r \to -\infty} g(r) |r|^{-q_{G,N}} < 0, \tag{1.4.9}$$

and $u \in C^2(\Omega) \cap C(\overline{\Omega} \setminus \{O\})$ *is a solution of* (1.1.1) *in* Ω *which coincides on* $\partial\Omega \setminus \{O\}$ *with a continuous function* φ *defined on whole* $\overline{\Omega}$. *Then the conclusion of Theorem 4.4 holds.*

Definition. A domain D of $\mathbb{R}^k$ or of a k-dimensionnal Riemannian manifold is admissible if there exist two positive constants C and R such that, for any $P \in \partial D$ and $\rho \in (0,R]$, there holds

$$\text{mes}\big(B(P,\rho) \cap {}^cD\big) \geq C\,\text{mes}(B(P,\rho)). \tag{1.4.10}$$

PROPOSITION 4.1. *Let* (M,g) *be a Riemannian manifold with metric tensor* g *and Laplacian* Δ_M, S *an admissible domain of* M , $q > 1$ *and* $\lambda_1(S)$ *the first eigenvalue of* $-\Delta_M$ *in* $W_0^{1,2}(S)$. *Then*

(i) if $h > 0$ *and* $\lambda \in \mathbb{R}$ *there exists a unique positive function* $u_h \in C(\overline{S}) \cap C^2(S)$ *which satisfies*

$$\begin{cases} -\Delta_M u_h - \lambda u_h + u_h^q = 0 & \text{in} \quad S, \\ \qquad\qquad\quad u_h = h & \text{on} \quad \partial S, \end{cases} \tag{1.4.11}$$

and the mapping $h \mapsto u_h$ *is increasing.*

(ii) if $h = 0$, *for any* $\lambda > \lambda_1(S)$ *(resp.* $\lambda \leq \lambda_1(S)$*) there exists a unique positive function (resp. there exists no function)* $w \in C(\overline{S}) \cap W_0^{1,2}(S)$ *satisfying* (1.4.11).

Before proving this result we need

LEMMA 4.9. *Let* (M,g), S *and* q *be as in Proposition 4.1 and* $\ell \in L^\infty(S)$. *If* u *and* v *belong to* $C(\overline{S}) \cap W_{\text{loc}}^{1,2}(S)$ *with* $u > 0$, $v \geq 0$ *in* S *and if they satisfy*

$$-\Delta_M u - \ell u + u^q \geq 0 \quad \text{and} \quad -\Delta_M v - \ell v + v^q \leq 0 \tag{1.4.12}$$

in S *with* $0 \leq v \leq u$ *on* ∂S, *then* $v \leq u$ *in* S.

Proof. For $\varepsilon > 0$, we set $u_\varepsilon = u + 2\varepsilon$, $v_\varepsilon = v + \varepsilon$. Then $(v_\varepsilon - u_\varepsilon)^+$ vanishes in a neighborhood of ∂S and we have

$$\frac{\Delta_M u_\varepsilon}{u_\varepsilon} + \ell - \frac{2\varepsilon\ell}{u_\varepsilon} - (u_\varepsilon - 2\varepsilon)^q \leq 0 \quad \text{and} \quad -\frac{\Delta_M v_\varepsilon}{v_\varepsilon} - \ell + \frac{\varepsilon\ell}{v_\varepsilon} + (v_\varepsilon - \varepsilon)^q \leq 0 \tag{1.4.13}$$

in weak sense. Therefore

$$\int_S \left(\left(\frac{\Delta_M u_\varepsilon}{u_\varepsilon} - \frac{\Delta_M v_\varepsilon}{v_\varepsilon} \right) - \ell\varepsilon \left(\frac{2}{u_\varepsilon} - \frac{1}{v_\varepsilon} \right) \right.$$
$$\left. - \left(\frac{(u_\varepsilon - 2\varepsilon)^q}{u_\varepsilon} - \frac{(v_\varepsilon - \varepsilon)^q}{v_\varepsilon} \right) \right) (v_\varepsilon^2 - u_\varepsilon^2)^+ \, dv_g \leq 0. \tag{1.4.14}$$

Since

$$\int_S \left(\frac{\Delta_M u_\varepsilon}{u_\varepsilon} - \frac{\Delta_M v_\varepsilon}{v_\varepsilon} \right) (v_\varepsilon^2 - u_\varepsilon^2)^+ dv_g$$
$$= \int_{S*} \left(\left| \nabla u_\varepsilon - \frac{u_\varepsilon}{v_\varepsilon} \nabla v_\varepsilon \right|^2 + \left| \nabla v_\varepsilon - \frac{v_\varepsilon}{u_\varepsilon} \nabla u_\varepsilon \right|^2 \right) dv_g \tag{1.4.15}$$

where $S*$ is some regular subdomain of S which contains the support of $(v_\varepsilon - u_\varepsilon)^+$, $\left|\dfrac{2\varepsilon}{u_\varepsilon} - \dfrac{\varepsilon}{v_\varepsilon}\right| \leq 3$, and $u > 0$ in S, we deduce from Lebesgue's theorem that

$$\lim_{\varepsilon \to 0} \int_S \ell \varepsilon \left(\frac{2}{u_\varepsilon} - \frac{1}{v_\varepsilon}\right)\left(v_\varepsilon^2 - u_\varepsilon^2\right)^+ dv_g = 0. \tag{1.4.16}$$

In the same way

$$\lim_{\varepsilon \to 0} \int_S \left(\frac{(u_\varepsilon - 2\varepsilon)^q}{u_\varepsilon} - \frac{(v_\varepsilon - \varepsilon)^q}{v_\varepsilon}\right)\left(v_\varepsilon^2 - u_\varepsilon^2\right)^+ dv_g \tag{1.4.17}$$

$$= \int_S \left(u^{q-1} - v^{q-1}\right)\left(v^2 - u^2\right)^+ dv_g \leq 0.$$

Therefore $(v - u)^+$ is zero.

Proof of Proposition 4.1. The uniqueness is a consequence of Lemma 4.9, and in fact, in case (ii) there exists no nonzero solution if $\lambda \leq \lambda_1(S)$. In case (i) the fact that $h \mapsto u_h$ is nondecreasing is also a consequence of Lemma 4.9 (and the strong maximum principle implies that $h \mapsto u_h$ is increasing). Existence in both cases (i) and (ii) with $\lambda > \lambda_1(S)$ is obtained my minimizing the functional

$$v \mapsto J(v) = \frac{1}{2} \int_S \left(|\nabla v|^2 + \frac{2}{q+1}|v|^{q+1} - \lambda v^2\right) dv_g, \tag{1.4.18}$$

This functional is defined on the v such that $v - h \in W_0^{1,2}(S) \cap L^{q+1}(S)$, $h \geq 0$. The minimizing procedure gives rise to a nontrivial nonnegative solution with the desired regularity, from the classical elliptic equations theory in admissible domains ([GT]). Existence in the case (i) can also be obtained by noticing that $a = \min\left(h, \lambda^{1/(q-1)}\right)$ (resp. $b = \max\left(h, \lambda^{1/(q-1)}\right)$) is a subsolution (resp. a supersolution) for (1.4.11).

Proof of Theorem 4.5. Step 1. Reduction of the equation to an autonomous system. As in the proof of Theorems 4.3-4.4, the problem is reduced to prove that any nonnegative function U which belongs to $C^2(\Omega) \cap C(\overline{\Omega} \setminus \{O\})$ and satisfies

$$-\Delta U + aU^{q_{G,N}} = 0 \quad \text{in} \quad \Omega \tag{1.4.19}$$

for some $a > 0$, and vanishes on $\partial\Omega \setminus \{O\}$ is identically 0. For the sake of simplicity we denote $q = q_{G,N}$. From Lemma 4.4, $|x|^{2/(q-1)}U(x)$ remains bounded in $\overline{\Omega} \setminus \{O\}$. If we set

$$y(t,\sigma) = r^{2/(q-1)}U(r,\sigma), \quad t = \ln(1/r), \quad \sigma \in S^{N-1}, \tag{1.4.20}$$

and $T_\rho = \ln(1/\delta_0)$, the function y is bounded, satisfies

$$y_{tt} + \left(\frac{2q+2}{q-1} - N\right)y_t + \frac{2}{q-1}\left(\frac{2q}{q-1} - N\right)y + \Delta_{S^{N-1}}y - ay^q = 0 \qquad (1.4.21)$$

in $[T_\rho,\infty)\times G$ and vanishes on $[T_\rho,\infty)\times\partial G$. Since $[T_\rho,\infty)\times\partial G$ is also admissible, we know from [GT] that $y(t,.)$ remains bounded in $C^\beta(\overline{G})\cap W_0^{1,2}(G)$, independently of t.

Step 2. We claim that $y(t,.) \underset{t\to\infty}{\to} 0$ in the $C^\gamma(\overline{G})\cap\left(W_0^{1,2}(G)\text{-weak}\right)$-topology for $0 < \gamma < \beta$. For $k > 0$ large enough $k\sqrt[\beta]{\psi_G}$ is a supersolution for (1.4.20) and is larger than y. For T positive let y^T be the solution of

$$y_{tt}^T + \left(\frac{2q+2}{q-1} - N\right)y_t^T + \frac{2}{q-1}\left(\frac{2q}{q-1} - N\right)y^T + \Delta_{S^{N-1}}y^T - a(y^T)^q = 0 \qquad (1.4.22)$$

on $\left(T_\rho, T_\rho + T\right)\times G$ which vanishes on $\left(T_\rho, T_\rho + T\right)\times\partial G$ and is subject to the boundary conditions $y^T(T_\rho,.) = y^T(T_\rho + T,.) = k\sqrt[\beta]{\psi_G(.)}$. From Lemma 4.9, we have

$$k\sqrt[\beta]{\psi_G} \geq y^T(t,\sigma) \geq y(t,\sigma) \quad \text{on } \left[T_\rho, T+T_\rho\right]\times G, \qquad (1.4.23)$$

which implies in particular that

$$y^{\tilde{T}} \leq y^T \quad \text{on } \left[T_\rho, T+T_\rho\right]\times G, \qquad (1.4.24)$$

for $\tilde{T} \geq T$. From local interior estimates, y^T remains bounded in $W_0^{1,2}(G)$ on $\left[T_\rho + 1, T + T_\rho - 1\right]$, uniformly with respect to T. When T tend to infinity, y^T decreases and converges in the $C_{loc}^\gamma\left((T_\rho,\infty)\times\overline{G}\right)$-topology to some Y which is a weak nonnegative solution of (1.4.21) on $(T_\rho,\infty)\times G$ vanishing on $(T_\rho,\infty)\times\partial G$ and such that $Y(t,.)$ belongs to $W_0^{1,2}(G)$ for any $t > T_\rho$. The main point is to prove that Y is decreasing:

For $\tau_0 > \tau > 0$, the function $t \mapsto x(t,.) = y^{T+\tau_0}(t+\tau,.)$ is defined in the cylinder $\left(T_\rho - \tau, T_\rho + T + \tau_0 - \tau\right)\times G$ where it solves the equation (1.4.21). When $t = T_\rho$ and $t = T_\rho + T$, we have

$$x(T_\rho,.) = y^{T+\tau_0}(T_\rho + \tau,.) \leq k\sqrt[\beta]{\psi_G(.)} \quad \text{and} \quad x(T_\rho + T,.) = y^{T+\tau_0}(t_\rho + T + \tau,.) \leq k\sqrt[\beta]{\psi_G(.)}$$

respectively from (1.4.23). Consequently

$$x(t,.) = y^{T+\tau_0}(t+\tau,.) \leq y^T(t,.), \qquad (1.4.25)$$

and letting T go to infinity yields

$$Y(t+\tau,\sigma) \le Y(t,\sigma) \qquad (1.4.26)$$

on $(T_\rho,\infty) \times G$. When t goes to infinity, $Y(t,.)$ converges in the $\left(W_0^{1,2}(G)-\text{weak}\right)-\cap C^\gamma(\overline{G})$-topology to some nonnegative function $w(.)$. Let ξ be any C^∞ nonnegative function, not identically zero, with compact support in $(0,1)$ and $\varphi \in W_0^{1,2}(G)$; then

$$\int_0^1 \int_G Y(t+\tau,\sigma)\left(\xi_{tt}(\tau) + \left(\frac{2q+2}{q-1} - N\right)\xi_t(\tau) \right.$$
$$\left. \left(+\frac{2}{q-1}\left(\frac{2q}{q-1} - N\right)\xi(\tau)\right)\varphi(\sigma)\mathrm{d}\sigma\mathrm{d}\tau \right) \qquad (1.4.27)$$
$$+\int_0^1 \int_G \xi(\tau)\left(Y(t+\tau,\sigma)\Delta_{S^{N-1}}\varphi - aY^q(t+\tau,\sigma)\varphi\right)\mathrm{d}\sigma\mathrm{d}\tau = 0.$$

Since $\displaystyle\int_0^1 \xi_t(\tau)\mathrm{d}\tau = \int_0^1 \xi_{tt}(\tau)\mathrm{d}\tau = 0$, we get

$$\int_G \left(w\Delta_{S^{N-1}}\varphi + \frac{2}{q-1}\left(\frac{2q}{q-1} - N\right)w\varphi - aw^q\varphi \right)\mathrm{d}\sigma = 0. \qquad (1.4.28)$$

From the assumption on G, $\dfrac{2}{q-1}\left(\dfrac{2q}{q-1} - N\right)$ is equal to $\lambda_1(G)$; since w is nonnegative and belongs to $C^\gamma(\overline{G}) \cap W_0^{1,2}(G)$, we deduce from Proposition 4.1-(ii) that it is identically zero. Since

$$0 \le y(t,\sigma) \le Y(t,\sigma) \qquad (1.4.29)$$

on $(T_\rho,\infty) \times G$, we conclude that

$$\lim_{t\to\infty} \|y(t,.)\|_{C^\gamma(\overline{G})} = 0. \qquad (1.4.30)$$

Step 3. End of the proof. For any $\varepsilon > 0$, we denote by u_ε the solution of

$$\begin{cases} -\Delta_{S^{N-1}}u_\varepsilon - \dfrac{2}{q-1}\left(\dfrac{2q}{q-1} - N\right)u_\varepsilon + u_\varepsilon^q = 0 & \text{in } G, \\[2ex] u_\varepsilon = \varepsilon & \text{on } \partial G, \end{cases} \qquad (1.4.31)$$

(see Proposition 4.1). Since $\dfrac{2}{q-1}\left(\dfrac{2q}{q-1}-N\right)=\lambda_1(G)$, $\lim\limits_{\varepsilon\to 0}\|u_\varepsilon\|_{C(\overline{G})}=0$. If we define now the function Ψ_ε by

$$\Psi_\varepsilon(t,\sigma)=u_\varepsilon(\sigma)+e^{-v_G(t-T_\rho)}\|y(T_\rho,.)\|_{L^\infty(G)}. \tag{1.4.32}$$

Then it is a supersolution of (1.4.21) in $(T_\rho,\infty)\times G$, which implies that the function $(t,\sigma)\mapsto\left(y-\Psi_\varepsilon\right)^+(t,\sigma)$ has a compact support in $(T_\rho,\infty)\times G$. From lemma 4.9 it is identically 0. Letting ε go to 0 yields $y(t,.)\le e^{-v_G(t-T_\rho)}\|y(T_\rho,.)\|_{L^\infty(G)}$ in $[T_\rho,\infty)\times\overline{G}$. Going back to U implies that it is uniformly bounded in Ω, and thus continuous in $\overline{\Omega}$. Since it vanishes on $\partial\Omega$, it is zero by the maximum principle.

Remark 4.5. If we assume that Ω admits an asymptotic cone $C_O(G)$ at O and that g has precisely the corresponding critical growth (1.4.9), the result of theorem 4.5 still applies provided $q_{G,N}\ne(N+2)/(N-2)$ and the approximation of $\partial\Omega\setminus\{O\}$ by the tangent cone is good enough. More precisely $\partial\Omega\setminus\{O\}$ is $C^{2,\gamma}$ for some $0<\gamma<1$, and there exists a smooth family of $C^{2,\gamma}$ diffeomorphisms Φ_ε from $\mathcal{H}_{O,\varepsilon^{-1}}(S_\varepsilon(O)\cap\partial\Omega)$ into $C_O(G)\cap S_1(O)=G$ satisfying

$$\|\Phi_r-I\|_{C^2}+r\|\partial\Phi_r/\partial r\|_{C^1}+r^2\|\partial^2\Phi_r/\partial r^2\|_{C^1}\le Cr^\theta, \tag{1.4.33}$$

for some $\theta>0$. The technique developped in [FV] consists in studying the behaviour at O of the solution U of (1.4.19) in $\Omega_\rho=\Omega\cap B(O,\rho)$, which is equivalent to study the asymptotics when t goes to infinity of the solution y of an equation of type (1.4.21) in a non-cylindrical domain $\Theta=\left\{(t,\sigma)\in(\ln(1/\rho),\infty)\times S^{N-1}:\ \sigma\in G_t\right\}$ where $t=\ln(1/r)$ and $G_t=\left\{\sigma\in S^{N-1}:(e^{-t},\sigma)\in\Omega_r\right\}$.

In the second direction the first results concerning the size of the removable set are due to Sheu [Sh]. By using a deep combination of probability theory and linear analysis, a general answer to the boundary removability problem for equation (1.1.1) has been given recently by Dynkin and Kuznetzov [DK1], unfortunately their method only works in the case $1<q\le 2$. We give below an overlook of their results.

Definition. Let $\Omega\subset\mathbb{R}^N$ be a bounded domain and $q>1$. We say that a compact subset $\Gamma\subset\partial\Omega$ is removable for the equation

$$-\Delta u+|u|^{q-1}u=0 \tag{1.4.34}$$

if any function u which belongs to $C^2(\Omega) \cap C(\overline{\Omega} \setminus \Gamma)$, solves (1.4.34) in Ω and vanishes on $\partial\Omega \setminus \Gamma$ is identically 0.

THEOREM 4.6. *Let* $\Omega \subset \mathbb{R}^N (\, N > 1)$ *be a bounded domain with a smooth boundary,* $1 < q \le 2$ *and* $\Gamma \subset \partial\Omega$ *be a compact subset. Then the following statements are equivalent:*
(i) Γ *is removable for equation* (1.4.34).
(ii) If $\mu \in \mathcal{M}(\partial\Omega)$ *is nonnegative and satisfies*

$$\int_\Omega \left(\int_\Gamma P(x,y) d\mu(y) \right)^q \rho(x) dx < \infty, \tag{1.4.35}$$

then $\int_K d\mu(y) = 0$.
(iii) If $\mu \in \mathcal{M}(\partial\Omega)$ *is nonnegative and satisfies*

$$\int_{\partial\Omega} \left(\int_\Gamma |x - y|^{2/q+1-N} d\mu(y) \right)^q \rho(x) dS(x) < \infty, \tag{1.4.36}$$

then $\int_K d\mu(y) = 0$.

Remark 4.6. The previous results are still valid if Δ is replaced by a second order strongly elliptic operator L with smooth coefficients. If fact, for such an operator, the corresponding Poisson kernel P_L has the same behaviour near $\partial\Omega$, namely

$$K_L^{-1} \rho(x) |x - y|^{-N} \le P_L(x,y) \le K_L \rho(x) |x - y|^{-N}. \tag{1.4.37}$$

([Ma2]). Therefore the condition (1.4.35) or (1.4.36) is independent of L.

Remark 4.7. It is interesting to express this type of result in terms of Poisson capacities. If $\Omega \subset \mathbb{R}^N$ is a bounded domain with a smooth boundary and $\Gamma \subset \partial\Omega$, we can define the p-Poisson capacity of $\Gamma (\, 1 < p < \infty)$ by

$$\mathrm{Cap}_p(\Gamma)$$
$$= \sup \left\{ \int_\Gamma d\mu(y) : \mu \in \mathcal{M}^+(\partial\Omega) , \int_\Omega \left(\int_\Gamma P(x,y) d\mu(y) \right)^{p/(p-1)} \rho(x) dx \le 1 \right\}. \tag{1.4.38}$$

Therefore (ii) means that the p'-Poisson capacity of Γ is zero.

In term of Hausdorff dimension Theorem 4.5 implies

COROLLARY 4.1. *Let* Ω *and q be as in Theorem* 4.5 *and* $\Gamma \subset \partial\Omega$ *with Hausdorff dimension* d. *If* $((N-1)q - (N+1))/(q-1) = d$, Γ *is removable for equation* (1.4.34).

Remark 4.8. In [Sh] Sheu proved that this result is valid for any $1 < q < \infty$.

4-1-5 Strong singularities

As in 4.1.2 we denote Ω a bounded domain of $\mathbb{R}^N$ with a C^3 boundary containing O and u is a solution of

$$-\Delta u + |u|^{q-1} u = 0 \quad \text{in } \Omega, \tag{1.5.1}$$

such that $u = \zeta$ on $\partial\Omega \setminus \{O\}$ where $\zeta \in \partial\Omega$ and $1 < q < (N+1)/(N-1)$. Let $T_O \partial\Omega$ be the tangent plane to $\partial\Omega$ at O ; we assume that

$$T_O \partial\Omega = \left\{ x = (x_1, x_2, \ldots, x_N) = (x', x_N) \in r^N : x_N = 0 \right\}. \tag{1.5.2}$$

There exist an open neighbourhood G of O and a C^3 real-valued function φ defined on $G \cap T_O \partial\Omega$ such that

$$G \cap \partial\Omega = \left\{ (x', \varphi(x')) : x' \in G \cap T_O \partial\Omega \right\}. \tag{1.5.3}$$

Moreover φ and $\nabla\varphi$ vanish at O. We set $y = \Phi(x)$ with $y_i = x_i$ if $1 \leq i < N$ and $y_N = x_N - \varphi(x')$, we can suppose that Φ is a diffeomorphism from G to $\tilde{G} = \Phi(G)$ and

$$\Phi(G \cap r^{N-1} \times r^+) = G \cap \Omega. \tag{1.5.4}$$

Let $z \in C^3(\overline{\Omega} \cap \overline{B}(O, 2R))$ be an harmonic lifting of ζ in $\Omega \cap B(O, 2R)$. If we set

$$u(x) - z(x) = \tilde{u}(y), \quad z(x) = \tilde{z}(y), \quad \zeta(x) = \tilde{\zeta}(y), \tag{1.5.5}$$

with $x = \varphi^{-1}(y)$ and $y \in G \cap \mathbb{R}^{N-1} \times \mathbb{R}^+$, then $\tilde{u}$ satisfies

$$\begin{cases} u_{x_i} - z_{x_i} = \tilde{u}_{y_i} - \varphi_{x_i} \tilde{u}_{y_N} & (\forall\, 1 \leq i \leq N-1), \\[4pt] u_{x_N} - z_{x_N} = \tilde{u}_{y_N}, \\[4pt] u_{x_i x_i} - z_{x_i x_i} = \tilde{u}_{y_i y_i} - 2\varphi_{x_i} \tilde{u}_{y_i y_N} - \varphi_{x_i x_i} \tilde{u}_{y_N} + \varphi_{x_i}^2 \tilde{u}_{y_N y_N} & (\forall\, 1 \leq i \leq N-1), \\[4pt] u_{x_N x_N} - z_{x_N x_N} = \tilde{u}_{y_N y_N}. \end{cases} \tag{1.5.6}$$

and it is a solution of

$$\Delta \tilde{u} + |\nabla\varphi|^2 \tilde{u}_{y_N y_N} - 2\nabla\varphi . \nabla\tilde{u}_{y_N} - \tilde{u}_{y_N} \Delta\varphi - (\tilde{u} + \tilde{z})|\tilde{u} + \tilde{z}|^{q-1} = 0, \tag{1.5.7}$$

in $\tilde{G} \cap \mathbb{R}^{N-1} \times \mathbb{R}^+$ which vanishes on $\tilde{G} \cap \mathbb{R}^{N-1} \times \{0\} \setminus \{O\}$.

We recall that $\mathcal{E}$ is the set of solutions of

$$-\Delta_{S^{N-1}}\omega - \frac{2}{q-1}\left(\frac{2q}{q-1}-N\right)\omega + |\omega|^{q-1}\omega = 0 \qquad (1.5.8)$$

on S^{N-1} and $\mathcal{E}^+$ the set of the solutions of the same equation (1.5.8) on S_+^{N-1} which vanishes on the equator $\partial S_+^{N-1} = S^{N-2}$; by convention $S_+^{N-1} = \overline{B}(O,1) \cap \mathbb{R}^{N-1} \times \mathbb{R}^+$. In order $\mathcal{E}^+$ be non-trivial we have to assume $1 < q < (N+1)/(N-1)$. With the above notations we prove

THEOREM 4.7. *If Ω, q and $\tilde{u}$ are as above, there exists a compact connected component $\mathcal{F}$ of $\mathcal{E}^+$ such that*

$$\lim_{r \to 0} \operatorname{dist}_{C^2}(r^{2/(q-1)}\tilde{u}(r,.),F) = \inf\left\{\left\|r^{2/(q-1)}\tilde{u}(r,.) - \omega(.)\right\|_{C^2(S_+^{N-1})} : \omega \in \mathcal{F}\right\} = 0. \quad (1.5.9)$$

Proof. Step 1. Reduction of the equation to a quasi-autonomous sytem. Let ∇_σ be the tangential gradient on S^{N-1} identified with its covariant derivative via the imbedding $S^{N-1} \subset \mathbb{R}^N$, $\vec{n}$ the normal outward unit vector to S^{N-1} and $(\vec{e}_1, \vec{e}_2, ..., \vec{e}_N)$ the canonical basis in $\mathbb{R}^N$. Then

$$\begin{cases}
\nabla\tilde{u} = \tilde{u}_r\,\vec{n} + r^{-1}\nabla_\sigma\tilde{u}\,, \\[2mm]
\tilde{u}_{y_N} = \tilde{u}_r <\vec{n}.\vec{e}_N> + r^{-1} <\vec{e}_N.\nabla_\sigma\tilde{u}>\,, \\[2mm]
\nabla\tilde{u}_{y_N} = u_{rr} <\vec{n}.\vec{e}_N>\vec{n} + r^{-1}\nabla_\sigma\left[\tilde{u}_r <\vec{n}.\vec{e}_N>\right] \\[2mm]
\qquad -\left(r^{-2} <\vec{e}_N.\nabla_\sigma\tilde{u}> - r^{-1} <\vec{e}_N.\nabla_\sigma\tilde{u}_r>\right)\vec{n} + r^{-2}\nabla_\sigma\left[<\vec{e}_N.\nabla_\sigma\tilde{u}>\right], \\[2mm]
\nabla\varphi = \varphi_r\,\vec{n} + r^{-1}\nabla_\sigma\varphi\,, \\[2mm]
\Delta\varphi = \varphi_{rr} + (N-1)r^{-1}\varphi_r + r^{-2}\Delta_{S^{N-1}}\varphi\,.
\end{cases} \qquad (1.5.10)$$

Therefore (1.5.7) reads as

$$\tilde{u}_{rr} + \frac{N-1}{r}\tilde{u}_r + \frac{1}{r^2}\Delta_{S^{N-1}}\tilde{u} - (\tilde{u}+\tilde{z})|\tilde{u}+\tilde{z}|^{q-1}$$

$$-\left(\tilde{u}_r <\vec{n}.\vec{e}_N> + \frac{1}{r}<\nabla_\sigma \tilde{u}.\vec{e}_N>\right)\Delta\varphi$$

$$-2\varphi_r\left(\tilde{u}_{rr} <\vec{n}.\vec{e}_N> - \frac{1}{r^2}<\nabla_\sigma \tilde{u}.\vec{e}_N> + \frac{1}{r}<\nabla_\sigma \tilde{u}_r.\vec{e}_N>\right)$$

$$+|\nabla\varphi|^2\left(\tilde{u}_{rr} <\vec{n}.\vec{e}_N>^2 - \frac{1}{r^2}<\nabla_\sigma \tilde{u}.\vec{e}_N><\vec{n}.\vec{e}_N> + \frac{1}{r}<\nabla_\sigma \tilde{u}_r.\vec{e}_N><\vec{n}.\vec{e}_N>\right)$$

$$+|\nabla\varphi|^2\left(\frac{1}{r}<\nabla_\sigma\left[\tilde{u}_r <\vec{n}.\vec{e}_N>\right].\vec{e}_N> + \frac{1}{r^2}<\nabla_\sigma\left[<\nabla_\sigma \tilde{u}.\vec{e}_N>\right].\vec{e}_N>\right) \quad (1.5.11)$$

$$-\frac{2}{r^2}<\nabla_\sigma\varphi.\nabla_\sigma\left[\tilde{u}_r <\vec{n}.\vec{e}_N>\right]> - \frac{2}{r^3}<\nabla_\sigma\varphi.\nabla_\sigma\left[<\nabla_\sigma \tilde{u}.\vec{e}_N>\right]> = 0.$$

Since

$$\vec{\nabla}_\sigma\left[\tilde{u}_r <\vec{n}.\vec{e}_N>\right] = \tilde{u}_r \vec{\nabla}_\sigma\left[<\vec{n}.\vec{e}_N>\right] + <\vec{n}.\vec{e}_N> \vec{\nabla}_\sigma \tilde{u}_r, \quad (1.5.12)$$

the following expression is derived from (1.5.11)

$$r^2\tilde{u}_{rr}\left(1 - 2\varphi_r <\vec{n}.\vec{e}_N> + |\nabla\varphi|^2 <\vec{n}.\vec{e}_N>^2\right)$$

$$+ r\tilde{u}_r\left(N-1-r<\vec{n}.\vec{e}_N>\Delta\varphi\right.$$

$$+ r|\nabla\varphi|^2\left(<\nabla_\sigma <\vec{n}.\vec{e}_N>.\vec{e}_N> - 2<\nabla_\sigma\varphi.\nabla_\sigma <\vec{n}.\vec{e}_N>>\right)\Big)$$

$$+ <\nabla_\sigma \tilde{u}.\vec{e}_N>\left(-r\Delta\varphi + 2\varphi_r - |\nabla\varphi|^2 <\vec{n}.\vec{e}_N>\right)$$

$$+ r<\nabla_\sigma \tilde{u}.\vec{e}_N>\left(2\varphi_r + 2|\nabla\varphi|^2 <\vec{n}.\vec{e}_N>\right) - 2<\vec{n}.\vec{e}_N><\nabla_\sigma \tilde{u}.v\varphi> \quad (1.5.13)$$

$$+ <\nabla_\sigma\left[<\nabla_\sigma \tilde{u}.\vec{e}_N>\right].\left(|\nabla\varphi|^2 \vec{e}_N - 2r^{-1}\nabla_\sigma\varphi\right)>$$

$$+ \Delta_{S^{N-1}}\tilde{u} - r^2(\tilde{u}+\tilde{z})|\tilde{u}+\tilde{z}|^{q-1} = 0.$$

We now set

$$v(t,\sigma) = r^{2/(q-1)}\tilde{u}(r,\sigma), \quad \alpha(t,\sigma) = r^{2/(q-1)}\tilde{z}(r,\sigma), \quad t = \ln(1/r); \quad (1.5.14)$$

then v satisfies

$$(1+\varepsilon_1)v_{tt} + \left(2\frac{q+1}{q-1} - N + \varepsilon_2\right)v_t + \left(\frac{2}{q-1}\left(\frac{2q}{q-1} - N\right) + \varepsilon_3\right)v$$

$$+\Delta_{S^{N-1}}v + <\nabla_\sigma v.\vec{\varepsilon}_4> + <\nabla_\sigma v_t.\vec{\varepsilon}_5> + <\nabla_\sigma\left(<\nabla_\sigma v.\vec{e}_N>\right).\vec{\varepsilon}_6> \qquad (1.5.15)$$

$$-(v+\alpha)|v+\alpha|^{q-1} = 0$$

in $(\ln(1/R), \infty) \times S_+^{N-1}$, vanishes on $(\ln(1/R), \infty) \times \partial S_+^{N-1}$; as for the functions ε_j $(1 \le j \le 6)$ they are defined by

$$\varepsilon_1 = -2 <\vec{n}.\vec{e}_N> \varphi_r + |\nabla\varphi|^2 <\vec{n}.\vec{e}_N>^2, \qquad (1.5.16)$$

$$\varepsilon_2 = r\Delta\varphi <\vec{n}.\vec{e}_N> - r|\nabla\varphi|^2 <\nabla_\sigma\left(<\vec{n}.\vec{e}_N>\right).\vec{e}_N>$$

$$+2<\nabla_\sigma\varphi.\nabla_\sigma\left(<\vec{n}.\vec{e}_N>\right)> -2\frac{q+3}{q-1}<\vec{n}.\vec{e}_N> \varphi_r + \frac{q+3}{q-1}|\nabla\varphi|^2 <\vec{n}.\vec{e}_N>^2, \qquad (1.5.17)$$

$$\varepsilon_3 = 4\frac{q+1}{q-1}\varphi_r <\vec{n}.\vec{e}_N> +2\frac{q+1}{q-1}|\nabla\varphi|^2 <\vec{n}.\vec{e}_N> + \frac{2r}{q-1}<\vec{n}.\vec{e}_N> \Delta\varphi$$

$$+\frac{2r}{q-1}|\nabla\varphi|^2 <\nabla_\sigma\left(<\vec{n}.\vec{e}_N>\right).\vec{e}_N> + \frac{4}{q-1}<\nabla_\sigma\varphi.\nabla_\sigma\left(<\vec{n}.\vec{e}_N>\right)>, \qquad (1.5.18)$$

$$\vec{\varepsilon}_4 = \left(-r\Delta\varphi + 2\frac{q+1}{q-1}\varphi_r - \frac{q+3}{q-1}|\nabla\varphi|^2 <\vec{n}.\vec{e}_N>\right)\vec{e}_N + \frac{4}{q-1}r^{-1}\nabla_\sigma\varphi, \qquad (1.5.19)$$

$$\vec{\varepsilon}_5 = \left(2\varphi_r - 2|\nabla\varphi|^2 <\vec{n}.\vec{e}_N>\right)\vec{e}_N - 2<\vec{n}.\vec{e}_N> r^{-1}\nabla_\sigma\varphi, \qquad (1.5.20)$$

$$\vec{\varepsilon}_6 = |\nabla\varphi|^2 \vec{e}_N + 2r^{-1}\nabla_\sigma\varphi. \qquad (1.5.21)$$

Step 2. Uniform estimates. From the choice of φ we have

$$|\varphi(r,.)| \le Cr^2 = Ce^{-2t}, \qquad\qquad |\nabla\varphi(r,.)| \le Cr = Ce^{-t}, \qquad (1.5.22)$$

which implies that

$$\left\|\varepsilon_j(t,.)\right\|_{L^\infty(S_+^{N-1})} \le Ce^{-t} \qquad\qquad (\forall\, 1 \le j \le 6). \qquad (1.5.23)$$

Since $|\alpha(t,.)| \le Ce^{-2t/(q-1)}$, we deduce from the Agmon-Douglis-Nirenberg estimates ([GT]) that

$$\|v\|_{W^{2,p}\left((T-1,T+1)\times S_+^{N-1}\right)} \leq C = C(p,N,q) \tag{1.5.24}$$

for any $T \geq 2 + \ln(1/R)$ and $p < \infty$. Moreover $\left|D^2\varphi(r,,.))\right| + \left|D^3\varphi(r,,.))\right| \leq C$, therefore, differentiating the ε_j with respect to t or taking its covariant derivatives, yields

$$\left\|(\varepsilon_j)_t(t,.)\right\|_{L^\infty(S_+^{N-1})} + \left\|\nabla_\sigma\varepsilon_j(t,.)\right\|_{L^\infty(S_+^{N-1})} \leq Ce^{-t}, \tag{1.5.25}$$

and, in the same way,

$$\left\|\alpha_t(t,.)\right\|_{L^\infty(S_+^{N-1})} + \left\|\nabla_\sigma\alpha(t,.)\right\|_{L^\infty(S_+^{N-1})} \leq Ce^{-2t/(q-1)}. \tag{1.5.26}$$

Therefore the Schauder estimates imply that for any $\gamma \in (0,1)$, there exists $C_\gamma > 0$ such that for any $t \geq 2 + \ln(1/R)$, there holds

$$\|v(t,.)\|_{C^2(S_+^{N-1})} + \|v_t(t,.)\|_{C^1(S_+^{N-1})} + \|v_{tt}(t,.)\|_{C(S_+^{N-1})}$$
$$+ \sup_{0<h\leq 1} \frac{\|v_{tt}(t+h,.) - v_{tt}(t,.)\|_{C(S_+^{N-1})}}{h^\gamma} \leq C_\gamma. \tag{1.5.27}$$

Step 3. Energy estimates: we claim that

$$\int_{\ln(1/R)}^{\infty} \int_{S_+^{N-1}} \left(v_t^2 + v_{tt}^2 + |\nabla_\sigma v_t|^2\right) d\sigma dt < \infty. \tag{1.5.28}$$

If we multiply (1.5.15) by v_t and integrate over S_+^{N-1}, we obtain

$$\left(2\frac{q+1}{q-1} - N + \varepsilon_2 - \frac{1}{2}(\varepsilon_1)_t\right)\int_{S_+^{N-1}} v_t^2 d\sigma - \frac{1}{2}(\varepsilon_3)_t \int_{S_+^{N-1}} v^2 d\sigma$$

$$= \frac{d}{dt}\left[\int_{S_+^{N-1}}\left(\frac{1}{2}|\nabla_\sigma v|^2 + \frac{1}{q+1}|v+\alpha|^{q+1}\right.\right.$$

$$\left.\left. - \frac{1}{2}\left(\frac{2}{q-1}\left(\frac{2q}{q-1} - N\right) + \varepsilon_3\right)v^2 - \frac{1}{2}(1+\varepsilon_1)v_t^2\right)d\sigma\right]$$

$$- \int_{S_+^{N-1}}(v+\alpha)|v+\alpha|^{q-1}\alpha_t d\sigma$$

$$- \int_{S_+^{N-1}} v_t\left(<\nabla_\sigma v.\vec{\varepsilon_4}> - <\nabla_\sigma v_t.\vec{\varepsilon_5}> - <\nabla_\sigma\left(<\nabla_\sigma v.\vec{e_N}>\right).\vec{\varepsilon_6}\right)d\sigma. \tag{1.5.29}$$

Integrating over $(\ln(1/R),\infty)$ and using (1.5.23), (1.5.25)-(1.5.27) yields

$$\int_{\ln(1/R)}^{\infty} \int_{S_+^{N-1}} v_t^2 \, d\sigma dt < \infty.$$

(1.5.30)

Multiply (1.5.15) by v_{tt} and integrating over S_+^{N-1} gives

$$(1+\varepsilon_1)\int_{S_+^{N-1}} v_{tt}^2 \, d\sigma + \int_{S_+^{N-1}} |\nabla_\sigma v_t|^2 \, d\sigma$$

$$= \frac{d}{dt}\left[\int_{S_+^{N-1}}\left(\frac{1}{2}\left(N-2\frac{q+1}{q-1}+\varepsilon_2\right)v_t^2\right.\right.$$

$$\left.\left.-\left(\frac{2}{q-1}\left(\frac{2q}{q-1}-N\right)+\varepsilon_3\right)vv_t + <\nabla_\sigma v.\nabla_\sigma v_t>\right)d\sigma\right]$$

$$+\int_{S_+^{N-1}}\left(\frac{1}{2}(\varepsilon_2)_t v_t^2 + \left(\frac{2}{q-1}\left(\frac{2q}{q-1}-N\right)+\varepsilon_3\right)v_t^2 + (\varepsilon_3)_t vv_t\right)d\sigma$$

(1.5.31)

$$+\int_{S_+^{N-1}} v_{tt}\left(<\nabla_\sigma v.\vec{\varepsilon_4}>+<\nabla_\sigma v_t.\vec{\varepsilon_5}>+<\nabla_\sigma\left(<\nabla_\sigma v.\vec{e_N}>\right).\vec{\varepsilon_6}\right)d\sigma$$

$$-q\int_{S_+^{N-1}} v_t(v_t+\alpha_t)|v+\alpha|^{q-1}d\sigma + \frac{d}{dt}\left[\int_{S_+^{N-1}} v_t(v+\alpha)|v+\alpha|^{q-1}d\sigma\right].$$

Integrating (1.5.31), using (1.5.23), (1.5.25)-(1.5.27), (1.5.29) infers the final estimate (1.5.28).

Step 4. End of the proof. From (1.5.27) and (1.5.28) we have

$$\lim_{t\to\infty}\int_{S_+^{N-1}}\left(v_t^2+v_{tt}^2+|\nabla_\sigma v_t|^2\right)(t,\sigma)d\sigma = 0$$

(1.5.32)

and the positive trajectory of v, say $\mathcal{T}$, which is relatively compact in $C^2(S_+^{N-1})$, has a non empty limit set at infinity. Any element ω of this limit set is a solution of (1.5.8) and therefore belongs to the same connected component of $\mathcal{E}^+$ which is the set of the solutions of (1.5.8) on S_+^{N-1} vanishing on ∂S_+^{N-1}.

In some particular cases the limit set of the trajectory of v is reduced to a single element.

THEOREM 4.8. *Suppose that* Ω, *ũ and* q *are as in Theorem* 4.7 *and suppose also that one the following hypotheses is fulfilled*
H1- $N=2$,
H2- $u^-(x) = o(|x|^{-2/(q-1)})$ *near* O, *or*
H3- $(N+2)/N \leq q < (N+1)/(N-1)$.
Then there exists $\omega \in \mathcal{E}^+$ *such that* $r^{2/(q-1)}\tilde{u}(r,.)$ *converges to* ω *in the* $C^2(S_+^{N-1})$-*topology when r goes to* 0.

Proof. If $N = 2$, $\mathcal{E}^+$ is the set of ω which satisfies

$$\begin{cases} \dfrac{d^2\omega}{d\sigma^2} + \dfrac{2}{q-1}\left(\dfrac{2q}{q-1} - N\right)\omega - |\omega|^{q-1}\omega = 0 & \text{on} \quad (0,\pi), \\[2mm] \omega(0) = \omega(\pi) = 0. \end{cases} \tag{1.5.33}$$

This set is totally disconnected from the proof of Proposition 2.3.

If $u^-(x) = o(|x|^{-2/(q-1)})$ near O, any element ω of the limit set of the trajectory is nonnegative and the strong maximum principle implies that either it is identically 0 or positive and, in both cases, unique from Proposition 4.1 (ii).

If $(N+2)/N \le q < (N+1)/(N-1)$, then

$$N - 1 = \lambda_1(S_+^{N-1}) < \frac{2}{q-1}\left(\frac{2q}{q-1} - N\right) \le \lambda_2(S_+^{N-1}) = 2N. \tag{1.5.34}$$

It is classical that any solution of (1.4.1) keeps a constant sign, which implies that $\mathcal{E}^+$ is totally disconnected. We recall the proof in a more general framework where G is a domain of $\mathbb{R}^N$, L a second order strongly elliptic operator in divergence form

$$L\zeta = \sum_{i,j} \frac{\partial}{\partial x_j}\left(a_{i,j}\frac{\partial\zeta}{\partial x_i}\right), \tag{1.5.35}$$

with associated bilinear form

$$Q_L(\mathbf{X}, \mathbf{Y}) = \sum_{i,j} a_{i,j} X_i Y_j. \tag{1.5.36}$$

We have the classical Courant-Hilbert characterisation of the second eigenvalue of -L in $W_0^{1,2}(G)$ (see [CR])

$$\lambda_2(G) = \min_{(G',G'')} \, \max\big(\lambda_1(G'), \lambda_1(G'')\big), \tag{1.5.37}$$

where G' and G'' run over the set of non-empty disjoint subdomains of G. If $\lambda_1(G) < \lambda \le \lambda_2(G)$ and $q > 0$, any solution of

$$-Lw - \lambda w + |w|^{q-1} w = 0 \tag{1.5.38}$$

in $W_0^{1,2}(G)$ has a constant sign. If it were not true, consider G^+ (resp. G^-) a connected component of the support of w^+ (resp. w^-). Then

$$0 = \int_{G^+} Q_L(\nabla w, \nabla w)dx - \lambda\int_{G^+} w^2 dx + \int_{G^+} w^{q+1} dx > (\lambda_1(G^+) - \lambda)\int_{G^+} w^2 dx, \tag{1.5.39}$$

$$0 = \int_{G^-} Q_L(\nabla w, \nabla w)dx - \lambda \int_{G^-} w^2 dx + \int_{G^-} (-w)^{q+1} dx > (\lambda_1(G^-) - \lambda)\int_{G^-} w^2 dx \quad (1.5.40)$$

Since (1.5.29), (1.5.30) implies that $\lambda > \max(\lambda_1(G^+), \lambda_1(G^-))$, this contradicts (1.5.37).

Remark 4.9. If u_i $(i = 1,2)$ are two solutions of (1.5.1) in Ω, continuous in $\overline{\Omega} \setminus \{O\}$, satisfying

$$\lim_{r \to 0} r^{2/(q-1)} \tilde{u}_i(r,.) = \omega^+(.) \quad (1.5.41)$$

where ω^+ is the unique positive element of $\mathcal{E}^+$ and if they coincide on $\partial\Omega \setminus \{O\}$ with a continuous function defined on whole $\partial\Omega$, then $u_1 = u_2$. Actually the maximum principle implies that they are bounded below on Ω by some negative constant - M. Moreover, since for any $\varepsilon > 0$, the function $(1 + \varepsilon)u_i + M$ is a super solution, we deduce that $|u_1 - u_2| \le M$ and we conclude by monotonicity.

4-1-6 Weak singularities

In this section we keep the notations u, $\tilde{u}$, Ω and G of the previous one.

THEOREM 4.9. *Suppose that* u *is a solution of* (1.5.1) *as in Theorem 4.7, such that* $\lim_{r \to 0} r^{2/(q-1)} \tilde{u}(r,.) = 0$ *uniformly on* S_+^{N-1} *and one of the three following conditions be fulfilled*

H1- $u^-(x) = O\left(|x|^{-2/(q-1)+\delta}\right)$ *near* O, *for some* $\delta > 0$,

H2- $N = 2$ *and* $\partial\Omega$ *is locally a straight line near* O,
H3- $2/(q-1)$ *is not an integer.*
Then
 (i) either u *can be extended to* $\overline{\Omega}$ *as a continuous function, or*
 (ii) there exist an integer $k \in [N-1, 2/(q-1))$ *and a non-zero spherical harmonic* ψ *of degree* $k + 2 - N$, *vanishing on* ∂S_+^{N-1} *such that*

$$\lim_{r \to 0} r^k \tilde{u}(r,.) = \psi(.) \quad (1.6.1)$$

in $C^2(S_+^{N-1})$.

The techniques for proving this results are very close to the ones of Theorem 2.4, therefore we shall abreviate the proof.

Proof. Step 1. There exists $M > 0$ and $\varepsilon > 0$ such that

$$\left\| |y|^{2/(q-1)} \tilde{u}(y) \right\| \le M|y|^{\varepsilon}. \quad (1.6.2)$$

We define v from ũ via the transformation (1.5.14). If estimate (1.6.2) is not true, we set $\rho(t) = \|v(t,.)\|_{L^\infty(S^{N-1}_+)}$ and introduce the slowly decaying function η defined in Lemma 2.1. Then $w = v / \eta$ is a solution of

$$(1+\varepsilon_1)w_{tt} + \left(2\frac{q+1}{q-1} - N + \varepsilon_2 + 2\frac{\eta'}{\eta}\right)w_t$$

$$+ \left(\frac{2}{q-1}\left(\frac{2q}{q-1} - N\right) + \varepsilon_3 + \frac{\eta''}{\eta} + \left(2\frac{q+1}{q-1} - N\right)\frac{\eta'}{\eta}\right)w + \Delta_{S^{N-1}}w$$

$$+ < \vec{\nabla}_\sigma w.\vec{\varepsilon}_4 + \frac{\eta'}{\eta}\vec{\varepsilon}_5 > + < \vec{\nabla}_\sigma w_t.\vec{\varepsilon}_5 > + < \vec{\nabla}_\sigma\left(< \vec{\nabla}_\sigma w.\vec{e}_N >\right).\vec{\varepsilon}_6 >$$

$$- \eta^{q-1}(w + \alpha\eta^{-1})|w + \alpha\eta^{-1}|^{q-1} = 0. \tag{1.6.3}$$

Since $\alpha(t,.) = O(e^{-2t/(q-1)})$, w is bounded; the elliptic and energy estimates as in the proof of Theorem 2.4 imply that the positive trajectory $\mathcal{T}_w$ of w is relatively compact in $C^2(S^{N-1}_+)$ and contains a non-zero element ω which is a solution of

$$-\Delta_{S^{n-1}}\omega = \frac{2}{q-1}\left(\frac{2q}{q-1} - N\right)\omega \tag{1.6.4}$$

on S^{N-1}_+, vanishing on ∂S^{N-1}_+. Consequently $\frac{2}{q-1}\left(\frac{2q}{q-1} - N\right)$ is an eigenvalue of $-\Delta_{S^{N-1}_+}$ in $W^{1,2}_0(S^{N-1}_+)$ and ω a corresponding eigenfunction.

If H1 holds ω must be positive and $\frac{2}{q-1}\left(\frac{2q}{q-1} - N\right) = \lambda_1(S^{N-1}_+) = N - 1$ which means that

$q = (N+1)/(N-1)$ and no singularity can occur from Theorem 4.3.

If H2 holds we perform a reflection through $\partial\Omega$. We apply Theorem 2.5 which gives a complete classification of the isolated singularities of the solutions of (1.5.1) and asserts that (1.6.2) should hold if $\lim_{r\to 0} r^{2/(q-1)}u(r,.) = 0$.

If H3 holds, $\frac{2}{q-1}\left(\frac{2q}{q-1} - N\right)$ is not an eigenvalue of $-\Delta_{S^{N-1}_+}$ in $W^{1,2}_0(S^{N-1}_+)$.

In any case we have a contradiction.

Step 2. Let k_q be the largest integer smaller than $2/(q-1)$. Then

$$\|\tilde{u}(r,.)\|_{W^{1,2}_0(S^{N-1}_+)} \leq Mr^{-k_q}, \tag{1.6.5}$$

for some $M > 0$. The technique of the proof is exactly the same as the one of Theorem 2.4 and this estimate is transformed in the same way into a uniform estimate, precisely

$$\left\| \tilde{u}(r,.) \right\|_{L^{\infty}(S_+^{N-1})} \le Mr^{-k_q} . \tag{1.6.6}$$

If we set

$$\tilde{u}(r,\sigma) = r^{-k_q} y(t,\sigma), \quad t = \ln(1/r), \tag{1.6.7}$$

and $\delta_q = 2/(q-1) - k_q$, then y satisfies

$$
\begin{aligned}
&(1+\varepsilon_1)y_{tt} + \left(2\frac{q+1}{q-1} - N + \varepsilon_2 - 2\delta_q \right)y_t \\
&\quad + \left(\frac{2}{q-1}\left(\frac{2q}{q-1} - N \right) + \varepsilon_3 + \delta_q^2 - \left(2\frac{q+1}{q-1} - N \right)\delta_q \right)y + \Delta_{S^{N-1}} y \\
&\quad + <\vec{\nabla}_\sigma y.\vec{\varepsilon}_4 - \delta_q \vec{\varepsilon}_5 > + <\vec{\nabla}_\sigma y_t.\vec{\varepsilon}_5 > + <\vec{\nabla}_\sigma \left(<\vec{\nabla}_\sigma y.\vec{e}_N > \right).\vec{\varepsilon}_6 > \\
&\quad - e^{-\delta_q(q-1)t}(y+\alpha e^{\delta_q t})\left| y+\alpha e^{\delta_q t} \right|^{q-1} = 0.
\end{aligned}
\tag{1.6.8}
$$

Consequently there exists an eigenfunction Φ of $-\Delta_{S_+^{N-1}}$ in $W_0^{1,2}(S_+^{N-1})$ corresponding to the eigenvalue $\lambda_{k_q-N+2}(S_+^{N-1}) = k_q(k_q+2-N)$, such that

$$\lim_{t\to\infty} y(t,.) = \Phi(.), \tag{1.6.9}$$

in $C^2(S_+^{N-1})$, which means

$$\lim_{r\to 0} r^{k_q}\tilde{u}(r,.) = \Phi(.). \tag{1.6.10}$$

Step 3. End of the proof. If the spherical harmonic Φ in (1.6.9) is zero, (1.6.6) is still valid with δ_q replaced by $\delta_q - 1$ and (1.6.10) by

$$\lim_{r\to 0} r^{k_q-1}\tilde{u}(r,.) = \Phi'(.), \tag{1.6.11}$$

for some spherical harmonic Φ' of degree k_q+1-N. Continuing this process, then either there exist an integer $k \in [N-1, 2/(q-1))$ and a nonzero spherical harmonic ψ of degree $k+2-N$ vanishing on ∂S_+^{N-1} such that (1.6.1) holds or

$$\lim_{r\to 0} r^{N-1}\tilde{u}(r,.) = 0 \tag{1.6.12}$$

in $C^2(S_+^{N-1})$. Consequently $u(x) = o(P(x,O))$ near O. We deduce from the maximum principle that

$$|u(x)| \leq \max_{y \in \partial\Omega} |\zeta(y)| = M, \tag{1.6.13}$$

and we conclude as in Theorem 4.3 that u can be extended by continuity in $\overline{\Omega}$.

As an example we have a classification of the nonnegative solutions of (1.5.1) with a single boundary singularity.

COROLLARY 4.2. *Let* $1 < q < (N+1)/(N-1)$ *and* $u \in C^3(\Omega) \cap C(\overline{\Omega} \setminus \{O\})$ *be a nonnegative solution of* (1.5.1) *in* Ω *which coincides on* $\partial\Omega \setminus \{O\}$ *with a continuous function* $\zeta \in C^3(\partial\Omega)$. *Then:*
  (i) either

$$\lim_{r \to 0} r^{2/(q-1)} \tilde{u}_i(r,.) = \omega^+(.), \tag{1.6.14}$$

where ω^+ *is the unique positive element of* $\mathcal{E}^+$,
  (ii) either there exists a positive spherical harmonic ψ *of degree* 1 *such that*

$$\lim_{r \to 0} r^{N-1} \tilde{u}(r,.) = \psi \tag{1.6.15}$$

and u *solves*

$$\int_{\Omega} \left(-u\Delta\xi + u^q\xi \right) dx = -c_\psi \frac{\partial\xi}{\partial\nu}(0) \quad \left(\forall\xi \in C_0^{1,1}(\overline{\Omega}) \right), \tag{1.6.16}$$

for some $c_\psi > 0$,
  (iii) or u *is continuous in* $\overline{\Omega}$.

4-2 Solutions with boundary blow-up

4-2-1 Uniform boundary blow-up: the smooth case

Let Ω be a domain of $\mathbb{R}^N$ and g a continuous real valued function; a function $u \in C^1(\Omega)$ is called a **large solution** of

$$-\Delta u + g(u) = 0 \tag{2.1.1}$$

in Ω, if it satisfies the equation and if

$$\lim_{x \to a} u(x) = \infty \quad (\forall a \in \partial\Omega). \tag{2.1.2}$$

If u is a large solution and is compact, it is clear that

$$\lim_{\rho(x)\to 0} u(x) = \infty, \tag{2.1.3}$$

where $\rho(x) = \text{dist}(x, \partial\Omega)$, otherwise there would exist a sequence $\{x_n\} \subset \Omega$ such that $x_n \to a$ for some $a \in \partial\Omega$ and some real number M such that $u(x_n) \to M$. This would contradict (2.1.2).

The main problems concerning those large solutions are their existence and their uniqueness. The following result are essentially due to Keller [Ke] and Osserman [Os]

THEOREM 4.10. *Suppose that Ω is bounded, $\partial\Omega$ satisfies the Wiener regularity criterion, g is nondecreasing and*

$$\int_\alpha^\infty \left(\int_\alpha^s g(\sigma)d\sigma \right)^{-1/2} ds < \infty \tag{2.1.4}$$

for some $\alpha > 0$. Then there exists at least one large solution of (2.1.1) *in Ω.*

Proof. We have seen in Theorem 2.6 that assumption (2.1.3) implies the local *a priori* upper bound for any solution of (2.1.1) in Ω, in the sense that there exists a decreasing function J defined on $(0,\infty)$ with value in $(0,\infty)$, such that $\lim_{r\to 0} J(r) = \infty$, with the property that if v is a solution of (2.1.1) in Ω, then the following upper estimate holds,

$$v(x) \leq J(\rho(x)) \qquad (\forall x \in \Omega). \tag{2.1.5}$$

For any $n \in \mathbb{N}_*$, there exists $u_n \in C(\overline{\Omega}) \cap C^1(\Omega)$ solution of (2.1.1) in Ω which takes the value n on $\partial\Omega$. Since $n \mapsto u_n$ is nondecreasing and u_n satisfies (2.1.5) the sequence $\{u_n\}$ converges in the $C^1_{loc}(\Omega)$-topology to a large solution u of (2.1.1).

Remark 4.10. The above existence result is still valid if we suppose that g is no longer nondecreasing but is bounded above by a continuous nondecreasing function $\tilde{g}$ which satisfies (2.1.4).

Definition. Let g be a nondecreasing continuous function satisfying (2.1.4) and Ω be a general domain of $\mathbb{R}^N$, then there exist two sequences of smooth domains Ω_n, Ω'_n such that, for any $n \in \mathbb{N}_*$ $\overline{\Omega}_n \subset \Omega_{n+1} \subset \Omega \subset \overline{\Omega} \subset \overline{\Omega}'_{n+1} \subset \Omega'_n$ with $\bigcup_{n\geq 0} \Omega_n = \Omega$ and $\bigcap_{n\geq 0} \Omega'_n = \overline{\Omega}$. For any n we call u_{Ω_n} (resp $u_{\Omega'_n}$) a large solution of (2.1.1) in Ω_n (resp Ω'_n). The sequence $\{u_{\Omega_n}\}$ (resp. $\{u_{\Omega'_n}\}$) is nonincreasing (resp. nondecreasing) and converges to some u_M^Ω (resp. u_m^Ω). The function u_M^Ω is the largest of all the solutions of (2.1.1) in Ω and is called the **maximal solution** of (2.1.1) in Ω. The function u_m^Ω is smaller than any large solution of (2.1.1) in Ω and is called the **minimal solution** of (2.1.1) in Ω.

Remark 4.11. If Ω satisfies the Wiener Criterion, the maximal solution is a large solution. The large solution $u = \tilde{u}_m^\Omega$ which is constructed in Theorem 4.10 is the **minimal large solution.** It has the property of being the smallest of all the large solutions of (2.1.1). Moreover $u_m^\Omega \leq \tilde{u}_m^\Omega$, but equality may not hold. It is also clear that if Ω and Ω' are general bounded domains of $\mathbb{R}^N$ such that $\overline{\Omega} \subset \Omega'$, then $u_M^{\Omega'} \leq u_m^\Omega$.

The following result is an adaptation of a result due to Iscoe ([Is]).

THEOREM 4.11. *Suppose that Ω is a bounded domain, starshapped with respect to O and g is nondecreasing and satisfies*

$$k^2 h(k) g(\rho) \leq g(h(k)\rho + \ell(k)), \ (\forall \rho \in \mathbb{R}, \ k > 1) \ (resp. (\forall \rho > 0, \ k > 1)), \ (2.1.6)$$

where h (resp. ℓ) is a continuous function defined on $(1,\infty)$ with limit 1 (resp. 0) at 1. Then there exists at most one large solution (resp. one nonnegative large solution) of (2.1.1) in Ω.

Proof. Suppose u and v are two large solutions and let $k > 1$. We define u_k by

$$u_k(x) = h(k)u(kx) + \ell(k) \qquad \left(\forall x \in \Omega_k = \mathcal{H}_{O,k^{-1}}(\Omega)\right), \tag{2.1.7}$$

where the function u_k satisfies

$$-\Delta u_k + k^2 h(k) g\left(\frac{u_k - \ell(k)}{h(k)}\right) = 0. \tag{2.1.8}$$

But $k^2 h(k) g\left(\dfrac{u_k - \ell(k)}{h(k)}\right) \leq g(u_k)$ and $\overline{\Omega}_k \subset \Omega$. By the maximum principle in Ω_k, $u_k \geq v$. Letting k go to 1 yields $u \geq v$. But $v \geq u$ in the same way and the proof is complete.

Example. (i) If $g(r) = |r|^{q-1} r$, $q > 1$, we can take $h(k) = k^{2/(q-1)}$ and (2.1.6) is verified for any $\rho \in \mathbb{R}$ (in fact u_k satisfies the same equation as u).

(ii) If $g(r) = e^{\alpha r}$, $\alpha > 0$, we can take $h(k) = 1$ and $\ell(k) = \dfrac{2}{\alpha}\ln(k)$ and the equation (2.1.1) is invariant under the transformation (2.1.7).

We give below a simple proof of uniqueness (see [BM] and [Ve8]) when $\partial\Omega$ is C^2 and $g(r) = |r|^{q-1} r$, $q > 1$. We shall see later on a generalisation due to Marcus and Véron ([MV1]) of this uniqueness result in the case where $\partial\Omega$ is locally the graph of a continuous function .

THEOREM 4.12. *Let Ω be a bounded domain of $\mathbb{R}^N$ with a C^2 boundary $\partial\Omega$ and g a continuous real valued function such that*

$$\lim_{r \to \infty} r^{-q} g(r) = 1, \tag{2.1.9}$$

for some $q > 1$. *We assume also that* $r \mapsto g(r) / r$ *is nondecreasing on* $\mathbb{R}_*^+$. *Then there is a unique positive large solution of* (2.1.1) *in* Ω.

Proof. Existence comes from Theorem 4.10 and Remark 4.10. For uniqueness we shall prove that any large solution u satisfies

$$\lim_{\rho(x) \to 0} \rho^{2/(q-1)}(x) u(x) = \left(2(q+1) / (q-1)^2 \right)^{1/(q-1)}. \tag{2.1.10}$$

Since $\partial \Omega$ is C^2, it satisfies both the interior and exterior sphere conditions: there exists $R_0 > 0$ such that for any $P \in \partial \Omega$, the spheres $B(P - R_0 \vec{v}, R_0)$ and $B(P + R_0 \vec{v}, R_0)$ which are tangent at $\partial \Omega$ at P are included respectively in Ω and $\overline{\Omega}^c$, $\vec{v}$ being the normal unit outward vector to $\partial \Omega$ at P. For $a > 0$ we call $v^{a,R}$ and $w^{a,c,R}$, $c > 1$, the radial solutions of

$$\begin{cases} -\Delta v^{a,R} + a(v^{a,R})^q = 0 & \text{in } B(O,R), \\[2mm] \displaystyle\lim_{|x| \to R} v^{a,R}(x) = \infty, \end{cases} \tag{2.1.11}$$

and

$$\begin{cases} -\Delta w^{a,c,R} + a(w^{a,c,R})^q = 0 & \text{in } \Gamma_O(R,cR), \\[2mm] \displaystyle\lim_{|x| \to R} w^{a,c,R}(x) = \infty, \\[2mm] w^{a,c,R}(x) = 0 & \text{for } |x| = cR, \end{cases} \tag{2.1.12}$$

respectively, where $\Gamma_O(R,cR) = B(O,cR) \setminus B(O,R)$; $v^{a,R}$ is the unique large solution from Theorem 4.11, as for $w^{a,c,R}$, it is obtained from approximation by solutions with constant values on the boundary. Since u is a large solution and g satisfies (2.1.9), for any $\varepsilon > 0$ there exists $R_\varepsilon \in (0, R_0]$ such that

$$1 - \varepsilon \le -\left(u^{-q} \Delta u \right)(x) \le 1 + \varepsilon, \quad \text{for } 0 < \rho(x) \le R_\varepsilon. \tag{2.1.13}$$

If $P \in \partial \Omega$ with coordinates x_P, the coordinates of $P - R_\varepsilon \vec{v}$ and $P + R_\varepsilon \vec{v}$ are called y_{P,R_ε} and z_{P,R_ε} respectively. From the construction, for any $\varepsilon > 0$, $R \in (0, R_\varepsilon] \subset (0, R_0]$ and $P \in \partial \Omega$, there holds

$$w^{1+\varepsilon,c,R}(x - (P + R \vec{v})) \le u(x) \le v^{1-\varepsilon,R}(x - (P - R \vec{v})) \tag{2.1.14}$$

220

for any x such that $x - P = \lambda \vec{v}$ for some $-R \leq \lambda < 0$. Moreover, for any $x \in \Omega$ with $\rho(x) \leq R \leq R_\varepsilon$, there exists $P = P(x)$ such that the relation (2.1.14) is valid.

Step 1. We claim that

$$\lim_{|x| \to R} (R - |x|)^{2/(q-1)} v^{a,R}(x) = \left(\frac{2(q+1)}{(q-1)^2 a} \right)^{1/(q-1)}. \tag{2.1.15}$$

Because $v^{a,R}(x) = (aR^2)^{-1/(q-1)} v^{1,1}(x/R)$ we can suppose $R = a = 1$. If $\delta > 1$, we define V_δ in $B(O,1)$ by $V_\delta(x) = V_\delta(r) = C(\delta^2 - r^2)^{-2/(q-1}$ and

$$\Delta V_\delta - V_\delta^q = \frac{4C}{q-1} (\delta^2 - r^2)^{-2q/(q-1)} \left(N\delta^2 - \left(N - 2\frac{q+1}{q-1} \right) r^2 - \frac{q-1}{4} C^{q-1} \right). \tag{2.1.16}$$

For the choice of C, we take $C = \left(\frac{8(q+1)}{(q-1)^2} \right)^{1/(q-1)}$ if $q \geq (N+2)/(N-2)$, or

$C = \left(\frac{4N}{(q-1)\delta^2} \right)^{1/(q-1)}$ if $1 < q < (N+2)/(N-2)$. Clearly the right-hand side of (2.1.16) is nonnegative which implies that $v^{1,1} \geq V_\delta$ in $B(O,\delta)$. Letting δ go to 1 yields

$$\liminf_{r \to 1} (1-r)^{2/(q-1)} v^{1,1}(x) \geq \min \left(\left(\frac{2(q+1)}{(q-1)^2} \right)^{1/(q-1)}, \left(\frac{N}{q-1} \right)^{1/(q-1)} \right) = \gamma. \tag{2.1.17}$$

Let a be a boundary point of $B(O,1)$ and v_k be

$$v_k(x) = k^{2/(q-1)} v^{1,1}(a + k(x-a)) \tag{2.1.18}$$

$\left(\forall x \in a + k^{-1}(B(O,1) \setminus a) = B((1-k^{-1})a, k^{-1}) \right)$; then v_k satisfies

$$-\Delta v_k + v_k^q = 0 \tag{2.1.19}$$

in $B((1-k^{-1})a, k^{-1})$. For $R > 0$ fixed, there exists $k_0 > 0$ such that, for any $k \in (0, k_0]$, there holds

$$C^{-1} \left(\operatorname{dist}(x, \partial B_{k,R}) \right)^{-2/(q-1)} \leq v_k(x) \leq C \left(\operatorname{dist}(x, \partial B_{k,R}) \right)^{-2/(q-1)} \tag{2.1.20}$$

where $B_{k,R} = B(a,R) \cap B((1-k^{-1})a, k^{-1})$. Let H be the half-space whose boundary contains a; H is the limit of $B((1-k^{-1})a, k^{-1})$ when k goes to 0. From (2.1.20) and the classical *a priori* elliptic equations estimates we deduce that $\{v_k\}$ is compact in the $C_{\text{loc}}^2(H)$-topology.

Therefore there exist a subsequence $\{k_n\}$ going to 0 and a function v solution of $-\Delta v + v^q = 0$ in H such that $v_{k_n} \to v$ in the $C^2_{loc}(H)$-topology. Moreover, if t denotes the distance function to ∂H and $x = (\tilde{x},t) \in \partial H \times \mathbb{R}^+_*$ is the coordinate in H, (2.1.20) implies

$$C^{-1}t^{-2/(q-1)} \le v(\tilde{x},t) \le Ct^{-2/(q-1)}. \tag{2.1.21}$$

Since for any $\varepsilon > 0$ the two functions

$$t \mapsto \left(\frac{2(q+1)}{(q-1)^2}\right)(t+\varepsilon)^{-2/(q-1)} \quad \text{and} \quad t \mapsto \left(\frac{2(q+1)}{(q-1)^2}\right)(t-\varepsilon)^{-2/(q-1)} \tag{2.1.22}$$

satisfies the same equation as v, we deduce by the maximum principle, (2.1.21) and letting ε go to 0 that $v(\tilde{x},t) = \left(\frac{2(q+1)}{(q-1)^2}\right)t^{-2/(q-1)}$ for any $t > 0$. Therefore $v_k(\tilde{x},t)$ goes to $\left(\frac{2(q+1)}{(q-1)^2}\right)t^{-2/(q-1}$ and in particular we get

$$\lim_{r \to 1}(1-r)^{2/(q-1)}v^{1,1}(ra) = \left(\frac{2(q+1)}{(q-1)^2}\right)^{1/(q-1)}, \tag{2.1.23}$$

uniformly with respect to a, since v is radial. This implies the claim.

Step 2. We claim that

$$\lim_{|x| \to R}(|x|-R)^{2/(q-1)}w^{a,c,R}(x) = \left(\frac{2(q+1)}{(q-1)^2 a}\right)^{1/(q-1)}, \tag{2.1.24}$$

for $c > 0$ small enough. As in Step 1 we can replace $w^{a,c,R}$ by $w^{1,c,1}$. Since there exists an *a priori* estimate of the following form

$$w^{1,c,1}(x) \le C(|x|-1)^{-2/(q-1)}, \tag{2.1.25}$$

valid for any $x \in \Gamma_0(1,c)$, the problem is reduced to find a lower estimate of the same type, the remaining of the proof being as above. Denote $W_\delta(x) = W_\delta(r) = C\left(r^2 - \delta^2\right)^{-2/(q-1}$ for $0 < \delta < 1$. Then

$$\Delta W_\delta - W_\delta^q = \frac{4C}{q-1}(r^2-\delta^2)^{-2q/(q-1)}\left(2\frac{q+1}{q-1}r^2 - N(r^2-\delta^2) - \frac{q-1}{4}C^{q-1}\right). \tag{2.1.26}$$

If c is chosen such that $\left(2\frac{q+1}{q-1} - N\right)c^2 - N > 0$, we can take $\delta \in (0,1)$ and $C > 0$ in order that the right-hand side of (2.1.26) be nonnegative in $\Gamma_0(1,c)$. For $D > 0$ the function

$W_\delta - D$ is still a subsolution of (2.1.19) in $\Gamma_O(1,c)$ and vanishes for $|x| = c$. Using the maximum principle and letting δ go to 1 yields

$$\liminf_{r \to 1} (r-1)^{2/(q-1)} w^{1,c,1}(x) = \gamma' > 0 \tag{2.1.27}$$

which finally implies (2.1.24).

Step 3. End of the proof. From (2.1.13)-(2.1.14) and Steps 1-2

$$\lim_{\rho(x) \to 0} \rho(x)^{2/(q-1)} u(x) = \left(\frac{2(q+1)}{(q-1)^2} \right)^{1/(q-1)} \tag{2.1.28}$$

is satisfied by any large solution u of (2.1.1) in Ω. Assume now that u and $\tilde{u}$ are two such positive large solutions and set $u_\varepsilon = (1+\varepsilon)u$ for $\varepsilon > 0$, then

$$-\Delta u_\varepsilon + (1+\varepsilon)g(u_\varepsilon / (1+\varepsilon)) = 0, \tag{2.1.29}$$

and

$$0 = -\frac{\Delta u_\varepsilon}{u_\varepsilon} + \frac{g(u_\varepsilon(1+\varepsilon)^{-1})}{u_\varepsilon(1+\varepsilon)^{-1}} \le -\frac{\Delta u_\varepsilon}{u_\varepsilon} + \frac{g(u_\varepsilon)}{u_\varepsilon}. \tag{2.1.30}$$

Proceeding as in Chapter 2, we obtain

$$-\left(\frac{\Delta \tilde{u}}{\tilde{u}} - \frac{\Delta u_\varepsilon}{u_\varepsilon} \right)(\tilde{u}^2 - u_\varepsilon^2)^+ + \left(\frac{g(\tilde{u})}{\tilde{u}} - \frac{g(u_\varepsilon)}{u_\varepsilon} \right)(\tilde{u}^2 - u_\varepsilon^2)^+ \le 0; \tag{2.1.31}$$

integrating by part and using the fact that $(\tilde{u} - u_\varepsilon)^+$ has compact support in Ω, we deduce that $\tilde{u} \le u_\varepsilon = (1+\varepsilon)u$ for any $\varepsilon > 0$, and $\tilde{u} \le u$. In the same way $\tilde{u} \ge u$, which ends the proof.

Remark 4.12. As we have seen it, the uniqueness lies upon the fact that all the large solutions have the same boundary behaviour. In [BM] the assumption on the function g which implies this uniqueness is essentially weaker that the one of Theorem 4.12, namely

(H-1) g is C^1 on $\mathbb{R}^+$, vanishes at 0 and is positive on $\mathbb{R}^+_*$,

(H-2) $g(\beta t) \le \beta^{1+\mu} g(t)$, $\forall \beta \in (0,1)$, $\forall t \ge t_0 / \beta$, for some $t_0 \ge 1$ and $\mu > 1$,

(H-3) $g(\beta t) \le \beta g(t)$, $\forall \beta \in (0,1)$, $\forall t \ge 0$.

In [Ve8] the uniqueness is obtained under the assumption that $g(t) = t^q$, $q > 1$, with a general elliptic operator and with forcing term $f(x)\,u^q$ in (2.1.1). Moreover, the boudedness assumption on Ω is replaced by a boundedness assumption on $\partial\Omega$.

4-2-2 Uniform boundary blow-up: the irregular case

In this Section we restrict ourself to study the uniform boundary blow-up problem for the equation

$$-\Delta u + |u|^{q-1} u = 0 \tag{2.2.1}$$

where $q > 1$, and u is a positive function in a domain Ω of $\mathbb{R}^N$. If $\partial\Omega$ satisfies the Wiener regularity criterion it is possible to construct a large solution of (2.1.1). However the above techniques for proving uniqueness by showing an estimate of type (2.1.28) are no longer valid. In [LG1], Le Gall proved a first uniqueness result for irregular domains by means of probability techniques. His result is the following

THEOREM 4.13. *Assume that Ω is a bounded domain of $\mathbb{R}^N$, regular in the probabilistic sense , which satisfies, for some $\theta \in (0,1)$ and $r_0 > 0$,*

$$c_{1,2}(\Omega^c \cap B(y,r)) \geq \theta\, c_{1,2}(B(y,r)), \ \forall y \in \partial\Omega, \ \forall r \in (0,r_0]. \tag{2.2.2}$$

Then there one and only one large of

$$-\Delta u + |u|u = 0. \tag{2.2.3}$$

In the above statement, the notion of regularity in a probabilitic sense would be too long to explain but it is satisfied in particular if $\partial\Omega$ is Lipschitz (and (2.2.2) is unuseful) or if $N = 2$ and Ω is any simply connected domain. In fact the maximal and the minimal large solutions have both a probabilistic interpretation. The next result due to Marcus and Véron ([MV1] ,[MV2]) is an extension, by purely analytic means, of Le Gall's uniqueness result.

Definition. Let Ω be a domain of $\mathbb{R}^N$, with a compact boundary. We say that Ω satisfies the local graph property if for every boundary point P there exist a neighborhood U_P of P, a set of coordinates $\xi = (\xi_1,...,\xi_N)$ obtained from x by rotation and a function $F_P \in C(\mathbb{R}^{N-1})$ such that $U_P \cap \Omega = U_P \cap G_u(F_P)$ where

$$G_u(F_P) = \left\{ \xi\colon \xi_N < F_P(\xi_1,...,\xi_N) \right\}. \tag{2.2.4}$$

The class of domains Ω satisfying the local graph property will be denoted by C_{gr}. If the condition holds with $F_P \in C^{0,1}(\mathbb{R}^{N-1})$, we shall say that Ω is of class $C_{gr}^{0,1}$.

THEOREM 4.14. *Let Ω be a domain of class C_{gr}. Then the equation* (2.2.1) *admits at most one large solution.*

Proof. Let P be a boundary point of Ω and suppose that the neighborhood U_P in the above definition is an open, bounded cylindrical domain centered at P, with axis parallel to the ξ_N-axis:

$$U_P = \left\{ \eta = (\eta', \eta_N),\ |\eta'| < R_P,\ |\eta_N| < T_P \right\} \tag{2.2.5}$$

where $\eta = \xi - P$ and $\eta' = (\eta_1, \dots, \eta_{N-1})$. We can also suppose that $\partial\Omega$ is bounded away from the top and the bottom of the cylinder U_P and that $\partial\Omega \cap \overline{U_P} = \overline{\partial\Omega \cap U_P}$. We denote

$$\Theta = U_P \cap \Omega, \quad \Gamma_1 = U_P \cap \partial\Omega, \quad \Gamma_2 = \partial U_P \cap \overline{\Omega}. \tag{2.2.6}$$

Step 1. Suppose that there exists a large solution u of (2.2.1) (necessarily positive) then we claim that there exists a positive function $v \in C^2(\Theta)$ such that

$$-\Delta v + v^q = 0 \tag{2.2.7}$$

in Θ which satisfies

$$\begin{cases} \text{(i)} & v(x) \to \infty \text{ locally uniformly as } x \to \Gamma_1, \\ \text{(ii)} & v(x) \to 0 \text{ locally uniformly as } x \to \Gamma_2, \end{cases} \tag{2.2.8}$$

where "$v(x) \to 0$ (or ∞) locally uniformly as x tends to Γ_j" means that, for every compact subset A contained in the relative interior of Γ_j, there holds $v(x) \to 0$ (or ∞) as $\operatorname{dist}(x, A) \to 0$. Set $\tilde{F}_P(\eta') = F_P(\eta' + P')$ and let $\{f_n\}$ be an increasing sequence of smooth positive functions defined in the closed $(N-1)$-ball $D_P = \left\{ \eta' = (\eta_1, \dots, \eta_{N-1}) : |\eta'| \le R_P \right\}$ which converges uniformly in D_P to $\tilde{F}_P$. We set $\Theta_n = \left\{ \eta : |\eta'| < R_P,\ -T_P < \eta_N < f_n(\eta') \right\}$ and denote $v_{n,k} \in C(\Theta_n)$ a positive solution of (2.2.7) in Θ_n which satisfies

$$\begin{cases} v_{n,k}(\eta) = k & \text{for } \eta \in \Gamma_{1,n}, \\ v_{n,k}(\eta', \eta_N) = 0 & \text{for } \eta \in \Gamma_{2,n} \text{ such that } -T_P \le \eta_N \le f_n(\eta') - n^{-1}, \\ v_{n,k}(\eta', \eta_N) = n\big(\eta_N - f_n(\eta') + n^{-1}\big)k \\ \qquad\qquad \text{for } |\eta'| = R_P,\ f_n(\eta') - n^{-1} \le \eta_N \le f_n(\eta'). \end{cases} \tag{2.2.9}$$

For n fixed the sequence $\{v_{n,k}\}$ is increasing in k and converges, as k goes to infinity, to some function v_n which is continuous in Θ_n where it satisfies (2.2.7) and

$$\begin{cases} \text{(i)} \quad v_n(\eta) \to \infty \text{ loc. uniformly as } \eta \to \Gamma_{1,n}, \\ \text{(ii)} \quad v(\eta) \to 0 \text{ loc. uniformly as } \eta \to \Gamma_{2,n}^* = \{\eta \in \Gamma_{2,n}: \eta_N \le f_n(\eta') - n^{-1}\}. \end{cases} \tag{2.2.10}$$

The sequence $\{v_n\}$ is decreasing and converges in the $C_{loc}(\Theta)$-topology to a continuous function v which satisfies (2.2.7) in Θ. Therefore v satisfies (2.2.8)-(ii). The main problem is to prove that it also satisfies (2.2.8)-(i). Indeed, if w is a large solution of (2.2.7) in the cylinder U_P (it exists from Theorem 4.10), $v_n + w$ satisfies

$$-\Delta(v_n + w) + (v_n + w)^q \ge 0 \tag{2.2.11}$$

in Θ_n and dominates u on $\partial\Theta_n$. Therefore $v_n + w \ge u$ in Θ_n and consequently $v + w \ge u$ in Θ, which implies (2.2.8)-(ii) and the claim, since w is continuous in U_P.

Step 2. We claim that if u is a large solution of (2.2.1) in Ω and v is the function defined in Step 1, then

$$\frac{u(x)}{v(x)} \to 1 \quad \text{locally uniformly as } \quad x \to \Gamma_1. \tag{2.2.12}$$

Denote $\hat{\Theta}_h = \{\eta \in \mathbb{R}^N: (\eta', \eta_N + h) \in \Theta\}$, $h > 0$. For any function f defined in Θ, we set $f_h(\eta) = f(\eta', \eta_N + h)$ if $\eta \in \hat{\Theta}_h$. If h is chosen such that $\overline{\hat{\Theta}_h} \subset \Omega$ and w is a large solution of (2.2.7) in the cylinder U_P, then w is defined in Θ and $v_h + w_h$ is a supersolution of (2.2.7) in $\hat{\Theta}_h$. From the above construction $v_h + w_h \ge u$ in $\hat{\Theta}_h$. On the other hand, by (2.2.8), $v_{-h} \le u$ on $\partial(\hat{\Theta}_{-h} \cap \Omega)$, which implies $v_{-h} \le u$ in $\hat{\Theta}_{-h} \cap \Omega$. Letting h go to 0 yields $v \le u \le v + w$ in Θ. Since w is continuous in U_P and v satisifies (2.2.8)-(i), (2.2.12) follows.

Step 3. End of the proof. Assume u_1 and u_2 are two large solutions of (2.2.1) in Ω, then (2.2.12) implies that for any boundary point P there exists a neighborhood U_P such that

$$\frac{u_1(x)}{u_2(x)} \to 1 \tag{2.2.13}$$

as $\text{dist}(x, \Gamma_1) \to 0$, where $\Gamma_1 = \overline{U}_P \cap \partial\Omega$. Since $\partial\Omega$ is compact, (2.2.13) holds as $\text{dist}(x, \partial\Omega) \to 0$. We conclude as in Theorem 4.12 by using the maximum principle.

In the particular case of equation (2.2.1) it is possible to extend Theorem 4.10 to domains Ω whose boundary does not satisfy the Wiener regularity criterion.

Definition. A domain Ω of $\mathbb{R}^N$ is said to satisfy the exterior segment property if there exists a compact segment K such that for any boundary point P there exists a segment K_P with one extremity P and isometric to K contained in Ω^c.

226

It is clear that if Ω is of class C_{gr} it satisfies the exterior segment property.

THEOREM 4.15. *Suppose that Ω is a domain of $\mathbb{R}^N$ with a compact boundary which satisfies the exterior segment property and $1 < q < (N-1)/(N-3)$, with no condition if $N = 3$. Then there exists a maximal large solution of (2.2.1).*

Proof. Assume P is a boundary point, the condition $1 < q < (N-1)/(N-3)$ ensures the existence and uniqueness in an apropriate class of a positive solution of

$$-\Delta v_{P,c} + v_{P,c}^q = c\mu_{K_P} \tag{2.2.14}$$

in $\mathbb{R}^N$ for any $c > 0$ where μ_{K_P} is the 1-dimensional Dirac measure with support K_P; such a solution exists from Lemma 2.5 and Remark 2.10 (the fact that the equation is considered in $\mathbb{R}^N$ instead of a bounded domain does not affect the result since the solution decays to 0 at infinity by the Osserman-Keller estimate). We consider a sequence of smooth domains Ω_n such that $\overline{\Omega}_n \subset \Omega_{n+1} \subset \Omega$ with $\bigcup_{n \geq 0} \Omega_n = \Omega$ and, for any n, denote u_n the maximal solution of (2.2.1) in Ω_n. From the maximum principle $u_n \geq v_{P,c}$ for any $c > 0$ and $P \in \partial\Omega$, which implies $u_n \geq v_{P,\infty}$. Since the sequence $\{u_n\}$ is decreasing and converges to some $u_M^\Omega = u$ which solves (2.2.1) in Ω, there also holds $u \geq v_{P,\infty}$ in Ω for any $P \in \partial\Omega$. In order to prove that $\lim_{x \to P} v_{P,\infty}(x) = \infty$, we take $P = O$ and $K_P = K_O = \{x = (x',x_N): x'=0, \ 0 \leq x_N \leq 1\}$. Let S_+^{N-1} be the upper hemisphere of the unit sphere in $\mathbb{R}^N$, N the North Pole with coordinates $(0,1)$ and ω a positive solution of

$$\Delta_{S^{N-1}}\omega + \frac{2}{q-1}\left(\frac{2q}{q-1} - N\right)\omega - \omega^q = 0 \quad \text{in} \quad S_+^{N-1}\setminus\{N\}, \tag{2.2.15}$$

which vanishes on the equator ∂S_+^{N-1} and satisfies

$$\lim_{\substack{x \to N \\ x \in S_+^{N-1}}} |x - N|^{2/(q-1)}\omega(x) = \left(\frac{2}{q-1}\left(\frac{2q}{q-1} - N + 1\right)\right)^{1/(q-1)} \tag{2.2.16}$$

(we write $x = (1,\sigma) = \sigma$ on S_+^{N-1} for simplicity), such a solution exists from Chapter 2 since $1 < q < (N-1)/(N-3)$ and is in fact unique. We define $x_N \mapsto \psi(x_N)$ as being the solution of $\psi'' - \psi^q = 0$ on $(-1,1)$ which blows up at 1 and whose derivative vanishes at 0. Then, by the maximum principle, because $u + \psi$ is a super solution for (2.2.1) and

$$\lim_{x_N \to 1^-} (1 - x_N)^{2/(q-1)}\psi(x_N) = \left(\frac{2(q+1)}{(q-1)^2}\right)^{1/(q-1)}, \tag{2.2.17}$$

we have

$$(u + \psi)(x) \geq |x|^{-2/(q-1)} \omega(x / |x|) \tag{2.2.18}$$

on $\Omega \cap \left\{ x = (x', x_N) : 0 < x_N < 1 \right\}$. This implies $\lim\limits_{x \to O} u(x) = \infty$ and the proof is complete.

Remark 4.13. By using an improved version of the techniques developped in the proof of Theorem 4.12, it is possible to describe the boundary blow-up near a wedge-like point ([MV1]) : For $0 \leq k \leq N - 2$ we consider a k-dimensional subspace Π_k of $\mathbb{R}^N$,

$$\Pi_k = \left\{ x = (x_1, \ldots, x_N) : x_j = 0, \ j = 1, \ldots, N - k \right\} \tag{2.2.19}$$

and denote by $(r, \sigma, z) \in \mathbb{R}^+ \times S^{N-k-1} \times \Pi_k$ the cylindrical coordinates in $\mathbb{R}^N$ with axis Π_k. Given a subdomain S in S^{N-k-1} whose boundary is piecewise C^2 we set

$$W_S(\Pi_k) = \left\{ (r, \sigma, z) : r > 0, \ \sigma \in S, \ z \in \mathbb{R}^k \right\}. \tag{2.2.20}$$

Then the large solution of (2.2.1) in $W_S(\Pi_k)$ is unique and is explicity given by

$$U_{S, \Pi_k}(r, \sigma, z) = r^{-2/(q-1)} \omega(\sigma), \tag{2.2.21}$$

where ω is the positive solution of

$$\Delta_{S^{N-k-1}} \omega + \frac{2}{q-1} \left(\frac{2q}{q-1} - N + k \right) \omega - \omega^q = 0 \tag{2.2.22}$$

in S which blows-up on ∂S (the large solution of (2.2.22) in S).

If Ω is a Lipchitz and piecewise C^2 domain in $\mathbb{R}^N$, O a boundary point such that $\partial \Omega$ is tangent to $W_S(\Pi_k)$ locally at O, then the large solution u of (2.2.1) in Ω satisfies

$$\lim_{x \to O} \frac{u(x)}{U_{S, \Pi_k}(x)} = 1. \tag{2.2.23}$$

4-3 Boundary trace of positive solutions

4-3-1 The global trace result

The problem under consideration in this section deals with the description of the boundary behaviour of any positive solution of

$$-\Delta u + g(u) = 0 \tag{3.1.1}$$

in the unit ball $B(0,1)$ of $\mathbb{R}^N$. In the next statement, we recall that $(r,\sigma) \in \mathbb{R}_*^+ \times S^{N-1}$ are the spherical coordinates in $\mathbb{R}^N \setminus \{O\}$.

THEOREM 4.16. *Let g be a continuous real valued function defined on $\mathbb{R}^+$ such that g(r) keeps a constant sign for r large enough and $u \in C^1(B(O,1))$ be a nonnegative solution of (3.1.1) in $B(0,1)$. Then $\int_{S^{N-1}} u(r,\sigma)d\sigma$ admits a limit $\ell \in [0,\infty]$ as r goes to 1. If ℓ is finite, u is integrable in $B(O,1)$, g(u) satisfies*

$$\int_{B(O,1)} |g(u)|(1-|x|)dx < \infty, \tag{3.1.2}$$

and there exists a nonnegative Radon measure $\mu \in \mathcal{M}(\partial B(O,1))$ with total mass ℓ such that

$$\lim_{r \to 1} u(r,.) = \mu, \tag{3.1.3}$$

weakly in $\mathcal{M}(\partial B(O,1))$. Moreover the following equality holds for any $\zeta \in C_0^{1,1}(\overline{B}(O,1))$

$$\int_{B(O,1)} (u\Delta\zeta - \zeta g(u))dx = \int_{\partial B(O,1))} \frac{\partial \zeta}{\partial \nu} d\mu. \tag{3.1.4}$$

Proof. Step 1. We claim that $\int_{S^{N-1}} u(r,\sigma)d\sigma$ admits a limit $\ell \in [0,\infty]$
Case 1: suppose that there exists a nonnegative k such that g(r) is nonnegative for any $r \geq k$. We write $g_1(r) = \chi_{[0,k)}(r)g(r)$, $g_2(r) = \chi_{[k,\infty)}(r)g(r)$ and get

$$\overline{u}_{rr} + \frac{N-1}{r}\overline{u}_r - \overline{g_1(u(r))} = \overline{g_2(u(r))} \tag{3.1.5}$$

where the overlining $^-$ means that the spherical average over S^{N-1} is taken. Therefore

$$\frac{d}{dr}\left(r^{N-1}\overline{u}(r) - \int_0^r \overline{g_1(u)}(s)s^{N-1}ds \right) \geq 0. \tag{3.1.6}$$

If we set $X(r) = r^{1-N}\int_0^r s^{N-1}\overline{g_1(u)}(s)ds$, then X can be extended as a continuous function on $[0,1]$. We encounter two possibilities:

(I)- If $r^{N-1}(\overline{u}_r(r) - X(r)) < 0$ on $(0,1)$. Then $r \mapsto \overline{u}(r) - \int_0^r X(s)ds$ is decreasing and, by continuity, there exists a real number m such that $m = \overline{u}(1) - \int_0^1 X(s)ds$.

229

(II)- If $r_0^{N-1}\big(\overline{u}_r(r_0) - X(r_0)\big) \geq 0$ for some $r_0 \in (0,1)$, then $r \mapsto \overline{u}(r) - \int_0^r X(s)ds$ is nondecreasing on $[r_0,1)$ and there exists $m \in \mathbb{R} \cup \{+\infty\}$ such that $m = \lim_{r \to 1} \overline{u}(r) - \int_0^1 X(s)ds$.

In both cases there exits $\ell \in [0,\infty]$ such that

$$\lim_{r \to 1} \overline{u}(r) = \ell. \tag{3.1.7}$$

Case 2: suppose that there exists a nonnegative k such that $g(r)$ is nonpositive for $r \geq k$. With the above notations (3.1.6) becomes

$$\frac{d}{dr}\left(r^{N-1}\overline{u}(r) - \int_0^r \overline{g_1(u)}(s)s^{N-1}ds \right) \leq 0, \tag{3.1.8}$$

which implies that the inequality

$$r^{N-1}\overline{u}(r) - \int_0^r \overline{g_1(u)}(s)s^{N-1}ds \leq 0 \tag{3.1.9}$$

holds on $(0,1)$. Therefore $r \mapsto \overline{u}(r) - \int_0^r X(s)ds$ is nonincreasing which again implies (3.1.7) for some $\ell \in [0,\infty)$.

Step 2. We suppose now that $\ell \in [0,\infty)$. It is clear that u is integrable in $B(O,1)$ and, from (3.1.5), we have

$$r^{N-1}\overline{u}_r(r) - \int_0^r s^{N-1}\overline{g_1(u)}(s)ds = \int_0^r s^{N-1}\overline{g_2(u)}(s)ds. \tag{3.1.10}$$

Multiplying by r^{1-N} and integrating on $(0, r)$ yields

$$\int_0^r s^{1-N}\int_0^s t^{N-1}\overline{g_2(u)}(t)dt = \overline{u}(r) - u(O) - \int_0^r s^{1-N}\int_0^s t^{N-1}\overline{g_1(u)}(t)dt. \tag{3.1.11}$$

The right-hand side of (3.1.11) admits a finite limit when r goes to 1, since g_1 is bounded and g_2 keeps a constant sign, therefore

$$\int_0^1 s^{1-N}\int_0^s t^{N-1}\overline{|g(u)|}(t)dt < \infty. \tag{3.1.12}$$

By Fubini's theorem $\int_{1/2}^{1} t^{N-1}(1-t)|\overline{g(u)}|(t)dt < \infty$, which means that (3.1.2) is valid. Let $\{r_n\}$ be a sequence going to 1 and $\mu \in \mathcal{M}(\partial B(O,1))$ a nonnegative measure such that $\lim_{n\to\infty} u(r_n,.) = \mu$, weakly in $\mathcal{M}(\partial B(O,1))$, then the total mass of μ is ℓ. If $\zeta \in C_0^{1,1}(\overline{B}(O,1))$ we set $\zeta_n(x) = \zeta(x / r_n)$ and we get

$$\int_{B(O,r_n)} (u\Delta\zeta_n - g(u)\zeta_n)dx = \int_{\partial B(O,r_n)} \frac{\partial \zeta_n}{\partial v} udS(x). \tag{3.1.13}$$

But $\Delta\zeta_n(x) = r_n^{-2}(\Delta\zeta)(x / r_n)$ and

$$\int_{\partial B(O,r_n)} \frac{\partial \zeta_n}{\partial v} udS(x) = r_n^{-1}\int_{\partial B(O,r_n)} \frac{\partial \zeta}{\partial v}(x / |x|)udS$$

$$\tag{3.1.14}$$

$$= r_n^{N-2}\int_{S^{N-1}} \frac{\partial \zeta}{\partial v}(1,\sigma)u(r_n\sigma)d\sigma$$

Moreover, for some $K > 0$ independent of n, there holds $|\zeta_n(x)| \leq K(1 - |x_n|)$ for any $x \in B(O,r_n)$. Consequently

$$\left|\left(\zeta_n g(u)\chi_{B(O,r_n)}\right)(x)\right| \leq K|g(u)|\chi_{B(O,r_n)}(1 - |x| / r_n) \leq K|g(u)|\chi_{B(O,1)}(1 - |x|). \tag{3.1.15}$$

From Lebesgue's theorem (3.1.4) is derived, which also implies that (3.1.3) holds.

Remark 4.14. If $g(r) \leq 0$ for r large enough then $\ell < \infty$. If g is nondecreasing and $\ell < \infty$, then u is the unique solution of (3.1.4) from Theorem 4.2. The above trace result can also be extended if the ball B(O, 1) is replaced by a smooth bounded domain of $\mathbb{R}^N$.

The important cases of equation (3.1.1) are when $g(r) = \pm r^q$ with $q > 1$ but in the remaining part of this section we shall just consider the equation

$$\Delta u + u^q = 0 \tag{3.1.16}$$

in a smooth bounded domain Ω of $\mathbb{R}^N$. We recall that $G(x,y)$ is the Green kernel in $\Omega \times \Omega$, $P(x,y)$ the Poisson kernel in $\Omega \times \partial\Omega$ and if $\mu \in \mathcal{M}(\partial\Omega)$, its Poisson potential m_μ is defined by

$$m_\mu(x) = \int_{\partial\Omega} P(x,y)d\mu(y) \tag{3.1.17}$$

We call ρ the distance function to $\partial\Omega$.

PROPOSITION 4.2. *Let μ be a Radon measure on $\partial\Omega$ and assume that there exists a nonnegative function $\phi \in L^1(\Omega)$ with $\phi \geq m_\mu$, $\int_{B(O,1)} \phi^q \rho(x)dx < \infty$, and satisfies a.e. in Ω*

$$\phi(x) \geq m_\mu(x) + \int_{B(O,1)} G(x,y)\phi^q(y)dy. \qquad (3.1.18)$$

Then there exists a nonnegative function $u \in L^1(B(O,1))$ with $\int_{B(O,1)} u^q \rho(x)dx < \infty$ such that

$$u(x) = m_\mu(x) + \int_{B(O,1)} G(x,y)u^q(y)dy \qquad (3.1.19)$$

for almost all $x \in \Omega$.

Proof. We first notice that (3.1.18) is the weak formulation for

$$\begin{cases} -\Delta\phi \geq \phi^q & \text{in } \Omega, \\ \phi_{|\partial\Omega} \geq \mu & \text{on } \partial\Omega. \end{cases} \qquad (3.1.20)$$

We consider a sequence $\{u_n\}$ satisfying

$$\begin{cases} u_0 = m_\mu, \\ u_n(x) = m_\mu(x) + \int_{B(O,1)} G(x,y)u_{n-1}^q(y)dy, \ n \geq 1. \end{cases} \qquad (3.1.21)$$

Then we claim that

$$u_0 \leq u_1 \leq \ldots \leq u_n \leq \phi, \qquad (3.1.22)$$

for every $n \in \mathbb{N}$. It is clear that $u_0 \leq u_1$. This implies that $u_0^q \leq u_1^q$ and $u_1 \leq u_2$. By induction $u_{n-1} \leq u_n$. Because $\phi \geq m_\mu$, we deduce from (3.1.18) that

$$\phi(x) \geq m_\mu(x) + \int_{B(O,1)} G(x,y)m_\mu^{\ q}(y)dy = u_1(x) \qquad (3.1.23)$$

which again implies

$$\phi(x) \geq m_\mu(x) + \int_{B(O,1)} G(x,y)u_1^q(y)dy = u_2(x) \qquad (3.1.24)$$

(always with (3.1.18)). By induction $u_n \leq \phi$, and (3.1.22) holds. By the monotone convergence theorem $u_n \to u$ in $L^1(\Omega)$ and $u_n^q \to u^q$ in $L^1(\Omega, \rho dx)$ when n goes to infinity. Moreover the two convergences occur a.e. in Ω. From the induction formula (3.1.21), we obtain (3.1.19) which actually means that u is a weak solution of (3.1.16) with boundary value μ in the sense of Theorem 4.16.

Remark 4.15. The range of applications of Proposition 4.2 has not yet been completely explored, however, if we take $\Omega = B(O,1)$ and $\mu = c\delta_P$ where P belongs to $\partial B(O,1)$), we obtain

$$0 \le (m_{c\delta_P})^q(x) \le c^q K^q \frac{\rho^q(x)}{|x - P|^{qN}},$$
(3.1.25)

where $K > 0$ depends on N. If we assume that $1 < q < (N+1)/(N-1)$ we deduce that

$$0 \le \int_{B(O,1)} G(x,y)(m_{c\delta_P})^q dy \le c^q M \rho(x) |x - P|^{-N}.$$
(3.1.26)

Consequently the function

$$\phi(x) = cm_{\delta_P}(x) + 2^q c^q \int_{B(O,1)} G(x,y) m_{\delta_P}^q(y) dy$$
(3.1.27)

satisfies the assumption of Proposition 4.2 for k small enough. Therefore there is a $c_0 > 0$ such that for any $c \in [0, c_0]$ there exists a solution of (3.1.16) in $B(O,1)$ with measure boundary value $c\delta_P$, which means that

$$\int_{B(O,1)} \left(u \Lambda \zeta + \zeta u^q \right) dx = c \frac{\partial \zeta}{\partial \nu}(P)$$
(3.1.28)

for any $\zeta \in C_0^{1,1}(\overline{B}(O,1))$.

4-3-2 The local trace result and the trace elements

Let u be a positive solution of

$$-\Delta u + u^q = 0$$
(3.2.1)

$(q > 1)$ in $B(O,1)$, then Theorem 4.16 applies and $\int_{S^{N-1}} u(r,\sigma) d\sigma$ admits a limit $\ell \in [0,\infty]$ as r goes to 1. However this limit may be infinite in many cases and this does not give any information on the boundary behaviour of u. The notion of trace elements has been introduced by LeGall ([LG2]) in the case $q = N = 2$; the general notion that we present here is due to Marcus and Véron ([MV2], [MV3]). The basic result which gives the basis of this theory is the following

THEOREM 4.17. *If* u *is a positive solution of* (3.2.1) *the following alternative holds for every* $\omega \in \partial B(O,1)$:

(i) either for every relative neighbourhood U *of* ω *on* $\partial B(O,1)$

$$\lim_{r \to 1} \int_U u(r,\sigma) d\sigma = \infty,$$
(3.2.2)

(ii) or there exists a relatively open neighbourhood U *of* ω *on* $\partial B(O,1)$ *such that for every* $\zeta \in C_0^\infty(U)$

$$\lim_{r \to 1} \int_U u(r,\sigma)\zeta(\sigma)\mathrm{d}\sigma = \ell(\zeta), \tag{3.2.3}$$

where ℓ *is a positive linear functional on* $C_0^\infty(U)$.

In order to prove this result we need the following lemma

LEMMA 4.10. *Let* U *be a smooth domain on* S^{N-1}, φ_U *the first eigenfunction of* $-\Delta_{S^{N-1}}$ *in* $W_0^{1,2}(U)$ *normalized by*

$$0 \le \varphi_U \le \sup_U \varphi_U = 1, \tag{3.2.4}$$

and $\alpha > (q+1)/(q-1)$, *then we have the following alternative:*
(i) either

$$\int_0^1 \int_U u^q \varphi_U^\alpha (1-r)\mathrm{d}\sigma\mathrm{d}r = \infty, \tag{3.2.5}$$

and in that case

$$\lim_{r \to 1} \int_U (u\varphi_U^\alpha)(r,\sigma)\mathrm{d}\sigma = \infty, \tag{3.2.6}$$

(ii) or

$$\int_0^1 \int_U u^q \varphi_U^\alpha (1-r)\mathrm{d}\sigma\mathrm{d}r < \infty, \tag{3.2.7}$$

and in that case, for any $C^2(U)$-*function* ζ *which satisfies*

$$0 \le \zeta \le k\varphi_U^\alpha, \quad \left|\Delta_{S^{N-1}}\zeta\right| \le k\varphi_U^{\alpha-2}, \tag{3.2.8}$$

for some $k > 0$, *the following limit exists*

$$\lim_{r \to 1} \int_U (u\zeta)(r,\sigma)\mathrm{d}\sigma = \ell(\zeta), \tag{3.2.9}$$

and the mapping $\zeta \mapsto \ell(\zeta)$ *is a nondecreasing linear functional defined on the set of* $C^2(U)$-*functions which satisfy* (3.2.8).

Proof. Step 1. Suppose that $\alpha > (q+1)/(q-1)$, then the following inequality holds

$$\int_U \left|\Delta_{S^{N-1}} \varphi_U^\alpha\right|^{q/(q-1)} (\varphi_U^\alpha)^{-1/(q-1)} d\sigma = K = K(\alpha, q, U) < \infty. \tag{3.2.10}$$

In fact, from Hopf boundary lemma, we have

$$\varphi_U^{\alpha/(q-1)} \geq C\left(\text{dist}(\sigma, \partial U)\right)^{\alpha/(q-1)} \tag{3.2.11}$$

for some $C > 0$, where the distance is the geodesic distance on the sphere. Moreover

$$\Delta_{S^{N-1}} \varphi_U^\alpha = -\lambda_U \alpha \varphi_U^\alpha + \alpha(\alpha-1)\varphi_U^{\alpha-2}\left|\nabla \varphi_U\right|^2, \tag{3.2.12}$$

where λ_U is the eigenvalue corresponding to φ_U. Therefore

$$\left|\Delta_{S^{N-1}} \varphi_U^\alpha\right|^{q/(q-1)} \leq C'\left(\text{dist}(\sigma, \partial U)\right)^{q(\alpha-2)/(q-1)} \tag{3.2.13}$$

for any $\sigma \in U$. By combination of (3.2.12) and (3.2.13) we obtain

$$\left|\Delta_{S^{N-1}} \varphi_U^\alpha\right|^{q/(q-1)} (\varphi_U^\alpha)^{-1/(q-1)} \leq C''\left(\text{dist}(\sigma, \partial U)\right)^{(q(\alpha-2)-\alpha)/(q-1)}. \tag{3.2.14}$$

Since $\alpha > (q+1)/(q-1)$, we have $(q(\alpha-2)-\alpha)/(q-1) = \alpha - 2q/(q-1) > -1$ and estimate (3.2.10) follows.

Step 2. Reduction of the equation.
Case 1, $N > 2$. We set classicaly

$$s = r^{N-2}/(N-2), \qquad u(r, \sigma) = r^{2-N}v(s, \sigma), \tag{3.2.15}$$

and v satisfies

$$s^2 v_{ss} + (N-2)^{-2}\Delta_{S^{N-1}}v - C(N)s^{N(N-2)-q}v^q, \tag{3.2.16}$$

where $C(N) = (N-2)^{(4-N)/(N-2)-q}$. Consequently

$$s^2 \frac{d^2}{ds^2}\int_U v\varphi_U^\alpha d\sigma + (N-2)^{-2}\int_U v\Delta_{S^{N-1}}\varphi_U^\alpha d\sigma - C(N)s^{N/(N-2)-q}\int_U v^q \varphi_U^\alpha d\sigma = 0. \tag{3.2.17}$$

Moreover

$$\left|\int_U v\Delta_{S^{N-1}}\varphi_U^\alpha d\sigma\right| \leq \left(\int_U v^q \varphi_U^{\alpha\,q}d\sigma\right)^{1/q}\left(\int_U \left|\Delta_{S^{N-1}}\varphi_U^\alpha\right|^{q/(q-1)}\varphi_U^{-\alpha/(q-1)}d\sigma\right)^{q/(q-1)}. \tag{3.2.18}$$

Using Step 1,

$$\left|\int_U v\Delta_{S^{N-1}}\varphi_U^\alpha d\sigma\right| \leq K^{q/(q-1)}\left(\int_U v^q \varphi_U^\alpha d\sigma\right)^{1/q}, \tag{3.2.19}$$

235

and finally we obtain the following inequalities:

$$s^2 \frac{d^2}{ds^2} \int_U v\varphi_U^\alpha d\sigma \overset{+}{\underset{-}{}} (N-2)^{-2} K^{(q-1)/q} \left(\int_U v^q \varphi_U^\alpha d\sigma \right)^{1/q}$$

$$-C(N)s^{N/(N-2)-q} \int_U v^q \varphi_U^\alpha d\sigma \overset{\geq}{\underset{\leq}{}} 0. \tag{3.2.20}$$

Case 2, $N = 2$. We set

$$r = e^{-t}, \qquad u(r,\sigma) = v(t,\sigma), \tag{3.2.21}$$

and v satisfies

$$v_{tt} + v_{\sigma\sigma} - e^{-2t} v^q = 0 \tag{3.2.22}$$

for $(t,\sigma) \in \mathbb{R}^+ \times S^1$. Consequently

$$\frac{d^2}{dt^2} \int_U v\varphi_U^\alpha d\sigma \overset{+}{\underset{-}{}} K^{(q-1)/q} \left(\int_U v^q \varphi_U^\alpha d\sigma \right)^{1/q} - e^{-2t} \int_U v^q \varphi_U^\alpha d\sigma \overset{\geq}{\underset{\leq}{}} 0. \tag{3.2.23}$$

Step 3. End of the proof when $N > 2$.
(i) We first assume that (3.2.5) holds; this means that

$$\int_\delta^{1/(N-2)} \int_U v^q \varphi_U^\alpha (1 - (N-2)s) d\sigma ds = \infty, \tag{3.2.24}$$

for some $\delta \in (0, (N-2)^{-1}]$ and (3.2.20) implies

$$s^2 \frac{d^2}{ds^2} \int_U v\varphi_U^\alpha d\sigma \geq C(N)\delta^{N/(N-2)-q} \int_U v^q \varphi_U^\alpha d\sigma$$

$$- (N-2)^{-2} K^{(q-1)/q} \left(\int_U v^q \varphi_U^\alpha d\sigma \right)^{1/q} \tag{3.2.25}$$

on $\left[\delta, (N-2)^{-1}\right)$. Consequently the differential inequality

$$\frac{d^2}{ds^2} \int_U v\varphi_U^\alpha d\sigma \geq A \int_U v\varphi_U^\alpha d\sigma - B, \tag{3.2.26}$$

holds for some positive constants A and B. Let $X(\tau)$ be $\left(\int_U v(s)\varphi_U^\alpha d\sigma \right)$, then

$$X(s) \geq X(\delta) + (s-\delta)X'(\delta) + A \int_\delta^s \int_\delta^\eta \left(\int_U v^q \varphi_U^\alpha d\sigma \right) d\tau d\eta - 2^{-1}B(s-\delta)^2. \tag{3.2.27}$$

236

But the function

$$s \mapsto \int_{\delta}^{s}\int_{\delta}^{\eta}\left(\int_{U} v^{q}\varphi_{U}^{\alpha}d\sigma\right)d\tau d\eta = \int_{\delta}^{s}(s-\tau)\left(\int_{U} v^{q}\varphi_{U}^{\alpha}d\sigma\right)d\tau \qquad (3.2.28)$$

tends to infinity as s goes to 1/(N-2). Therefore

$$\lim_{s \to (N-2)^{-1}} \int_{U} v(s)\varphi_{U}^{\alpha}d\sigma = \infty, \qquad (3.2.29)$$

which is (3.2.6).

(ii) Let us suppose that (3.2.7) holds, it means

$$\int_{\delta}^{1/(N-2)} \int_{U} v^{q}\varphi_{U}^{\alpha}(1-(N-2)s)d\sigma ds < \infty, \qquad (3.2.30)$$

and, as above, the inequality

$$\frac{d^{2}}{ds^{2}}\int_{U} v\varphi_{U}^{\alpha}d\sigma - A'\int_{U} v\varphi_{U}^{\alpha}d\sigma + B' \le 0, \qquad (3.2.31)$$

holds on $\left[\delta,(N-2)^{-1}\right)$ for some positive constants A' and B'. Consequently the function

$s \mapsto Y(s) = X(s) - A'\int_{\delta}^{s}\int_{\delta}^{\eta}\left(\int_{U} v^{q}\varphi_{U}^{\alpha}d\sigma\right)d\tau d\eta + 2^{-1}B'(s-\delta)^{2}$ is concave, $\lim_{s \to (N-2)^{-1}} X(s)$ exists,

is finite and nonnegative. If $\zeta \in C^{2}(U)$ and satisfies (3.2.8), then

$$s^{2}\frac{d^{2}}{ds^{2}}\int_{U} v\zeta d\sigma + (N-2)^{-1}\int_{U} v\Delta_{S^{N-1}}\zeta d\sigma - C(N)s^{N/(N-2)-q}\int_{U} v^{q}\zeta d\sigma = 0. \quad (3.2.32)$$

Since

$$\left|\int_{U} v\Delta_{S^{N-1}}\zeta d\sigma\right| \le k\int_{U} v\varphi_{U}^{\alpha-2}d\sigma \le k\left(\int_{U} v\varphi_{U}^{\alpha}d\sigma\right)^{1/q}\left(\int_{U} \varphi_{U}^{\alpha-2q/(q-1)}d\sigma\right)^{(q-1)/q} \qquad (3.2.33)$$

and $0 \le \int_{U} v^{q}\zeta d\sigma \le k\int_{U} v^{q}\varphi_{U}^{\alpha}d\sigma$, there exist $A'' > 0$ and $B'' > 0$ such that

$$s \mapsto \int_{U} v(s)\zeta d\sigma - kA''\int_{\delta}^{s}\int_{\delta}^{\eta}\int_{U} v^{q}\zeta d\sigma d\tau d\eta + 2^{-1}kB''(s-\delta)^{2} \qquad (3.2.34)$$

is concave on $\left[\delta,(N-2)^{-1}\right)$. From the integrability estimate (3.2.30) and (3.2.8) the function $s \mapsto \int_{U} v(s,.)\zeta d\sigma$ admits a finite limit as s goes to $1/(N-2)$. With the unknown u it means that $r \mapsto \int_U u(r,.)\zeta d\sigma$ also admits a limit at 1 and this limit, that we denote $\ell(\zeta)$, defines a nondecreasing linear functional on the space of $C^2(U)$ functions satisfying (3.2.8). Therefore this functional can be extended as a measure on U.

Step 4. End of the proof when N = 2.
(i) If we assume (3.2.5), we have in the t variable

$$\int_0^\infty \int_U v^q \varphi_U^\alpha (1 - e^{-t}) e^{-t} d\sigma dt = \infty, \tag{3.2.35}$$

and the blow-up problem occurs at t = 0, therefore

$$\int_0^1 \int_U v^q \varphi_U^\alpha d\sigma dt = \infty. \tag{3.2.36}$$

From (3.2.23)

$$\frac{d^2}{dt^2} \int_U v \varphi_U^\alpha d\sigma \geq A \int_U v^q \varphi_U^\alpha d\sigma - B \tag{3.2.37}$$

is valid for some A, B > 0. A double integration yields

$$\lim_{t \to 0} \int_U v(t,.) \varphi_U^\alpha d\sigma = \infty, \tag{3.2.38}$$

with the help of (3.2.36), which is (3.2.6).
(ii) If we assume (3.2.7) we have

$$\int_0^1 \int_U v^q \varphi_U^\alpha d\sigma dt < \infty, \tag{3.2.39}$$

and, from (3.2.23),

$$\frac{d^2}{dt^2} \int_U v \varphi_U^\alpha d\sigma \leq A \int_U v^q \varphi_U^\alpha d\sigma + B. \tag{3.2.40}$$

The same type of concavity argument as the one of Step 3 implies that $t \mapsto \int_U v(t,.) \varphi_U^\alpha d\sigma$ admits a finite limit at 0 which also means that $r \mapsto \int_U u(r,.) \varphi_U^\alpha d\sigma$ admits a finite limit at $r = 1$. In the same way as in Step 3, for any $\zeta \in C^2(U)$ satisfying (3.2.8), there exists $\ell(\zeta) = \lim_{r \to 1} \int_U u(r,.) \zeta d\sigma$, defining ℓ as a nonnegative measure on U.

The proof of Theorem 4.17 is an immediate consequence of Lemma 4.10. Thank to this result we state the following definition:

Definition. Let $\omega \in \partial B(0,1)$. We shall say that ω is a regular point relative to u if the statement (ii) of Theorem 4.17 holds. The set of regular point, which is a relatively open subset, will be denoted by $\mathcal{R} = \mathcal{R}(u)$. We denote by $S = S(u)$ the set $\partial B(0,1) \setminus \mathcal{R}$. The elements of S are called singular relatively to u. The statement (ii) of Theorem 4.17 implies that there exists a nonnegative Radon measure μ on $\mathcal{R}$ such that

$$\lim_{r \to 1} \int_U u(r,\sigma)\zeta(\sigma)d\sigma = \int_U \zeta d\mu \tag{3.2.41}$$

for any $\zeta \in C_0(\mathcal{R})$. The couple (S, μ) will be called **the trace of** u.

In the case $1 < q < (N+1)/(N-1)$ it is possible to give a minimum blow-up estimate of u for any $a \in S$. If a is a boundary point and $c > 0$ we denote $v_{c,a}$ the solution of

$$\begin{cases} -\Delta v_{c,a} + (v_{c,a})^q = 0 & \text{in } B(O,1), \\ \qquad\quad v_{c,a} = c\delta_a & \text{on } \partial B(O,1). \end{cases} \tag{3.2.42}$$

When c tends to infinity, $v_{c,a}$ increases and converges to $v_{\infty,a}$ which is a solution of (3.2.1) in $B(O,1)$ and satisfies

$$\lim_{x \to a} |x-a|^{2/(q-1)} v_{\infty,a}(x) = \omega_a\big(x-a/|x-a|\big), \tag{3.2.43}$$

in the sense of Theorem 4.8, where ω_a is the unique positive solution of

$$-\Delta_{S^{N-1}}\omega_a - \frac{2}{q-1}\left(\frac{2q}{q-1} - N\right)\omega_a + \omega_a^q = 0 \tag{3.2.44}$$

in $S_{a+}^{N-1} = S^{N-1} \cap \{y: \ <y,a> \ <0\}$, vanishing on the equator ∂S_{a+}^{N-1}.

THEOREM 4.18. *Let* $1 < q < (N+1)/(N-1)$, u *be a nonnegative solution of* (3.2.1) *with trace* (S,μ) *and* $a \in S$. *Then* $u(x) \geq v_{\infty,a}(x)$ *for any x in* $B(O,1)$. *In particular*

$$\liminf_{\substack{x \to a \\ x \in C}} |x-a|^{2/(q-1)} u(x) > 0 \tag{3.2.45}$$

where C *is any cone interior to* $B(O,1)$ *with vertex* a.

Proof. From Theorem 4.17, for any $\varepsilon \in (0,1)$

$$\lim_{r \to 1} \int_{D_\varepsilon} u(r,\sigma)d\sigma = \infty \tag{3.2.46}$$

where D_ε is the geodesic ball on S^{N-1} of radius ε and center a. We set $\theta = 1 - \varepsilon$, define u_θ in $B(O, 1/\theta)$ by

$$u_\theta(x) = \theta^{2/(q-1)} u(\theta x);\qquad (3.2.47)$$

then function u_θ satisfies (3.2.1). If $\eta \in (0, \pi)$ we set

$$M_{\varepsilon,\eta} = \int_{D_\eta} u(1 - \varepsilon, \sigma) d\sigma \qquad (3.2.48)$$

and, for any $\eta \in (0, \pi)$ and $m > u(O)$, there exists $\varepsilon = \varepsilon(\eta, m) \in (0, 1)$ such that $M_{\varepsilon,\eta} = m$. Let w_η be the solution of (3.2.1) in $B(O, 1)$ subject to the following boundary conditions

$$w_\eta(x) = \begin{cases} 0 & \text{if } x \notin D_\eta, \\ u_\theta(x) & \text{if } x \in D_\eta. \end{cases} \qquad (3.2.49)$$

From the maximum principle $w_\eta \leq u_\theta$. Since $\lim_{\eta \to 0} \varepsilon = 0$, $\lim_{\eta \to 0} \theta = 1$ and $\lim_{\eta \to 0} w_{\eta|\partial B(O,1)} = m\delta_a$ weakly in the sense of measures on $\partial B(O, 1)$, the set of functions $\{w_\eta\}$ remains bounded in the Marcinkiewicz spaces $M^{(N+1)/(N-1)}(B(O,1), \rho dx) \cap M^{N/(N-1)}(B(O,1), dx)$ where $\rho(x) = 1 - |x|$. Therefore w_η converges to $v_{m,a}$ and $v_{m,a} \leq u$ from Theorem 4.1. Letting m tend to infinity yields

$$u(x) \geq v_{\infty,a}(x), \qquad (3.2.50)$$

for any $x \in \partial B(O, 1)$.

Remark 4.16. When $q \geq (N + 1)/(N - 1)$ the situation is more complicated since isolated points are removable singularities for solutions of (3.2.1) in the sense of Theorem 4.3. However it is possible to prove that if $q > 1$ and S has a non-empty relative interior int (S), then for any $\omega \in \text{int}(S)$ there holds

$$\lim_{r \to 1} (1 - r)^{2/(q-1)} u(r, \omega) = \left(\frac{2(q+1)}{(q-1)^2} \right)^{1/(q-1)}, \qquad (3.2.51)$$

uniformly on any compact subset $K \subset \text{int}(S)$.

As a consequence of Theorem 4.18 we have the uniqueness of the complete solutions of (3.2.1), that is the positive solutions u which satisfy

$$\int_0^1 u^{(q-1)/2}(\gamma(t)) dt = \infty, \qquad (3.2.52)$$

for any $\gamma \in C^{0,1}\left([0,1], \overline{B}(O,1)\right)$ with $\gamma([0,1)) \subset B(O,1)$ and $\gamma(1) \in \partial B(O,1)$.

COROLLARY 4.3. *Suppose that* $1 < q < (N+1)/(N-1)$, *then there exists one and only one complete solution of* (3.2.1) *in* $B(O,1)$.

Proof. We first notice that $(q-1)/2 < 1$, which implies that for any measurable subset $U \subset \partial B(O,1)$ and any, $T \in (0,1)$ there holds

$$\int_0^T \int_U u^{(q-1)/2}(t,\sigma)d\sigma dt \leq (T \operatorname{mes}(U))^{(3-q)/2} \left(\int_0^T \int_U u(t,\sigma)d\sigma dt \right)^{(q-1)/2}, \qquad (3.2.53)$$

and the completeness assumption yields $\int_0^1 \int_U u(t,\sigma)d\sigma dt = \infty$. Since U is arbitrary, $S = \partial B(O,1)$ and u is a large solution of (3.2.1) in $B(O,1)$ and therefore is unique.

Remark 4.17. The most interesting case for the notion of completeness deals with the case $q = (N+2)/(N-2)$. In that case, if u is a complete solution of (3.2.1) the new metric $g_u = u^{4/(N-2)}g_0$ defines a complete metric on the unit ball $B(O,1)$. If we set $g_H = (4/(1-|x|^2)^2 g_0$, the Poincaré representation of the N-dimensional hyperbolic space consists in the identification

$$\mathbb{H}^N = \left(B(O,1), g_H\right). \qquad (3.2.54)$$

If $v(x) = u(x)(2/(1-|x|^2)^{-(N-2)/2}$, the metric $g_v = v^{4/(N-2)}g_H$ is complete and has scalar curvature -1 (see [RRV] for details about this type of problems).

When $1 < q < (N+1)/(N-1)$, the understanding of the trace problem is essentially complete since we have

THEOREM 4.19. *Assume* $1 < q < (N+1)/(N-1)$, S *is a closed subset of* $\partial B(O,1)$ *and* μ *a nonnegative Radon measure on* $\mathcal{R} = \partial B(0,1)\backslash S$. *Then there exist one and only one nonnegative solution* u *of* (3.2.1) *with trace* (S, μ).

This result has been obtained by Le Gall ([LG2]) in the case q = N = 2 and in the general case by Marcus and Véron ([MV3]); their proof needs the construction of a minimal and a maximal solution of (3.2.1) with trace element (S, μ), and a local scaling comparison argument. When $q \geq (N+1)/(N-1)$ the situation is more complicated, but, given a trace (S, μ), we can construct a corresponding solution provided μ and S satisfy some admissibility and compatibility condition.

5 Singularities of quasilinear elliptic equations involving the p-Laplace or the mean curvature operator

The aim of this chapter is to present some important singularity results for the following types of nonlinear equations

$$-\Delta_p u + g(u) = -\text{div}\left(|\nabla u|^{p-2}\nabla u\right) + g(u) = 0 \tag{1}$$

with $g(u) = \pm|u|^{q-1}u$ and $q > p - 1 > 0$ or

$$-H(u) - \lambda u = -\text{div}\left(\nabla u / \sqrt{1 + |\nabla u|^2}\right) - \lambda u = 0. \tag{2}$$

with $\lambda > 0$. The first equation is the natural extension of the Emden-Fowler one, as for the second, called the capillarity equation, it plays a fundamental role in describing the shape of a falling liquid drop.

5-1 The p-Laplace equation

5-1-1 $C^{1,\alpha}$-regularity results for degenerate elliptic equations

Let Ω be a domain of $\mathbb{R}^N$. We consider a vector valued function $A = \left(A_j\right)$, continuous from $\Omega \times \mathbb{R} \times \mathbb{R}^N$ into $\mathbb{R}^N$ and C^1 from $\Omega \times \mathbb{R} \times \mathbb{R}^N \setminus \{0\}$ and a real valued function B defined in $\Omega \times \mathbb{R} \times \mathbb{R}^N$, measurable in Ω and separately continuous in $\mathbb{R} \times \mathbb{R}^N$. We also suppose that A satisfies the following growth and ellipticity conditions:

H-1 $\quad A(x,r,0) = 0,$

H-2 $\quad \sum_{i,j} \dfrac{\partial A_j}{\partial p_i}(x,r,q)\xi_i\xi_j \geq \gamma(\delta + |q|^2)^{(p-2)/2}|\xi|^2, \quad \forall(x,r,q,\xi) \in \Omega \times \mathbb{R} \times \mathbb{R}^N_* \times \mathbb{R}^N \setminus \{O\},$

$\quad$ for some $p > 1$, $\gamma > 0$ and $\delta \in [0,1]$.

H-3 $\quad \sum_{i,j} \left|\dfrac{\partial A_j}{\partial p_i}(x,r,q)\right| \leq \Gamma(\delta + |q|^2)^{(p-2)/2}, \ \forall(x,r,q) \in \Omega \times \mathbb{R} \times \mathbb{R}^N_* \setminus \{O\}$, for some $\Gamma > 0$.

H-4 $\quad \sum_{i,j} \left|\dfrac{\partial A_j}{\partial x_i}(x,r,q)\right| + \sum_j \left|\dfrac{\partial A_j}{\partial r}(x,r,q)\right| \leq \Gamma(\delta + |q|^2)^{(p-2)/2}|q|, \ \forall(x,r,q) \in \Omega \times \mathbb{R} \times \mathbb{R}^N.$

H-5 $\quad |B(x,r,p)| \leq \Gamma' (1+|q|^2)^{p/2}, \quad \forall(x,r,q) \in \Omega \times \mathbb{R} \times \mathbb{R}^N$ for some $\Gamma' \geq 0$.

The following regularity result is due to Tolksdorf [To1] under this general form and to Ural'Ceva[Ua], Ulhenbeck[Ul] and Lewis [Le1] and [Le2] under a much less general one.

THEOREM 5.1. *Let* $u \in W^{1,p}_{loc}(\Omega) \cap L^{\infty}_{loc}(\Omega)$ *be a weak solution of*

$$-\text{div}(A(x,u,\nabla u)) + B(x,u,\nabla u) = 0 \tag{1.1.1}$$

in Ω. *We suppose that* $a \in \Omega$ *is such that* $B(a,3) \subset \Omega$. *Then there exist* $C > 0$ *and* $\alpha \in (0,1]$ *depending on* N, p, γ, Γ *and* $\|u\|_{L^{\infty}(Ba,3R)}$ *such that*

$$|\nabla u(x)| \leq C, \tag{1.1.2}$$

$$|\nabla u(x) - \nabla u(x')| \leq C|x-x'|^{\alpha}, \tag{1.1.3}$$

for any $(x, x') \in B(a,1)$.

When the coefficient δ in H-2-H-4 is positive, the equation is not degenerate and $\alpha = 1$ from the classical elliptic equations theory ([GT]). The strict comparison principle is usually not valid for degenerate equations except for non degenerate points or p-harmonic functions. The following result has been proved by Tolksdorf in [To2].

PROPOSITION 5.1. *Assume that* Ω *is an open domain of* $\mathbb{R}^N$, *H-1--H-4 is fulfilled, and the functions* u_i $(i=1, 2)$ *belong to* $W^{1,p}_{loc}(\Omega) \cap L^{\infty}_{loc}(\Omega)$ *and satisfy* $u_1 \geq u_2$ *and*

$$-\text{div}(A(x,u_i,\nabla u_i)) = 0 \tag{1.1.4}$$

in Ω. *If* u_1 *is not equal to* u_2, *then* $u_1 > u_2$ *in* Ω.

5-1-2 Singular p-harmonic functions

Let Ω be a bounded domain of $\mathbb{R}^N$, $\alpha = (\alpha_1, \alpha_2, ..., \alpha_N) \in \mathbb{N}^N$, containing O , we shall say that a function u belonging to $W^{1,p}(\Omega)$ is p-harmonic in Ω or $\Omega^* = \Omega \setminus \{O\}$ if

$$-\text{div}(|\nabla u|^{p-2} \nabla u) = 0 \tag{1.2.1}$$

in weak sense in Ω or Ω^*. Any p-harmonic function is locally $C^{1,\alpha}$ from Chapter 1 and Theorem 5.1. From Serrin's results in Chapter 1, we know that if u is bounded below, p-harmonic in Ω^* and singular at O, there exists $k > 1$ such that

$$k^{-1} \le \frac{u(x)}{\mu_p(x)} \le k \tag{1.2.2}$$

near O, where μ_p is the fundamental solution of the p-Laplace operator in $\mathbb{R}^N \setminus \{O\}$, that is

$$\mu_p(x) = \begin{cases} C(p,N)|x|^{(p-N)/(p-1)} & \text{if } 1 < p < N, \\ C(N)Ln(1/|x|) & \text{if } p = N, \end{cases} \tag{1.2.3}$$

with $C(p,N) = \dfrac{p-1}{N-p}\left|S^{N-1}\right|^{-1/(p-1)}$, $C(N) = \left|S^{N-1}\right|^{-1/(N-1)}$. Moreover μ_p solves

$$-\text{div}\left(\left|\nabla\mu_p\right|^{p-2}\nabla\mu_p\right) = \delta_O, \tag{1.2.4}$$

in the sense of distributions in $\mathbb{R}^N$; if $p > N$, the fundamental solution of the p-Laplace equation is bounded and therefore less interesting. In fact estimate (1.2.2) can be made much more precise in the case of p-harmonic functions and the following result has been proved by Kichenassamy and Véron [KV].

THEOREM 5.2. *Let* $1 < p \le N$ *and* u *be a p-harmonic function in* Ω^* *such that* $u(x)/\mu_p(x)$ *remains bounded in some neighborhood of* O. *Then there exists a real number* γ *such that*

$$u - \gamma\mu_p \in L^\infty_{loc}(\Omega). \tag{1.2.5}$$

Moreover, when $\gamma \ne 0$, *the following relation holds*

$$\lim_{x \to O} |x|^{(N-p)/(p-1)+|\sigma|}D^\sigma(u - \gamma\mu_p) = 0, \tag{1.2.6}$$

for all multi-indices $\sigma = (\sigma_1, \sigma_2, ..., \sigma_N) \in \mathbb{N}^N$ *with* $|\sigma| = \sigma_1 + \sigma_2 + ... + \sigma_N \ge 1$, $|\nabla u|^{p-1}$ *is locally integrable in* Ω *and* u *satisfies the following equation (even if* $\gamma = 0$)

$$-\text{div}\left(|\nabla u|^{p-2}\nabla u\right) = \gamma|\gamma|^{p-2}\delta_O \tag{1.2.7}$$

in the sense of distributions in Ω.

We may assume that $\overline{B}(O,1) \subset \Omega$ and we have the following *a priori* estimates

LEMMA 5.1. *Under the assumptions of Theorem 5.1 there exist two constants* $C = C(N,p,u) > 0$ *and* $\alpha = \alpha(N,p) \in (0,1)$ *such that for any* (x, x') *satisfying* $0 < |x| \le |x'| \le 1$, *we have*

$$|\nabla u(x)| \le C|x|^{-1}(\mu_p(x)+1), \tag{1.2.8}$$

$$|\nabla u(x) - \nabla u(x')| \le C|x - x'|^\alpha |x|^{-1-\alpha}(\mu_p(x)+1). \tag{1.2.9}$$

Proof. For $0 < a < 1/2$ and $0 < |x| \le 1$, we set $x = ay/(1+a)$ and $v(y) = u(ay/(1+a))/\mu(a)$. The function v is p-harmonic in $\{y: 1/2 \le |y| \le 3\}$ and we deduce from the classical regularity theory for degenerate equations ([Le1], [Le2] and [Ua], see Theorem 5.1) that

$$\|\nabla v\|_{C^\alpha(\{y: \, 2/3 \le |y| \le 2\})} \le C \tag{1.2.10}$$

where α and C depend on N,p and N,p,u respectively. Replacing v by u yields ((1.2.8)-(1.2.9).

Remark 5.1. In the case $p = N$ the above estimates are not optimal since the term $\mu_N(x)+1$ is not necessary. However, improvement of this estimate requires a scaling over the quantity $u - \gamma\mu_N$ which we do not yet known to be bounded. Moreover the pointwise estimate over $u(x)/\mu_p(x)$ can be replaced by an integral one, namely

$$\int_{r \le |x| \le 2r} |u(x)|^p dx \le C \begin{cases} r^{(p^2-N)/(p-1)} & \text{if } p < N, \\ (r\ln(1/r))^N & \text{if } p = N. \end{cases} \tag{1.2.11}$$

As a direct consequence of Proposition 5.1, we have

LEMMA 5.2. *Let* G *be an open domain in* $\mathbb{R}^N$ *which does not contain* 0 *and u is p-harmonic in* G *with p > 1. If* $v = u/\mu_p$ *or* $v = u - \mu_p$ *achieves its maximum in* G, *then v is constant.*

Proof of Theorem 5.2. Step 1. The case $1 < p < N$. We define γ^+ and γ^- by

$$\gamma^+ = \limsup_{x\to 0} u(x)/\mu_p(x), \qquad \gamma^- = \liminf_{x\to 0} u(x)/\mu_p(x). \tag{1.2.12}$$

If $\gamma^+ = \gamma^- = 0$, $\lim_{x\to 0} u(x)/\mu_p(x) = 0$. So we can assume that $\gamma^+ > 0$ (or $\gamma^- < 0$ in the same way). Let β be the maximum of u on $\partial B(O,1)$. The function $u_\beta = u - \beta$ satisfies $\limsup_{x\to 0} u_\beta(x)/\mu_p(x) = \gamma^+$ and its maximum on $\partial B(O,1)$ is 0; for the sake of simplicity, we shall denote it u and, for $r \in [0,1]$, we set

$$\tilde\gamma(r) = \sup_{r \le |x| \le 1} u(x)/\mu_p(x). \tag{1.2.13}$$

The function $\tilde{\gamma}$ is nonnegative. Moreover, if there exist some $r \in (0,1]$ and some y with $r < |y| < 1$ such that $\tilde{\gamma}(r) = u(y)/\mu_p(y)$, then $u(x) = \tilde{\gamma}(r)\mu_p(x)$ in $\Gamma_r = \{x: r \leq |x| \leq 1\}$ from Lemma 5.2 and in fact $\tilde{\gamma} = 0$ on $[r,1]$. Therefore the function $\tilde{\gamma}$ is nonincreasing and we have

$$\tilde{\gamma}(r) = \sup_{|x|=r} u(x)/\mu_p(x), \tag{1.2.14}$$

and

$$\lim_{r \to 0} \tilde{\gamma}(r) = \gamma^+. \tag{1.2.15}$$

Let x_r be such that $|x_r| = r$ and $\tilde{\gamma}(r) = u(x_r)/\mu_p(x_r)$. We define u_r on $\Lambda_r = \{\xi: 0 < |\xi| < 1/r\}$ by

$$u_r(\xi) = u(r\xi)/\mu_p(r). \tag{1.2.16}$$

Since u_r is p-harmonic, we deduce from Lemma 5.1 that the following estimates hold:

$$\begin{cases} \text{(i)} & |u_r(\xi)| \leq C\mu_p(\xi), \\[2mm] \text{(ii)} & |\nabla u_r(\xi)| \leq C|\xi|^{(1-N)/(p-1)}, \\[2mm] \text{(iii)} & |\nabla u_r(\xi) - \nabla u_r(\xi')| \leq C|\xi - \xi'|^\alpha \left(\min|\xi|,|\xi'|\right)^{(1-N)/(p-1)-\alpha}, \end{cases} \tag{1.2.17}$$

where C is independent of r. From the Arzela-Ascoli Theorem, there exist a sequence $\{r_n\}$ going to 0 and a p-harmonic function v defined in $\mathbb{R}^N \setminus \{O\}$ such that the sequence $\{u_{r_n}\}$ converges to v in the $C^1_{loc}(\mathbb{R}^N \setminus \{O\})$-topology. Moreover we have

$$\frac{u_r(\xi)}{\mu_p(\xi)} = \frac{u(\xi)}{\mu_p(r)\mu_p(\xi)} = \frac{u(\xi)}{\mu_p(r\xi)}\frac{\mu_p(r\xi)}{\mu_p(r)\mu_p(\xi)} = \frac{u(r\xi)}{\mu_p(r)\mu_p(1)}, \tag{1.2.18}$$

which implies that $u_r(\xi)/\mu_p(\xi) \leq \gamma^+/\mu_p(1)$. We can also suppose that $\{\xi_{r_n}\}$ converges to some $\xi_0 \in S^{N-1}$; therefore

$$v(\xi_0)/\mu_p(\xi_0) = \gamma^+/\mu_p(1) \quad \text{and} \quad v(\xi)/\mu_p(\xi) \leq \gamma^+/\mu_p(1). \tag{1.2.19}$$

We deduce from Lemma 5.2 that $v(\xi)/\mu_p(\xi) = \gamma^+/\mu_p(1)$ and $\lim_{r \to 0} u_r(\xi) = v(\xi)$ in the above C^1_{loc}- topology on $\mathbb{R}^N \setminus \{O\}$. In the particular case where $\xi \in S^{N-1}$, we have (with $x = r\xi$)

$$\lim_{x \to O} u(x) / \mu_p(x) = \gamma^+ = \gamma. \tag{1.2.20}$$

In order to prove the boundedness of $u - \gamma\mu_p$, we consider

$$\begin{cases} \text{(i)} \quad v_\varepsilon^+(x) = (\gamma + \varepsilon)\mu_p(x) - (\gamma + \varepsilon)\mu_p(1) + \sup_{|y|=1} u(y), \\[2mm] \text{(ii)} \quad v_\varepsilon^-(x) = (\gamma - \varepsilon)\mu_p(x) - (\gamma - \varepsilon)\mu_p(1) + \inf_{|y|=1} u(y), \end{cases} \tag{1.2.21}$$

for $\varepsilon > 0$ and v_ε^+ and v_ε^- are both p-harmonic functions in $B(O,1)$. Since $(u - v_\varepsilon^+)^+$ and $(v_\varepsilon^- - u)^+$ vanish in a neighborhood of O and near $\partial B(O,1)$, we deduce that $v_\varepsilon^-(x) \le u(x) \le v_\varepsilon^+(x)$. Letting ε go to 0 yields

$$\inf_{|x|=1} u(x) - \gamma\mu_p(1) \le u - \gamma\mu_p \le \sup_{|x|=1} u(x) - \gamma\mu_p(1) \tag{1.2.22}$$

in $B(O,1)\backslash\{O\}$. From (1.2.20) there also holds

$$\lim_{r \to 0} \nabla u_r(\xi) = \gamma / \mu_p(1)\nabla\mu_p(\xi), \tag{1.2.23}$$

uniformly on every compact subset of $\mathbb{R}^N\backslash\{O\}$. If we take $|\xi| = 1$ and set $x = r\xi$, we have (1.2.6) with $|\sigma| = 1$. This implies also that ∇u_r never vanishes on $\overline{\Gamma} = \{\xi : 1/2 \le |\xi| \le 2\}$ for $0 < r \le r_0$. Therefore the equation satisfies by u_r is nondegenerate and u_r is C^∞. Using the same device as in Lemma 5.1, we deduce that

$$\left|D^\sigma u_r(\xi)\right| \le C|\xi|^{(p-N)/(p-1)-|\sigma|} \tag{1.2.24}$$

holds for any multi-indices σ with $|\sigma| \ge 2$, which gives (1.2.6). Moreover

$$\int_{|x|>r} |\nabla u|^{p-2} \nabla u . \nabla\phi dx = -\int_{|x|=r} \phi|\nabla u|^{p-2} \frac{\partial u}{\partial \nu} dS(x) \tag{1.2.25}$$

for any $\phi \in C_0^1(\Omega)$ and $r \in (0,1)$. Because

$$\nabla u(x) \approx -\left(N|S^{N-1}|\right)^{-1/(p-1)} |x|^{(1-N)/(p-1)} x / |x|, \quad \text{as} \quad x \to O, \tag{1.2.26}$$

we obtain

$$\int_\Omega |\nabla u|^{p-2} \nabla u . \nabla\phi dx = \gamma|\gamma|^{p-2} \phi(O). \tag{1.2.27}$$

Step 2. The case p = N. In the same way as in Step 1 we define γ^{+} and γ^{-} and we assume that $\gamma^{+} = \gamma = \limsup\limits_{x \to O} u(x) / \mu_{N}(x) > 0$. We can modify u in such a way that the radial function $\tilde{\gamma}$ defined by

$$\tilde{\gamma}(r) = \sup_{r \le |x| \le 1/2} u(x) / \mu_{N}(x), \tag{1.2.28}$$

satisfies $\tilde{\gamma}(1/2) = 0$, is nonincreasing and such that $\lim\limits_{r \to 0} \tilde{\gamma}(r) = 0$. We set again

$$u_{r}(\xi) = u(r\xi) / \mu_{N}(\xi) \text{ and } \Lambda_{r} = \left\{ \xi : 0 < |\xi| < 1/(2r) \right\}.$$

Then u_{r} is N-harmonic on Λ_{r} and the following estimates are satisfied

$$\begin{cases} |u_{r}(\xi)| \le C\left(1 + |Ln|\xi|| / Ln(1/r)\right), \\ |\nabla u_{r}(\xi)| \le C\left(1 + |Ln|\xi|| / Ln(1/r)\right)|\xi|^{-1}, \\ |\nabla u_{r}(\xi) - \nabla u_{r}(\xi')| \le C|\xi - \xi'|^{\sigma}\left(1 + |Ln|\xi|| / Ln(1/r)\right)|\xi|^{-1-\sigma}, \end{cases} \tag{1.2.29}$$

for some C > 0 and any $0 < |\xi| \le |\xi'| \le (1/2r)$. Moreover $1/(Ln(1/r)) \le 1/Ln2$ since r $< 1/2$. By the classical compactness argument, there exist a N-harmonic function v and a sequence $\left\{ r_{n} \right\}$ going to 0 such that $\left\{ u_{r_{n}} \right\}$ converges to v in the C^{1}_{loc}-topology of $\mathbb{R}^{N} \setminus \left\{ O \right\}$ and v satisfies

$$\begin{cases} |v(\xi)| \le C, \\ |\nabla v(\xi)| \le C|\xi|^{-1}, \\ |\nabla v(\xi) - \nabla v(\xi')| \le C|\xi - \xi'|^{\sigma}|\xi|^{-1-\sigma}, \end{cases} \tag{1.2.30}$$

for $0 < |\xi| \le |\xi'|$. Since v is bounded, it can be extended to $\mathbb{R}^{N}$ as a N-harmonic function $\tilde{v}$ in $\mathbb{R}^{N}$ which still satisfies (1.2.30). From a classical Liouville type result due to Reshetniak ([Re]), $\tilde{v}$ is constant . Moreover, we know that there exists $\xi_{r} \in S^{N-1}$ such that $\tilde{\gamma}(r) = u(r\xi_{r}) / \mu_{N}(r)$. Since we can also suppose that $\left\{ \xi_{r_{n}} \right\}$ converges to some $\xi_{0} \in S^{N-1}$, this implies

$$\gamma = \lim_{n \to \infty} \tilde{\gamma}(r_{n}) = v(\xi_{0}) = v(\xi), \tag{1.2.31}$$

and $\lim\limits_{r \to 0} u_{r}(\xi) = \gamma$. If we take in particular $|\xi| = 1$ and $|x| = 1$, we get $\lim\limits_{x \to O} u(x) / \mu_{N}(x) = \gamma$.

In order to prove the boundedness of $u - \gamma\mu_N$, we set $v_r(\xi) = u(r\xi) - \gamma\mu_N(r)$. It is clear that this function is bounded on any compact subset of $\Lambda_r = \{\xi: 0 < |\xi| < 1/(2r)\}$, since $\mu_N(r\xi) = \mu_N(r) + \mu_N(\xi)$ and

$$\gamma\mu_N(\xi) - |u(r\xi) - \gamma\mu_N(r\xi)| \le v_r(\xi) \le \gamma\mu_N(\xi) + |u(r\xi) - \gamma\mu_N(r\xi)|. \tag{1.2.32}$$

From Theorem 5.1, the following estimates hold

$$\|\nabla v_r\|_{C^\sigma (x:\ 1/2 \le |x| \le 2)} \le C\|v_r\|_{L^\infty (x:\ 1/4 \le |x| \le 3)}, \tag{1.2.33}$$

which implies the following improvement of Lemma 5.1:

$$\begin{cases} |\nabla u(x)| \le C|x|^{-1}, \\ |\nabla u(x) - \nabla u(x')| \le C|x - x'|^\sigma |x|^{-1-\sigma}, \end{cases} \tag{1.2.34}$$

for $0 < |x| \le |x'| \le 1$. If we go back to the function v_r, we have

$$\begin{cases} |\nabla v_r(\xi)| \le C|\xi|^{-1}, \\ |\nabla v_r(\xi) - \nabla v_r(\xi')| \le C|\xi - \xi'|^\sigma |\xi|^{-1-\sigma}, \end{cases} \tag{1.2.35}$$

for $0 < |\xi| \le |\xi'| \le (1/2r)$. We prove (1.2.6) in the following way. Since the function $u - \gamma\mu_N$ is bounded in $\overline{B}(O, 1/2)$, it achieves its maximum ; if the maximum is achieved in $B(O, 1/2)$ then $u - \gamma\mu_N$ is constant otherwise we assume that this maximum is achieved either for $|x| = 1/2$ or for $|x| = 0$, and in that case

$$\sup_{0 < |x| \le 1/2} (u(x) - \gamma\mu_N(x)) = \limsup_{x \to O} (u(x) - \gamma\mu_N(x)). \tag{1.2.36}$$

Case 1. We assume that (1.2.36) holds and denote λ the value of the supremum. We define $\lambda(r)$ as being

$$\lambda(r) = \sup_{r \le |x| \le 1/2} (u(x) - \gamma\mu_N(x)) = \sup_{|x| = r} (u(x) - \gamma\mu_N(x)) = u(x_r) - \gamma\mu_N(x_r), \tag{1.2.37}$$

in which formula $|x_r| = r$. Since the set of functions $\{v_r\}$ is relatively compact in the C^1_{loc}-topology of $\Lambda_r = \{\xi: 0 < |\xi| < 1/(2r)\}$, there exist some $v \in C^1_{loc}(\mathbb{R}^N \setminus \{O\})$ and a subsequence $\{v_{r_n}\}$ with $r_n \to 0$ such that $v_{r_n} \underset{n \to \infty}{\to} v$ in the $C^1_{loc}(\mathbb{R}^N \setminus \{O\})$-topology. Moreover we can assume that $\xi_{r_n} = x_{r_n} / r_n \to \xi_0 \in S^{N-1}$ as n goes to infinity. We deduce from (1.2.37) that

$$u(r_n \xi_{r_n}) - \gamma \mu_N(r_n \xi_{r_n}) \underset{n \to \infty}{\longrightarrow} \lambda. \tag{1.2.38}$$

Therefore

$$v(\xi) \le \gamma \mu_N(\xi) + \lambda \quad \text{and} \quad v(\xi_0) = \gamma \mu_N(\xi_0) + \lambda, \tag{1.2.39}$$

which implies that $v = \gamma \mu_N + \lambda$ from Lemma 5.2. Therefore v_r converges to $\gamma \mu_N + \lambda$, and this means

$$\lim_{x \to 0} \left(u(x) - \gamma \mu_N(x) \right) = \lambda, \tag{1.2.40}$$

$$\lim_{x \to 0} |x| \nabla u(x) = \gamma \nabla \mu_N(x / |x|). \tag{1.2.41}$$

The remaining of the proof of (1.2.6) is the same as in the case $1 < p < N$.

Case 2. We assume $\sup\limits_{0 < |x| \le 1/2} (u(x) - \gamma \mu_N(x)) = \sup\limits_{|x| = 1/2} (u(x) - \gamma \mu_N(x))$ and we perform the scaling transformation $v_r(\xi) = u(r\xi) - \gamma \mu_N(r)$. There exist a N-harmonic function v and a sequence $\{r_n\}$ going to 0 such that $\{v_{r_n}\}$ converges to v in the $C^1_{loc}(\mathbb{R}^N \setminus \{O\})$. Moreover (1.2.32) implies

$$C_1 \le v(\xi) - \gamma \mu_N v(\xi) \le C_2. \tag{1.2.42}$$

Consider now the point where $u - \gamma \mu_N$ achieves its maximum in $\mathbb{R}^N \setminus \{O\}$. If this supremum is achieved at some ξ_0, then $u - \gamma \mu_N$ is identically equal to some constant λ, which means that

$$\lim_{r_n \to 0} \left(u(r_n \xi) - \gamma \mu_N(r_n \xi) \right) = \lambda. \tag{1.2.43}$$

Therefore, for $\varepsilon > 0$, there exists n_0 such that for $n \ge n_0$

$$\gamma \mu_N(r_n \xi) + \lambda - \varepsilon \le u(r_n \xi) \le \gamma \mu_N(r_n \xi) + \lambda + \varepsilon, \tag{1.2.44}$$

for $|\xi| = 1$. Applying the maximum principle in the spherical shell $\{x : r_n < |x| < r_{n_0}\}$, we deduce that $\gamma \mu_N + \lambda - \varepsilon \le u \le \gamma \mu_N + \lambda + \varepsilon$. Since ε is arbitrary, it implies again that $u - \gamma \mu_N = \lambda$. Therefore we are left with the case where this maximum is achieved, either at O or at infinity. If the maximum is achieved at infinity, we use the conformal invariance of the N-harmonic equation ([Re]) and perform the inversion $\theta: x \to x / |x|^2$. Consequently we are left with the case where $\tilde{v}$ which is v or $v \circ \theta$ achieves its maximum at O. If $v = \tilde{v}$,

$$\lim_{\xi \to 0} \left(v(\xi) - \gamma \mu_N(\xi) \right) = \nu, \tag{1.2.45}$$

for some constant ν; this implies

$$\lim_{\xi \to 0} \lim_{r_n \to 0} \left(u(r_n \xi) - \gamma \mu_N(r_n \xi) \right) = v, \tag{1.2.46}$$

and the maximum principle yields again (1.2.40). If $\tilde{v} = v \circ \theta$, then

$$\lim_{|\xi| \to \infty} \left(v(\xi) - \gamma \mu_N(\xi) \right) = v, \tag{1.2.47}$$

and

$$\lim_{|\xi| \to \infty} \lim_{r_n \to 0} \left(u(r_n \xi) - \gamma \mu_N(r_n \xi) \right) = v. \tag{1.2.48}$$

For $\varepsilon > 0$, there exists $K > 0$ such that for $|\xi| = K$,

$$v - \varepsilon \leq \lim_{r_n \to 0} \left(u(r_n \xi) - \gamma \mu_N(r_n K) \right) \leq v + \varepsilon.$$

Since $\{r_n\} \to 0$, we deduce again (1.2.40)-(1.2.41). The remaining of the proof is unchanged.

Remark 5.2. It is easy to adapt the above techniques to the exterior problem: if u is a p-harmonic function $(1 < p < N)$ in an exterior domain G of $\mathbb{R}^N$ such that u / μ_p remains bounded for $|x|$ large enough, then there exists a real number γ such that

$$\lim_{|x| \to \infty} u(x) / \mu_p(x) = \gamma. \tag{1.2.49}$$

Remark 5.3. In the case $p > N$, it has been proved by Serrin [Se4] that any bounded p-harmonic function can be extended by continuity. By adapting the proof of Theorem 5.2, it is possible to prove that if u is p-harmonic and satisfies $|u(x)| \leq C|x|^{(p-N)/(p-1)}$, there exists some γ such that

$$\lim_{x \to 0} |x|^{(N-p)/(p-1)} u(x) = \gamma. \tag{1.2.50}$$

The singular Dirichlet problem can be uniquely solved as a consequence of Theorem 5.2. Let Ω be a bounded domain in $\mathbb{R}^N$ containing O with a smooth boundary, $\Omega^* = \Omega \setminus \{O\}$, then the following result holds [KV].

COROLLARY 5.1. *Assume* $1 < p \leq N$, $g \in L^\infty(\Omega) \cap W^{1,p}(\Omega)$ *and* λ *is a real number. Then there exists a unique function* $u \in C^{1,\sigma}(\Omega^*)$ *with* $|\nabla u|^{p-1} \in L^1(\Omega)$, $\nabla u \in L^p(\Omega \setminus B(O,r))$ *for* $r > 0$ *small enough and* $u / \mu_p \in L^\infty(\Omega)$, *satisfying*

$$\begin{cases} -\mathrm{div}\left(|\nabla u|^{p-2}\nabla u\right) = \lambda\delta_0 & in \quad \mathcal{D}'(\Omega), \\ \qquad\qquad u = g & on \quad \partial\Omega. \end{cases} \tag{1.2.51}$$

Another consequence of Theorem 5.2 is the following type of singular Liouville result

COROLLARY 5.2. *Assume* $1 < p \leq N$ *and* u *is p-harmonic in* $\mathbb{R}^N\setminus\{O\}$ *such that*

$$|u(x)| \leq a\left|\mu_p(x)\right| + b \quad (\forall x \in \mathbb{R}^N\setminus\{O\}), \tag{1.2.52}$$

for some constants a *and* b. *Then there exist two constants* γ *and* λ *such that*

$$u(x) = \gamma\mu_p(x) + \lambda. \tag{1.2.53}$$

Proof. Step 1. The case $p = N$. Because the equation is invariant under the conformal transformation $x \mapsto \theta(x) = x/|x|^2$, there exist γ and γ' such that

$$\lim_{x\to O} u(x)/\mu_N(x) = \gamma \quad and \quad \lim_{|x|\to\infty} u(x)/\mu_N(x) = \gamma'. \tag{1.2.54}$$

Moreover

$$u - \gamma\mu_N \in L^\infty(B(O,1)) \quad and \quad u - \gamma\mu_N \in L^\infty(\mathbb{R}^N\setminus B(O,1)). \tag{1.2.55}$$

Assuming that γ or γ' is not zero, $\gamma > 0$ for example, we define

$$v_\varepsilon(x) = (1-\varepsilon)\gamma\mu_N(x) + (\gamma' - (1-\varepsilon)\gamma)\mu_N(R) - K, \tag{1.2.56}$$

where $K = \max\left(\left\|u - \gamma\mu_N\right\|_{L^\infty(B(O,1))}, \left\|u - \gamma'\mu_N\right\|_{L^\infty(\{x:\,|x|>1\})}\right)$, $R > 1$ and $\varepsilon > 0$. From the maximum principle, $u \geq v_\varepsilon$ in $B(O,R)\setminus\{O\}$. Letting ε go to zero yields

$$u(x) \geq \gamma\mu_N(x) + (\gamma' - \gamma)\mu_N(R) - K. \tag{1.2.57}$$

In the same way,

$$u(x) \leq \gamma\mu_N(x) + (\gamma' - \gamma)\mu_N(R) + K. \tag{1.2.58}$$

Letting R go to infinity gives $\gamma = \gamma'$. Since $u - \gamma\mu_N$ admits a finite limit, both at O and at infinity, we denote $\lambda = \lim_{|x|\to\infty}(u(x) - \gamma\mu_N(x))$. For any $\varepsilon > 0$ let $R(\varepsilon)$ be such that

$$\sup_{|x|\geq R(\varepsilon)}\left|u(x) - \gamma\mu_N(x) - \lambda\right| \leq \varepsilon. \tag{1.2.59}$$

If we compare u and $u_\varepsilon = \gamma\mu_N + \lambda + \varepsilon$ in $B(O,R(\varepsilon))\setminus\{O\}$ which are two p-harmonic functions with the same singularity at 0, we conclude that $u \le u_\varepsilon$. In the same way $u \ge \gamma\mu_N + \lambda - \varepsilon$. Letting ε tend to 0 yields $u = \gamma\mu_N + \lambda$.

Step 2. The case $1 < p < N$. From Theorem 5.2 and Lemma 5.2 there exists a real number γ such that

$$\lim_{|x|\to 0} u(x)/\mu_p(x) = \gamma \quad \text{and} \quad u - \gamma\mu_p \in L^\infty(\mathbb{R}^N) \tag{1.2.60}$$

and we have the following situation concerning the position of the point where the maximum of $u - \gamma\mu_p$ is achieved:

(i) either it is achieved at some interior point, therefore the theorem follows,
(ii) or the maximum is achieved at infinity,
(iii) or the maximum is achieved at O.

In case (i) we set $\lambda(r) = \sup_{|x|\le r}\big(u(x) - \gamma\mu_p(x)\big)$, for $r > 0$, and

$$v_r(\xi) = u(r\xi) - \gamma\mu_p(r). \tag{1.2.61}$$

The function λ is increasing, with limit $\lambda(\infty)$ at infinity. Moreover there exists $\xi_r \in S^{N-1}$ such that $v_r(\xi_r) = \lambda(r)$. Since v_r is p-harmonic in $\mathbb{R}^N\setminus\{O\}$ and satisfies

$$\begin{cases} \text{(i)} & |\nabla v_r(\xi)| \le C|\xi|^{-1}, \\ \text{(ii)} & |\nabla v_r(\xi) - \nabla v_r(\xi')| \le C|\xi|^{-1-\sigma}|\xi - \xi'|^{\sigma}, \end{cases} \tag{1.2.62}$$

for $r^{-1} \le |\xi| \le |\xi'|$, there exist a sequence $\{r_n\}$ going to infinity, a p-harmonic function v and a point $\xi_0 \in S^{N-1}$ such that $v_{r_n} \to v$ in the C^1_{loc}-topology and $\xi_{r_n} \to \xi_0$. Since

$$v_r(\xi) = u(r\xi) - \gamma\mu_p(r) \le \lambda(\infty) + \gamma(\mu_p(r\xi) - \mu_p(r)), \tag{1.2.63}$$

we deduce that $v(\xi) \le \lambda(\infty)$ and $v(\xi_0) = \lambda(\infty)$. Consequently $v = \lambda(\infty)$ and

$$\lim_{|x|\to\infty}\big(u(x) - \gamma\mu_p(x)\big) = \lim_{|x|\to\infty} u(x) = \lambda(\infty) = \lambda. \tag{1.2.64}$$

The proof of (1.2.64) in case (iii) is similar. In order to end the proof, we can suppose that $\gamma \ne 0$, positive for example, since we know from [Re] that any bounded p-harmonic function is constant. If we denote $u_\gamma(x) = \gamma\mu_p(x) + \lambda$, we have

$$(1-\varepsilon)u_\gamma(x) - |\lambda|\varepsilon \le u(x) \le (1+\varepsilon)u_\gamma(x) + |\lambda|\varepsilon \tag{1.2.65}$$

for $\varepsilon > 0$, from the maximum principle. Letting ε go to 0 implies that $u(x) = \gamma\mu_p(x) + \lambda$.

COROLLARY 5.3. *Any nonnegative N-harmonic function in* $\mathbb{R}^N\backslash\{O\}$ *is constant.*

There is a natural extension of the spherical harmonics to the p-harmonic framework. If we look for p-harmonic functions under the form $u(r,\sigma)=r^{-\beta}\omega(\sigma)$, then a straightforward computation yields

$$-\text{div}_{S^{N-1}}\left((\beta^2\omega^2+|\nabla\omega|^2)^{(p-2)/2}\nabla\omega\right)+\beta\theta(\beta^2\omega^2+|\nabla\omega|^2)^{(p-2)/2}\omega=0 \qquad (1.2.66)$$

on S^{N-1} with $\theta=N-1-(\beta+1)(p-1)$. Using a technique introduced by Tolksdorf [To2], it is possible to construct non-zero and changing sign solutions of (1.2.65) (see [Ve6]) with a high degree of symmetry corresponding to the tessalations of the sphere. However it does not give all the solutions. When $N=2$ the equation (1.2.66) reduces to

$$\left((\beta^2\omega^2+\omega_\sigma^2)^{(p-2)/2}\omega_\sigma\right)_\sigma-(1-(\beta+1)(p-1))\beta(\beta^2\omega^2+\omega_\sigma^2)^{(p-2)/2}\omega=0 \qquad (1.2.67)$$

on S^1. If we set $Y=\omega_\sigma/\omega$, then

$$\left(\frac{\beta^2}{Y^2+\beta^2}-\frac{\beta+1}{Y^2+\beta(\beta-\beta_0)}\right)Y_\sigma=1. \qquad (1.2.68)$$

This equation has been completely integrated by Kroll and Mazja in the case $\beta<0$ and independently by Kichenassamy and Véron [KV] in the case $\beta>0$. We summarize with

THEOREM 5.3. *Let* $N=2$ *and* $p>1$. *Then for each positive integer* k *there exists* $(\beta_{k,1},\omega_{k,1})$ *and* $(\beta_{k,2},\omega_{k,2})$ *where the* $\beta_{k,j}$ *(* $j=1,\ 2$ *) are real numbers with* $\beta_{k,1}<0$, $\beta_{k,2}>0$ *and* $\omega_{k,j}$ C^∞ *real valued functions with least period* $2\pi/k$, *such that the function*

$$x\mapsto u_{k,j}(x)=|x|^{-\beta_{k,j}}\omega_{k,j}(x/|x|) \qquad (1.2.69)$$

is p-harmonic in $\mathbb{R}^2$ *or* $\mathbb{R}^2\backslash\{O\}$; $\beta_{k,1}$ *is the root smaller or equal to* -1 *of*

$$(1-1/k)^2(\beta^2-\beta\beta_0)=(\beta+1)^2, \qquad (1.2.70)$$

$\beta_{k,2}$ *is the positive root of*

$$(1+1/k)^2(\beta^2-\beta\beta_0)=(\beta+1)^2. \qquad (1.2.71)$$

Moreover $(\beta_{k,j},\omega_{k,j})$ *is unique up to homothety and translation over* $\omega_{k,j}$.

5-2 Equations with absorption term

5-2-1 Removability results

Let Ω be a bounded domain of $\mathbb{R}^N$ containing O and $q > p\text{-}1 > 0$. If we look for solutions of the following equation in Ω

$$-\mathrm{div}\left(|\nabla u|^{p-2}\nabla u\right)+|u|^{q-1}u=0, \tag{2.1.1}$$

under the form $u(r)=\beta r^{\alpha}$, then

$$\alpha=-\frac{p}{q+1-p}\qquad\text{and}\qquad \beta^{q+1-p}=\left(\frac{p}{q+1-p}\right)^{p-1}\left(\frac{pq}{q+1-p}-N\right), \tag{2.1.2}$$

but the last term exists if and only if

$$\begin{cases}(\text{i})\quad p-1<q<\dfrac{N(p-1)}{N-p}&\text{if }\ 1<p<N,\\[2ex](\text{ii})\quad p-1<q&\text{if }\ p\ge N.\end{cases} \tag{2.1.3}$$

When $1 < p < N$ and $q \ge N(p-1)/N-p$ such a solution no longer exists. This corresponds to a very general removability result due to Vazquez and Véron [VV1], dealing with solutions of

$$-\mathrm{div}\left(|\nabla u|^{p-2}\nabla u\right)+g(u)=0, \tag{2.1.4}$$

in $\Omega^*=\Omega\setminus\{O\}$ where g is a real valued function satisfying some growth condition.

THEOREM 5.4. *Let* $1<p<N$ *and* g *be continuous and such that*

$$\liminf_{r\to\infty}\ g(r)r^{-N(p-1)/(N-p)}>0\ ,\qquad\limsup_{r\to-\infty}\ g(r)|r|^{-N(p-1)/(N-p)}<0. \tag{2.1.5}$$

If $u \in L^{\infty}_{\mathrm{loc}}(\Omega^*)\cap W^{1,p}_{\mathrm{loc}}(\Omega^*)$ *is a weak solution of (2.1.4) in* Ω^*, *then* u *can be extended as a* $C^{1,\alpha}(\Omega)$-*solution of the same equation in whole* Ω.

As in the semilinear case, the problem is to obtain a local uniform estimate on u in Ω.

PROPOSITION 5.2. *Suppose that* $1<p<N$ *and* $q\ge N(p-1)/N-p$. *If* $u \in L^{\infty}_{\mathrm{loc}}(\Omega^*)\cap W^{1,p}_{\mathrm{loc}}(\Omega^*)$ *with* $\mathrm{div}\left(|\nabla u|^{p-2}\nabla u\right)\in L^1_{\mathrm{loc}}(\Omega^*)$, *satisfies*

$$-\mathrm{div}\left(|\nabla u|^{p-2}\nabla u\right)+au^q\le C\quad a.e.\ on\quad \{x\in\Omega:\ u(x)>0\}, \tag{2.1.6}$$

for some positive constants a *and* C, *then* $u^+ \in L^\infty_{loc}(\Omega)$.

Proof. Without any loss of generality we may suppose that $\overline{B}(O,1) \subset \Omega$.
Step 1. We claim that there exists a positive constant $A = A(N, p, q)$ such that

$$u(x) \le A|x|^{-p/(q+1-p)}, \quad \forall \ 0 < |x| \le 1/2. \tag{2.1.7}$$

We pick $x_0 \in B(O, 1/2)$ and define, for $p' = p/(p-1)$,

$$v(x) = \lambda \left(2^{-p'}|x_0|^{p'} - |x - x_0|^{p'} \right)^{-q/(q+1-p)} + \mu \tag{2.1.8}$$

in $B(x_0, |x_0|/2)$. We set $R = |x_0|/2$, $r = |x - x_0|$ and $v(x) = v(r)$. Because the following identity holds

$$-\mathrm{div}\left(|\nabla v|^{p-2}\nabla v\right) = -|v'|^{p-2}\left((p-1)v'' + \frac{N-1}{r}v' \right), \tag{2.1.9}$$

a straightforward calculation implies that there exists some positive constant $c_1 = c_1(N, p, q)$ such that

$$-\mathrm{div}\left(|\nabla v|^{p-2}\nabla v\right) \ge -c_1 \lambda^{p-1} R^{p'} \left(R^{p'} - r^{p'} \right)^{-pq/(q+1-p)}. \tag{2.1.10}$$

If we set $v = \min(1, 2^{1-1/q})$, choose

$$\lambda = \frac{1}{v}\left(\frac{c_1}{a}\right)^{1/(q+1-p)} R^{p'/(q+1-p)} \quad \text{and} \quad \mu \ge \left(\frac{C}{a}\right)^{1/q}, \tag{2.1.11}$$

we deduce that the inequality

$$-\mathrm{div}\left(|\nabla v|^{p-2}\nabla v\right) + av^q \ge C, \tag{2.1.12}$$

holds in $B(x_0, |x_0|/2)$. From the maximum principle $u \le v$, and in particular, if $x = x_0$,

$$u(x_0) \le v(x_0) = \frac{1}{v}\left(\frac{c_1}{a}\right)^{1/(q+1-p)} R^{-p/(q+1-p)} + \mu. \tag{2.1.13}$$

This implies (2.1.7).

Step 2. We claim that for every $\eta \in \mathcal{D}^+(\Omega^*)$ and $\mu \ge \left(\frac{C}{a}\right)^{1/q}$, we have

$$\left\| \eta \nabla(u - \mu)^+ \right\|_{L^p(\Omega)} \le p \left\| (u - \mu)^+ \nabla \eta \right\|_{L^p(\Omega)}. \tag{2.1.14}$$

Let ρ be a nonnegative bounded C^1 function vanishing on $(-\infty,0)$, increasing on $(0,\infty)$. If $\eta \in \mathcal{D}^+(\Omega^*)$ and $\mu \geq \left(\dfrac{C}{a}\right)^{1/q}$, we have

$$\int_\Omega \left(|\nabla u|^{p-2}\nabla u.\nabla(\eta^p\rho(u-\mu)) + (au^q - u)\eta^p\rho(u-\mu)\right)dx \leq 0. \qquad (2.1.15)$$

Therefore

$$\int_\Omega \eta^p |\nabla(u-\mu)|^p \rho'(u-\mu)dx$$
$$+p\int_\Omega \eta^{p-1}\rho(u-\mu)|\nabla(u-\mu)|^{p-2}\nabla(u-\mu).\nabla\eta dx \leq 0. \qquad (2.1.16)$$

Letting $\rho(r)$ converge to $\mathrm{sign}^+(r)$ and using Hölder's inequality yields (2.1.14).

End of the proof. We consider $\zeta_n \in \mathcal{D}^+(\Omega^*)$ such that

$$\zeta_n(x) = \begin{cases} 0 & \text{if } |x| < 1/(2n) \text{ or } |x| > 2/3, \\ 1 & \text{if } 1/n < |x| < 1/2. \end{cases} \qquad (2.1.17)$$

with $0 \leq \zeta_n \leq 1$ and $|\nabla\zeta_n| \leq Cn$. Let ρ be as in Step 2 and $\mu \geq \max\left\{\left(\dfrac{C}{a}\right)^{1/q}, \max_{1/2\leq|x|\leq2/3} u(x)\right\}$. If we set $\Delta_n = \{x: 1/(2n) \leq |x| \leq 1/n\}$ and $X_n = -\int_\Omega \rho(u-\mu)|\nabla u|^{p-2}\nabla u.\nabla\zeta_n dx$, then

$$0 \leq \int_\Omega (au^q - C)\rho(u-\mu)dx \leq X_n. \qquad (2.1.18)$$

Moreover, by Hölder's inequality, $|X_n| \leq c\|\rho\|_{L^\infty}\, n^{1-N/p}\left\|\nabla(u-\mu)^+\right\|_{L^p(\Delta_n)}^{p-1}$. Using Step 2 yields

$$\left\|\nabla(u-\mu)^+\right\|_{L^p(\Delta_n)}^{p-1} \leq p\left\|(u-\mu)^+\nabla\zeta_{2n}\right\|_{L^p}^{p-1} \text{ and finally}$$

$$|X_n| \leq Cn^{(p^2-N)/p}\left\|(u-\mu)^+\right\|_{L^p(\Delta_{2n})}^{p-1}. \qquad (2.1.19)$$

From step 1 we get

$$\left\|(u-\mu)^+\right\|_{L^p(\Delta_{2n})}^{p-1} \leq Cn^{\frac{p^2-p}{q+1-p}-N(1-\frac{1}{p})}, \qquad (2.1.20)$$

and because $\dfrac{p^2-p}{q+1-p}-N(1-\dfrac{1}{p})+\dfrac{p^2-N}{p}=\dfrac{pq}{q+1-p}-N=0$, X_n remains bounded.
Letting n go to infinity and $\rho(r)$ to $sign^+(r)$ implies that

$$\int_\Omega (au^q-\mu)sign^+(u-\mu)dx < \infty. \tag{2.1.21}$$

Case 1. Suppose that $q \ge p$ (which means $p^2 \ge N$), then

$$\left\| (u-\mu)^+ \right\|^{p-1}_{L^p(\Delta_{2n})} \le \left\| (u-\mu)^+ \right\|^{p-1}_{L^q(\Delta_{2n})} \left| \Delta_{2n} \right|^{(p-1)(p^{-1}-q^{-1})} \tag{2.1.22}$$

and $\lim\limits_{n\to\infty} X_n = 0$ from (2.1.19).

Case 2. Suppose that $q < p$ (which means $p^2 < N$), then

$$\left\| (u-\mu)^+ \right\|_{L^p(\Delta_{2n})} \le \left\| (u-\mu)^+ \right\|^{q/p}_{L^q(\Delta_{2n})} \left\| (u-\mu)^+ \right\|^{1-q/p}_{L^\infty(\Delta_{2n})} \tag{2.1.23}$$

and

$$\left| X_n \right| \le Cn^{\frac{p^2-N}{p}+(1-\frac{q}{p})\frac{p^2-p}{q+1-p}}, \tag{2.1.24}$$

and finally $\lim\limits_{n\to\infty} X_n = 0$.

In any case $\displaystyle\int_{B(O,1/2)} (au^q-C)sign^+(u-\mu)dx = 0$ and $u \le \mu$ in $B(O,1/2)$.

Remark 5.4. A general *a priori* estimate of Keller-Osserman type for any solution u of (2.1.4) in a N-dimensionnal domain G is due to Vazquez [Va1]. If we suppose that g is nondecreasing (for simplicity) and satisfies

$$\int^{\infty} \frac{ds}{\sqrt[p]{sg(s)}} ds < \infty, \tag{2.1.25}$$

then for any compact $K \subset G$ there exists $C(K) > 0$ such that for any u satisfying (2.1.4), we have $u(x) \le C(K)$ in K.

Proof of Theorem 5.5. From Proposition 5.2, u is locally bounded in Ω. From Chapter 1 and Serrin's results, u can be extended as a solution of (2.1.4) and the regularity follows .

Remark 5.5. The previous removability results extends to a singular set lying on a d-dimensionnal complete and compact C^2 submanifold Σ of $\mathbb{R}^N$. If we assume $1 < p < N-d$ and

$$\liminf_{r\to\infty} g(r)r^{-(N-d)(p-1)/(N-p-d)} > 0 \,, \quad \limsup_{r\to-\infty} g(r)|r|^{-(N-d)(p-1)/(N-p-d)} < 0, \quad (2.1.26)$$

any solution u of (2.1.4) in a some open subset $\Omega\setminus\Sigma$ can be extended as a regular solution of the same equation in Ω. The technique is a simple adaptation of [Ve10]. However it appears that there exists no general result of Baras and Pierre type [BP1] giving a necessary and sufficient condition for the removability of any closed subset in terms of some capacity estimates.

5-2-2 The singular Dirichlet problem

We recall that the μ_p defined in (1.2.3) are the fundamental solutions of the p-Laplace equations in $\mathbb{R}^N$. Given a bounded domain Ω containing O, we consider the singular problem (2.1.1) in the range of values p and q where Theorem 5.5 does not apply and in which there exist solutions in the sense of distributions.

THEOREM 5.5. *Let* p *and* q *be subject to the following condition*

$$\begin{cases} \text{(i)} \quad p-1 \le q < \dfrac{N(p-1)}{N-p} & \text{if } 1 < p < N, \\[2mm] \text{(ii)} \quad p-1 \le q & \text{if } p = N. \end{cases} \qquad (2.2.1)$$

If u *is a solution of (2.1.1) in* $\Omega^* = \Omega\setminus\{O\}$ *such that* u / μ_p *remains bounded near O, there exists a real number* γ *such that*

$$\lim_{x\to O} \frac{u(x)}{\mu_p(x)} = \gamma. \qquad (2.2.2)$$

Furthermore

$$\left| u(x) - \gamma\mu_p(x) \right| \le C(1 + |x|^N |\mu_p(x)|^{q+1}) \qquad (\forall x \in \Omega), \qquad (2.2.3)$$

$$\lim_{x\to O}\left(\frac{x}{|x|} |x|^{(N-1)/(p-1)} \nabla u(x) \right) = -\gamma \left| S^{N-1} \right|^{-1/(p-1)}, \qquad (2.2.4)$$

and u *satisfies*

$$-\mathrm{div}\left(|\nabla u|^{p-2} \nabla u \right) + |u|^{q-1} u = |\gamma|^{p-2} \gamma \qquad (2.2.5)$$

in $\mathcal{D}'(\Omega)$.

The following regularity results extend Lemma 5.1

LEMMA 5.3. *Suppose that $1 < p \le q+1$ and let u be a solution of (2.1.1) in $B^*(O,1) = B(O,1)\setminus\{O\}$ such that*

$$|u(x)| \le C_1|x|^{-\delta}, \tag{2.2.6}$$

for some $C_1 > 0$ and δ with $0 \le \delta \le p/(q+1-p)$ ($0 \le \delta < \infty$ if $q = p\text{-}1$). Then there exist a positive constant C and $\alpha \in (0,1)$ such that

$$|\nabla u(x)| \le C|x|^{-1-\delta}, \tag{2.2.7}$$

$$|\nabla u(x) - \nabla u(x')| \le C|x|^{-1-\delta-\alpha}|x-x'|^{\alpha}, \tag{2.2.8}$$

for any (x,x') satisfying $0 < |x| \le |x'| \le 1/2$.

Proof. We first define $\Gamma = \{y \in \mathbb{R}^N : 1 < |y| < 7\}$ and $\Gamma^* = \{y \in \mathbb{R}^N : 2 < |y| < 6\}$. For $0 < |x| < 1/2$ we have $2\beta \le |x| \le 3\beta$ for some $\beta \in (0, 1/16)$. Setting $u_\beta(\xi) = \beta^\delta u(\beta\xi)$ for $\xi \in \Gamma$, then

$$-\mathrm{div}\left(\left|\nabla u_\beta\right|^{p-2}\nabla u_\beta\right) + \beta^{\delta(p-1-q)+p}\left|u_\beta\right|^{q-1}u_\beta = 0 \tag{2.2.9}$$

in Γ. Because $\delta(p-1-q)+p \ge 0$ and $\left|u_\beta(\xi)\right| \le C_1|\xi|^{-\delta} \le C_1$, we deduce from Theorem 5.1 , as in Lemma 5.1, that (2.2.7) and (2.2.8) hold.

Remark 5.6. If we replace (2.2.6) by

$$|u(x)| \le C_1\left(\left|\ln(|x|)\right|+1\right), \tag{2.2.10}$$

then (2.2.7) and (2.2.8) are replaced by

$$\begin{cases} |\nabla u(x)| \le C\left(\left|\ln(|x|)\right|+1\right)|x|^{-1}, \\ |\nabla u(x) - \nabla u(x')| \le C\left(\left|\ln(|x|)\right|+1\right)|x|^{-1-\alpha}|x-x'|^{\alpha}, \end{cases} \tag{2.2.11}$$

but these estimates are not optimal as we shall see it later on.

LEMMA 5.4. *Let G be a domain in $\mathbb{R}^N$, $p > 1$, $c \in L^\infty_{loc}(G)$, and let u and v be two functions belonging to $C^1(G)$ and satisfying*

$$-\mathrm{div}\left(|\nabla u|^{p-2}\nabla u\right) + cu \le 0, \tag{2.2.12}$$

$$-\mathrm{div}\left(|\nabla v|^{p-2}\nabla v\right) + cv \ge 0, \tag{2.2.13}$$

in weak sense in G and $\nabla v(x) \neq 0$ $(\forall x \in G)$. If $u \leq v$ in G and if there exists a point
$x_0 \in G$ *such that* $u(x_0) = v(x_0)$, *then* $u \equiv v$.

Proof. By the mean value theorem

$$|\nabla u|^{p-2} \frac{\partial u}{\partial x_i} - |\nabla v|^{p-2} \frac{\partial v}{\partial x_i} = \sum_j \alpha_{ij}(x_0) \frac{\partial(u-v)}{\partial x_i} \quad \text{in} B(x_0, \delta), \text{ where}$$

$$\alpha_{ij}(x_0) = \left|t_i \nabla u + (1-t_i)\nabla v\right|^{p-4}\left(\delta_{ij}\left|t_i \nabla u + (1-t_i)\nabla v\right|^2 - \right.$$

$$\left. + (p-2)\left(t_i \frac{\partial u}{\partial x_i} + (1-t_i)\frac{\partial v}{\partial x_i}\right)\left(t_i \frac{\partial u}{\partial x_j} + (1-t_i)\frac{\partial v}{\partial x_j}\right)\right). \tag{2.2.14}$$

But the graphs of u and v are tangent at x_0 and $\nabla v(x_0) \neq 0$, then

$$\alpha_{ij}(x_0) = |\nabla v(x_0)|^{p-4}\left(\delta_{ij}|\nabla v(x_0)|^2 + (p-2)\frac{\partial v(x_0)}{\partial x_i}\frac{\partial v(x_0)}{\partial x_j}\right). \tag{2.2.15}$$

The matrix $\left(\alpha_{ij}(x_0)\right)$ is positive definite because it is the Hessian of the strictly convex

function $(X_1, \ldots, X_n) \mapsto \frac{1}{p}\left(\sum_i X_i^2\right)^{p/2}$. Consequently $\left(\alpha_{ij}(x)\right)$ is also positive definite in

some $B(x_0, \delta')$. If we set $w = v - u$, then

$$-\sum_{ij} \frac{\partial}{\partial x_i}\left(\alpha_{ij}\frac{\partial w}{\partial x_i}\right) + c^+ w \geq 0 \tag{2.2.16}$$

holds in $B(x_0, \delta')$ and therefore w is identically 0. We conclude by a classical connexity
argument since the coincidence set of u and v is closed.

Remark 5.7. Lemma 5.2 is also valid as a consequence of Lemma 5.4: if u is a solution
of (2.1.1) in G and $u - \mu$ (resp. u/μ) achieves a nonnegative maximum (resp. a
maximum) in G, then it is constant in G.

LEMMA 5.5. *Suppose that (2.1.1) holds and let R be any positive number. Then for any*
real numbers α *and* γ *there exists a unique function* $\psi = \psi^{\alpha, \gamma}$ *in* $C^1((0,R])$ *satisfying*

$$\begin{cases} -\left(r^{N-1}|\psi_r|^{p-2}\psi_r\right)_r + r^{N-1}|\psi|^{q-1}\psi = 0 \quad \text{in} (0,R), \\[2ex] \lim_{r\to 0}\frac{\psi(r)}{\mu_p(r)} = \gamma, \quad \psi(R) = \alpha. \end{cases} \tag{2.2.17}$$

Furthermore

$$\lim_{r \to 0} \frac{\psi_r(r)}{\mu_{p_r}(r)} = \gamma,$$

(2.2.18)

the function $(\alpha, \gamma) \mapsto \psi^{\alpha, \gamma}$ *is nonincreasing separately in* α *and* γ *and*

$$\left|\psi^{\alpha, \gamma} - \psi^{\alpha', \gamma}\right| \leq \left|\alpha - \alpha'\right|$$

(2.2.19)

holds for any real numbers α *and* α'.

Proof. Step 1. Existence. If $\gamma = 0$, the problem is regular and the solution is unique and obtained by minimization. Therefore we can assume that γ is nonzero, positive for example. For $\varepsilon > 0$, set ψ_ε the solution in $\Gamma_\varepsilon = \{x:\ 0 < |x| < \varepsilon\}$ of

$$\begin{cases} -\mathrm{div}\left(\left|\nabla\psi_\varepsilon\right|^{p-2}\nabla\psi_\varepsilon\right) + \left|\psi_\varepsilon\right|^{q-1}\psi_\varepsilon = 0 & \text{in} \quad \Gamma_\varepsilon, \\ \qquad\qquad\qquad \psi_\varepsilon = \alpha \quad \text{for} \quad |x| = R, \\ \qquad -\frac{\partial\psi_\varepsilon}{\partial\nu}(x) = \gamma\mu_{\mathrm{pr}}(\varepsilon) \quad \text{for} \quad |x| = \varepsilon. \end{cases}$$

(2.2.20)

This function ψ_ε is unique and obtained also by minimization. By the maximum principle

$$-|\alpha| \leq \psi_\varepsilon \leq \gamma\mu_p + k, \quad (k = \alpha^+ + \gamma|\mu_p(R)|).$$

(2.2.21)

Consequently $\{\psi_\varepsilon\}_\varepsilon$ remains locally bounded in $C^1\left(B^*(O,R)\right) \cap C\left(\overline{B}^*(O,R)\right)$ $\cap W^{1,p}_{\mathrm{loc}}\left(\overline{B}^*(O,R)\right)$ by Theorem 5.1, and there exist a sequence $\{\varepsilon_n\}$ and a function ψ which solves (2.1.1) in $B^*(O,R)$, takes the value α for $|x| = R$, such that $\psi_{\varepsilon_n} \to \psi$ in the topology of local uniform convergence in $C^1\left(B^*(O,R)\right) \cap C\left(\overline{B}^*(O,R)\right) \cap W^{1,p}_{\mathrm{loc}}\left(\overline{B}^*(O,R)\right)$.

The last point is to prove that ψ satisfies $\lim_{r \to 0} \psi(r)/\mu_p(r) = \gamma$. We shall only treat the case $1 < p < N$, the case $p = N$ being similar. We introduce the new variable and unknown

$$s = r^{(p-N)/(p-1)}, \quad \psi_\varepsilon(r) = v_\varepsilon(s);$$

(2.2.22)

then v satisfies

$$\begin{cases} \left(\left|v_{\varepsilon s}\right|^{p-2}v_{\varepsilon s}\right)_s - \left(\dfrac{p-1}{N-p}\right)^p s^{-p((N-1)/(N-p))}v_\varepsilon^q = 0 \quad \text{on} \quad (S,\varepsilon^{(p-N)/(p-1)}), \\[2mm] v_\varepsilon(S) = \alpha, \qquad v_{\varepsilon s}(\varepsilon^{(p-N)/(p-1)}) = \gamma, \end{cases} \tag{2.2.23}$$

where $S = R^{(p-N)/(p-1)}$. Moreover $\left|v_\varepsilon(s)\right|^q \leq Cs^q$. Therefore

$$\gamma^{p-1} - \left|v_{\varepsilon s}\right|^{p-2}v_{\varepsilon s}(s) \leq C \int_S^{\varepsilon^{(p-N)/p-1)}} s^{q-p(N-1)/(N-p)}ds$$

$$= \frac{C(N-p)}{N(p-1)}\left(s^{q-\frac{N(p-1)}{N-p}} - \varepsilon^{N-q\frac{N-p}{p-1}}\right) \tag{2.2.24}$$

and

$$\left|v_{\varepsilon s}\right|^{p-2}v_{\varepsilon s}(s) \geq \gamma^{p-1} - Ms^{q-\frac{N(p-1)}{N-p}}. \tag{2.2.25}$$

Consequently

$$v_\varepsilon(s) \geq s\gamma\left(1 - M's^{\frac{q}{p-1}-\frac{N}{n-p}}\right), \tag{2.2.26}$$

with M' independent of ε. This yields $\liminf\limits_{r\to 0}\psi_\varepsilon(r)/\mu_p(r) \geq \gamma$, and finally $\lim\limits_{r\to 0}\psi(r)/\mu_p(r) = \gamma$, by using transformation (2.2.22). For the derivative, we have also

$$\left|v_{\varepsilon s}\right|^{p-2}v_{\varepsilon s}(s) \leq \gamma^{p-1} + Ms^{q-\frac{N(p-1)}{N-p}}. \tag{2.2.27}$$

Combining (2.2.25) and (2.2.27) and returning to ψ implies (2.2.18). Actually, we have a stronger estimate, namely

$$\left|\psi_r(r) - \gamma\mu_{pr}(r)\right| \leq Cr^{\frac{1-N}{p-1}+N+q\frac{p-N}{p-1}}. \tag{2.2.28}$$

Step 2. Uniqueness. If $\gamma = 0$ uniqueness is classical. Therefore let us suppose that $\gamma > 0$ for example and that ψ and $\tilde{\psi}$ are two solutions of (2.2.17).

Case 1. Suppose that $\alpha \geq 0$. For $\delta > 0$, $(1+\delta)\psi$ is a supersolution and $(\tilde{\psi} - (1+\delta)\psi)^+$ has compact support in $\overline{B}^*(O,R)$. Therefore it vanishes identically. Letting δ go to 0 yields $\tilde{\psi} \leq \psi$. In the same way $\psi \leq \tilde{\psi}$.

Case 2. Suppose $\alpha < 0$. From the maximum principle and the behaviour of ψ and $\tilde{\psi}$ near O, we have $\alpha = \min\limits_{0<|x|\leq R}\psi(x) = \min\limits_{0<|x|\leq R}\tilde{\psi}(x)$. For $\delta > 0$, $(1+\delta)\psi - \delta\alpha$ is a super solution which dominates $\tilde{\psi}$ both in a neighborhood of O and on $\partial B(O,R)$ therefore, it dominates $\tilde{\psi}$ in $\overline{B}^*(O,R)$. Letting δ go to 0 and reversing the role of ψ and $\tilde{\psi}$ yields $\psi = \tilde{\psi}$.

Remark 5.8. It is easy to check that the solution ψ constructed in Lemma 5.5 is a solution of the singular Dirichlet problem

$$\begin{cases} -\mathrm{div}\left(|\nabla\psi|^{p-2}\nabla\psi\right) + |\psi|^{q-1}\psi = |\gamma|^{p-2}\gamma\delta_O & \text{in } \mathcal{D}'(B(O,R)), \\ \psi(R) = \alpha & \text{on } \partial(B(O,R)). \end{cases} \tag{2.2.29}$$

Proof of Theorem 5.5. Case 1. $1 < p < N$. We proceed as in Theorem 5.2 and define

$$\gamma^+ = \limsup_{x\to O} \frac{u(x)}{\mu_p(x)}, \qquad\qquad \gamma^- = \liminf_{x\to O} \frac{u(x)}{\mu_p(x)}. \tag{2.2.30}$$

If $\gamma^+ = \gamma^- = 0$, we denote U the solution of (2.2.1) in $B(O,1)$ with $u = U$ on $\partial B(O,1)$. For any $\varepsilon > 0$, we have $|u - U| \leq \varepsilon\mu_p$ from the maximum principle. Therefore u is equal to U.

From now we may suppose that $(\gamma^+)^2 + (\gamma^-)^2 > 0$ and for definiteness $\gamma = \gamma^+ > 0$. Set

$$\tilde{\gamma}(r) = \sup_{|x|=r} \frac{u(x)}{\mu_p(x)}; \tag{2.2.31}$$

from Remark 5.7, $\lim_{r\to 0} \tilde{\gamma}(r) = \gamma$. We choose $\xi_r \in S^{N-1}$ such that

$$\frac{u(r\xi_r)}{\mu_p(r)} = \tilde{\gamma}(r), \tag{2.2.32}$$

and consider the functions $\xi \mapsto u_r(\xi) = \dfrac{u(r\xi)}{\mu_p(r)}$, defined on $\left\{\xi:\ 0 < |\xi| < r_0/r\right\}$, for some $r_0 \in (0,1)$. Then the following estimate holds

$$|u_r(\xi)| \leq |u(r\xi)/\mu_p(r)| \leq C|\mu_p(\xi)|$$

and

$$-\mathrm{div}\left(|\nabla u_r|^{p-2}\nabla u_r\right) + r^{N-q(N-p)/(p-1)}|u_r|^{q-1}u_r = 0 \text{ in } (B(O,r_0/r)). \tag{2.2.33}$$

From Lemma 5.3 we have

$$\begin{cases} \left|\nabla u_r(\xi)\right| \le C|\xi|^{(1-N)/(p-1)}, \\ \left|\nabla u_r(\xi) - \nabla u_r(\xi')\right| \le C|\xi - \xi'||\xi|^{(1-N)/(p-1)-\alpha}, \end{cases} \tag{2.2.34}$$

for $0 < |\xi| \le |\xi'| < r / r_0$. Furthermore

$$\frac{u_r(\xi)}{\mu(\xi)} \le \frac{\tilde{\gamma}(r|\xi|)}{\mu(1)}, \qquad \frac{u_r(\xi_r)}{\mu(\xi_r)} \le \frac{\tilde{\gamma}(r)}{\mu(1)}. \tag{2.2.35}$$

Consequently there exist a sequence $\{r_n\}$ and a function v, p-harmonic in $\mathbb{R}^N \backslash \{O\}$, such that $u_{r_n} \to v$ in the $C^1_{loc}(\mathbb{R}^N \backslash \{O\})$-topology. Assuming that $\lim_{r_n \to 0} \xi_{r_n} = \xi_0$, we deduce from (2.2.35) that

$$\frac{v(\xi)}{\mu(\xi)} \le \frac{\gamma}{\mu(1)} \qquad \text{and} \qquad \frac{v(\xi_0)}{\mu(\xi_0)} = \frac{\gamma}{\mu(1)}. \tag{2.2.36}$$

From lemma 5.4, it follows that $v(\xi) = \gamma \mu_p(\xi) / \mu_p(1)$. Because the sequence $\{r_n\}$ is arbitrary, we have $\lim_{r \to 0} u_r(\xi) = \gamma \mu_p(\xi)/\mu_p(1)$. Taking in particular $|\xi| = 1$, then

$$\lim_{x \to O} \frac{u(x)}{\mu_p(x)} = \gamma. \tag{2.2.37}$$

Moreover, the local uniform convergence of ∇u_r, implies

$$\lim_{x \to O} \left(|x| \frac{\nabla u(x)}{\mu_p(x)} \right) = \gamma \frac{\nabla \mu_p(\xi)}{\mu_p(1)}, \tag{2.2.38}$$

by taking $|\xi| = 1$. Estimate (2.2.3) is derived from (2.2.37) as in Lemma 5.5 and the fact that (2.2.5) holds is a straightforward consequence of (2.2.37)-(2.2.38).

Case 2. $p = N$. We introduce u_r as before and get

$$|u_r(\xi)| \le C(1 + \ln|\xi|/\ln(1/r)) \tag{2.2.39}$$

$$|\nabla u_r(\xi)| \le C(1 + \ln|\xi|/\ln(1/r))|\xi|^{-1} \tag{2.2.40}$$

$$|\nabla u_r(\xi) - \nabla u_r(\xi')| \le C(1 + \ln|\xi|/\ln(1/r))|\xi|^{-1-\alpha}|\xi - \xi'|^{\alpha} \tag{2.2.41}$$

for $0 < |\xi| \le |\xi'| < r_0 / r$. Therefore there exist a sequence $\{r_n\}$ and a function v, N-harmonic in $\mathbb{R}^N \backslash \{O\}$, such that $u_{r_n} \to v$ in the $C^1_{loc}(\mathbb{R}^N \backslash \{O\})$-topology. Moreover v is uniformly bounded. From Chapter 1 v can be extended to $\mathbb{R}^N$ as a N-harmonic function. From Corollary 5.2, v is constant and because

$$\lim_{r \to 0} \tilde{\gamma}(r_n) = \gamma = v(\xi_0) \tag{2.2.42}$$

for some $\xi_0 = \lim_{r_n \to 0} \xi_{r_n}$ on the unit sphere, we get $v = \gamma$. This implies

$$\lim_{r \to 0} u(r\xi)/\mu_N(\xi) = \gamma. \tag{2.2.43}$$

As in the first case we easily obtain

$$u - \gamma\mu_N \in L^\infty_{loc}(\Omega). \tag{2.2.44}$$

In order to prove the gradient estimate, we set

$$v_r(\xi) = u(r\xi) - \gamma\mu_N(r). \tag{2.2.45}$$

Since

$$-\mathrm{div}\left(\left|\nabla v_r\right|^{N-1}\nabla v_r\right) + r^N\left|v_r + \gamma\mu_N\right|^{q-1}(v_r + \gamma\mu_N) = 0 \tag{2.2.46}$$

in $B^*(O, r/r_0)$ and

$$\left|v_r(\xi)\right| \le C + \gamma\left|\mu_N(\xi)\right|, \tag{2.2.47}$$

from (2.2.44), we deduce that

$$\left\|\nabla v_r\right\|_{C^\alpha(\xi:\ 1/2 \le |\xi| \le 3/2)} \le C. \tag{2.2.48}$$

Returning to u it means that

$$\left|\nabla u(x)\right| \le C|x|^{-1}, \tag{2.2.49}$$

$$\left|\nabla u(x) - \nabla u(x')\right| \le C|x|^{-1-\alpha}|x - x'|^\alpha, \tag{2.2.50}$$

for $0 < |x| \le |x'| \le 1$. Applying also Lemma 5.3 with $\delta = 0$ gives

$$\left|\nabla v_r(\xi)\right| \le C|\xi|^{-1}, \tag{2.2.51}$$

$$\left|\nabla u(\xi) - \nabla u(\xi')\right| \le C|\xi|^{-1-\alpha}|\xi - \xi'|^\alpha, \tag{2.2.52}$$

for $0 < |\xi| \le |\xi'| \le 1/r$. We define now

$$\lambda(r) = \max_{r \le |x| \le r_1} \big(u(x) - \gamma\mu_N(x)\big), \tag{2.2.53}$$

where r_1 is chosen such that

$$\gamma\mu_N(r_1) > \sup_{0 < r < 1} \lambda(r). \tag{2.2.54}$$

Clearly the function $x \mapsto w(x) = \gamma\mu_N(x) + \lambda(r)$ is nonnegative in $B^*(0, r_1)$ where it satisfies

$$-\mathrm{div}\big(|\nabla w|^{p-1}\nabla w\big) + |w|^{q-1}w \ge 0. \tag{2.2.55}$$

From Lemma 5.4, the maximum of w is achieved either on $|x| = r_1$ or on $|x| = r$.

Case 1. Suppose that $\lambda(r) = \max\limits_{|x| = r}\big(u(x) - \gamma\mu_N(r)\big)$ for some $r \in (0, r_1)$. Therefore this relation holds for any $r \in (0, r_1)$. Since λ decreases with r, $\tilde{\lambda} = \lim\limits_{r \to 0} \lambda(r)$. Therefore there are a sequence $\{r_n\}$ and a function z, N-harmonic in $\mathbb{R}^N \setminus \{O\}$, such that $v_{r_n} \to z$ in the $C^1_{\mathrm{loc}}(\mathbb{R}^N \setminus \{O\})$-topology. Because the function z satisfies the same estimate (2.2.47) as v_r, we know from Corollary 5.3 that $z(\xi) = \alpha\mu_N(\xi) + \beta$. This, in turn, implies that

$$\lim_{r_n \to 0}\big(u(r_n\xi) - \gamma\mu_N(r_n)\big) = \alpha\mu_N(\xi) + \beta \tag{2.2.56}$$

holds uniformly on the compact subsets of $\mathbb{R}^N \setminus \{O\}$. Taking $\xi = \xi_{r_n} \in S^{N-1}$ in (2.2.57) defined from the fact that

$$\lambda(r_n) = u(r_n\xi_{r_n}) - \gamma\mu_N(r_n), \tag{2.2.57}$$

we deduce $\tilde{\lambda} = \beta$. It follows that

$$\lim_{r_n \to 0}\big(u(r_n\xi) - \gamma\mu_N(r_n\xi)\big) = (\alpha - \gamma)\mu_N(\xi) + \tilde{\lambda}, \tag{2.2.58}$$

and $\alpha = \gamma$, since the left-hand side of (2.2.58) is uniformly bounded. Consequently

$$\lim_{r \to 0}\big(u(r\xi) - \gamma\mu_N(r)\big) = \gamma\mu_N(\xi) + \tilde{\lambda}, \tag{2.2.59}$$

and the gradient estimate follows.

Case 2. $\lambda(r) = \max\limits_{|x| = r_1}\big(u(x) - \gamma\mu_N(r)\big)$ for any $r \in (0, r_1)$. Defining

$$\lambda_0(r) = \max_{0 < |x| < r}\big(u(x) - \gamma\mu_N(r)\big) = \max_{|x| = r}\big(u(x) - \gamma\mu_N(r)\big) \tag{2.2.60}$$

(from Remark 5.7 the maximum cannot be achieved at a interior point), then $\lambda_0(r) \to \lambda_0$ as r goes to 0. In the same way as in Case 1, we prove that

$$\lim_{r \to 0}\left(u(r\xi) - \gamma\mu_N(r)\right) = \gamma\mu_N(\xi) + \lambda_0. \tag{2.2.61}$$

Equation (2.2.5) follows from (2.2.4).

By combining the techniques of Lemma 5.5 and Theorem 5.5, it is easy to give an existence and uniqueness result for the singular Dirichlet problem

$$\begin{cases} -\mathrm{div}\left(|\nabla u|^{p-2}\nabla u\right) + |u|^{q-1}u = \lambda\delta_0 & \text{in} \quad \mathcal{D}'(\Omega), \\ u = h & \text{on} \quad \partial\Omega. \end{cases} \tag{2.2.62}$$

THEOREM 5.6. *Let* (2.2.1) *hold and* Ω *be a bounded domain of* $\mathbb{R}^N$ *containing* 0, *with a* C^1 *boundary*, $h \in C^1(\partial\Omega)$, $\lambda \in \mathbb{R}$. *Then there exists a unique* $u \in C^{1,\alpha}(\Omega^*) \cap L^q(\Omega)$ *such that* $|\nabla u|^{p-1} \in L^1(\Omega)$ *and* $u/\mu_p \in L^\infty(B(O,r))$ *for some* $0 < r < 1$, *which satisfies* (2.2.62). *Furthermore*

$$\limsup_{x \to 0} |x|^{-N}\mu_p^{1-q}(x)\left|u(x) - \lambda^{1/(p-1)}\mathrm{sgn}(\lambda)\mu_p(x)\right| < \infty, \tag{2.2.63}$$

$$\left|\nabla u(x) - \lambda^{1/(p-1)}\mathrm{sgn}(\lambda)\nabla\mu_p(x)\right| = o\left(|x|^{(1-N)/(p-1)}\right), \tag{2.2.64}$$

near O.

5-2-3 Classification of nonnegative singularities

Hereafter we assume that Ω is an open subset of $\mathbb{R}^N$ containing O, $\Omega^* = \Omega \setminus \{O\}$ and

$$1 < p \leq N, \qquad p - 1 < q < \frac{N(p-1)}{N-p}, \tag{2.3.1}$$

($p - 1 < q$ if $p = N$) and we denote

$$\gamma_{N,p,q} = \left(\left(\frac{p}{q+1-p}\right)^{p-1}\left(\frac{pq}{q+1-p} - N\right)\right)^{1/(p+q-1)}. \tag{2.3.2}$$

The following result provides a complete classification of the isolated singularities of the positive solutions of (2.1.1).

THEOREM 5.7. *Suppose that* $u \in L^\infty_{loc}(\Omega^*)$ *is a nonnegative solution of* (2.1.1) *in* Ω^*. *Then one of the following three possibilities must occur:*

(i) $\quad \lim_{x \to 0} |x|^{p/(p+q-1)} u(x) = \gamma_{N,p,q},$

(ii) there exists a positive number γ such that $\lim\limits_{x \to O} u(x) / \mu_p(x) = \gamma$ and

$$-\mathrm{div}\left(|\nabla u|^{p-2}\nabla u\right) + u^q = \gamma^{p-1}\delta_O \qquad in \ \ \mathcal{D}'(\Omega) ,$$

(iii) $u(x)$ admits a finite limit when x goes to O and the resulting function is a solution of (2.1.1) in Ω.

The next lemma is a Harnack inequality for nonnegative solutions of (2.1.1) (we can suppose that Ω contains the unit ball of $\mathbb{R}^N$).

LEMMA 5.6. *Let $0 < p-1 < q$ and u be a nonnegative solution of (2.1.1) in Ω^*. Then there existsa positive constant $C = C(N, p, q)$ such that*

$$\max_{|x|=r} u(x) \leq C \min_{|x|=r} u(x). \tag{2.3.3}$$

Proof. We write (2.1.1) under the following form

$$-\mathrm{div}\left(|\nabla u|^{p-2}\nabla u\right) + b^p u^{p-1} = 0 \tag{2.3.4}$$

where $b^p = u^{q+1-p}$. If $y \in B^*(O,1/2)$, $\rho = |y|/2$, $B(y,\rho) \subset \Omega^*$. By a result of Trudinger ([Tr]), there exists a constant C_1 depending on N, p and $\rho\|b\|_{L^\infty(B(y,\rho))}$, such that

$$\max_{|x-y|\leq\rho/3} u(x) \leq C_1 \min_{|x-y|\leq\rho/3} u(x). \tag{2.3.5}$$

Note that by Proposition 5.2-Step 1

$$\rho\|b\|_{L^\infty(B(y,\rho))} \leq C_2 \tag{2.3.6}$$

where C_2 is independent of y. Hence it is the same with C_1 in (2.3.5). If x_1 and x_2 are two points such that $0 < |x_1| = |x_2| \leq 1/2$, we can join them by at most 10 connected balls of radius $|x_1|/6$ and centers on $\partial B(O,|x_1|)$. Consequently (2.3.3) follows from (2.3.5) with $C = C_1^{10}$.

LEMMA 5.7. *Suppose that (2.3.1) holds and let $\psi \in C^1((0,1])$ be a nonnegative solution of*

$$-\left(r^{N-1}|\psi_r|^{p-2}\psi_r\right) + r^{N-1}\psi^q = 0 \qquad in \quad (0,1], \tag{2.3.7}$$

such that $\psi(r)/\mu_p(x)$ is unbounded in $(0,1/2]$. Then there exists a positive constant C depending only on N, p, q such that

$$\left|r^{p/(p+q-1)}\psi(r) - \gamma_{N,p,q}\right| \leq Cr^\tau \qquad on \quad (0,1/2], \tag{2.3.8}$$

where τ is the positive root of the polynomial equation

$$X^2 + \left(N - \frac{p(q+1)}{q+1-p}\right)X + \frac{p}{p-1}\left(N - \frac{pq}{q+1-p}\right) = 0.$$ (2.3.9)

Proof. Step 1. We claim that there is a positive constant α such that

$$\alpha r^{-p/(q+1-p)} \le \psi(r) \le \alpha^{-1} r^{-p/(q+1-p)} \quad \text{in } (0,1].$$ (2.3.10)

To prove it, suppose first that $1 < p < N$ and consider again the change of variable and unknown of Lemma 5.5

$$s = r^{(p-N)/(p-1)}, \qquad \phi(s) = \psi(r).$$ (2.3.11)

Then

$$-\left(|\phi_s|^{p-2}\phi_s\right)_s + \left(\frac{p-1}{N-p}\right)^p s^{-p(N-1)/(N-p)}\phi^q = 0 \quad \text{in } [1,\infty).$$ (2.3.12)

Clearly ϕ_s is increasing and
  (i) either $\lim\limits_{s\to\infty} \phi_s(s) = \beta < \infty$,
  (ii) or $\lim\limits_{s\to\infty} \phi_s(s) < \infty$.

Because $\psi_r(r) = \dfrac{p-N}{p-1} r^{(1-N)/(p-1)} \phi_s(s)$, the first case implies that $\lim\limits_{r\to 0} r^{(N-p)/(p-1)}\psi(r) = \beta$, which contradicts the assumption. Therefore (ii) holds, $\phi(s) > 0$, $\phi(s) \le s\phi_s(s)$ for s large enough and

$$\left(\phi_s^{p-1}\right)_s \le s^{q-p(N-1)/(N-p)}\phi_s^q.$$ (2.3.13)

We set $a(s) = \phi_s^{p-1}(s)$ and obtain by integration

$$\frac{p-1}{q+1-p}(a(s))^{(p-1-q)/(p-1)} \le \left(\frac{p-1}{N-p}\right)^p\left(\frac{N(p-1)}{N-p} - q\right)^{-1} s^{q-N(p-1)/(N-p)},$$ (2.3.14)

or, equivalently,

$$\phi_s(s) \ge C_{N,p,q} s^{(N(p-1)/(N-p)-q)/(q+1-p)}.$$ (2.3.15)

Integrating again gives the left-hand side inequality (2.3.10). The right-hand side is just Proposition 5.2-Step 1.
In the case $p = N$ we proceed similarly by setting

$$s = \ln(1/r), \quad \psi(r) = \phi(s). \tag{2.3.16}$$

Step 2. There holds

$$\lim_{r \to 0} r^{p/(q+1-p)} \psi(r) = \gamma_{N,p,q}. \tag{2.3.17}$$

To prove it, we consider the function

$$r \mapsto v(r) = r^{p/(p+q-1)} \psi(r). \tag{2.3.18}$$

By step 1, v is bounded from above and from below and

$$\left| rv_r - \frac{p}{q+1-p} v \right|^{p-2} \left(r^2(p-1)v_{rr} + r\left(N-1-2\frac{p(q-1)}{q+1-p} \right)v_r \right.$$
$$\left. + \frac{p^2 q - Np(q+1-p)}{(q+1-p)^2} v \right) = v^q \tag{2.3.19}$$

holds in $(0,1)$ except at the points r such that $rv_r - \dfrac{p}{q+1-p} v = 0$.

Suppose first that $r \mapsto v(r)$ is monotone on some interval $\left(0, r^*\right)$, and set $v_0 = \lim\limits_{r \to 0} v(r)$ and
$v_k(r) = v(r/k)$. Applying Lemma 5.3 gives $\left| v_{kr}(r) \right| \leq Cr^{-1}$ and
$\left| v_{kr}(r) - v_{kr}(r) \right| \leq Cr^{-1-\alpha} |r - r'|^\alpha$ for $0 < r \leq r' \leq k/2$. Up to some subsequence $\{k_n\}$ going
to infinity with n , $\lim\limits_{n \to \infty} v_{k_n} = \tilde{v}$, $\lim\limits_{n \to \infty} v_{k_n} = \tilde{v}_r$ and $\tilde{v}$ is a solution of (2.3.19) on $(0,\infty)$.
Clearly $\tilde{v} = v_0$ wich is a positive solution of

$$\left(\frac{p}{q+1-p} v_0 \right)^{p-2} \left(\frac{p^2 q - Np(q+1-p)}{(q+1-p)^2} v_0 \right) = v_0^q, \tag{2.3.20}$$

which implies $v_0 = \gamma_{N,p,q}$ and (2.3.17) holds.

In order to complete the proof of (2.3.17) it suffices to prove that $r \mapsto v(r)$ is
monotone on some interval $\left(0, r^*\right)$. Suppose that this is not so, then there exist points
$0 < r_1 < r_2 < 1$ such that $v(r_1) < v(r) < v(r_2)$ for $r_1 < r < r_2$ and $v(r_1)$ is a local minimum,
$v(r_2)$ is a local maximum. Since v is C^2 in a neighborhood of r_1 and of r_2, $v_{rr}(r_1) \geq 0$ and
$v_{rr}(r_2) \leq 0$. Therefore, we deduce from (2.3.19) that

$$\left(\frac{p}{q+1-p} v(r_1) \right)^{p-2} \left(\frac{p^2 q - Np(q+1-p)}{(q+1-p)^2} v(r_1) \right) \leq (v(r_1))^q, \tag{2.3.21}$$

and $v(r_1) \geq \gamma_{N,p,q}$. Similarly $v(r_2) \leq \gamma_{N,p,q}$, which contradicts the inequality $v(r_1) < v(r_2)$.

Step 3. Set $\gamma = \gamma_{N,p,q}$ and $\psi(r) = \gamma r^{-p/(q+1-p)}(1+z(r))$. Then $z(r)$ tends to 0 with r and

$$\gamma^{p-1}\left|rz_r - \frac{p}{q+1-p}(z+1)\right|^{p-2}\left(r^2(p-1)z_{rr}\right.$$

$$\left.+r\left(N-1-2\frac{q(p-1)}{q+1-p}z_r + \frac{p^2q - Np(q+1-p)}{(q+1-p)^2}(1+z)\right)\right) = \gamma^q(1+z)^q. \qquad (2.3.22)$$

By the scaling argument of Step 2, $rz_r(r) \to 0$ and $r^2 z_{rr}(r) \to 0$ if r goes to 0. Therefore

$$\left|1+z+\frac{q+1-p}{p}rz_r\right|^{p-2} = 1+(p-2)\left(z - \frac{q+1-p}{p}rz_r\right) + O(z^2 + r^2 z_r^2).$$

Because $(1+z)^q = 1+qz+O(z^2)$, (2.3.22) becomes

$$r^2(p-1)z_{rr} + r\left(N(p-1)-1+2p - \frac{p^2q}{q+1-p}\right)z_r$$

$$+p\left(N - \frac{pq}{q+1-p}\right)z = O\left(z^2 + r^2 z_r^2 + r^4 z_{rr}^2\right). \qquad (2.3.23)$$

Consider now the homogeneous equation

$$r^2(p-1)y_{rr} + r\left(N(p-1)-1+2p - \frac{p^2q}{q+1-p}\right)y_r + p\left(N - \frac{pq}{q+1-p}\right)y = 0. \qquad (2.3.24)$$

Since $N - \dfrac{pq}{q+1-p} < 0$ and

$$\Delta = \left(N - \frac{p(q+1)}{q+1-p}\right)^2 - 4\frac{p}{p-1}\left(N - \frac{pq}{q+1-p}\right) > 0, \qquad (2.3.25)$$

equation (2.3.9) admits two roots τ and τ' with $\tau' < 0 < \tau$ and equation (2.3.24) admits two linearly independent solutions, $r \mapsto r^{\tau}$ and $r \mapsto r^{\tau'}$. By standard comparison techniques, for any $\varepsilon > 0$ we have $z(r) = O(r^{\tau-\varepsilon})$. In order to estimate the derivatives of z we set $t = \ln(1/r)$ and $\tilde{z}(t) = z(r)$. With this change of variable equation (2.3.23) becomes

$$r^2(p-1)\tilde{z}_{tt} - \left((N-1)(p-1)-1+2p - \frac{p^2q}{q+1-p} + a(t)\right)\tilde{z}_t$$

$$+p\left(N - \frac{pq}{q+1-p} + b(t)\right)\tilde{z} = 0, \qquad (2.3.26)$$

where a and b are bounded functions which tend to 0 when t goes to infinity. From elliptic equations regularity estimates, we obtain

$$\left|\tilde{z}_t(t)\right| + \left|\tilde{z}_{tt}(t)\right| \le C_\alpha \max_{|t-\sigma|\le\alpha} \left|\tilde{z}(\sigma)\right|, \tag{2.3.27}$$

or equivalently

$$\left|rz_r(r)\right| + \left|r^2 z_{rr}(r)\right| \le C_\alpha \max_{re^{-\alpha}\le\sigma re^\alpha} \left|z(\sigma)\right|. \tag{2.3.28}$$

Since the right-hand side is bounded by $Cr^{\tau-\varepsilon}$, we can rewrite (2.3.23) under the form

$$r^2(p-1)z_{rr} + r\left(N(p-1)-1+2p-\frac{p^2 q}{q+1-p}\right)z_r$$
$$+p\left(N-\frac{pq}{q+1-p}\right)z = O\!\left(r^{2\tau-2\varepsilon}\right). \tag{2.3.29}$$

Setting $z(r) = r^\tau h(r)$, then

$$\left(r^\delta h_r(r)\right)_r = O\!\left(r^{\delta-2+\tau-2\varepsilon}\right), \tag{2.3.30}$$

where $\delta = N + 2\tau + \dfrac{(2p-1)(q+1-p)-pq^2}{(p-1)(q+1-p)}$. Notice that $\delta + \tau - 1 = \tau + \sqrt{\Delta} > 0$, if Δ is as in (2.3.25). Therefore $r \mapsto \left(r^\delta h_r(r)\right)_r$ is integrable on $(0,1)$ and $\lim_{r\to 0} r^\delta h_r(r) = C_0$, for some C_0. Since $h(r) = O(r^{-\varepsilon})$, C_0 is zero and there holds

$$\left|r^\delta h_r(r)\right| \le \int_0^r \sigma^{\delta-2+\tau-2\varepsilon} d\sigma = \frac{r^{\delta-1+\tau-2\varepsilon}}{\delta-1+\tau-2\varepsilon}. \tag{2.3.31}$$

Consequently $h_r(r) = O(r^{\tau-1-2\varepsilon})$, $\lim_{r\to 0} h(r)$ and

$$\lim_{r\to 0} r^\tau \left|r^{p/(q+1-p)}\psi(r) - \gamma_{N,p,q}\right| < \infty. \tag{2.3.32}$$

From the proof, it is clear that the constant C in (2.3.8) only depends on N, p and q.

Proof of Theorem 5.7. Case 1. u/μ_p is bounded near O. From Theorem 5.6 it admits a finite limit γ when x goes to O and (ii) holds. If $\gamma = 0$, we have (iii).

Case 2. u/μ_p is not bounded near O. Then there exists a sequence $\{x_n\}$ converging to 0 when n goes to infinity such that

$$\lim_{n\to\infty}\frac{u(x_n)}{\mu_p(x_n)}=\infty,\tag{2.3.33}$$

and, from Lemma 5.6,

$$\lim_{n\to\infty}\min_{y=|x_n|}\frac{u(y)}{\mu_p(y)}=\infty.\tag{2.3.34}$$

Step 1. We claim that

$$\lim_{x\to 0}\frac{u(x)}{\mu_p(x)}=\infty.\tag{2.3.35}$$

Indeed, for any $\gamma>0$ we introduce the solution u_γ of

$$\begin{cases}-\left(r^{N-1}\left|u_{\gamma r}\right|u_{\gamma r}\right)_r+r^{N-1}u_\gamma{}^q=0 \ \ \text{in} \ \ (0,1),\\[2mm] \lim_{r\to 0}u_\gamma(r)/\mu_p(r)=\gamma \ \ \text{and} \ \ u_\gamma(1)=0.\end{cases}\tag{2.3.36}$$

By (2.3.34), $u(x)\geq u_\gamma(x)$ for $|x|=|x_n|$ and n large enough. From the maximum principle $u(x)\geq u_\gamma(x)$ for any $0<|x|\leq 1$ and

$$\liminf_{x\to 0}\frac{u(x)}{\mu_p(x)}\geq\gamma.\tag{2.3.37}$$

Since γ is arbitrary, (2.3.35) follows.

Step 2. By the maximum principle $\gamma\mapsto u_\gamma$ is increasing and by Proposition 5.2-Step 1, $u_\gamma(r)\leq Ar^{-p/(p+q-1)}$. Therefore u_γ converges to some u_∞ which satisfies

$$\begin{cases}-\left(r^{N-1}\left|u_{\infty r}\right|^{p-2}u_{\infty r}\right)_r+r^{N-1}u_\infty^q=0 \ \ \text{on} \ \ (0,1),\\[2mm] \lim_{r\to 0}u_\infty(r)/\mu_p(r)=\infty \ \ \ \text{and} \ \ \ u_\infty(1)=0,\end{cases}\tag{2.3.38}$$

and from Lemma 5.7

$$\liminf_{x\to 0}|x|^{p/(q+1-p)}u(x)\geq\liminf_{r\to 0}r^{p/(q+1-p)}u(r)\geq\gamma_{N,p,q}.\tag{2.3.39}$$

Step 3. For any positive integer n we set $\alpha_n=\max_{|x|=1/n}u(x)$, and let v_n be the solution of

$$\begin{cases} -\left(r^{N-1}\left|v_{nr}\right|^{p-2}v_{nr}\right)_{r} + r^{N-1}v_{n}^{\ q} = 0 & \text{in } (1/n,1), \\ v_{n}(1/n) = \alpha_{n} \quad \text{and} \quad v_{n}(1) = \max_{|x|=1} u(x). \end{cases} \tag{2.3.40}$$

By the maximum principle, $u(x) \le v_{n}(x)$ for $1/n \le |x| \le 1$. Using again Proposition 5.2-Step 1, we get $v_{n}(r) \le A(r-1/n)^{-p/(p+q-1)}$. By Theorem 5.1, there exist a sequence $\{n_{k}\}$ going to infinity and a function v_{∞} such that $\lim_{n\to\infty} v_{n_{k}} = v_{\infty}$ in the $C^{1}_{loc}(\overline{B}^{*}(O,1))$-topology. Moreover v_{∞} satisfies

$$-\left(r^{N-1}\left|v_{\infty r}\right|^{p-2}v_{\infty r}\right)_{r} + r^{N-1}v_{\infty}^{\ q} = 0 \quad \text{in } (0,1), \tag{2.3.41}$$

and $u(x) \le v_{\infty}(x)$ for $0 < |x| \le 1$. From (2.3.39) $\liminf_{r\to 0} r^{p/(q+1-p)}v_{\infty}(r) \ge \gamma_{N,p,q}$ and, from Lemma 5.7,

$$\lim_{r\to 0} r^{p/(q+1-p)}v_{\infty}(r) = \gamma_{N,p,q}. \tag{2.3.42}$$

Consequently $\lim_{x\to O} |x|^{p/(p+q-1)}u(x) = \gamma_{N,p,q}$, and in fact

$$\left| |x|^{p/(p+q-1)}u(x) - \gamma_{N,p,q} \right| \le C|x|^{\tau} \quad \text{for } 0 < |x| \le 1/2, \tag{2.3.43}$$

which completes the proof.

The techniques introduced in Theorem 5.7 apply to give the asymptotic behaviour of any nonnegative solution of (2.1.1) in an exterior domain of $\mathbb{R}^{N}$.

THEOREM 5.8. *Suppose that* (2.3.1) *holds and let* u *be a nonnegative non-trivial solution of* (2.1.1) *in* $\Theta \supset \{x \in \mathbb{R}^{N} : |x| \ge 1\}$. *Then there exists a constant* $C' = C'(N,p,q)$ *such that*

$$\left| |x|^{p/(p+q-1)}u(x) - \gamma_{N,p,q} \right| \le C|x|^{\tau} \quad \text{for } |x| \ge 2, \tag{2.3.44}$$

where τ *is the negative root of the polynomial equation* (2.3.9).

5-2-4 Global solutions

If u is a solution of (2.1.1) in whole $\mathbb{R}^{N}$, then it is identically 0; this is an immediate consequence of Proposition 5.2-Step 1. The same holds if u is a solution in $\mathbb{R}^{N}\setminus\{O\}$ and $1 < p < N$, $q \ge N(p-1)/(N-p)$, since it can be extended as a solution in whole $\mathbb{R}^{N}$ as a consequence of Theorem 5.5;.

COROLLARY 5.4. *Suppose that* (2.3.1) *holds and* $u \in C^1(\mathbb{R}^N \setminus \{O\})$ *is a solution of* (2.1.1) *in* $\mathbb{R}^N \setminus \{O\}$ *such that* $u(x)/\mu_p(x)$ *remains bounded in a punctured neighborhood of* O. *Then* u *is a radial function and either it is identically* 0 *or there exists* $\gamma \neq 0$ *such that* u *is the solution of*

$$-\mathrm{div}\left(|\nabla u|^{p-2}\nabla u\right) + |u|^{q-1}u = |\gamma|^{p-2}\gamma \quad in \quad \mathcal{D}'(\mathbb{R}^N). \tag{2.4.1}$$

Proof. From Proposition 5.2-Step 1 and Lemma 5.3 there exists $C > 0$ such that

$$|u(x)| \leq C|x|^{-p/(q+1-p)}, \tag{2.4.2}$$

$$|\nabla u(x)| \leq C|x|^{-(q+1)/(q+1-p)}, \tag{2.4.3}$$

and

$$\lim_{x \to 0} \frac{u(x)}{\mu_p(x)} = \gamma, \tag{2.4.4}$$

for some $\gamma \in \mathbb{R}$. If $\gamma = 0$, u is solution in $\mathbb{R}^N$ and therefore identically 0. If $\gamma \neq 0$, say $\gamma > 0$, then we can compare u with the function $x \mapsto \tilde{u}(x) = \tilde{u}(r)$ which satisfies

$$\begin{cases} -\left(r^{N-1}|\tilde{u}_r|^{p-2}\tilde{u}_r\right) + r^{N-1}u^q = 0 \quad in \quad (0,\infty), \\ \lim_{r \to 0} \tilde{u}(r)/\mu_p(r) = \gamma. \end{cases} \tag{2.4.5}$$

Such a solution is constructed from approximation by the solutions u_n of the same equation on $(0,n)$, which vanish at $r = n$ and satisfy $\lim_{r \to 0} u_n(r)/\mu_p(r) = \gamma$. The blow-up estimate at 0 is kept since we have

$$0 \leq u_n(|x|) \leq u_{n+1}(|x|) \leq \tilde{u}(x) \leq \gamma\mu_p(x) + \delta \tag{2.4.6}$$

in $B^*(O,1)$. For $\varepsilon > 0$, $\tilde{u}_\varepsilon = (1+\varepsilon)\tilde{u}$ is a super solution of (2.1.1) which dominates u both at O and at infinity, therefore it is larger than u and letting ε go to 0 gives $\tilde{u} \geq u$. Since u is positive in a punctured neighborhood of O and goes to 0 at infinity, it is positive in $\mathbb{R}^N \setminus \{O\}$. Consequently, we can reverse the role of u and $\tilde{u}$ and prove that $(1+\varepsilon)u \geq \tilde{u}$ for any $\varepsilon > 0$, which finally yields $u = \tilde{u}$.

We now characterize all nonnegative solutions in $\mathbb{R}^N \setminus \{O\}$.

COROLLARY 5.5. *Suppose that* (2.3.1) *holds and* $u \in C^1(\mathbb{R}^N \setminus \{O\})$ *is a nonnegative solution of* (2.1.1) *in* $\mathbb{R}^N \setminus \{O\}$. *Then the following alternative holds:*
(i) *either* $u(x) = \gamma_{N,p,q}|x|^{-p/(q+1-p)} = u_s(x)$,

(ii) *or there exists a nonnegative real number γ such that* $u(x) = u(|x|)$ *is the unique solution of* (2.4.1) *for which* $u(x) / \mu_p(x)$ *remains bounded in a punctured neighborhood of* O.

Proof. From Theorem 5.8 and Corollary 5.4, we are left with the case when $u(x) / \mu_p(x)$ is not bounded near O and therefore u satisfies $\lim_{x \to O} |x|^{p/(q+1-p)} u(x) = \gamma_{N,p,q}$. Comparing u with $(1 \pm \varepsilon) u_s(x)$ $(\varepsilon > 0)$ and letting ε go to 0 gives yields (i).

5-3 Equations with source term

5-3-1 The Bidaut-Véron Lemma

As we have seen it in Chapter 3, the Brezis-Lions Lemma (Theorem 3.1) plays an important role for describing the isolated singularities of positive solutions of

$$-\Delta u = u^q \tag{3.1.1}$$

in a punctured N-dimensional domain when $1 < q < N/(N-2)$. This result has been extended to the p-Laplacian by Bidaut-Véron [B.V1]

THEOREM 5.9. *Let* $N > 1$ *and* $1 < p < N$. *If* $u \in C^0(\overline{B}^*(O,1)) \cap W^{1,p}_{loc}(\overline{B}^*(O,1))$ *with* $\Delta_p u \in L^1_{loc}(B^*(O,1))$ *in the sense of distributions in* $B^*(O,1)$, *is nonnegative and satisfies*

$$\operatorname{div}\left(|\nabla u|^{p-2} \nabla u\right) \le 0 \quad \textit{a.e. in } B(O,1), \tag{3.1.2}$$

then $u^{p-1} \in M^{N/(N-p)}_{loc}(\overline{B}(O,1))$, $|\nabla u|^{p-1} \in M^{N/(N-1)}_{loc}(\overline{B}(O,1))$ *and there exist a nonnegative constant* β *and a function* $g \in L^1(B(O,R))$ *such that*

$$-\operatorname{div}\left(|\nabla u|^{p-2} \nabla u\right) = g + \beta \delta_O \quad \textit{in } \mathcal{D}'(B(O,1)). \tag{3.1.3}$$

The proof of this result heavily relies on techniques introduced by Bénilan, Serrin and Trudinger. In the next four lemmas we keep the notations of Theorem 5.9.

LEMMA 5.8. *Let* $g(x) = -\Delta_p u(x)$ *a.e. in* $B(O,1)$. *Then* $g \in L^1_{loc}(B(O,R))$ *and for any function* $\eta \in C_0^\infty(B(O,1))$, $0 \le \eta \le 1$, $\eta(x) = 1$ *near* O, *there holds*

$$\int_{B(O,1)} g \eta^p dx \le \int_{B(O,1)} |\nabla u|^{p-2} \nabla u . \nabla \eta^p dx. \tag{3.1.4}$$

Moreover, for any $0 < \rho < 1$, *there exists* $c_\rho > 0$ *such that*

277

$$\int_{B_{\rho,\alpha,k}} |\nabla u|^p \, dx \le c_\rho \alpha \quad \text{for any } k \ge 0 \text{ and } \alpha > 0, \tag{3.1.5}$$

where $B_{\rho,\alpha,k} = \{x \in B(O,\rho) : k < u(x) < k + \alpha\}$.

Proof. From the definition of g we have

$$\int_{B(O,1)} g\zeta \, dx = \int_{B(O,1)} |\nabla u|^{p-2} \nabla u . \nabla \zeta \, dx dx, \tag{3.1.6}$$

for any $\zeta \in C_0^{0,1}(B^*(O,1))$. Set $p_{k,\alpha}(t) = \min\big((t-k)^+/\alpha, 1\big)$ for t and $k \ge 0$, and $\alpha > 0$. Let $0 < \rho < 1$ and $\varepsilon < \rho/2$. If $\eta \in C_0^\infty(B(O,1))$, $0 \le \eta \le 1$, $\eta(x) = 1$ in $B(O,\rho)$, we denote $\zeta_\varepsilon = \xi_\varepsilon \eta$ with $\xi_\varepsilon \in C^\infty(B(O,1))$, $\xi_\varepsilon(x) = 0$ for $|x| \le \varepsilon$, $\xi_\varepsilon(x) = 1$ for $|x| \ge 2\varepsilon$ and $|\nabla \xi_\varepsilon| \le C/\varepsilon$. We take $\zeta = (1 - p_{k,\alpha}(u))\zeta_\varepsilon^p$ in (3.1.6) and get

$$\int_{B(O,1)} g(1 - p_{k,\alpha}(u))\zeta_\varepsilon^p \, dx + \alpha^{-1} \int_{B_{1,\alpha,k}} |\zeta_\varepsilon \nabla u|^p \, dx$$
$$= \int_{B(O,1)} g(1 - p_{k,\alpha}(u))|\nabla u|^{p-2} \nabla u . \nabla(\zeta_\varepsilon^p) \, dx. \tag{3.1.7}$$

We first take $\alpha = 1$ and add the corresponding equalities with $k = 0, 1, \dots,$ n. Because $\sum_{k=0}^n (1 - p_{k,1}(t)) = (n + 1 - t)^+$, we get

$$\int_{\{u < n+1\}} g(n + 1 - u)\zeta_\varepsilon^p \, dx + \int_{\{u < n+1\}} |\zeta_\varepsilon \nabla u|^p \, dx$$
$$\le \int_{\{u < n+1\}} g(n + 1 - u)|\nabla u|^{p-2} \nabla u . \nabla(\eta^p) \, dx \tag{3.1.8}$$
$$+ p \int_{\{u < n+1\} \cap B(O,2\varepsilon)} (n + 1 - u)|\zeta_\varepsilon \nabla u|^{p-1} \nabla u . \nabla(\xi_\varepsilon) \, dx.$$

We consider $h > 0$; since $n + 1 - u(x) > (n+1)h/(h+1)$ a.e. on $\left\{x : u(x) < \dfrac{n+1}{h+1}\right\}$, we have, from inequality (3.1.8) and by using Hölder's inequality,

$$\frac{h}{h+1} \int_{\{u < (n+1)/(h+1)\}} g\zeta_\varepsilon^p \, dx + \frac{1}{n+1} \int_{\{u < n+1\}} |\zeta_\varepsilon \nabla u|^p \, dx$$
$$\le \int_{\{u < n+1\}} \left(1 - \frac{u}{n+1}\right) |\nabla u|^{p-2} \nabla u . \nabla(\eta^p) \, dx + p \int_{B(O,2\varepsilon) \cap \{u < n+1\}} |\zeta_\varepsilon \nabla u|^p |\nabla \xi_\varepsilon| \, dx \tag{3.1.9}$$
$$\le \int_{\{u < n+1\}} \left(1 - \frac{u}{n+1}\right) |\nabla u|^{p-2} \nabla u . \nabla(\eta^p) \, dx$$
$$+ (p-1)\beta^{p/(p-1)} \int_{\{u < n+1\}} |\zeta_\varepsilon \nabla u|^p \, dx + \beta^{-p} \int_{B(O,2\varepsilon)} |\nabla \xi_\varepsilon|^p \, dx,$$

for any $\beta > 0$. Now we can choose β in such a way that

$$\frac{h}{h+1}\int_{\{u<(n+1)/(h+1)\}} g\zeta_\varepsilon^p dx + \frac{1}{2(n+1)}\int_{\{u<n+1\}}|\zeta_\varepsilon \nabla u|^p dx$$

$$\leq \int_{\{u<n+1\}}\left(1-\frac{u}{n+1}\right)|\nabla u|^{p-2}\nabla u.\nabla(\eta^p)dx + M(n+1)^{p-1}C^{p-1}\varepsilon^{N-p}, \tag{3.1.10}$$

where $M = M(N,p)$. Now we let ε go to 0, n and h to infinity successively. From (3.1.10) and Fatou's lemma we deduce that $g\eta^p \in L^1(B(O,1))$ and (3.1.4) holds. Moreover, for any integer n,

$$\int_{B(O,\rho)\cap\{u<n+1\}}|\nabla u|^p dx \leq 2(n+1)\int_{B(O,1)}|\nabla u|^{p-1}|\nabla(\eta^p)|dx. \tag{3.1.11}$$

Taking $\alpha > 0$ and $k \geq 0$ arbitrary in (3.1.7), we deduce from (3.1.11) that the following inequality, in which $c' = c'(N,p,k,\alpha)$,

$$\frac{1}{\alpha}\int_{B_{1,u,\lambda}}\zeta_\varepsilon|\nabla u|^p dx \leq \int_{B(O,1)}|\nabla u|^{p-1}|\nabla(\eta^p)|dx + p\int_{B(O,2\varepsilon)\cap\{u<k+\alpha\}}|\nabla u|^{p-1}|\nabla\zeta_\varepsilon|dx$$

$$\leq \int_{B(O,1)}|\nabla u|^{p-1}|\nabla(\eta^p)|dx + c'\,\varepsilon^{(N-p)/p}\left(\int_{B(O,1)}|\nabla u|^{p-1}|\nabla(\eta^p)|dx\right)^{(p-1)/p} \tag{3.1.12}$$

holds. Letting ε go to 0 gives (3.1.5).

LEMMA 5.9. *For any* $\gamma \in (0, N(p-1)/(N-p))$, $u^\gamma \in L^1(B(O,R))$ *and there exists a positive constant* $C = C(\gamma, N, p, u)$ *such that for any* $\sigma \in (0,1)$,

$$\int_{B(O,\sigma)} u^\gamma dx \leq C\sigma^{N-(N-p)\gamma/(p-1)}. \tag{3.1.13}$$

Proof. We recall Serrin's test functions that have been used in the proof of Theorem 1.21: first we set $C_1 = 2\max_{|x|=1/2} u(x)$, $\bar{u} = u - C_1$ and for any $\sigma \in (0,1/2)$, $m(\sigma) = \min_{|x|=\sigma}\bar{u}(x)$. If $m(\sigma) > 0$, then

$$v(\sigma)(x) = \begin{cases} 0 & \text{if } \sigma < |x| < 1/2 \text{ and } \bar{u}(x) \leq 0, \text{ or if } |x| \geq 1/2, \\ \bar{u}(x) & \text{if } 0 \leq \bar{u}(x) \leq m(\sigma) \text{ and } \sigma < |x| < 1/2, \\ m(\sigma) & \text{if } \bar{u}(x) > m(\sigma) \text{ and } \sigma < |x| < 1/2, \text{ or if } |x| \leq \sigma. \end{cases} \tag{3.1.14}$$

Then $v(\sigma) \in C^0(\bar{B}(O,R)) \cap W^{1,p}(B(O,R))$. We take $\zeta = v(\sigma) - m(\sigma)\eta$ as a test function, where η is as in Lemma 5.8. Then

$$\int_{B(O,1)}|\nabla u|^{p-2}\nabla u.\nabla(v(\sigma))dx + \int_{B(O,1)}g(m(\sigma)-v(\sigma))dx = m(\sigma)K, \tag{3.1.15}$$

where K does not depend on σ,

$$K = \int_{B(O,1)} |\nabla u|^{p-2} \nabla u . \nabla \eta \, dx + \int_{B(O,1)} g(1-\eta) dx.$$

Because $g(m(\sigma) - v(\sigma)) \geq 0$, we obtain as in the proof of Theorem 1.21,

$$m(\sigma) \geq \int_{B(O,1)} |\nabla u|^{p-2} \nabla u . \nabla (v(\sigma)) dx = \int_{B(O,1)} |\nabla (v(\sigma))|^p dx$$

$$\geq m^p(\sigma) c_{1,p}(B(O,\sigma)) = m^p(\sigma) N^{-1} \left| S^{N-1} \right| \left(\frac{N-p}{p-1} \right)^{p-1} \sigma^{N-p}. \tag{3.1.16}$$

Dividing by $m(\sigma)$ yields

$$\min_{|x|=\sigma} u(x) \leq C_1 + C_2 \sigma^{(p-N)/(p-1)} \qquad (\forall \sigma \in (0,1/2]), \tag{3.1.17}$$

where the C_j do not depend on σ. Since u is a weak supersolution of the p-Laplace equation, the following form of the weak Harnack inequality holds ([Se1]): for any $\gamma \in (0, N(p-1)/(N-p))$, there exists $C = C(N,p,\gamma)$ such that for any ball $B(x_0, 3\rho) \subset B^*(O,1)$,

$$\rho^{-N/\gamma} \left(\int_{B(x_0, 2\rho)} u^\gamma dx \right)^{1/\gamma} \leq C \min_{x \in B((x_0,\rho))} u(x). \tag{3.1.18}$$

Combining with (3.1.17) and using the covering argument which has been introduced in the proof of Lemma 5.6 yields

$$\rho^{-N/\gamma} \left(\int_{\{3\sigma/4 < |x| < 5\sigma/4\}} u^\gamma dx \right)^{1/\gamma} \leq C \min_{|x|=\sigma} u(x) \quad (\forall \sigma \in (0,1/2]), \tag{3.1.19}$$

where C does not depend on σ. Clearly (3.1.13) follows from (3.1.17) and (3.1.19).

LEMMA 5.10. *The function* u *satisfies* (3.1.3).

Proof. We start with an estimate of $|\nabla u|^{p-1}$ near O. Let $0 < \rho < 1$ and $\delta > 0$ be fixed, then

$$\int_{B(O,\sigma)} |\nabla u|^{p-1} dx = \int_{B(O,\sigma)} \left| \frac{\nabla u}{(1+u)^{(\delta+1)/p}} \right|^{p-1} (1+u)^{(\delta+1)(p-1)/p} dx$$

$$\leq \left(\int_{B(O,\sigma)} \frac{|\nabla u|^p}{(1+u)^{\delta+1}} dx \right)^{(p-1)/p} \left(\int_{B(O,\sigma)} (1+u)^{(\delta+1)(p-1)} dx \right)^{1/p}, \tag{3.1.20}$$

for any $\sigma \in (0,\rho]$. From (3.1.5),

280

$$\int_{B(O,\sigma)} \frac{|\nabla u|^p}{(1+u)^{\delta+1}} dx \le \sum_{k=0}^{\infty} \int_{B_{\rho,1,k}} \frac{|\nabla u|^p}{(1+u)^{\delta+1}} dx \le c_\rho \sum_{k=0}^{\infty} \frac{1}{(1+k)^{\delta+1}} < \infty. \tag{3.1.21}$$

Take $\delta < p/(N-p)$, then $(\delta+1)(p-1) < N(p-1)/(N-p)$ and by Lemma 5.9 $|\nabla u|^{p-1}$ is integrable in $B(O,1)$. Moreover there exists $C_\delta > 0$ such that

$$\int_{B(O,\sigma)} |\nabla u|^{p-1} dx \le C_\delta \sigma^{1-\delta(N-p)/p}. \tag{3.1.22}$$

We can now define the distribution $T = -\mathrm{div}\left(|\nabla u|^{p-2} \nabla u\right) - g$ in $B(O,1)$ as being associated to a locally integrable function. Since its support is reduced to $\{O\}$ we have $T = \sum_{|\theta| \le m} \beta_\theta D^\theta \delta_O$ for some integer m. Consider now $\zeta \in C_0^\infty(B(O,1))$ such that $(-1)^{|\theta|} D^\theta \zeta(O) = \beta_\theta$ and $\zeta_\varepsilon(x) = \zeta(x/\varepsilon)$. Then

$$\langle T, \zeta_\varepsilon \rangle = \sum_{|\theta| \le m} \beta_\theta^2 \varepsilon^{-|\theta|} = \int_{B(O,1)} |\nabla u|^{p-2} \nabla u . \nabla \zeta_\varepsilon dx - \int_{B(O,1)} g \zeta_\varepsilon dx. \tag{3.1.23}$$

Because g is nonnegative, we get for ε small enough,

$$\sum_{|\theta| \le m} \beta_\theta^2 \varepsilon^{-|\theta|} \le C_\delta \varepsilon^{-\delta(N-p)/p}. \tag{3.1.24}$$

Since $\delta(N-p)/p < 1$, $\beta_\theta = 0$ for any $|\theta| \ge 1$. Finally, if we take $\eta \in C_0^\infty(BO,1))$ with $0 \le \eta(x) \le 1$ and $\eta(x) = 1$ near O, we have

$$\langle T, \eta^p \rangle = \beta_0 = \int_{B(O,1)} |\nabla u|^{p-2} \nabla u . \nabla(\eta^p) dx - \int_{B(O,1)} g \eta^p dx, \tag{3.1.25}$$

which implies $\beta_0 \ge 0$ and ends the proof.

In the next lemma we shall prove that we can take $\delta = 0$ in (3.1.22).

LEMMA 5.11. $u^{p-1} \in M^{N/(N-p)}(B(O,1))$ and $|\nabla u|^{p-1} \in M^{N/(N-1)}(B(O,1))$.

Proof. Let $0 < \gamma < (p-1)/p$; since $u \in W_{loc}^{1,p}(\overline{B}^*(O,1))$, $(1+u)^\gamma \in W_{loc}^{1,p}(\overline{B}^*(O,1))$ and

$$\nabla((1+u)^\gamma) = \gamma(1+u)^{\gamma-1} \nabla u. \tag{3.1.26}$$

Taking now $\delta = p-1-p\gamma$ in (3.1.21) implies that $\nabla((1+u)^\gamma) \in L_{loc}^p(\overline{B}^*(O,1))$. We also have $\gamma < (N-1)(p-1)/(N-p)$, therefore $(1+u)^\gamma \in L_{loc}^1(\overline{B}(O,1))$. For any $1 \le i \le N$, there holds

$$\frac{\partial}{\partial x_i}(1+u)^\gamma = \gamma(1+u)^{\gamma-1}\frac{\partial u}{\partial x_i} + \sum_{|\theta|\le m_i}\alpha_{i,\theta}^2 D^\theta \delta_0 \tag{3.1.27}$$

in the sense of distributions in $B(O,1)$. If $\zeta_\varepsilon(x) = \zeta(x/\varepsilon)$ is the test function of Lemma 5.10, with $(-1)^{|\theta|}D^\theta\zeta(O) = \alpha_{i,\theta}$, then

$$\left\langle \frac{\partial}{\partial x_i}((1+u)^\gamma),\zeta_\varepsilon \right\rangle = \sum_{|\theta_i|\le m}\alpha_{i,\theta}^2\varepsilon^{-|\theta|} + \gamma\int_{B(O,1)}(1+u)^{\gamma-1}\frac{\partial u}{\partial x_i}\zeta_\varepsilon dx$$
$$= -\int_{B(O,1)}(1+u)^\gamma\frac{\partial\zeta_\varepsilon}{\partial x_i}dx, \tag{3.1.28}$$

and

$$\sum_{|\theta|\le m_i}\alpha_{i,\theta}^2\varepsilon^{-|\theta|} \le C\left(\left\|(1+u)^{\gamma-1}\nabla u\right\|_{L^p(B(O,1/2))}\varepsilon^{Np/(p-1)} + \varepsilon^{N-1-\gamma(N-p)/(p-1)}\right). \tag{3.1.29}$$

Consequently all the $\alpha_{i,\theta}$ vanish and $(1+u)^\gamma \in W^{1,1}_{loc}(B(O,1))$. We follow now Bénilan's method: for any $k\ge 0$ and $\alpha > 0$, we have

$$p_{k,\alpha}(1+u) \le p_{k^\gamma,(k+\alpha)^\gamma - k^\gamma}((1+u)^\gamma), \tag{3.1.30}$$

a.e. in $B(O,1)$, since $\gamma < 1$. Let $0 < \rho < 1$, then Sobolev's inequality implies

$$\left\|p_{k,\alpha}(1+u)\right\|_{L^{N/(N-1)}(B(O,\rho))} \le C\left\|\nabla p_{k^\gamma,(k+\alpha)^\gamma - k^\gamma}((1+u)^\gamma)\right\|_{L^1(B(O,\rho))} \tag{3.1.31}$$

where C depends on N and p. We deduce from (3.1.26) that

$$\left\|p_{k,\alpha}(1+u)\right\|_{L^{N/(N-1)}(B(O,\rho))} \le \frac{C}{(k+\alpha)^\gamma - k^\gamma}\int_{B_{\rho,\alpha,k}}\left|\nabla(1+u)^\gamma\right|dx,$$
$$\le \frac{C\gamma(k+1)^{\gamma-1}}{(k+\alpha)^\gamma - k^\gamma}\int_{B_{\rho,\alpha,k}}\left|\nabla u\right|dx, \tag{3.1.32}$$
$$\le \frac{C}{\alpha}\int_{B_{\rho,\alpha,k}}\left|\nabla u\right|dx,$$

if $\alpha < 1$. Consequently

$$\text{meas}\left(\left\{x \in B(O,\rho): u(x) > k\right\}\right) \le Ck^{N(1-p)/(N-p)}, \tag{3.1.33}$$

$$\text{meas}\left(\left\{x \in B(O,\rho): |\nabla u(x)| > k\right\}\right) \le Ck^{N(1-p)/(N-1)}. \tag{3.1.34}$$

This completes the proof of the lemma and of Theorem 5.9.

The case $p = N$ can only be treated under a more restrictive hypothesis

THEOREM 5.10. *Let* $N > 1$, $u \in C^0(\overline{B}^*(O,1)) \cap W_{loc}^{1,N}(\overline{B}^*(O,1))$ *with* $\Delta_N u \in$ $iL_{loc}^1(B^*(O,1))$ *in the sense of distributions in* $B^*(O,1)$, *be nonnegative and satisfy*

$$\operatorname{div}\left(|\nabla u|^{N-2}\nabla u\right) \leq 0 \quad a.e. \ in \ B(O,1);$$ (3.1.35)

then $u^\gamma \in L^1(B(O,1))$ *for any* $\gamma > 0$ *and* $|\nabla u|^\gamma \in L^1(BO,1))$ *for any* $\gamma \in (0,N)$. *Moreover, if* $\lim_{x \to O} u(x) = \infty$, *there exist a nonnegative constant* β *and a function* $g \in L^1(B(O,R)$ *such that*

$$-\operatorname{div}\left(|\nabla u|^{N-2}\nabla u\right) = g + \beta\delta_O \quad in \ \mathcal{D}'(B(O,1)).$$ (3.1.36)

Proof. Let $g(x) = -\Delta_N u(x)$ for almost all x in $B(O,1)$. Since

$$c_{1,N}(B(O,\sigma)) = N^{-1}|S^{N-1}|(\ln(R/\sigma))^{1-N} ,$$ (3.1.37)

we deduce as in Lemma 5.9 that

$$\min_{|x|=\sigma} u(x) \leq C_1 + C_2|\ln\sigma| \qquad (\forall\sigma \in (0,1/2)).$$ (3.1.38)

Now the weak Harnack inequality is available for any $\gamma > 0$, and (3.1.13) is replaced by

$$\int_{B(O,\sigma)} u^\gamma dx \leq C\sigma^N|\ln\sigma|^\gamma,$$ (3.1.39)

for σ small enough and $C = C(\gamma,N,u)$. The function $u_k = \min(u,k)$ is super-N-harmonic in $B^*(O,1)$, therefore it follows from [LM] that, for any $x_0 \in B^*(O,1/2)$ and $0 < r < \rho < |x_0|$,

$$\int_{B(x_0,r)}\left|\nabla u_k/(1+u_k)\right|^N dx \leq (N/(N-1))^N N^{-1}|S^{N-1}|(\ln(\rho/r))^{1-N}.$$ (3.1.40)

Letting k go to infinity implies that (3.1.40) also holds with u_k replaced by u. By a covering argument, there exists C independent of σ such that

$$\int_{\{\sigma/2<|x|<3\sigma/2\}}\left|\nabla u/(1+u)\right|^N dx \leq C.$$ (3.1.41)

From Hölder's inequality, we obtain

$$\int_{\{\sigma/2<|x|<3\sigma/2\}}|\nabla u|^{\gamma}\,dx$$

$$\leq\left(\int_{\{\sigma/2<|x|\}}|\nabla u/(1+u)|^{N}\,dx\right)^{\gamma/N}\left(\int_{\{\sigma/2<|x|<3\sigma/2\}}(1+u)^{\gamma N/(N-\gamma)}\,dx\right)^{(N-\gamma)/N}, \tag{3.1.42}$$

for any $\gamma\in(0,N)$. Consequently, from (3.1.39), we deduce that for σ small enough there holds

$$\int_{\{\sigma/2<|x|<3\sigma/2\}}|\nabla u|^{\gamma}\,dx\leq C\sigma^{N-\gamma}|\ln\sigma|^{\gamma}. \tag{3.1.43}$$

Replacing σ by $2^{-\ell}\sigma$, we conclude by summation that $|\nabla u|^{\gamma}\in L^{1}(B(O,1))$, with the estimate

$$\int_{B(O,\sigma)}|\nabla u|^{\gamma}\,dx\leq C\sigma^{N-\gamma}|\ln\sigma|^{\gamma}\qquad(\forall\sigma\in(0,1/2)), \tag{3.1.44}$$

where $C=C(\gamma,N,u)$ and $\gamma\in(0,N)$.

We suppose now that $\lim_{x\to O}u(x)=\infty$. Then for any integer n, there exists $\varepsilon>0$ such that

$$\int_{B(O,2\varepsilon)\{u<n+1\}}(n+1-u)|\zeta_{\varepsilon}\nabla u|^{N-1}|\nabla\zeta_{\varepsilon}|\,dx=0. \tag{3.1.45}$$

Therefore the conclusion of Lemma 5.8 follows with

$$\int_{B_{\rho,\alpha,k}}|\nabla u|^{N}\,dx\leq c_{\rho}\alpha, \tag{3.1.46}$$

for $k\geq0$ and $\alpha>0$. In Lemma 5.10, (3.1.22) has to be replaced by

$$\int_{B(O,\sigma)}|\nabla u|^{N-1}\,dx\leq C\sigma|\ln\sigma|^{N-1}, \tag{3.1.47}$$

and (3.1.24) by

$$\sum_{|\theta|\leq m}\beta_{\theta}^{2}\varepsilon^{-|\theta|}\leq C|\ln\varepsilon|^{N-1}, \tag{3.1.48}$$

This, again, implies that $\beta_{\theta}=0$ for $|\theta|\geq1$. As for β_{0}, it is nonnegative as in Lemma 5.10.

5-3-2 Isolated singularities in the sub-critical case

Let Ω be an open subset of $\mathbb{R}^{N}$ containing O and $\Omega^{*}=\Omega\setminus\{O\}$. We consider the following equation in Ω^{*}, in the range of values of p and q in which Serrin's results (see Chapter 1) do not apply, that is $1<p<q-1$,

$$-\mathrm{div}\left(|\nabla u|^{p-2}\nabla u\right)=u^q, \qquad u>0. \tag{3.2.1}$$

If we look for solutions of (3.2.1) under the form $u(r)=\lambda r^\alpha$, then

$$\alpha=-\frac{p}{q+1-p} \quad \text{and} \quad \lambda^{q+1-p}=\left(\frac{p}{q+1-p}\right)^{p-1}\left(N-\frac{pq}{q+1-p}\right), \tag{3.2.2}$$

but the last term exists if and only if

$$1<p<N \quad \text{and} \quad q>\frac{N(p-1)}{N-p}. \tag{3.2.3}$$

By reference to the semilinear case, the subcritical case of equation (3.2.1) corresponds to

$$\begin{cases} \text{(i)} \quad p-1<q<\dfrac{N(p-1)}{N-p} \quad \text{if} \quad 1<p<N, \\[2mm] \text{(ii)} \quad p-1<q \qquad\qquad \text{if} \quad p\geq N. \end{cases} \tag{3.2.4}$$

Hereafter we suppose that $\overline{B}(O,1)\subset\Omega$. The following result is proved by combining techniques of [GaV] and [B.V1],

THEOREM 5.11. *Let* p *and* q *satisfy (3.2.4) and* $u\in C^1(\Omega^*)$ *be a nonnegative solution of (3.2.1) in* Ω^**. Then one of the two following possibilities must occur:*
(i) $\lim\limits_{x\to 0} u(x)$ *exists and* u *can be extended by continuity as a solution of (3.1.1) in* Ω,
(ii) *there exists* $\alpha>0$ *such that* $\lim\limits_{x\to 0} u(x)/\mu_p(x)=\alpha$ *and* u *satisfies*

$$-\mathrm{div}\left(|\nabla u|^{p-2}\nabla u\right)=u^q+\alpha^{p-1}\delta_o \qquad in \ \mathcal{D}'(\Omega). \tag{3.2.5}$$

Moreover

$$\lim_{x\to 0}\left(\frac{x}{|x|}|x|^{(N-1)/(p-1)}\nabla u(x)\right)=-\alpha\left|S^{N-1}\right|^{-1/(p-1)}, \tag{3.2.6}$$

and for $0<|x|\leq 1$, *we have*

$$-C\leq u(x)-\alpha\mu_p(x)\leq C(1+|x|^{(N(p-2)+p-(N-p)q)(p-1)}), \tag{3.2.7}$$

if $p<N$, $q\neq(N(p-2)+p)/(N-p)$,

$$-C\leq u(x)-\alpha\mu_p(x)\leq C(1+|\ln|x||), \tag{3.2.8}$$

if $p < N$, $q = (N(p-2)+p)/(N-p)$,

$$-C \leq u(x) - \alpha \mu_N(x) \leq C, \tag{3.2.9}$$

if $p = N$, *where* $C = C(N,p,q,u)$.

Proof. We write (3.2.1) under the following form

$$\mathrm{div}\left(|\nabla u|^{p-2}\nabla u\right) + du^{p-1} = 0, \tag{3.2.10}$$

where $d = u^{q+1-p}$.

Step 1. We claim that either $u(x)$ admits a finite limit at O, or there exists $C > 0$ such that

$$C^{-1} \leq u(x)/\mu_p(x) \leq C, \tag{3.2.11}$$

holds near O. From Lemma 5.9 and Theorem 5.10, $u^\gamma \in L^1(B(O,R))$ for any $\gamma \in (0, Np/(N-p))$ (no condition if $N = p$) Therefore there exists $\delta > 0$ such that $d \in L^{N/(p-\delta)}(B(O,1))$ and we deduce the claim from Theorem 1.21. Hereafter we suppose that (3.2.11) holds.

Step 2. There exists $\alpha > 0$ such that

$$\lim_{x \to O} \frac{u(x)}{\mu_p(x)} = \alpha. \tag{3.2.12}$$

We follow the proof of [GaV – Theorem 1.2]. Set $\alpha = \limsup_{x \to O} u(x)/\mu_p(x)$, then there exists a sequence $\{x_n\}$ tending to O when n goes to infinity such that $\alpha = \lim_{n \to \infty} u(x_n)/\mu_p(x_n)$. We set $\delta_n = |x_n|$, $\xi_n = x_n/\delta_n$ and we can suppose that $\alpha_n = u(x_n)/\mu_p(x_n) = \max_{|x|=\delta_n} u(x)/\mu_p(x)$. If we define u_n on $B(O,1/\delta_n)\setminus\{O\}$ by

$$u_n(\xi) = u(\delta_n \xi)/\mu_p(\delta_n), \tag{3.2.13}$$

then u_n satisfies

$$-\mathrm{div}\left(|\nabla u_n|^{p-2}\nabla u_n\right) = C_n u_n^q \tag{3.2.14}$$

in this punctured ball, with $C_n = \delta_n^p(\mu_p(\delta_n))^{q+1-p}$. From Lemma 5.3 and Remark 5.6, there holds

$$|\nabla u(x)| \leq C|x|^{-1}\mu_p(x), \tag{3.2.15}$$

$$|\nabla u(x) - \nabla u(x')| \leq C|x|^{-1-\alpha}|x-x'|\mu_p(x), \qquad (3.2.16)$$

for $0 < |x| \leq |x'| \leq 1$. Therefore $\{u_n\}$ remains bounded in the $C^{1,\alpha}_{loc}(\mathbb{R}^N \setminus \{O\})$-topology and there exists a subsequence $\{u_{n_k}\}$ and a p-harmonic function v such that $\{u_{n_k}\}$ converges to v in the $C^1_{loc}(\mathbb{R}^N \setminus \{O\})$-topology (notice that C_n tends to 0). Moreover v is nonnegative. Consequently, if $p = N$, v is constant with value α (we take $|\xi| = 1$ in (3.2.13)). For $\varepsilon > 0$ there exists n_O such that $\alpha_n \geq \alpha - \varepsilon$ for any $n \geq n_O$. Comparing u and $(\alpha - \varepsilon)\mu_N$ in the spherical shell $\{x: \delta_p < |x| < \delta_{n_O}\}$ yields $u \geq (\alpha - \varepsilon)\mu_N$. Therefore

$$\liminf_{x \to O} u(x)/\mu_N(x) \geq (\alpha - \varepsilon) \qquad (3.2.17)$$

and we get (3.2.12). In the case $1 < p < N$, v satisfies

$$c^{-1}\mu_p(x) \leq v(x) \leq c\mu_p(x), \qquad (3.2.18)$$

and therefore $v = \lambda\mu_p$ for some $\lambda > 0$. Clearly $\lambda = \alpha / \mu_p(1)$ and finally (3.2.17) holds as above. The proof of (3.2.5) and (3.2.6) follows as in Theorem 5.5.

Step 3. In order to prove (3.2.7)-(3.2.9), for any $\varepsilon > 0$ there exists $a > 0$ such that $u(x) \geq (\alpha - \varepsilon)\mu_p(x)$ in $B(O,a)$. Letting ε go to 0 yields $u(x) \geq \alpha\big(\mu_p(x) - \mu_p(\rho)\big)$ in $B^*(O,1)$. Moreover there is a $K > 0$ such that $u \leq K\big(\mu_p + 1\big)$ in $B^*(O,1)$. If we compare u with the radial solution of

$$\begin{cases} -\mathrm{div}\big(|\nabla u_\varepsilon|^{p-2}\nabla u_\varepsilon\big) = \big((K+1)\mu_p\big)^q & \text{in } B(O,1), \\[2mm] \lim_{r \to 0} u_\varepsilon(r)/\mu_p(r) = \alpha + \varepsilon, \\[2mm] u_\varepsilon(1) = k = \max_{|x|=1} u(x), \end{cases} \qquad (3.2.19)$$

then $u \leq u_\varepsilon$; as for u_ε, it is explicitely given by

$$u_\varepsilon(r) = \begin{cases} k + \alpha_\varepsilon C'(p,N)\int_r^1 s^{(1-N)/(p-1)}\big(1 + M_\varepsilon s^{N-q(N-p)/(p-1)}\big)^{1/(p-1)} ds & \text{if } p < N, \\[4mm] k + \alpha_\varepsilon C'(N)\int_r^1 s^{-1}\left(1 + N_\varepsilon \int_0^s t^{N-1}(\ln(1/t))^q dt\right)^{1/(N-1)} ds & \text{if } p = N, \end{cases} \qquad (3.2.20)$$

with $\alpha_\varepsilon = \alpha + \varepsilon$ and where M_ε and N_ε have finite limit when ε goes to 0. From the maximum principle $u \leq u_\varepsilon$, and the following estimates hold for u_ε

$$
\left| u_\varepsilon - \alpha_\varepsilon \mu_p \right| \leq
\begin{cases}
C(1 + r^N \mu_p^{q+1}(r)) & \text{if } p < N \text{ and } q \neq (N(p-2)+p)/(N-p), \\
C(1 + |\ln r|) & \text{if } p < N \text{ and } q = (N(p-2)+p)/(N-p), \qquad (3.2.21) \\
C & \text{if } p = N.
\end{cases}
$$

Letting ε go to 0 yields (3.2.7)-(3.2.9), which completes the proof.

We end this Section with a result of non-existence of positive solutions of (3.2.1) in an exterior domain ([B.V1]). We start with the following radial result due to [GaV]

LEMMA 5.12. *Let* $1 < p \leq N$ *and* $p - 1 < q \leq N(p-1)/(N-p)$ *(*$p - 1 < q$ *if* $p = N$*). Then there exists no positive radial solution of* (3.2.1) *in* $\{x : |x| > 1\}$.

Proof. If ψ is such a solution and $1 < p < N$, we perform the change of variable (2.3.11), $\phi(s) = \psi(r)$, $s = r^{(p-1)/(N-p)}$. Then

$$
-\left(|\phi_s|^{p-2} \phi_s \right)_s = \left(\frac{p-1}{N-p} \right)^p s^{-p(N-1)/(N-p)} \phi^q \qquad \text{in } (0,1]. \qquad (3.2.22)
$$

Since ϕ is concave it admits a nonnegative limit at 0. Let β be this limit. If β is nonzero, then (3.2.22) implies that there exists c_1 such that, for s small enough, there holds

$$
c_1 s^{-N(p-1)/(N-p)} \leq \left(|\phi_s|^{p-2} \phi_s \right)(s) \leq \frac{1}{c_1} s^{-N(p-1)/(N-p)}. \qquad (3.2.23)
$$

Hence $\phi_s > 0$ on $(0, s_0]$ and

$$
c_2 s^{-N/(N-p)} \leq \phi_s(s) \leq \frac{1}{c_2} s^{-N/(N-p)}, \qquad (3.2.24)
$$

which implies $\lim_{s \to 0} \phi(s) = -\infty$. Consequently $\lim_{s \to 0} \phi(s) = 0$, $\phi(s) \geq s \phi_s(s) > 0$ on some $(0, s_0]$ and (3.2.22) yields

$$
\left(\phi_s^{p-1} \right)_s + \left(\frac{N-p}{p-1} \right)^p s^{q-p(N-1)/(N-p)} \phi_s^q \leq 0. \qquad (3.2.25)
$$

Set $\phi_s^{p-1} = \rho$, then by integration,

$$
(\rho(s))^{-(q+1-p)/(p-1)} \geq c' \begin{cases} s^{q-N(p-1)/(N-p)} & \text{for } q < N(p-1)/(N-p), \\ \ln(1/s) & \text{for } q = N(p-1)/(N-p), \end{cases} \qquad (3.2.26)
$$

for some c' > 0. But $\phi_s^{p-1}(0) = \rho(0) \in (0,\infty]$ by concavity, contradiction. In the case p = N, we proceed similarly by setting $\tau = \ln(1/r)$, $\phi(\tau) = \psi(r)$.

THEOREM 5.12. *Let* p, q, N *be as in Lemma 5.12, then there exists no nonnegative non-zero solution of* (3.2.1) *in an exterior domain of* $\mathbb{R}^N$.

Proof. Let us suppose that such a solution exists. Without any loss of generality, we may suppose that u is a solution in $\{x: |x| > 1\}$ and is continuous up to the boundary. From the strict maximum principle, u is positive in $\{x: |x| \geq 1\}$. For $1 < n$ and $k > 0$ we consider a sequence of radial functions $\{u_{n,k}\}$ which satisfy

$$\begin{cases} -\mathrm{div}\left(|\nabla u_{n,k}|^{p-2} \nabla u_{n,k}\right) = |u_{n,k-1}|^{q-1} u_{n,k-1} & \text{in } \{x: 1 < |x| < n\}, \\[2mm] u_{n,k} = m = \min_{|x|=1} u(x) & \text{for } |x| = 1, \\[2mm] u_{n,k} = 0 & \text{for } |x| = n, \end{cases} \qquad (3.2.27)$$

with $u_{n,0} = 0$. From the maximum principle $u_{n,k} \leq u_{n,k+1} \leq u$. From (3.2.27), the set of functions $\left\{r^{N-1}|(u_{n,k})_r|^{p-2}(u_{n,k})_r\right\}_{k \geq 0}$ is equicontinuous on $[1,n]$ and therefore $\{u_{n,k}\}$ converges, when k goes to infinity, to some radial function u_n which belongs to $C^{1,\alpha}([1,n])$ and satisfies

$$\begin{cases} -\mathrm{div}\left(|\nabla u_n|^{p-2} \nabla u_n\right) = |u_n|^{q-1} u_n & \text{in } \{x: 1 < |x| < n\}, \\[2mm] u_n = m = \min_{|x|=1} u(x) & \text{for } |x| = 1, \\[2mm] u_n = 0 & \text{for } |x| = n, \end{cases} \qquad (3.2.28)$$

Moreover $0 < u_{n,k} \leq u_n \leq u$. We consider now the sequence $\{u_n\}_{n>1}$. From the previous estimates and the regularity results, there exist a subsequence $\{u_{n_k}\}$ and a function $\tilde{u}$ such that $\lim_{n_k \to \infty} u_{n_k} = \tilde{u}$ in the $C^1_{loc}(x: |x \geq 1|)$-topology. The function $\tilde{u}$ is a radial solution of (3.2.1) in $\{x: |x| > 1\}$, it is not zero since minorized by any $u_{n,k}$ and it is dominated by u. This contradicts Lemma 5.12.

5-3-3 Isolated singularities in the super-critical case

In the super-critical case $q \geq N(p-1)/(N-p)$ of equation (3.2.1), only radial results are known concerning the singularities or the asymptotic behavior of solutions : Guedda and Véron studied the positive case when $N(p-1)/(N-p) \leq q < (N(p-1)+p)/(N-p)$, in [GaV], but the full description, with no restriction on the sign of u and of the value of q, has been provided by Bidaut-Véron [B.V1]. Without too many details, we shall give an outline of the techniques and results. The radial form of (3.2.1) is

$$r^{1-N}\left(r^{N-1}|u_r|^{p-2}u_r\right)_r + |u|^{q-1}u = 0,\tag{3.3.1}$$

where r belongs to some interval I of the form $(0,1]$, $[1,\infty)$ or $(0,\infty)$. Let E be the classical energy function (with $p' = p/(p-1)$)

$$E_u(r) = \frac{|u_r|^p}{p'} + \frac{|u|^{q+1}}{q+1},\tag{3.3.2}$$

and $(E_u)_r = -(N-1)|u_r|^p/r$. Setting again $\phi(s) = u(r)$, $s = r^{(p-1)/(N-p)}$ (in the case $1 < p < N$), then ϕ satisfies (3.2.22) and we obtain a new energy function

$$F_\phi(s) = \frac{|\phi_s|^p}{p'} + \left(\frac{p-1}{N-p}\right)^p s^{-p(N-1)/(N-p)}\frac{|\phi|^{q+1}}{q+1},\tag{3.3.3}$$

with derivative $(F_\phi)_s = -\dfrac{p(N-1)}{N-p}\left(\dfrac{p-1}{N-p}\right)^p s^{-p(N-1)/(N-p)-1}\dfrac{|\phi|^{q+1}}{q+1}$. In the case $p = N$, the logarithm change of variable yields a somewhat similar expression. Consequently, for any $r_0 > 0$ and $(u_0,u_1)\in\mathbb{R}\times\mathbb{R}$, there exists a unique solution of (3.3.1) on $(0,\infty)$ with $u(r_0) = u_0$, $u_r(r_0) = u_1$.

The equation (3.3.1) can also be transformed into an autonomous equation by setting

$$\delta = p/(p+q-1), \qquad u(r) = r^{-\delta}w(t), \qquad t = \ln(1/r),\tag{3.3.4}$$

and w satisfies

$$\left(|w_t + \delta w|^{p-2}(w_t + \delta w)\right)_t - (N-\delta q)|w_t + \delta w|^{p-2}(w_t + \delta w) + |w|^{q-1}w = 0.\tag{3.3.5}$$

If we denote

$$y(t) = -r^{-(\delta+1)/(p-1)}|u_r|^{p-2}u_r,\tag{3.3.6}$$

then (w, y) is solution of an autonomous system, namely

$$\begin{cases} w_t = -\delta w + |y|^{(2-p)/(p-1)}y, \\ y_t = -|w|^{q-1}w + (N-\delta q)y. \end{cases}\tag{3.3.7}$$

Define now

$$\begin{cases} \text{(i)} \quad A = N - \delta(q+1) = ((N-p)q - (N(p-1)+p))/(q+1-p), \\ \text{(ii)} \quad B = N - \delta q = ((N-p)q - N(p-1))/(q+1-p), \end{cases} \qquad (3.3.8)$$

and

$$\begin{cases} \text{(i)} \quad V(w,y) = \dfrac{|y|^{p'}}{p'} - \delta wy - A\dfrac{\delta^{p-1}}{p}|w|^p + \dfrac{|w|^{q+1}}{q+1}, \\[3mm] \text{(ii)} \quad W(w,y) = \dfrac{|y|^{p'}}{p'} - Bwy + \dfrac{q}{q+1}A|B|^{(1-q)/q}B|y|^{(q+1)/q} + \dfrac{|w|^{q+1}}{q+1}, \end{cases} \qquad (3.3.9)$$

and finally a straightforward computation yields the following lemma.

LEMMA 5.13. *Set* $\mathbf{V}(t) = V(w(t),y(t))$ *and* $\mathbf{W}(t) = W(w(t),y(t))$, *then*

$$\mathbf{V}_t = AX(t), \qquad \mathbf{W}_t = AZ(t), \qquad (3.3.10)$$

where

$$\begin{cases} \text{(i)} \quad X = \left(|\delta w|^{p-2}\delta w - y\right)\left(\delta w - |y|^{(2-p)/(p-1)}y\right) \geq 0, \\[2mm] \text{(ii)} \quad Z = \left(w - |By|^{(1-q)/q}By\right)\left(|w|^{q-1}w - By\right) \geq 0. \end{cases} \qquad (3.3.11)$$

PROPOSITION 5.3. *Let* (w,y) *be a solution of* (3.3.7) *on* $[1,\infty)$. *If* w *is nonnegative, or if* $q \leq (N(p-1)+p)/(p-1)$, *then* (w,y) *remains uniformly bounded.*

Proof. In the case $q \leq (N(p-1)+p)/(p-1)$, the coefficient A is nonpositive, $\mathbf{V}$ is nonincreasing and, from Young's inequality,

$$\frac{|w(t)|^{q+1}}{q+1} - \delta^p \frac{|w(t)|^p}{p} \leq \mathbf{V}(1), \qquad (3.3.12)$$

for $t \geq 1$. When w remains positive, we just use a concavity argument as in Lemma 5.12.

PROPOSITION 5.4. *Suppose* $q \in (p-1,\infty) \setminus (N(p-1)+p)/(p-1)$ *and* w *is a bounded solution of* (3.3.7) *on* $[0,\infty)$. *Then* $\lim\limits_{t\to\infty} w_t(t) = 0$, $\lim\limits_{t\to\infty} w(t) = \ell$ *and*

$$\ell\left(|\ell|^{q+1-p} - \delta^{p-1}B\right) = 0. \qquad (3.3.13)$$

Proof. From Lemma 5.13, $\mathbf{V}$ and $\mathbf{W}$ are monotonous functions of t and they have finite limit at infinity. Since $A \neq 0$, the functions X and Z are integrable on $[1,\infty)$. If $1 < p \leq 2$ we have the following inequality

$$\left(|a|^{p-2}a - |b|^{p-2}b\right)(a - b) \geq c_p(a - b)^2\left(|a| + |b|\right)^{p-2}, \tag{3.3.14}$$

therefore

$$X(t) \geq c_p w_t^2(t)\left(\delta|w(t)| + |y(t)|^{1/(p-1)}\right)^{p-2}. \tag{3.3.15}$$

and because $\left(\delta|w(t)| + |y(t)|^{1/(p-1)}\right)^{2-p}$ is bounded, we have $\int_1^\infty w_t^2 dt < \infty$. From the equation w_t and w_{tt} are both bounded, and finally $\lim\limits_{t\to\infty} w_t(t) = 0$. From (3.3.9) and the monotonicity of $\mathbf{V}$ and $\mathbf{W}$, we deduce

$$\lim_{t\to\infty} \mathbf{V}(t) = \lim_{t\to\infty}\left(\frac{|w|^{q+1}}{q+1} - B\delta^{p-1}\frac{|w|^p}{p}\right)(t). \tag{3.3.16}$$

Therefore w admits a limit ℓ at infinity with

$$\lim_{t\to\infty} y(t) = \delta^{p-1}|\ell|^{p-2}\ell, \qquad \lim_{t\to\infty} y_t(t) = -|\ell|^{q-1}\ell + B\delta^{p-1}|\ell|^{p-2}\ell \tag{3.3.17}$$

and $\lim\limits_{t\to\infty} y_t(t) = 0$, which implies (3.3.13). If $p > 2$, $1 + 1/q < 2$ and (3.3.14 is replaced by

$$\left(|a|^{1/q-1}a - |b|^{1/q-1}b\right)(a - b) \geq c_q(a - b)^2\left(|a| + |b|\right)^{1/q-1}, \tag{3.3.18}$$

and

$$Z(t) \geq c_q y_t^2(t)\left(|w(t)|^q + |By(t)|\right)^{1/q-1}. \tag{3.3.19}$$

As above $\int_1^\infty y_t^2 dt < \infty$. Since $y_{tt} = -(q-1)|w|^{q-1}w_t + By_t$, y_{tt} is bounded; y_t goes to 0 at infinity and y admits a limit, so does w. The completion of the proof is as above.

With this Proposition it is possible to give a full characeterisation of the behaviour near O of any radial solution of (3.2.1); the detailed proofs are in [B.V1].

THEOREM 5.13. *Suppose* $q > (N(p-1) + p)/(N - p)$ *and* u *is any radial non trivial solution of (3.2.1) in* $B^*(O,1)$. *Then we have three possibilites near* O:
(i) $u(r) \equiv \pm\lambda_{N,p,q}r^{-q/(q+1-p)}$ *where*

$$\lambda_{N,p,q} = \left(\left(\frac{p}{q+1-p} \right)^{p-1} \left(N - \frac{pq}{q+1-p} \right) \right)^{1/(q+1-p)}, \qquad (3.3.20)$$

(ii) u *is regular at* O *with*

$$\lim_{r\to 0} u(r) = \alpha \neq 0, \qquad \lim_{r\to 0} r^{-1}\left(|u|^{p-2}u \right)(r) = -N^{-1}|\alpha|^{q-1}\alpha, \qquad (3.3.21)$$

(iii) $r^{p/(q+1-p)}u(r)$ *is not bounded,* u *oscillates and intersects the graphs* $\left(r, \pm\lambda_{N,p,q} r^{-q/(q+1-p)} \right)$, $r > 0$, *infinitely many times.*

THEOREM 5.14. *Suppose* $N(p-1)/(N-p) < q < (N(p-1)+p)/(N-p)$ *and* u *is any non-trivial solution of* (3.2.1) *in* $B^*(O,1)$. *Then we have three possibilites near* O:
(i) $u(r) \equiv \pm\lambda_{N,p,q} r^{-q/(q+1-p)}$ *where*

$$\lambda_{N,p,q} = \left(\left(\frac{p}{q+1-p} \right)^{p-1} \left(N - \frac{pq}{q+1-p} \right) \right)^{1/(q+1-p)}, \qquad (3.3.22)$$

(ii) u *is regular at* O *with* (3.3.20),
(iii) $r^{p/(q+1-p)}u(r)$ *is nonconstant and* $\displaystyle\lim_{r\to 0} r^{p/(q+1-p)}u(r) = \lambda_{N,p,q}$ *(or by symmetry* $\displaystyle\lim_{r\to 0} r^{p/(q+1-p)}u(r) = -\lambda_{N,p,q}$ *).*

Moreover all those solutions exist. There are two critical cases which correspond to $q = N(p-1)/(N-p)$ and $q = (N(p-1)+p)/(N-p)$. When $q = N(p-1)/(N-p)$, the blow-up rate of the fundamental solutions of the p-Laplace equation is equal to $p/(q+1-p)$ and it appears an additional logarithm damping term in the blow-up at O.

THEOREM 5.15. *Suppose* $q = (N(p-1)+p)/(N-p)$. *Then, to within exchange of* u *and* -u, *each radial nontrivial solution of* (3.2.1) *in* $B^*(O,1)$ *has one of the following forms*
(i) u *is regular near* O,
(ii) u *is singular near* O *and*

$$\lim_{r\to 0} r^{(N-p)/(p-1)}\left(\ln(1/r) \right)^{(N-p)/p(p-1)} u(r) = \left(\frac{N-p}{p}\left(\frac{N-p}{p-1} \right)^{p-1} \right)^{(N-p)/p(p-1)}. \qquad (3.3.23)$$

When $q = (N(p-1)+p)/(N-p)$, the renormalized system in (w,y) is conservative since the two energy functions $\mathbf{V}$ and $\mathbf{W}$ are constant and there exists some K such that

$$\frac{p-1}{p}|y(t)|^p - \delta w(t)y(t) + \frac{|w(t)|^{q+1}}{q+1} = K, \qquad (3.3.24)$$

for $t \in \mathbb{R}$.

THEOREM 5.16. *Suppose* $q = (N(p-1)+p)/(N-p)$. *Then, to within exchange of* u *and* -u, *each radial nontrivial solution of* (3.2.1) *in* $B^*(O,1)$ *has one of the following forms* :

  (i) u *is regular near* O,

 (ii) $u(r) \equiv \lambda_{N,p,q} r^{-q/(q+1-p)} = \lambda_{N,p,q} r^{-(N-p)/p}$,

 (iii) $r^{p/(q+1-p)} u(r)$ *is positive and almost periodic near* $r = 0$ *and oscillates between two values* a *and* b *with* $0 < a < \lambda_{N,p,q} < b$,

 (iv) $r^{p/(q+1-p)} u(r)$ *is almost periodic and oscillates near* $r = 0$ *between two values* b *and* -b *with* $0 < \lambda_{N,p,q} < b$; *it takes infinitely many times the values* $\lambda_{N,p,q}$ *and* $-\lambda_{N,p,q}$.

5-4 Singular solutions of the capillarity equation

5-4-1 Existence results

As we have seen it in Chapter 1, there exists no isolated singularity for the solutions of the mean curvature equation (see Theorem 1.20); for the capillarity equation in $\mathbb{R}^N \setminus \{O\}$,

$$\text{div}\left(\frac{\nabla v}{\sqrt{1+|\nabla v|^2}} \right) = Kv, \qquad (4.1.1)$$

the situation is the same when $K > 0$, but completely different when $K < 0$. In this last case the solution represents the surface of a pendent water drop. We shall not go into the very rich details of the theory of pendent water drops (see [CF3]) but present the very surprising results of existence and uniqueness of singular radial solutions of (4.1.1); these results are originaly due to Concus and Finn ([CF1], [CF2]) and have been simplified and improved by Bidaut-Véron [B.V3] . Assuming that K is a negative constant, we define u from v by

$$u(x) = \sqrt{-K/(N-1)}\, v\left(-\left(\sqrt{(N-1)/K}\right)x\right), \qquad (4.1.2)$$

and since v is radial, u satisfies

$$-\left(\frac{r^{N-1}u_r}{\sqrt{1+u_r^2}} \right)_r = (N-1)r^{N-1}u. \qquad (4.1.3)$$

In their first results, Concus and Finn [CF1] proved the existence of a local solution of (4.1.3) under the form

$$u(r) = -\frac{1}{r} + \frac{N+3}{2(N-1)}r^3 + r^3\varepsilon(r), \qquad (4.1.4)$$

where in fact $\varepsilon(r) = o(r)$. The technique is to seek a solution u in the form

$$u(r) = -\frac{1}{r} + \frac{N+3}{2(N-1)}r^3 + r^3\alpha_0(r), \qquad (4.1.5)$$

for $0 < r \le r_0$, with $\lim_{r \to 0}\alpha_0(r) = 0$. The key point is to introduce a vector valued operator $\mathbf{T}$ which transforms $\alpha = (\alpha_0, \alpha_1)$ into $\mathbf{T}\alpha = \big(T_0(\alpha_0, \alpha_1), T_1(\alpha_0, \alpha_1)\big)$ defined as follows: if α is a continuous vector valued function defined on $I_0 = [0, r_0]$, we denote

$$\begin{aligned}
\mathcal{F}(\alpha(r), r) &= \frac{2}{r^3} - 3\frac{N+3}{N-1}r + \frac{(N+2)(N+6)}{4}\alpha_0 r + 3(N+5)\alpha_1 r \\
&\quad + \frac{N-1}{r^3}\alpha_0 - \frac{N-1}{r}\gamma_1(1+\gamma_1^2) - (N-1)\gamma_0(1+\gamma_1^2)^{3/2},
\end{aligned} \qquad (4.1.6)$$

with

$$\begin{cases}
\text{(i)} \quad \gamma_0 = \gamma_0(\alpha_0(r), r) = -\frac{1}{r} + \frac{N+3}{2(N-1)}r^3 + \alpha_0(r)r^3, \\[2mm]
\text{(ii)} \quad \gamma_1 = \gamma_1(\alpha_1(r), r) = \frac{1}{r^2} + \frac{3(N+3)}{2(N-1)}r^2 + 3\alpha_1(r)r^2.
\end{cases} \qquad (4.1.7)$$

Then we set

$$\mathcal{S}[f](r) = \int_0^r f(\tau)\sin\left(\frac{\sqrt{N-1}}{2}\left(\frac{1}{\tau^2} - \frac{1}{r^2}\right)\right)d\tau, \qquad (4.1.8)$$

$$\mathcal{C}[f](r) = \int_0^r f(\tau)\cos\left(\frac{\sqrt{N-1}}{2}\left(\frac{1}{\tau^2} - \frac{1}{r^2}\right)\right)d\tau. \qquad (4.1.9)$$

If any function f defined on $(0, \infty)$ is written as

$$^{(q)}f(r) = r^{(N+q)/2}f(r), \qquad (4.1.10)$$

then T_0 and T_1 are defined by

$$T_0(\alpha_0, \alpha_1) = \frac{1}{\sqrt{N-1}}r^{-(N+8)/2}\mathcal{S}\big[^{(8)}\mathcal{F}(\alpha(.),.)\big], \qquad (4.1.11)$$

$$T_1(\alpha_0, \alpha_1) = -\frac{N+2}{6}T_0(\alpha_0, \alpha_1) + \frac{1}{3}r^{-(N+12)/2}\mathcal{C}\big[^{(8)}\mathcal{F}(\alpha(.),.)\big]. \qquad (4.1.12)$$

A lengthy and formal computation shows that for any C^2 function u defined by (4.1.5), the relations (4.1.3) and (4.1.13) below

$$\frac{d}{dr}\gamma_0(\alpha_0(r),r) = \gamma_1(\alpha_1(r),r),\tag{4.1.13}$$

are equivalent to

$$\mathbf{T}\alpha = \alpha.\tag{4.1.14}$$

The space $C(I_0,\mathbb{R}\times\mathbb{R})$ is endowed with the norm

$$\|\alpha\| = \|(\alpha_0,\alpha_1)\| = \max_{0\le r\le r_0}\left|p_0^{-1}(r)\alpha_0(r)\right| + \max_{0\le r\le r_0}\left|\alpha_1(r)\right|,\tag{4.1.15}$$

where p_0 is a nondecreasing positive function defined in $(0,1]$ such that

$$\lim_{r\to 0}p_0(r) = 0,\qquad \lim_{r\to 0}\frac{r^4}{p_0(r)} = 0.\tag{4.1.16}$$

Let $\mathcal{A}_M = \left\{\alpha\in C(I_0,\mathbb{R}\times\mathbb{R}): \|\alpha\|\le M\right\}$ for $M > 0$. The existence result in [CF1] is the following:

THEOREM 5.17. *There exist* M *and* $r_0 > 0$ *such that the mapping* $\mathbf{T}$ *is a strict contraction from* $\mathcal{A}_M$ *into itself. Consequently it admits a fixed point and therefore there exists a singular solution of* (4.1.3) *in* $B(O,r_0)$ *under the form* (4.1.5).

This result has been considerably simplified and extended by Bidaut-Véron [B.V2] who introduced in this problem the angle $\psi = \psi(r)$ between the tangent at the curve $(r,u(r))$ and the r-axis by setting

$$z(r) = \frac{u_r(r)}{\sqrt{1+u_r^2(r)}} = \sin\psi(r),\tag{4.1.17}$$

and z satisfies

$$z_r(r) + \frac{N-1}{r}z(r) = -(N-1)u(r),\tag{4.1.18}$$

subject to the boundary conditions

$$\lim_{r\to 0}z(r) = 1,\qquad \lim_{r\to 0}z_r(r) = 0.\tag{4.1.19}$$

The problem of finding a solution is therefore reduced to a fixed-point theorem on w for an operator $\mathbf{L}$ where

$$z(r) = 1 - r^4/2 + r^6 w(r) \tag{4.1.20}$$

and

$$(\mathbf{L}w)(r) = \frac{r^{-(N+8)/2}}{\sqrt{N-1}} \int_0^r \tau^{(N+2)/2} F(w(\tau),\tau) \sin\left(\frac{\sqrt{N-1}}{2}\left(\frac{1}{\tau^2} - \frac{1}{r^2}\right)\right) d\tau. \tag{4.1.21}$$

In this formula F and $\varPhi$ are defined by

$$F(w,r) = 2(N+2)r^2 + \frac{N(N-4)}{4} r^4 w + (N-1)w + \varPhi(1 - r^4/2 + r^6 w, r), \tag{4.1.22}$$

$$\varPhi(y,r) = (N-1)\left(\frac{y}{r^2} - \frac{y}{\sqrt{1-y^2}}\right). \tag{4.1.22}$$

Denoting $B_{M,r_0} = \left\{v \in C([0,r_0]) \ \|v\| = \max_{0 \le r \le r_0}|v(r)| \le M\right\}$, then $\mathbf{L}$ is a strict contraction in B_{M,r_0} for some M and r_0, and more precisely,

THEOREM 5.18. *Let* $M < M_0 = (N+8)/(3\sqrt{N-1})$. *Then for* $r_0 > 0$ *small enough the problem* (4.1.17)-(4.1.19) *admits a unique solution* z *belonging to* $C([0,r_0]) \cap C^2((0,r_0])$ *with*

$$\begin{cases} z(r) = 1 - r^4/2 + r^6 w(r), \\ |w(r)| \le M \quad in \quad [0,r_0]. \end{cases} \tag{4.1.23}$$

If we compare the two techniques, and transcript Concus and Finn fixed point theorem into the z variable then we see that they look for a w(r) with $z(r) = 1 - r^4/2 + r^8 w(r)$.

5-4-2 Global existence and uniqueness results

Some uniqueness results are immediate consequence of the contraction mapping principle; however, the class of uniqueness where the fixed point theorem takes place, are very restrictive. In [CF2] and [B.V2] improved uniqueness results are given. By a very delicate analysis of the graph of a solution, the following result is proved in [CF2]:

THEOREM 5.19. *Let* $u(x) = u(r)$ *be a solution of* (4.1.3) *in the punctured ball* $B^*(O,R)$, *then the following alternative holds :*

(i) either can be extended by continuity at O *so that the resulting function is a solution of* (4.1.3)
in the whole ball B(O,R),
(ii) or for any two constants

$$\lambda_0 > \frac{\pi + \sqrt{2}}{\sqrt{N-1}}, \qquad \lambda_1 > \sqrt{\frac{2}{N-1}}, \tag{4.2.1}$$

there holds

$$-\lambda_0 r < U(r) + |u(r)| < \lambda_1 r, \tag{4.2.2}$$

for r *small enough where* U *is the singular solution constructed in Theorem 5.19.*

Unfortunately estimate (4.2.2) is far from the class of uniqueness obtained through the fixed point method in Theorem 5.18. In [B.V2] global existence and uniqueness results are obtained in a widder class than the one of Theorem 5.18.

THEOREM 5.20. *Any solution* z *of* (4.1.17)-(4.1.19), *or equivalently any singular solution* u *of* (4.1.3) *admits a unique extension defined on* $\mathbb{R}^N \setminus \{O\}$.

Proof. We have only to consider the unknown z and set $X = (z, z_r)$ which satisfies a system of the form

$$X_r(r) = G(r, X(r)), \tag{4.2.3}$$

where G is a C^1 function defined on the open set $\Omega = (0, \infty) \times (-1,1) \times \mathbb{R}$. We may suppose that z is the maximal solution defined on some bounded interval $(0, R^*)$ with $R^* < \infty$.

Step 1. We claim that

$$|z_r(r)| < \sqrt{N-1} \min(r, \sqrt{2}), \quad \text{for } 0 < r < R^*. \tag{4.2.4}$$

If we set

$$g(r) = \frac{z_r^2(r)}{2(N-1)} + \frac{1 - z^2(r)}{2r^2} - \sqrt{1 - z^2(r)}, \tag{4.2.5}$$

then

$$g_r(r) = -\frac{z_r^2(r)}{r} - \frac{1 - z^2(r)}{r^3} < 0, \tag{4.2.6}$$

and

$$\left(r^2 g(r)\right)_r = -\left(\frac{N-2}{N-1}z_r^2(r) + 2r\sqrt{1-z^2(r)}\right) < 0. \tag{4.2.7}$$

But $\lim_{r\to 0} r^2 g(r) = 0$, therefore g is negative and

$$0 < \sqrt{1-z^2(r)} < 2r^2. \tag{4.2.8}$$

Because we can write $2g(r) = \dfrac{z_r^2(r)}{N-1} + \left(\dfrac{\sqrt{1-z^2(r)}}{r} - r\right)^2 - r^2$, we get (4.2.4).

Step 2. Since z_r is bounded, $z(r)$ admits a limit z_0 when r goes to R^*. From (4.2.6) the function g which is decreasing and bounded below by -1, admits a limit $\gamma > -1$. Therefore $z_0 \neq \pm 1$ and the solution can be extended to some interval on the right of R^*, contradiction.

The main achievement of $[B.\,V2]$ is the following geometric uniqueness result:

THEOREM 5.21. *There exists a unique solution* z*of* (4.1.17)-(4.1.19) *defined on* $(0,\infty)$ *such that* z *is nonincreasing near* $r = 0$, *or equivalently a unique solution* u *of* (4.1.13) *in* $\mathbb{R}^N \setminus \{O\}$ *such that* u *is concave near* O.

The proof is highly technical but the main idea is to prove that under this geometric assumption, the function w defined from u and z by (4.1.20) is bounded from above and from below by $M < M_0 = (N+8)/(3\sqrt{N-1})$ on some interval $(0, r_0]$. Therefore it falls into the scope of the local uniqueness result of the fixed point of the contractive mapping **L** introduced in Theorem 5.19.

5-4-3 Further extensions

A more general equation is treated in $[B.\,V3]$,

$$-\text{div}\left(\frac{\nabla u}{\sqrt{1+|\nabla u|^2}}\right) = f(u), \qquad \text{in } \mathbb{R}^N \setminus \{O\}, \tag{4.3.1}$$

(always in the radial case) with the singularity condition

$$\lim_{x\to O} u(x) = -\infty, \tag{4.3.2}$$

with a growth assumption on f, from a very different point of view: a singular solution can be seen as a limit of regular solutions u_n, that is of solution of

$$-\left(\frac{r^{N-1}u_{nr}}{\sqrt{1+u_{nr}^2}}\right)_r = r^{N-1}f(u_n) \quad \text{on } [0,\varepsilon_n), \tag{4.3.3}$$

$$u_n(0) = \alpha_n < 0, \qquad u_{nr}(0) = 0, \tag{4.3.4}$$

when α_n goes to $-\infty$. The problem is that ε_n goes to 0 because of the occurence of vertical points on the graph of $(r, u_n(r))$, closer and closer to 0. Therefore the notion of solution must be extended in a more general and geometric framework. A solution is a C^2-imbedded surface in $\mathbb{R}^N \times \mathbb{R}$ without self-intersection whose mean curvature at each point $(x, ux))$ is $-N^{-1}f(u(x))$. This leads to the parametric system

$$\begin{cases} \dfrac{d\psi}{ds} = -f(u) - (N-1)r^{-1}\sin\psi, \\[2mm] \dfrac{du}{ds} = \sin\psi, \\[2mm] \dfrac{dr}{ds} = \cos\psi. \end{cases} \tag{4.3.5}$$

In this formulation s is the arc-length of a vertical section of the hypersurface and ψ is the angle between the tangent at this section and the r-axis. When $\psi \in (0, \pi)$ the hypersurface can be represented by the graph of a function $u \mapsto r(u)$ and (4.3.5) is equivalent to

$$\frac{r_{uu}(u)}{(1+r_u^2(u))^{3/2}} - \frac{N-1}{r(u)}\frac{1}{\sqrt{1+r_u^2(u)}} = f(u). \tag{4.3.6}$$

With this formulation the regular solutions are represented by the solutions $u \mapsto r(u)$ of (4.3.6) on $(u_{0,n}, \infty)$ satisfying the initial conditions

$$\lim_{u \to u_{0,n}} r(u) = 0, \qquad \lim_{u \to u_{0,n}} r_u(u) = \infty. \tag{4.3.7}$$

By comparison with onduloids (or Delaunay surfaces) which are rotationnally symmetric surfaces with constant mean curvature, Concus and Finn proved in [CF3] that, when $N = 2$ and $f(u) = (1 - N)u$, the regular solutions u_n of (4.3.6)-(4.3.7) converges uniformly on any compact subset of $(-\infty, 0)$ to a singular solution of (4.3.6). In [B.V3] some existence and local behaviour results are obtained. In particular, if $u \mapsto |u|^{-1/(N-1)}f(u)$ is nondecreasing for large $|u|$, there exists a local singular solution of (4.3.1)-(4.3.2) under the form $u \mapsto r(u)$.

Remark 5.9. The existence of non-radial singular solutions to the capillarity type equations appears to be completely open.

6 Singularities of nonlinear parabolic equations

6-1 Singularities of linear or quasilinear equations

6-1-1 The heat equation

As in the stationary case, the simplest example of a singular solution of a linear parabolic equation is the heat kernel in $\mathbb{R}^N \times (0, \infty)$ which is given by

$$E(x, t) = (4\pi t)^{-N/2} e^{-|x|^2/4t}, \tag{1.1.1}$$

and satisfies the heat equation

$$u_t - \Delta u = 0 \tag{1.1.2}$$

in $\mathbb{R}^N \times (0, \infty)$, with a singular initial data δ_O. A natural way to construct many other singular solutions of is to use the self-similar variable $\eta = x / \sqrt{t}$ and to look for solutions of (1.1.2) under the form $u(x, t) = t^{-\alpha} f(x / \sqrt{t}) = t^{-\alpha} f(\eta)$. This leads to the following equation in $\mathbb{R}^N$

$$-\Delta f - \frac{1}{2} \eta \cdot \nabla f - \alpha f = -K^{-1} \mathrm{div}(K \nabla f) = 0, \tag{1.1.3}$$

with

$$K = K(\eta) = e^{|\eta|^2/4}. \tag{1.1.4}$$

As for the initial condition for u, it means that $\lim_{t \to 0} u(x, t) = 0$ for any $x \neq O$, which reads

$$\lim_{|\eta| \to \infty} |\eta|^{2\alpha} f(\eta) = 0. \tag{1.1.5}$$

Equation (1.1.3) can be understood as an eigenvalue problem in a weighted Lebesgue space

$$L_K^2(\mathbb{R}^N) = \left\{ \phi \in L_{loc}^2(\mathbb{R}^N) : \int_{\mathbb{R}^N} \phi^2(\eta) K(\eta) d\eta < \infty \right\}. \tag{1.1.6}$$

We also define the weighted Sobolev spaces

$$W_{K^\theta}^{m,2}(\mathbb{R}^N) = \left\{ \phi \in W_{loc}^{m,2}(\mathbb{R}^N) : \int_{\mathbb{R}^N} |D^\alpha \phi|^2 K^\theta(\eta) d\eta < \infty, \forall \alpha \in \mathbb{N}^N, |\alpha| \leq m \right\} \tag{1.1.7}$$

301

for some integer m and $\theta > 0$ and,

$$L^s_{K^\theta}(\mathbb{R}^N) = \left\{ \phi \in L^s_{loc}(\mathbb{R}^N) : \int_{\mathbb{R}^N} \phi^s(\eta) K^\theta(\eta) d\eta < \infty \right\}, \tag{1.1.8}$$

for $s \geq 1$. The spectral analysis of the operator $f \mapsto -K^{-1} \mathrm{div}(K\nabla f) = Lf$ is completely known ([EK] or [We2]) and we recall some elements of it that we shall use later on.

PROPOSITION 6.1. *Suppose* $N \geq 1$; *then :*

(i) the embedding of $W^{1,2}_{K^\theta}(\mathbb{R}^N)$ *into* $L^2_{K^\theta}(\mathbb{R}^N)$ *is compact for any* $\theta > 0$;

(ii) if $N > 2$, *the embedding of* $W^{1,2}_{K^\theta}(\mathbb{R}^N)$ *into* $L^{2N/(N-2)}_{K^{\theta N/(N-2)}}(\mathbb{R}^N)$ *is continuous ;*

(iii) if $N = 2$, *the embedding of* $W^{1,2}_{K^\theta}(\mathbb{R}^2)$ *into* $L^q_{K^{\theta q/2}}(\mathbb{R}^2)$ *is continuous ;*

(iv) if $N = 1$, *the embedding of* $W^{1,2}_{K^\theta}(\mathbb{R})$ *into the Hölder space* $C^{0,1/2}_{K^{\theta/2}}(\mathbb{R})$ *is continuous .*

The operator L whose domain $D(L) = \left\{ \phi \in L^2_K(\mathbb{R}^N) : L\phi \in L^2_K(\mathbb{R}^N) \right\}$ is contained in $W^{1,2}_K(\mathbb{R}^N)$, is self-adjoint in $L^2_K(\mathbb{R}^N)$.

PROPOSITION 6.2. *The eigenvalues of* L *are the numbers* $\lambda_k = (N+k-1)/2$, *for* $k > 1$. *The eigenspaces* $N(L - \lambda_k I)$ *are spanned by the* $D^\alpha(K^{-1})$, *where* $\alpha \in \mathbb{N}^N$, $|\alpha| = k - 1$ *and*

$$\dim N(L - \lambda_k I) = \binom{N+k-2}{N-1} = n_k. \tag{1.1.9}$$

6-1-2 Removable singularities for parabolic equations

Let Ω be a domain in $\mathbb{R}^N$ and K a compact subset of $\Omega \times (0,T)$, we shall say that K is a p-q null set if

$$\inf \left\{ \int_0^T \left(\int_{\mathbb{R}^N} \left(|\nabla v|^2 + v_t^- \right)^{p/2} dx \right)^{q/p} dt : v \in C^1_{K,P} \right\} = 0, \tag{1.2.1}$$

for $p, q \geq 2$, where $C^1_{K,P}$ is the set of functions v belonging to $C^1\left(\mathbb{R}^N \times [0,\infty)\right)$ such that $0 \leq v \leq 1$, $v \equiv 1$ in a neighbourhood of K and $v \equiv 0$ ouside the sphere $B(O,R)$ of $\mathbb{R}^N$.

LEMMA 6.1. *Suppose* K *is a* p-q *null set for some for some* $2 \leq p \leq N$ *and* $q \geq 2$. *Then* K *has zero Lebesgue* $(N+1)$-*dimensional measure and there exists a sequence* $\psi_n \in C^1_{K,P}$ *such that* $\psi_n \to 0$ *a.e. in* $\mathbb{R}^N \times (0,T)$ *when* n *tends to infinity.*

Proof. Let us define the functional $F_{p,q}$ by

302

$$F_{p,q}(v) = \int_0^T \left(\int_{\mathbb{R}^N} \left(|\nabla v|^2 + v_t^- \right)^{p/2} dx \right)^{q/p} dt. \tag{1.2.2}$$

There exists a sequence $\{\psi_n\}$ in $C^1_{K,P}$ such that $F_{p,q}(\psi^n) \to 0$ when n goes to infinity; this implies in particular that

$$\lim_{n\to\infty} F_{p,q}(\psi_n) = \lim_{n\to\infty} \left(\int_0^T \left(\int_{r^N} |\nabla \psi_n|^p dx \right)^{q/p} dt \right)^{1/q} = 0. \tag{1.2.3}$$

First suppose that $p < N$ and set $p^* = Np/(N-p)$. By Sobolev's inequality,

$$\left(\int_0^T \left(\int_{\mathbb{R}^N} |\nabla \psi_n|^p dx \right)^{q/p} dt \right)^{1/q} \geq c(p,N) \left(\int_0^T \left(\int_{\mathbb{R}^N} \psi_n^{p^*} dx \right)^{q/p^*} dt \right)^{1/q}, \tag{1.2.4}$$

and by Hölder's inequality,

$$\left(\int_0^T \left(\int_{\mathbb{R}^N} |\nabla \psi_n|^p dx \right)^{q/p} dt \right)^{1/q} \geq c(p,N) \begin{cases} T^{1/q-1/p'} \|\psi_n\|_{L^{p^*}(\mathbb{R}^N \times (0,T))} & \text{if } q \geq p^*, \\ \Sigma^{1/p^*-1/q} \|\psi_n\|_{L^q(\mathbb{R}^N \times (0,T))} & \text{if } q < p^*, \end{cases} \tag{1.2.5}$$

where $\Sigma = \text{meas}(B(O,R) \times (0,T))$. Therefore there exists a subsequence $\{\psi_{n_k}\}$ which converges to 0 a.e. in $\mathbb{R}^N \times (0,T)$. The case $p = N$ needs a simple adaptation of the above technique.

We consider the following equation in $\Omega \times (0,T)$

$$u_t - \Delta u - c(x,t)u = 0, \tag{1.2.6}$$

and we prove the following result due to Aronson [Ar1].

THEOREM 6.1. *Suppose* c *is locally bounded in* $\Omega \times (0,T)$, K *is a compact* p-q *null set in* $\Omega \times (0,T)$ *for some* $2 \leq p \leq N$ *and* $q \geq 2$, *and* u *is a solution of* (1.2.6) *in* $\Omega \times (0,T) \backslash K$ *such that*

$$u \in L^s(0,T;L^r(\Omega)) \cap C^{2,1}(\Omega \times (0,T) \backslash K) \tag{1.2.7}$$

for some

$$r \geq p/(p-2) \quad \text{and} \quad s \geq q/(q-2). \tag{1.2.8}$$

Then u *can be extended to* $\Omega \times (0,T)$ *so that the resulting function is a solution of* (1.2.6) *in whole* $\Omega \times (0,T)$.

Proof. Without any loss of generality, we may suppose that Ω is bounded and that u is continuous up to the lateral and lower boundaries of $\Omega \times (0,T)$. Let w be the solution of (1.2.6) in $\Omega \times (0,T)$ which coincides with u on the lateral and lower boundaries. From the parabolic equations regularity theory ([Fd]), the function w belongs to $C^{2,1}(\Omega \times (0,T)$ $\cap C(\overline{\Omega} \times [0,T])$. Set $v = u - w$, $z = \max(0, v - \delta)$ for some $\delta > 0$ and $\phi = (1+z)^\varepsilon - 1$, for $\varepsilon > 0$. If $\psi \in C^1_{K,P}$ and $\eta = 1 - \psi$, then η vanishes in a neighborhood of K and $0 \le \eta \le 1$. Because for any $0 < \tau < T$,

$$\int_0^\tau \int_\Omega \phi \eta^2 (v_t - \Delta v - cv)\,dxdt = 0, \qquad (1.2.9)$$

we obtain

$$\frac{1}{1+\varepsilon}\int_\Omega \eta^2 A\,dx\Big|_{t=\tau} + \varepsilon\int_0^\tau\int_\Omega \eta^2 B\,dxdt = \frac{1}{1+\varepsilon}\int_\Omega \eta^2\,dx\Big|_{t=0} + \frac{2}{1+\varepsilon}\int_0^\tau\int_\Omega \eta\eta_t A\,dxdt$$

$$\qquad (1.2.10)$$

$$-2\int_0^\tau\int_\Omega \phi\eta\nabla\eta.\nabla v\,dxdt + \int_0^\tau\int_\Omega c\phi v\eta^2\,dxdt,$$

where $A = (1+z)^{1+\varepsilon} - (1+\varepsilon)z$ and $B = (1+z)^{\varepsilon-1}|\nabla v|^2$. From Cauchy-Schwarz inequality

$$2\left|\int_0^\tau\int_\Omega \eta\phi\nabla\phi.\nabla\eta\,dxdt\right| \le \varepsilon\int_0^\tau\int_\Omega \eta^2 B\,dxdt + \frac{1}{\varepsilon}\int_0^\tau\int_\Omega |\nabla\eta|^2 C\,dxdt, \qquad (1.2.11)$$

where $C = (1+z)^{1+\varepsilon}$. Thus, since $0 \le \eta \le 1$, $1 \le A \le C$, $c\eta^2\phi v \le 0$ and $\eta = 1 - \psi$, we obtain

$$\int_\Omega \eta^2 A\,dx\Big|_{t=\tau} \le \int_\Omega \eta^2\,dx\Big|_{t=0} + k(\varepsilon)\int_0^\tau\int_\Omega \left(|\nabla\psi|^2 + \psi_t^-\right)C\,dxdt, \qquad (1.2.12)$$

and $k(\varepsilon)$ only depends on ε. If $p > 2$, we have to assume $\varepsilon \le \dfrac{r(p-2)}{p} - 1 = \varepsilon_1$ and we apply Hölder's inequality to the right-hand side of (1.2.12). Therefore

$$\int_\Omega \eta^2 A\,dx\Big|_{t=\tau} \le \int_\Omega \eta^2\,dx\Big|_{t=0}$$

$$+ k(\varepsilon)|\Omega|^{(\varepsilon_1-\varepsilon)/\varepsilon}\int_0^T\left(\int_{\Gamma_N}\left(|\nabla\psi|^2 + \psi_t^-\right)^{p/2}\,dx\right)^{2/p}\left(\int_\Omega C^{r/(1+\varepsilon)}\,dx\right)^{(1+\varepsilon)/r}\,dt. \qquad (1.2.13)$$

If $q > 2$, we assume $\varepsilon \le s(q-2)/q - 1 = \varepsilon_2$ and, again with Hölder's inequality,

$$\int_\Omega \eta^2 A\,dx\Big|_{t=\tau} \le \int_\Omega \eta^2 dx\Big|_{t=0}$$
$$+k(\varepsilon)|\Omega|^{(\varepsilon_1-\varepsilon)/\varepsilon}\,T^{(\varepsilon_2-\varepsilon)/\varepsilon}M\,F_{p,q}^2(\psi), \tag{1.2.14}$$

where $\displaystyle M=\left(\int_0^T\left(\int_\Omega C^{r/(1+\varepsilon)}dx\right)^{s/r}dt\right)^{(1+\varepsilon)/s}$. From the integrability assumption on u and the

definition of C, which is essentially $(v^+)^{1+\varepsilon}$, M is finite. Finally, integrating with respect to τ on (0,T) yields

$$\int_0^T\int_\Omega \eta^2 A\,dxdt \le T\int_\Omega \eta^2 dx\Big|_{t=0}+k(\varepsilon,p,q,r,s,|\Omega|,T)F_{p,q}^2(\psi). \tag{1.2.15}$$

If we replace ψ by ψ_n and η by $\eta_n=1-\psi_n$, we obtain

$$\int_0^T\int_\Omega A\,dxdt = \int_0^T\int_\Omega\left((1+z)^{1+\varepsilon}-(1+\varepsilon)z\right)dxdt \le T|\Omega|. \tag{1.2.16}$$

But, if z is not zero, by continuity we get $\displaystyle\int_0^T\int_\Omega A\,dxdt > T|\Omega|$, which is a contradiction.

Therefore z is identically zero and $v \le \delta$. In the same way $v \ge -\delta$. Since δ is arbitrary, u = w in $\overline{\Omega}\times[0,T]$.

Remark 6.1. By a simple adaptation of the proof it is possible to weaken the assumption on K and assume only that $K \subset \overline{\Omega}\times[0,T]$. In that case we can prove that u is continuous in any interior point $P \in \Omega\times(0,T)$ and then apply the above technique.

6-1-3 Isolated singularities for parabolic equations

We consider the equation

$$u_t - \sum_{i,j}\frac{\partial}{\partial x_j}\left(a_{ij}(x,t)\frac{\partial u}{\partial x_i}\right)=0 \tag{1.3.1}$$

in the cylinder $S = B(O,\rho)\times(0,T]\subset \mathbb{R}^N\times\mathbb{R}^+$, and we assume that the matrix $\left(a_{ij}(.,.)\right)$ is measurable and almost everywhere symmetric in $\overline{S}$ where the ellipticity conditions

$$\lambda^{-1}\sum_i \xi_i^2 \le \sum_{i,j}a_{ij}(x,t)\xi_i\xi_j \le \lambda\sum_i \xi_i^2, \tag{1.3.2}$$

holds, for some $\lambda \ge 1$ and any $\xi = (\xi_1,...,\xi_N)\in\mathbb{R}^N$. The fonction u is a weak solution of (1.3.1) in $S_\eta = B(O,\rho)\times(\eta,T]$, $\eta > 0$, if u belongs to the space $C\left([\eta,T]\times L^2(B(O,\rho))\right)$

$\cap L^2\big((\eta, T); W^{1,2}(B(O, \rho))\big)$ and for any $t \in [\eta, T]$ and any function $\phi \in C^1(\mathbb{R}^N \times \mathbb{R}^+)$ with compact support in $B(O, \rho)$, we have

$$\int_{B(O,\rho)} u(x,t)\phi(x,t)dx + \int_\eta^T \int_{B(O,\rho)} \left(-u(x,\tau)\phi_t(x,\tau) + \sum_{i,j} a_{ij} \frac{\partial u}{\partial x_i} \frac{\partial \phi}{\partial x_j} \right) dx d\tau \tag{1.3.3}$$
$$- \int_{B(O,\rho)} u(x,\eta)\phi(x,\eta)dx.$$

The maximum principle applied to u yields

$$\min_{\Gamma_\eta} u(y,s) \leq u(x,t) \leq \max_{\Gamma_\eta} u(y,s), \tag{1.3.4}$$

for any $(x,t) \in S_\eta$, where $\Gamma_\eta \in B(O,\rho) \times \{0\} \cup \partial B(O,\rho) \times [\eta, T]$. The existence and uniqueness theorem for parabolic equations applies. Consequently, for any $u_0 \in C_0(\overline{B}(O,\rho))$, there exists a unique weak solution of (6.2.1) in S_η, which vanishes in the $W_0^{1,2}$-sense on the lateral boundary $\partial B(O,\rho) \times [\eta, T]$ and takes the initial value u_0 for $t = 0$. Moreover u is continuous in $\overline{S}_\eta$ and

$$u(x,t) = \int_{B(O,\rho)} g(x,t,\xi,\eta)u_0(\xi)\, d\xi, \tag{1.3.5}$$

where g is the Green's function for (6.2.1) in S. Similarly the function ũ defined by

$$\tilde{u}(\xi,\eta) = \int_{B(O,R)} g(x,t,\xi,\eta)u_0(x)dx, \tag{1.3.6}$$

is the unique weak solution of the backward equation adjoint to (1.3.1) in $S \backslash S_t$

$$\tilde{u}_\eta + \sum_{i,j} \frac{\partial}{\partial \xi_j}\left(a_{ij}(\xi,\eta) \frac{\partial \tilde{u}}{\partial \xi_i} \right) = 0, \tag{1.3.7}$$

subject to the Dirichlet boundary conditions on $\partial B(O,\rho) \times [0,t]$ and $\tilde{u}(\xi,t) = u_0(\xi)$ in $B(O,\rho)$ as initial condition. The following properties of the Green's kernel can be found in [Ar3].

PROPOSITION 6.3. *With the previous notations, there holds*
(i) $g(x,t,\xi,\eta) \geq 0$,
(ii) *for* (x,t) *fixed and* $\varepsilon > 0$, $(\xi,\eta) \mapsto g(x,t,\xi,\eta)$ *is a continuous function in*
    $\overline{B}(O,\rho) \times [0, t - \varepsilon]$,
(iii) *for arbitrary* $\varepsilon > 0$, $(x,t) \mapsto g(x,t,0,0)$ *is a weak solution of* (6.2.1) *in* S_ε *and is*
continuous in $\overline{S}_\varepsilon$,
(iv) $g(x,t,0,0) = 0$ *for* $|x| = \rho$ *and* $t = 0$, $\lim_{t \to 0} g(x,t,0,0) = 0$ *for* $x \neq 0$ *and*

$$\limsup_{(x,t)\to(0,0)} g(x,t,0,0) = \infty,$$

(v) $\int_{B(O,\rho)} g(x,t,0,0)\,dx \le 1$ *and* $\lim_{t\to 0}\int_{B(O,\rho)} g(x,t,0,0)h(x)dx = h(0)$ *for any*

$h \in C(\overline{B}(O,\rho))$.

THEOREM 6.2. *Suppose that* $u \in C(\overline{B}(O,\rho)\times(0,T])\cap L^{\infty}(0,T;B(O,\rho)\setminus B(O,\delta))$ *for any* $\delta \in (0,\rho)$ *is a nonnegative weak solution of* (6.2.1) *in* S_η *for every* $\eta \in (0,T)$. *We suppose also there exists a function* $\psi \in C(\Gamma)$ *with* $\Gamma = \Gamma_0$ *such that* $u(x,t) = \psi(x,t)$ *on* $\partial B(O,\rho)\times(0,T]$,

$$\lim_{t\to 0^+}\int_{\delta\le|x|\le\rho}|u(x,t)-\psi(x,0)|dx = 0 \tag{1.3.8}$$

for every $\delta \in (0,\rho)$ *and that there exists a weak solution* $\overline{u} \in C(\overline{S})$ *which coincides with* ψ *on* Γ. *Then there exists* $\alpha \ge 0$ *such that the equality*

$$u(x,t) = \alpha g(x,t,0,0) + \overline{u}(x,t) \tag{1.3.9}$$

holds in S.

Proof. Step 1. we claim that there exists $A > 0$ such that

$$\int_{B(O,\rho)} u(x,t)dx \le A, \tag{1.3.10}$$

for every $t \in (0,T]$. Because for $\theta \in (0,T)$, $u \in C(\overline{S}_\theta)\cap L^{\infty}(0,T;B(O,\rho)\setminus B(O,\rho/2))$, there exist $A_1 = A_1(\theta) > 0$ and $A_2 > 0$ such that the inequalities

$$\int_{B(O,\rho)} u(x,t)dx \le A_1(\theta) \tag{1.3.11}$$

for $t \in [\theta,T]$, and

$$\int_{\rho/2\le|x|\le\rho} u(x,t)dx \le A_2 \tag{1.3.12}$$

for $t \in [0,T]$, hold. Let $v = v(x)$ be a smooth function which takes the value 1 on $\overline{B}(O,\rho/2)$, vanishes for $|x| \ge 3\rho/4$ and such that $0 \le v \le 1$. For any $\eta > 0$, the function $\hat{u}$ defined by

$$\hat{u}(x,t) = \int_{B(O,\rho)} g(x,t,\xi,\eta)v(\xi)u(\xi,\eta)dx \tag{1.3.13}$$

is continuous in $\overline{S}_\eta$ and is the unique weak solution of (1.3.1) in S_η which coincides with v on $\overline{B}(O,\rho)\times\{\eta\}$ and vanishes on $\partial B(O,\rho)\times[\eta,T]$. From the maximum principle, $u \ge \hat{u}$ in $\overline{S}_\eta$. In particular there holds

$$u(x,t) \geq \int_{B(O,\rho/2)} g(x,t,\xi,\eta)u(\xi,\eta)\,d\xi \qquad\qquad (1.3.14)$$

in S_η, and therefore

$$\int_{B(O,\rho)} u(x,t)dx \geq \int_{B(O,\rho/2)} \int_{B(O,\rho)} g(x,t,\xi,\eta)dx u(\xi,\eta)\,d\xi \qquad\qquad (1.3.15)$$

for $t > \eta$. Now

$$(\xi,\eta) \mapsto \tilde{u}(\xi,\eta) = \int_{B(O,\rho)} g(x,t,\xi,\eta)v(x)dx \qquad\qquad (1.3.16)$$

is continuous in $\overline{B}(O,\rho)\times[0,t]$ and is the unique weak solution of (1.3.7) in $B(O,\rho)\times(0,t)$ with $\tilde{u} = v$ on $\overline{B}(O,\rho)\times\{t\}$ and $\tilde{u} = 0$ on $\partial B(O,\rho)\times[0,t]$. But $\tilde{u}(\xi,t) = 1$ in $B(O,\rho/2)$, therefore there exists $\theta = \theta(N,T,\rho,\lambda) > 0$ such that $\tilde{u}(\xi,\eta) \geq 1/2$ if $(\xi,\eta) \in \overline{B}(O,\rho/2)\times[t-\theta,t]$. Hence

$$\int_{B(O,\rho)} g(x,t,\xi,\eta)dx \geq 1/2, \qquad\qquad (1.3.17)$$

if $(\xi,\eta) \in \overline{B}(O,\rho/2)\times[t-\theta,t]$. Consequently, for $\eta \leq \theta$,

$$\int_{B(O,\rho/2)} u(\xi,\eta)d\xi \leq 2\int_{B(O,\rho)} u(x,\theta)dx \leq 2A_1(\theta) \qquad\qquad (1.3.18)$$

and (1.3.10) holds with $A = 2A_1(\theta) + A_2$.

Step 2. From Step 1 and $u \geq 0$ there exists β such that

$$0 \leq \beta = \limsup_{t\to 0} \int_{B(O,\rho)} u(x,t)dx \leq A, \qquad\qquad (1.3.19)$$

and, from the continuity of $\tilde{u}$ in $\overline{S}$, $\int_{B(O,\rho)} |\overline{u}(x,t)|dx \leq B$ and

$$\gamma = \int_{B(O,\rho)} \psi(x,0)dx = \lim_{t\to 0} \int_{B(O,\rho)} \overline{u}(x,t)dx, \qquad\qquad (1.3.20)$$

for some B and γ. Set $\alpha = \beta - \gamma$ and

$$w(x,t) = u(x,t) - \overline{u}(x,t) - \alpha g(x,t,0,0); \qquad\qquad (1.3.21)$$

next we prove that $w = 0$ in S.

Let $h = h(\xi)$ be continuous in $B(O,\rho)$ and denote

$$I(\tau) = \int_{B(O,\rho)} h(\xi)w(\tau,\xi)d\xi. \qquad\qquad (1.3.22)$$

We claim that for any sequence $\{\tau_n\}$ with $\lim_{n \to \infty} \tau_n = 0$ and $\lim_{n \to \infty} \int_{B(O,\rho)} u(\xi, \tau_n) d\xi = \beta$, we have $\lim_{n \to \infty} I(\tau_n) = 0$. In order to prove it, write

$$I(\tau) = h(0) \int_{B(O,\rho)} (u - \overline{u}) d\xi - \alpha \int_{B(O,\rho)} hg d\xi$$

$$+ \int_{B(O,\rho)} (h(\xi) - h(0))(u - \overline{u}) d\xi, \tag{1.3.23}$$

$$= h(0) I_1(\tau) - \alpha I_2(\tau) + I_3(\tau).$$

Then $\lim_{n \to \infty} I_1(\tau_n) = 0$ and $\lim_{\tau \to 0} I_2(\tau) = h(0)$ are derived from the properties of the Green kernel. Moreover, from (1.3.10),

$$I_3(\tau) \leq (A + B) \max_{|\xi| \leq \rho} |h(\xi) - h(0)| + 2 \max_{|\xi| \leq \rho} |h| \int_{\sigma \leq |\xi| \leq \rho} |u - \overline{u}| d\xi. \tag{1.3.24}$$

For $\varepsilon > 0$ we can choose $\sigma = \sigma(\varepsilon)$ such that the first term in the right-hand side of (1.3.24) is smaller than $\varepsilon / 2$. Because

$$\int_{\sigma \leq |\xi| \leq \rho} |u - u| d\xi \leq \int_{\sigma \leq |\xi| \leq \rho} |u - \psi(\xi, 0)| d\xi + \int_{\sigma \leq |\xi| \leq \rho} |\psi(\xi, 0) - u| d\xi, \tag{1.3.25}$$

which goes to 0 with τ, there exists a $\tau(\varepsilon) > 0$ such that the second term in the right-hand side of (1.3.24) is smaller than $\varepsilon / 2$. Therefore $\lim_{\tau \to 0} I_3(\tau) = 0$ and $\lim_{n \to \infty} I(\tau_n) = 0$.

For any $\tau > 0$, the function w which belongs to $C(\overline{S}_t)$, vanishes on the lateral boundary and is a weak solution of (1.3.1) in S_τ, can be written under the following form

$$w(x,t) = \int_{B(O,\rho)} g(x,t,\xi,\tau) w(\xi,\tau) d\xi \tag{1.3.26}$$

in S_τ, and this can be re-written as

$$w(x,t) = \int_{B(O,\rho)} g(x,t,\xi,0) w(\xi,\tau) d\xi$$

$$+ \int_{B(O,\rho)} (g(x,t,\xi,\tau) - g(x,t,\xi,0)) w(\xi,\tau) d\xi; \tag{1.3.27}$$

$$= J_1(\tau) + J_2(\tau).$$

For (x,t) fixed and τ bounded away from t, $(\xi,\tau) \mapsto g(x,t,\xi,\tau)$ is a continuous function. Therefore $\lim_{n \to \infty} J_1(\tau_n) = 0$. In view of (1.3.10) and $\int g d\xi \leq 1$, there holds

$$|J_2(\tau)| \leq (A + B + \alpha) \max_{|\xi| \leq \rho} |g(x,t,\xi,\tau) - g(x,t,\xi,0)|, \tag{1.3.28}$$

and consequently $\lim_{\tau \to 0} J_2(\tau) = 0$. Since the representation (1.3.26) holds for all $\tau < t$, it holds also for any positive t for $\tau_n \to 0$. Therefore w (x,t) is zero for any $(x,t) \in S$. Finally, since

$$\limsup_{(x,t) \to (0,0)} g(x,t,0,0) = +\infty \qquad (1.3.29)$$

it follows that $u \geq 0$, which completes the proof.

Remark 6.2. From the above proof, it is clear that the choice of the sequence $\{\tau_n\}$ is arbitrary and in fact

$$\lim_{\tau \to 0} \int_{B(O,\rho)} u(x,\tau)dx = \beta. \qquad (1.3.30)$$

Moreover the sign assumption on u can be dropped provided there exists some $A \geq 0$ such that

$$\int_{B(O,\rho)} |u(x,t)|dx \leq A \qquad (1.3.31)$$

for every $t \in (0,T]$. In that case (1.3.30) holds for some β and (1.3.9) for some α.

6-1-4 The Aronson and Serrin's results

Let Ω be a bounded domain of $\mathbb{R}^N$, T a positive real number and $Q = \Omega \times (0,T)$. We consider the following second order quasilinear parabolic equation

$$u_t = \mathrm{div}\,\mathcal{A}(x,t,u,\nabla u) + \mathcal{B}(x,t,u,\nabla u) = \sum_{j=1}^{N} \frac{\partial}{\partial x_j} \mathcal{A}_j(x,t,u,\nabla u) + \mathcal{B}(x,t,u,\nabla u), \quad (1.4.1)$$

where $\mathcal{A}$ (resp. $\mathcal{B}$) is a vector valued (resp. real valued) function of (x,t,r,p) which belongs to $\Omega \times (0,T) \times \mathbb{R} \times \mathbb{R}^N$. The functions $\mathcal{A}$ and $\mathcal{B}$ are supposed to be measurable in (x,t) for all values of (r,p). We define the space $L^{p,q}(Q)$ for p and $q \geq 1$ by

$$L^{p,q}(Q) = \left\{ w \in L^1_{loc}(Q) : \int_0^T \left(\int_\Omega |w|^p dx \right)^{q/p} dt < \infty \right\}, \qquad (1.4.2)$$

with the corresponding norm

$$\|w\|_{L^{p,q}} = \left(\int_0^T \left(\int_\Omega |w|^p dx \right)^{q/p} dt \right)^{1/q}. \qquad (1.4.3)$$

The structural assumptions on $\mathcal{A}$ and $\mathcal{B}$ are the following:

310

$$p.\mathcal{A}(x,t,r,p) = \sum_j p_j.\mathcal{A}_j(x,t,r,p) \geq a|p|^2 - b^2 r^2 - f^2, \tag{1.4.4}$$

$$|\mathcal{A}(x,t,r,p)| \leq \bar{a}|p| + e|r| + h, \tag{1.4.5}$$

$$|\mathcal{B}(x,t,r,p)| \leq c|p| + d|r| + g. \tag{1.4.6}$$

Here a and $\bar{a}$ are positive constants and the functions b, c, d, e, h, and g are supposed to belong to some $L^{p,q}$ spaces, namely

$$b,c,e,f,h \in L^{p,q}(Q), \text{ with } p > 2 \text{ and } \frac{N}{2p} + \frac{1}{q} < \frac{1}{2}, \tag{1.4.7}$$

$$d,g \in L^{p,q}(Q), \text{ with } p > 1 \text{ and } \frac{N}{2p} + \frac{1}{q} < 1. \tag{1.4.8}$$

The admissible weak solutions of (1.4.1) are functions w in $L^{2,\infty}_{loc}(Q)$ with $\nabla w \in L^{2,2}_{loc}(Q)$ such that

$$\iint_Q \left(-u\varphi_t + \nabla\varphi.\mathcal{A}(x,t,u,\nabla u)\right)dxdt = \iint_Q \left(\varphi\mathcal{B}(x,t,u,\nabla u)\right)dxdt, \tag{1.4.9}$$

holds for all $\varphi \in C^1_0(Q)$. Defining the parabolic boundary Γ of Q as being $\partial\Omega \times [0,T) \cup \Omega \times \{0\}$, and because u may not be continuous up to Γ, we shall say that $u \leq M$ on Γ if for any $\varepsilon > 0$ there exists a neighborhood of Γ in wich $u \leq M + \varepsilon$ a.e. In this Section we present without proof some of the deep results obtained by Aronson and Serrin in [AS].

THEOREM 6.3. *Let u be a weak solution of (1.4.1) in Q such that $u \leq M$ on Γ. Then almost everywhere in Q there holds*

$$u(x,t) \leq M + Ck, \tag{1.4.10}$$

where C depends only on T and meas(Ω) *and the structural assumptions, while*

$$k = \left(\|b\| + \|d\|\right)|M| + \left(\|f\| + \|g\|\right), \tag{1.4.11}$$

where the norms of the various coefficients are taken in there respective $L^{p,q}$ spaces, provided conditions (1.4.7),(1.4.8) be fulfilled.

Their method can also give local bound without any boundary assumption. Let $(\bar{x},\bar{t}) \in \Omega$ and define

$$R(\rho) = \left\{x \in \mathbb{R}^N : |x_j - \bar{x}_j| < \rho/2,\ 1 \leq j \leq N\right\} \text{ and } Q(\rho) = R(\rho) \times (\bar{t} - \rho^2, \bar{t}) \tag{1.4.12}$$

In the next result u need not be in $L^{2,\infty}_{loc}(Q)$ but only in $L^{2,2}_{loc}(Q)$

THEOREM 6.4. *Let* u *be a weak solution of* (1.4.1) *in* Q. *Suppose that the set* $Q(3\rho)$ *is included in* Q. *Then for almost all* (x,t) *in* $Q(\rho)$ *we have*

$$|u(x,t)| \le C\left(\rho^{-(N+2)/2}\|u\|_{L^{2,2}(Q(3\rho))} + k\rho^{\theta}\right),\tag{1.4.13}$$

where C *depends only on the structural assumptions and* ρ *while*

$$k = \|f\| + \|g\| + \|h\|,\tag{1.4.14}$$

and θ *is a positive number such that*

$$p \ge \frac{2}{1-\theta} \quad and \quad \frac{N}{2p} + \frac{1}{q} \le \frac{1-\theta}{2},\tag{1.4.15}$$

for the $L^{p,q}$ *space to which belong* b, c, f *and* h, *and*

$$p \ge \frac{1}{1-\theta} \quad and \quad \frac{N}{2p} + \frac{1}{q} \le 1-\theta,\tag{1.4.16}$$

for the coefficients d *and* g.

A consequence of this type of estimate is the parabolic Harnack inequality. Define

$$Q^*(\rho) = R(\rho) \times (\bar{t} - 8\rho^2, \bar{t} - 7\rho^2),\tag{1.4.17}$$

then there holds,

THEOREM 6.5. *Let* u *be a nonnegative weak solution of* (1.4.1) *in* Q. *Suppose that the set* $Q(3\rho)$ *is included in* Q. *Then*

$$\max_{Q^*(3\rho)} u \le C \min_{Q(3\rho)}\left(u + k\rho^{\theta}\right),\tag{1.4.18}$$

where C *depends only on the structural assumptions and* ρ *while* k *is as in* (1.4.14).

As in the elliptic case, the Harnack inequality implies the Hölder continuity of the weak solutions of (1.4.1). The following pointwise form of the Harnack inequality shows more precisely the dependence of the oscillation coefficient in (1.4.18).

THEOREM 6.6. *Let* u *be a nonnegative weak solution of* (1.4.1) *in* Q. *Suppose that* Ω' *is a convex subdomain of* Ω *which has a positive distance* δ *from the boundary of* Ω. *Then for all* (x,s) *and* (y,t) *in* $Q' = \Omega \times (0,T)$ *with* $0 < s < t < T$, *we have*

312

$$u(y,s) + k \leq (u(x,t) + k)\exp C\left(\frac{|x-y|^2}{t-s} + \frac{t-s}{R} + 1\right), \tag{1.4.19}$$

where C depends only on the structural assumptions and ρ, while k is as in (1.4.14) and $R = \min(1, s, \delta^2)$.

As a consequence of Theorem 6.6 we have

COROLLARY 6.1. *Let* u *be a nonnegative weak solution of* (1.4.1) *in Q. If*

$$M = \sup_{0 < t < T} \int_\Omega u(x,t)dx < \infty, \tag{1.4.20}$$

then for any compact subset Ω' *of* Ω *and* t *small enough, we have*

$$u(x,t) \leq CMt^{-N/2} + Ckt^{\theta/2} \qquad (\forall x \in \Omega'), \tag{1.4.21}$$

where C *depends only on the structural assumptions and* ρ, *while* k *is as in* (1.4.14).

With Theorem 6.6 it is also possible to derive a lower estimate for nonnegative solutions of (1.4.1) with a non-removable isolated singularity. Without restriction, we shall suppose that $O \in \Omega$ and we have

THEOREM 6.7. *Let* u *be a nonnegative weak solution of* (1.4.1) *in Q such that*

$$\liminf_{t \to 0} \int_{|x|^2 \leq \alpha t} u(x,t)dx > 0 \tag{1.4.22}$$

for some $\alpha > 0$. *Suppose that* Ω' *is a convex subdomain of* Ω *which has a positive distance* δ *from the boundary of* Ω *and contains* O *in its interior. Then there exists a positice constant* C *such that for any* t *small enough and* x *in* Ω' *there holds*

$$u(x,t) + k \geq \mathbb{C}t^{-N/2} \exp\left(-2C|x|^2 / t\right), \tag{1.4.23}$$

and C *depends on the structural assumptions,* T *and* δ *and* k *is as above.*

6-2 Isolated singularities: the absorption case

6-2-1 Removable singularities

Let Ω be an open domain of $\mathbb{R}^N$ containing 0, $p > 1$ be a real number and u belonging to $C^{2,1}(\Omega \times (0,T)) \cap C(\Omega \times [0,T) \setminus \{(O,0)\})$ be a solution of

$$u_t - \Delta u + u|u|^{p-1} = 0 \tag{2.1.1}$$

in $\Omega \times (0,T)$, $0 < T \leq \infty$, satisfying the initial condition

$$u(x,0) = 0 \qquad\qquad (2.1.2)$$

in $\Omega \setminus \{O\}$. The singularity problem is to determine the behaviour of $u(x,t)$ when (x,t) tends to $(0,0)$. In this Section we study the removability case due to Brezis and Friedman [BF] which corresponds to $p \geq (N+2)/N$.

The following *a priori* estimate of Keller-Osserman type applies to the parabolic equation

LEMMA 6.2. *Assume that* $p > 1$ *and* u *belonging to* $C(\Omega \times [0,T) \setminus \{(O,0)\})$ $\cap\, C^{2,1}(\Omega \times (0,T))$ *is a solution of (2.1.1) in* $\Omega \times (0,T)$ *vanishing on* $\Omega \setminus \{O\} \times \{0\}$. *Then if* $\rho > 0$ *is such that* $B(O, 2\rho) \subset \Omega$ *there exist positive constant* $0 < T_1 < T$ *and* $C = C(N,p)$ *such that*

$$|u(x,t)| \leq C\left(|x|^2 + t\right)^{-1/(p-1)} \;\text{ for } (x,t) \in B(O,\rho) \times (0,T_1) \setminus \{(O,0)\}). \qquad (2.1.3)$$

Proof. Consider $0 < T_1 < T$ and $x_0 \in \mathbb{R}^N$ with $0 < |x_0| < \rho$. We fix $0 < R < |x_0|$ and set

$$G = \left\{ (x,t)\colon |x - x_0| < R^2 + t,\ 0 < t < T_1 \right\}. \qquad (2.1.4)$$

We can always suppose that $G \subset \Omega \times (0,T)$ if T_1 is small enough. Defining U by

$$U(x,t) = \frac{C(R^2 + t)^{\theta/2}}{(R^2 - r^2 + t)^{\theta}}, \qquad\qquad (2.1.5)$$

with $\theta = 2/(p-1)$, $r = |x - x_0|$ and C a positive constant to be determined, then U is well defined in G where it satisfies

$$U_t - \Delta U + U^p = \frac{\theta}{2} \frac{C(R^2 + t)^{\theta/2-1}}{(R^2 - r^2 + t)^{\theta}} - \frac{4C(\theta+1)r^2(R^2 + t)^{\theta/2}}{(R^2 - r^2 + t)^{\theta+2}}$$
$$- \frac{C(2N+1)\theta(R^2 + t)^{\theta/2}}{(R^2 - r^2 + t)^{\theta+1}} + \frac{C^p(R^2 + t)^{\theta p/2}}{(R^2 - r^2 + t)^{\theta p}}. \qquad (2.1.6)$$

Since $\theta + 2 = \theta p$, there holds

$$U_t - \Delta U + U^p \geq 0 \quad \text{in } G, \qquad\qquad (2.1.7)$$

if $C^{p-1} \geq \max((2N+1)\theta, 4\theta(\theta+1))$. Since U dominates u on the lateral boundary of G the same holds in whole G and in particular

$$u(x_0,t) \le U(x_0,t) = \frac{C}{(R^2+t)^{\theta/2}}.\tag{2.1.8}$$

Letting R go to $|x_0|$ and replacing u by -u yields (2.1.3). Brezis and Friedman main result is the following

THEOREM 6.8. *Let u be an element of* $C^{2,1}(\Omega \times (0,T]) \cap C(\Omega \times [0,T] \setminus \{(O,0)\})$ *satisfying* (2.1.1)-(2.1.2) *in* $\Omega \setminus \{O\}$ *and solution of* (2.1.1) *in* $\Omega \times (0,T)$. *If* $p \ge (N+2)/N$ *then u can be extended as an element of* $C^{2,1}(\Omega \times [0,T])$.

Proof. The proof given here follows the one of [BF], another one will be derived in Section 6.3 as a consequence of the classification of isolated singularities of solutions of (2.1.2).
Step 1. We claim that

$$\int_0^T \int_{B(O,\rho)} |u(x,t)|^p dxdt < \infty,\tag{2.1.9}$$

if ρ is as in Lemma 6.2. Since $p \ge (N+2)/N$, we have

$$\int_0^T \int_{B(O,\rho)} |u(x,t)|dxdt < \infty.\tag{2.1.10}$$

With an adaptation of the elliptic Kato inequality it is easy to check that the following inequality

$$|u|_t - \Delta|u| + |u|^p \le 0\tag{2.1.11}$$

holds in the sense of distributions in $\Omega \times (0,T)$. Let $\zeta \in \mathcal{D}(\Omega \times [0,T])$ such that $0 \le \zeta \le 1$ and $\zeta = 1$ on $B(O,\rho) \times [0,T/2]$. Let $\eta \in C^\infty([0,\infty))$ be a truncation function such that $\eta' \ge 0$ and

$$\eta(t) = \begin{cases} 1 & \text{on} \quad [2,\infty), \\ 0 & \text{on} \quad [0,1]. \end{cases}\tag{2.1.12}$$

We set $\eta_k(t) = \eta(kt)$ for any integer k and take $(x,t) \mapsto \eta_k(|x|^2+t)\zeta(x,t) = \phi_k(x,t)$ as a test function. Then

$$\int_0^T \int_\Omega \left(-|u|(\phi_k)_t - |u|\eta_k\Delta\phi_k + |u|^p \phi_k\right)dxdt \le 0,\tag{2.1.13}$$

or equivalently,

$$\int_0^T \int_\Omega |u|^p \phi_k \, dxdt \le \int_0^T \int_\Omega \big(|u|(\phi_k)_t + |u| \Delta \phi_k \big) dxdt. \tag{2.1.14}$$

We define $D_k = \left\{ (x,t) : k^{-1} \le |x|^2 + t \le 2k^{-1} \right\}$. Since

$$(\phi_k)_t = \eta_k{}' \zeta + \eta_k \zeta_t, \quad \Delta \phi_k = \zeta \Delta \eta_k + 2\nabla \eta_k . \nabla \zeta + \eta_k \Delta \zeta, \tag{2.1.15}$$

there holds

$$|(\phi_k)_t| \le C\big((k+1)\chi_{D_k} + \chi_{{}^c D_k}\big) \ \text{ and } \ |\Delta \phi_k| \le C\big((k+1)\chi_{D_k} + \chi_{{}^c D_k}\big), \tag{2.1.16}$$

for some $C > 0$. Therefore

$$\int_0^T \int_\Omega |u|^p \phi_k \, dxdt \le Ck \iint_{D_k} |u| \, dxdt + C. \tag{2.1.17}$$

From Lemma 6.2 and $p \ge (N+2)/N$,

$$\iint_{D_k} |u| \, dxdt \le C \iint_{D_k} \big(|x|^2 + t\big)^{-N/2} dxdt \le Ck^{N/2} \operatorname{meas}(D_k), \tag{2.1.18}$$

which implies (2.1.9).

Step 2. The function u satisfies (2.1.1) in the sense of distributions in $\Omega \times [0,T)$. If $\zeta \in \mathcal{D}\big(\Omega \times [0,T)\big)$ and $(x,t) \mapsto \eta_k(|x|^2 + t)\zeta(x,t) = \phi_k(x,t)$, then

$$\int_0^T \int_\Omega \big(-u(\phi_k)_t - u\Delta \phi_k + |u|^{p-1} u \phi_k \big) dxdt = 0. \tag{2.1.19}$$

It is therefore sufficient to check that

$$\lim_{k \to \infty} \int_0^T \int_\Omega u\zeta(\eta_k)_t \, dxdt = \lim_{k \to \infty} \int_0^T \int_\Omega u\zeta \Delta \eta_k \, dxdt = \lim_{k \to \infty} \int_0^T \int_\Omega u\nabla \eta_k . \nabla \zeta \, dxdt. \tag{2.1.20}$$

In fact all of the three terms in (2.1.20) are dominated by $Ck \iint_{D_k} |u| \, dxdt$ and

$$\left| \int_0^T \int_\Omega u\zeta(\eta_k)_t \, dxdt \right| \le Ck \iint_{D_k} |u| \, dxdt \le Ck \left(\iint_{D_k} |u|^p \, dxdt \right)^{1/p} |D_k|^{1/p'},$$

$$\tag{2.1.21}$$

$$\le C'' \left(\iint_{D_k} |u|^p \, dxdt \right)^{1/p'},$$

since $|D_k| = C' k^{-(1+N/2)}$ and $(1 + N/2)/p' \ge 1$. Consequently, we have (2.1.20), which implies

316

$$\int_0^T \int_\Omega \left(-u\zeta_t - u\Delta\zeta + |u|^{p-1} u\zeta \right) dxdt = 0, \tag{2.1.22}$$

which is the claim.

Step 3. End of the proof.

We set $M = \max\limits_{\partial B(O,\rho)\times[O,T]} u(x,t)+1$ and define $\tilde{u}$ by

$$\tilde{u} = \begin{cases} u(x,t) & \text{if } t>0, \\ 0 & \text{if } t<0. \end{cases} \tag{2.1.23}$$

Then $\tilde{u}$ satisfies $\tilde{u}_t - \Delta\tilde{u} + |\tilde{u}|^{p-1}\tilde{u} = 0$ in the sense of distributions in $\Omega\times(-T,T)$. Therefore the function $(\tilde{u}-M)^+$ is subcaloric in $\Omega\times(-T,T)$ and vanishes in a neighborhood of the parabolic boundary $\partial\Omega\times(-T,T)\cup\Omega\times\{-T\}$. It is therefore identically 0 and $u \leq M$. In the same way u is uniformly minorized in $\Omega\times(0,T)$. By the parabolic estimate theory u is a strong solution and can be extended by continuity into a function which belongs to $C^{2,1}(\Omega\times[0,T))$.

The theory of parabolic capacities as developped in [BP2] gives much more precise removability results for solutions of (2.1.1)

Definition. Let $1 < p < \infty$, g be an open subset of $\mathbb{R}^N$, $T > 0$ and $G \times (0,T) = \Xi$, then

$$W_p^{2,1}(\Xi) = \left\{ u \in L^p(\Xi): \frac{\partial u}{\partial t}, \frac{\partial u}{\partial x_i}, \frac{\partial^2 u}{\partial x_i \partial x_j} \in L^p(\Xi), \ \forall\, 1 \leq i,j \leq N \right\}, \tag{2.1.24}$$

is a Banach space under the norm

$$\|u\|_{W_p^{2,1}} = \left(\iint_\Xi \left(\left|\frac{\partial u}{\partial t}\right|^p + \sum_i \left|\frac{\partial u}{\partial x_i}\right|^p + \sum_i \left|\frac{\partial^2 u}{\partial x_i \partial x_j}\right|^p \right) dxdt \right)^{1/p}. \tag{2.1.25}$$

If K is a compact subset of Ξ, we set

$$c_{2,1,p}(K) = \inf\left\{ \|v\|_{W_p^{2,1}} : v \in C_0^\infty(\Xi), v \geq 1 \text{ in a neighborhood of } K \right\}. \tag{2.1.26}$$

The Sobolev-Besov capacity of a compact subset K of G is defined for a non integer order of differentiation α by

$$c_{\alpha,p}(K) = \inf\left\{ \|v\|_{W^{\alpha,p}} : v \in C_0^\infty(G), v \geq 1 \text{ in a neighborhood of } K \right\} \tag{2.1.27}$$

where $W^{\alpha,p}(G)$ is the fractional Sobolev-Besov space of order α ([Ma1]). In fact we need not go further in that direction since the Sobolev-Besov capacity on G can be understood as the trace of a parabolic capacity in Ξ via the following estimates

PROPOSITION 6.4. *Set* $G = \mathbb{R}^N$ *and* $\Xi = \mathbb{R}^N \times (0,T)$. *Then there exists* $0 < k_1 < k_2$ *such that for any compact subset* F *in* G *and any* $t_o \in (0,T)$, *there holds*

$$k_1 c_{2-2/p,p}(F) \leq c_{2,1,p}(F \times \{t_0\}) \leq k_2 c_{2-2/p,p}(F).\tag{2.1.28}$$

For any $0 < t_1 < t_2 < T$, *there exist positive constants* ℓ_1 *and* ℓ_2 *such that for any subset* F *of* G *such that the relation*

$$\ell_1 c_{2,p}(F) \leq c_{2,1,p}(F \times [t_0,t_1]) \leq \ell_2 c_{2,p}(F)\tag{2.1.29}$$

holds.

The following general removability result is proved in [BP2] by using an approriate extension to the parabolic case of the duality technique of Sections 2.2.2 and 2.2.3.

THEOREM 6.9. *Let* K *be a compact subset of* Ξ, F *a compact subset of* G *and* $p > 1$. *Then any function* u *which satisfies*

$$\begin{cases} u \in L^p_{loc}(\Xi \setminus K), \\ \dfrac{\partial u}{\partial t} - \Delta u + |u|^{p-1} u = 0 \text{ in the sense of distributions in } \Xi \setminus K, \end{cases}\tag{2.1.30}$$

satisfies also

$$\begin{cases} u \in L^p_{loc}(\Xi), \\ \dfrac{\partial u}{\partial t} - \Delta u + |u|^{p-1} u = 0 \text{ in the sense of distributions in } \Xi, \end{cases}\tag{2.1.31}$$

if and only if $c_{2,1,p'}(K) = 0$. *Any solution* u *of* (2.1.31) *which satsifies also*

$$\lim_{t \to 0} u(.,t) = 0 \text{ essentially in the sense of measure in } G\backslash F\tag{2.1.32}$$

satisfies in fact

$$\lim_{t \to 0} u(.,t) = 0 \text{ essentially in the sense of measure in } G\tag{2.1.33}$$

if and only if $c_{2/p,p'}(F) = 0$.

6-2-2 Measure as initial data

Let Ω be a bounded open domain of $\mathbb{R}^N$ with a smooth boundary $\partial\Omega$, then the initial value Cauchy-Dirichlet problem for equation (2.1.1) with a measure as initial condition can be formulated in the following way: find a function u belonging to $C^{2,1}\left(\overline{\Omega}\times(0,\infty)\right)$ such that

$$u_t - \Delta u + |u|^{p-1}u = 0 \quad \text{in}\quad \Omega\times(0,\infty), \tag{2.2.1}$$

$$u = 0 \quad \text{in}\quad \partial\Omega\times(0,\infty), \tag{2.2.2}$$

$$u(x,0) = \mu \in \mathcal{M}_b(\Omega), \tag{2.2.3}$$

where $\mathcal{M}_b(\Omega)$ is the set of bounded measures in Ω. As for the initial condition (2.2.3) it must be understood in the sense of trace, that is

$$\lim_{t\to 0}\int_\Omega u(x,t)\zeta(x)dx = \int_\Omega \zeta(x)d\mu(x), \tag{2.2.4}$$

for any $\zeta \in C_0(\overline{\Omega})$. Brezis and Friedman ([BF]) proved the first existence result for this problem.

THEOREM 6.10. *Suppose* $1 < p < (N+2)/N$ *and* $\mu \in \mathcal{M}_b(\Omega)$. *Then there exists a unique solution* $u \in C^{2,1}\left(\overline{\Omega}\times(0,\infty)\right)$ *of* (2.2.1)-(2.2.3) *in* $\Omega\times(0,\infty)$.

Later on Baras and Pierre ([BP2]) extended this result to more general measures

THEOREM 6.11. *Suppose* $p > 1$ *and* $\mu \in \mathcal{M}_b(\Omega)$. *Then there exists a unique solution* u *belonging to* $C^{2,1}\left(\overline{\Omega}\times(0,\infty)\right)$*of* (2.2.1)-(2.2.3) *in* $\Omega\times(0,\infty)$ *if and only if* μ *does not charge the compact subsets* $F \subset \Omega$ *with* $c_{2/p,p'}(F) = 0$.

For a more general nondecreasing nonlinearity there is a simple proof of existence which is essentially due to Moutoussamy and Véron ([MV1]). We recall that $E(x,t)$ is the heat kernel in $\mathbb{R}^N$ defined by (1.1.1). We denote $E^\Omega(x,t,y)$ the fundamental solution of the heat equation in Ω with Dirichlet conditions on $\partial\Omega$ and δ_y as initial condition and $(S^\Omega(t))_{t\geq 0}$ the corresponding semi-group. Clearly

$$0 \leq E^\Omega(x,t,y) \leq E(x,t) \qquad (\forall(x,y,t)\in \Omega\times\Omega\times(0,\infty)). \tag{2.2.5}$$

The solution of the heat equation in $\Omega\times(0,\infty)$ with Dirichlet conditions on $\partial\Omega$ and $\mu \in \mathcal{M}_b(\Omega)$ as initial condition is

$$E^\Omega_\mu(x,t) = \int_\Omega E^\Omega(x,t,y)d\mu(y), \tag{2.2.6}$$

for (x,t) in $\Omega \times (0,\infty)$. If g is a continuous nondecreasing function vanishing at 0, we shall say that $\mu \in \mathcal{M}_b(\Omega)$ is g-admissible if

$$\int_0^T \int_\Omega \left| g(E_{|\mu|}^\Omega(x,t)) \right| dxdt < \infty. \tag{2.2.7}$$

THEOREM 6.12. *Let g be as above and $\mu \in \mathcal{M}_b(\Omega)$ be a g-admissible measure. Then there exists a unique $u \in C^{2,1}\left(\overline{\Omega} \times (0,\infty)\right)$ satisfying $g(u) \in L^1_{loc}\left(\overline{\Omega} \times [0,\infty)\right)$ and*

$$u_t - \Delta u + g(u) = 0 \quad in \quad \Omega \times (0,\infty), \tag{2.2.8}$$

$$u = 0 \quad on \quad \partial\Omega \times (0,\infty), \tag{2.2.9}$$

$$u(x,0) = \mu \quad in \quad \Omega, \tag{2.2.10}$$

where (2.2.10) has to be understood in the sense (2.2.4).

Proof. Existence: We first assume that μ is nonnegative. If k and n are two nonzero integers, we set $g_k(r) = \text{sign}(r)\min(|g(r)|, k)$ and $u^n = u^{n,k}$ the solution of

$$u_t^n - \Delta u^n + g_k(u^n) = 0 \quad in \quad \Omega \times (0,\infty), \tag{2.2.11}$$

$$u^n = 0 \quad on \quad \partial\Omega \times (0,\infty), \tag{2.2.12}$$

$$u^n(x,0) = E_\mu^\Omega \ (x,1 = \! \mid n) \quad in \quad \Omega. \tag{2.2.13}$$

We can express u^n with the help of the semi-group $(S^\Omega(t))_{t \geq 0}$ via the following formula

$$u_n(.,t) = E_\mu^\Omega(.,t+1/n) - \int_0^t S^\Omega(t-s)g_k(u_n(.,s))ds, \tag{2.2.14}$$

and clearly
$$0 \leq u_n(x,t) \leq E_\mu^\Omega(x,t+1/n). \tag{2.2.15}$$

Since g_k is bounded, the set $\{g_k(u_n)\}$ remains bounded in $L^\infty\left(0,T;L^\infty(\Omega)\right)$ independently of n, for any $T > 0$. From [BHV] we know that the operator

$$\phi \mapsto K\phi, \quad with \quad K\phi(t) = \int_0^t S^\Omega(t-s)\phi(s)ds, \tag{2.2.16}$$

is compact from $L^\infty(0,T;L^\infty(\Omega))$ into $C([0,T];L^\infty(\Omega))$. Consequently there exist a sequence $\{n_m\}$ and a function $h \in L^\infty(0,\infty;L^\infty(\Omega))$ such that $\{Kg_k(u_{n_m})\}$ converges to Kh in $L^\infty_{loc}(0,\infty;L^\infty(\Omega))$. If we go to the limit in (2.2.14), we can define U^k by

$$U^k(.,t) = E^\Omega_\mu(.,t) - Kh(t), \tag{2.2.17}$$

and $h(x,t) = g_k(U^k(x,t))$ a.e. in $\Omega \times (0,\infty)$. Therefore U^k satisfies (2.2.8)-(2.2.10) with g replaced by g_k. Since $0 \le g_k(r) \le g_{k'}(r)$ for $r > 0$ and $k > k' > 0$, we have

$$0 \le U^k \le U^{k'} \text{ for } k > k' > 0, \text{ and } 0 \le g_k(U^k) \le g(U^k) \le g(E^\Omega_\mu). \tag{2.2.18}$$

Let u be the decreasing limit of U^k when k goes to infinity. Because the relation

$$\int_0^\infty \int_\Omega \left(U^k(-\zeta_t - \Delta\zeta) + g_k(U^k)\zeta \right) dxdt = \int_\Omega \zeta(x,0)d\mu(x) \tag{2.2.19}$$

holds for any $\zeta \in C^{2,1}(\overline{\Omega} \times [0,\infty))$ vanishing on $\partial\Omega \times [0,\infty)$ and for t large enough, and because $g_k(U^k(x,t))$ converges to $g(u(x,t))$ a.e. in $\Omega \times (0,\infty)$, we deduce that the following equality follows

$$\int_0^\infty \int_\Omega \left(u(-\zeta_t - \Delta\zeta) + g(u)\zeta \right) dxdt = \int_\Omega \zeta(x,0)d\mu(x). \tag{2.2.20}$$

This finally implies that (2.2.8)-(2.2.10) hold.

For a general measure μ, we construct, in the same way $u^n = u^{n,k}$, which satisfies

$$0 \le |u^n(x,t)| \le E^\Omega_{|\mu|}(x,t+1/n), \tag{2.2.21}$$

and U^k such that (2.2.19) holds for any $\zeta \in C^{2,1}(\overline{\Omega} \times [0,\infty))$ vanishing on $\partial\Omega \times [0,\infty)$. Since

$$U^k(.,t) = E^\Omega_\mu(.,t) - \int_0^t S^\Omega(t-s)g_k(U^k(.,s))ds, \tag{2.2.22}$$

we deduce from the compactness result ([BHV]) and $0 \le |g_k(U^k)| \le |g(U^k)| \le g(E^\Omega_{|\mu|})$ that the set $\{U^k - E^\Omega_\mu\}$ is relatively compact in $L^\infty_{loc}([0,\infty);L^1(\Omega))$. If u is any function limit of a subsequence $\{U^{k_n}\}$ in the $L^\infty_{loc}(0,\infty;L^1(\Omega))$-topology and a.e. in $\Omega \times (0,\infty)$, u belongs to $C^1(0,\infty;L^1(\Omega))$ and (2.2.20) is verified, which in turn implies (2.2.8)-(2.2.10) .

Uniqueness. Suppose u and u^* are two solutions of the same equation with the same initial data μ, then $w = u - u^*$ satisfies

321

$$\int_0^\infty \int_\Omega \big(w(-\zeta_t - \Delta\zeta) + (g(u) - g(u^*))\zeta \big) dxdt = 0, \tag{2.2.23}$$

where $\zeta \in C^{2,1}(\overline{\Omega} \times [0,\infty))$ vanishes on $\partial\Omega \times [0,\infty)$ and for t large enough. Equivalently, with $f = g(u^*) - g(u)$,

$$w_t - \Delta w = f \quad \text{in} \quad \Omega \times (0,\infty), \tag{2.2.24}$$

with vanishing data on $\partial\Omega \times [0,\infty) \cup \Omega \times \{0\}$. An approximation technique [BF] (see Lemma 4.1 for the elliptic case) yields

$$\int_0^T \int_\Omega f \, \mathrm{sgn}(w) dxdt \geq \int_0^T \int_\Omega |w| dxdt. \tag{2.2.25}$$

From this we deduce that w vanishes since $f \, \mathrm{sgn}(w) \leq 0$.

Remark 6.3. It is possible to weaken the assumption on the measure μ by writting its Lebesgue's decomposition $\mu = \mu_S + \mu_R$ where μ_S (resp. μ_R) is the singular part (resp. the regular part) and in assuming that g is nondecreasing, vanishes at O, satisfies the Δ_2-condition as in Theorem 4.2 and that

$$\int_0^T \int_\Omega \left| g(E_{|\mu_s|}^\Omega(x,t)) \right| dxdt < \infty, \tag{2.2.26}$$

holds. The existence proof is an adaptation of the one of Theorem 4.2.

6-2-3 The very singular solutions

Let $p > 1$ be a real number and $u \in C^{2,1}(\mathbb{R}^N \times [0,T)\backslash\{O,0)\}$ a solution of

$$u_t - \Delta u + u|u|^{p-1} = 0 \qquad \text{in} \quad \mathbb{R}^N \times (0,T), \tag{2.3.1}$$

$0 < T \leq \infty$, satisfying the initial condition

$$u(x,0) = 0 \qquad \text{in} \quad \mathbb{R}^N \backslash \{O\}. \tag{2.3.2}$$

The problem is to determine the behaviour of $u(x,t)$ when (x,t) tends to $(0,0)$. Concerning the problem (2.3.1)-(2.3.2), Brezis, Peletier and Terman [BPT] first proved the existence of a very singular solution W under the following form:

$$W(x,t) = t^{-1/(p-1)} f(|x| / \sqrt{t}), \tag{2.3.3}$$

where f is the unique positive solution of the following differential equation:

$$-f_{yy} - \left(\frac{N-1}{y} + \frac{y}{2}\right)f_y - \frac{1}{p-1}f + f^p = 0 \quad \text{on} \quad (0,+\infty), \tag{2.3.4}$$

with the boundary conditions

$$f_y(0) = 0 \ , \ \lim_{y\to\infty} y^{2/(p-1)}f(y) = 0. \tag{2.3.5}$$

Their methods is essentially based on a delicate phase space analysis. But it gives also very precise asymptotic expansion, namely

$$f(y) = Ae^{-y^2/4}y^{2/(p-1)-N}\left[(1-(2/(p-1)-N)(2/(p-1)-2)y^{-2} + \circ(y^{-2})\right]. \tag{2.3.6}$$

In fact, if we look for particular solution of (2.3.1)-(2.3.2) under the form

$$u(x,t) = t^{-1/(p-1)}\omega(\eta) \quad \text{with} \quad \eta = x/\sqrt{t} \ , \tag{2.3.7}$$

then ω satisfies

$$-\Delta\omega - \frac{1}{2} < \nabla\omega, y > + \omega|\omega|^{p-1} - \frac{1}{p-1}\omega = 0 \tag{2.3.8}$$

in $\mathbb{R}^N$, or equivalently

$$-K^{-1}\text{div}(K\nabla\omega) + \omega|\omega|^{p-1} - \frac{1}{p-1}\omega = 0, \tag{2.3.9}$$

with $K = K(\eta) = e^{|\eta|^2/4}$. For $\lambda \in \mathbb{R}$, define the set

$$\mathcal{E}_\lambda = \left\{\omega \in H^1(K,\mathbb{R}^N) \cap L^{p+1}(K,\mathbb{R}^N) \text{ satisfying (2.3.11) below }\right\} \tag{2.3.10}$$

$$-K^{-1}\text{div}(K\nabla\omega) + \omega|\omega|^{p-1} - \lambda\omega = 0. \tag{2.3.11}$$

Then $\mathcal{E}_\lambda$ has the following structure.

PROPOSITION 6.5. *Let* p > 1, *then*
  (i) *if* $\lambda \le N/2$, $\mathcal{E}_\lambda$ *is reduced to the zero function ;*
  (ii) *if* $N/2 < \lambda \le (N+1)/2$, $\mathcal{E}_\lambda$ *contains three elements: the zero function and* $\pm\omega$ *where* ω *is the unique positive solution of* (2.3.11) *belonging to* $H^1_K \cap L^{p+1}_K$;
  (iii) *if* $\lambda > (N+1)/2$, $\mathcal{E}_\lambda$ *contains the three preceeding elements and some changing sign solutions which are not invariant under* O(N).

Proof. Step 1. The first assertion is a consequence of the fact that the first eigenvalue of the operator L in H^1_K is N/2 (see the definition in Section 6.1.1). For the second assertion,

323

we use a device introduced by Véron [Ve4] for proving first that ω is a radial function; we write (2.3.11) in spherical coordinates $(\rho,\sigma) \in (0,\infty) \times S^{N-1}$

$$\omega_{\rho\rho} + \left(\frac{N-1}{\rho} + \frac{\rho}{2}\right)\omega_\rho + \frac{1}{\rho^2}\Delta_{S^{N-1}}\omega + \lambda\omega - |\omega|^{p-1}\omega = 0, \qquad (2.3.12)$$

where $\Delta_{S^{N-1}}$ is the Laplace-Beltrami operator on the unit sphere. Let $\overline{\omega}$ be the spherical average of w; then

$$\overline{\omega}_{\rho\rho} + \left(\frac{N-1}{\rho} + \frac{\rho}{2}\right)\overline{\omega}_\rho + \lambda\overline{\omega} - \overline{\omega|\omega|^{p-1}} = 0. \qquad (2.3.13)$$

If we set $\xi = \omega - \overline{\omega}$, the function ξ satsfies

$$K^{-1}\mathrm{div}(K\nabla\xi) + \lambda\xi - \left(\omega|\omega|^{p-1} - \overline{\omega|\omega|^{p-1}}\right) = 0, \qquad (2.3.14)$$

and this implies

$$-\int_{\mathbb{R}^N}|\nabla\xi|^2 K d\eta + \lambda\int_{\mathbb{R}^N}\xi^2 K d\eta - -\int_{\mathbb{R}^N}\left(\omega|\omega|^{p-1} - \overline{\omega|\omega|^{p-1}}\right)(\omega - \overline{\omega})K d\eta. \qquad (2.3.15)$$

Moreover

$$\int_{S^{N-1}}\left(\omega|\omega|^{p-1} - \overline{\omega|\omega|^{p-1}}\right)(\omega - \overline{\omega})d\sigma = \int_{S^{N-1}}\left(\omega|\omega|^{p-1} - \overline{\omega}|\overline{\omega}|^{p-1}\right)(\omega - \overline{\omega})d\sigma$$
$$\geq 2^{1-p}\int_{S^{N-1}}|\omega - \overline{\omega}|^{p+1}d\sigma. \qquad (2.3.16)$$

and

$$\int_{\mathbb{R}^N}\left(\xi K^{-1}\right)K d\eta = \int_0^\infty\left(\int_{S^{N-1}}\xi d\sigma\right)\rho^{N-1}d\rho = 0. \qquad (2.3.17)$$

But (2.3.17) means that ξ is orthogonal to the first eigenfunction of L in L_K^2. As $(N+1)/2$ is the second eigenvalue, there holds

$$\int_{\mathbb{R}^N}|\nabla\xi|^2 K d\eta \geq \frac{N+1}{2}\int_{\mathbb{R}^N}\xi^2 K d\eta. \qquad (2.3.18)$$

Therefore,

$$\left(\lambda - \frac{N-1}{2}\right)\int_{\mathbb{R}^N}\xi^2 K d\eta - 2^{1-p}\int_{\mathbb{R}^N}|\xi|^{p+1}K d\eta \geq 0, \qquad (2.3.19)$$

324

which implies that ξ is zero and ω is radial.

Step 2. Let $H^1_{K,rad}$ be the space of radial functions over $\mathbb{R}^N$ which belong to H^1_K. From the computation of the spectrum of L in H^1_K, it is clear that the second eigenvalue in $H^1_{K,rad}$ is $1+N/2$ with a one dimensional eigenspace generated by $\eta \mapsto \left(|\eta|^2 - N\right)e^{-|\eta|^2/4}$. *We claim that there exists no changing sign element in* $H^1_{K,rad} \cap \mathcal{E}_\lambda$ *when* $\lambda \leq 1+N/2$. In fact, if such a solution ω exists, we denote a the first zero of ω on $(0,\infty)$ and let Ω_a (resp. Ω_a^c) be the open ball with center 0 and radius a (resp. its complementary in $\mathbb{R}^N$). If λ_a (resp. μ_a) is the first eigenvalue of L in $H^1_{K,rad}(\Omega_a)$ (resp. $H^1_{K,rad}(\Omega_a^c)$) with zero boundary conditions, then by the Courant Min-Max principle ([CH]), we have

$$1+N/2 \leq \max(\lambda_a, \mu_a), \tag{2.3.20}$$

which implies that one of the two following inequalities holds

$$\int_{\Omega_a} |\nabla\omega|^2 K d\eta \geq (1+N/2)\int_{\Omega_a} \omega^2 K d\eta, \tag{2.3.21}$$

$$\int_{\Omega_a^c} |\nabla\omega|^2 K d\eta \geq (1+N/2)\int_{\Omega_a^c} \omega^2 K d\eta. \tag{2.3.22}$$

Therefore, multiplying (2.3.11)) by $K\omega$ and integrating the resulting relation on Ω_a or on Ω_a^c implies that one of the two following inequalities is valid

$$\int_{\Omega_a}\left(\left(1+\frac{N}{2}-\lambda\right)\omega^2 + |\omega|^{p+1}\right)K d\eta \leq 0, \tag{2.3.23}$$

$$\int_{\Omega_a^c}\left(\left(1+\frac{N}{2}-\lambda\right)\omega^2 + |\omega|^{p+1}\right)K d\eta \leq 0, \tag{2.3.24}$$

consequently ω is zero.

Step3. End of the proof. As there exists only one positive element in $\mathcal{E}_\lambda$ (see [EK]), we have (ii). For the last assertion, we construct a non-radial element of $\mathcal{E}_\lambda$ as in [Ve4], just by constructing it on $\mathbb{R}^N_+ = \mathbb{R}^{N-1} \times (0,\infty)$ - in the space $H^1_K(\mathbb{R}^N_+)$, vanishing on the boundary - and this is possible since $\lambda > 1+N/2$ which is the first eigenvalue of the operator L in $H^1_K(\mathbb{R}^N_+)$ with zero boundary conditions, and then to reflect it through $\mathbb{R}^{N-1} \times \{0\}$.

The following result which extends, with a different argument adapted from [KV], a previous result due to Oswald ([Os]) points out the role of the positive very singular solution as a majorant for any solution of (2.3.1)-(2.3.2).

PROPOSITION 6.6. *Let* $1 < p < (N+2)/N$, u *be a solution of* (2.3.1)-(2.3.2) *and* W *be the unique positive very singular solution of the equation. Then there exists a positive constant* C *such that*

$$|u(x,t)| \leq CW(x,t) \qquad \left(\forall (x,t) \in \mathbb{R}^N \times [0,\infty) \setminus \{(O,0)\} \right). \qquad (2.3.25)$$

Proof. Step 1. By considering specific solutions of (2.3.1) independent of x, under the form $t \mapsto (1/(p-1))^{1/(p-1)} (t-\tau)^{1/(p-1)}$ for $\tau > 0$ and letting τ go to 0, we see that the following estimate holds in $\mathbb{R}^N \times (0,T)$

$$|u(x,t)| \leq (1/(p-1))^{1/(p-1)} t^{-1/(p-1)} = C_* t^{-1/(p-1)}. \qquad (2.3.26)$$

For $\varepsilon > 0$, let Ω_ε be the set of x in $\mathbb{R}^N$ such that $|x| > \varepsilon$. For $\delta > 0$ we consider the following Cauchy -Dirichlet problem in $\Omega_\varepsilon \times (0,\infty)$

$$\begin{cases} v_t^{\varepsilon,\delta} - \Delta v^{\varepsilon,\delta} + (v^{\varepsilon,\delta})^p = 0 & \text{in } \Omega_\varepsilon \times (0,\infty), \\ v^{\varepsilon,\delta} = C_*(t+\delta)^{-1/(p-1)} & \text{in } \partial\Omega_\varepsilon \times (0,\infty), \\ v^{\varepsilon,\delta} = 0 & \text{in } \Omega_\varepsilon \times \{0\}. \end{cases} \qquad (2.3.27)$$

A straightforward adaptation of the proof of Lemma 6.2 yields the following *a priori* estimate

$$v^{\varepsilon,n}(x,t) \leq C(p,N)(1/((|x|-\varepsilon)^{2/(p-1)} + t^{1/(p-1)})) \quad \text{in } \Omega_\varepsilon \times (0,\infty), \qquad (2.3.28)$$

and (2.3.26) extends as follows

$$v^{\varepsilon,\delta}(x,t) \leq C_* t^{-1/(p-1)} \qquad \left(\forall (x,t) \in \Omega_\varepsilon \times (0,\infty) \right). \qquad (2.3.29)$$

Moreover $(\varepsilon,\delta) \mapsto v^{\varepsilon,\delta}$ is nonincreasing in ε and δ, and from uniqueness we have

$$v^{\varepsilon/\sqrt{k},\delta/k}(x,t) = k^{1/(p-1)} v^{\varepsilon,\delta}(\sqrt{k}x,kt), \qquad (2.3.30)$$

for any $k > 0$. When δ goes to 0, $v^{\varepsilon,n}$ converges uniformly on any compact subset of $\Omega_\varepsilon \times [0,\infty)$ to some function v^ε which satisfies the same equation (2.3.1) in $\Omega_\varepsilon \times (0,\infty)$, with zero initial value on $\Omega_\varepsilon \times \{0\}$ and such that

$$v^\varepsilon(x,t) = C_* t^{-1/(p-1)} \qquad \left(\forall (x,t) \in \partial\Omega_\varepsilon \times (0,\infty) \right). \qquad (2.3.31)$$

From the scaling property (2.3.30), v^ε satifies:

$$v^{\varepsilon/\sqrt{k}}(x,t) = k^{1/(p-1)} v^\varepsilon(\sqrt{k}x,kt) \qquad (2.3.32)$$

326

Comparing u and v^ε on $\Omega_\varepsilon \times (0,T)$ implies that $|u| \le v^\varepsilon$ holds from (2.3.26) and (2.3.31). When goes to 0, that v^ε decreases and converges uniformly on any compact subset of $\mathbb{R}^N \backslash \{O\} \times (0,\infty)$ to some solution v of (2.3.1) in $\mathbb{R}^N \backslash \{O\} \times (0,\infty)$ which vanishes on $\mathbb{R}^N \backslash \{O\} \times \{0\}$; the property (2.3.32) for v^ε implies that v is invariant under the previous changes of scale, that is

$$v(x,t) = k^{1/(p-1)} v(\sqrt{k}x, kt) \tag{2.3.33}$$

for any $k > 0$. Therefore v is a similarity solution and can be written under the following form:

$$v(x,t) = t^{-1/(p-1)} \psi(x/\sqrt{t}) \tag{2.3.34}$$

where ψ is a positive radial function solution of

$$-\psi_{yy} - \left(\frac{N-1}{y} + \frac{y}{2}\right)\psi_y - \frac{1}{p-1}\psi + \psi^p = 0 \qquad \text{on } (0,\infty). \tag{2.3.35}$$

Moreover ψ satisfies

$$\lim_{y\to\infty} y^{2/(p-1)} \psi(y) = 0. \tag{2.3.36}$$

Step 2. We claim that there exists a constant $C > 0$ such that

$$\psi \le Cf, \tag{2.3.37}$$

where f is defined from W by (2.3.3)-(2.3.5). From [BPT] we have the very precise asymptotic expansion (2.3.6) which reads

$$f(y) = e^{-y^2/4} y^{2/(p-1)-N}\left(1 + O(y^{-2})\right) \qquad \text{as } y \to \infty \tag{2.3.38}$$

and implies, by a direct integration at infinity,

$$f_y(y) / f(y) = -\frac{1}{2} y(1 + o(1)) \qquad \text{as } y \to \infty. \tag{2.3.3}$$

Because ψ satisfies (2.3.35)-(2.3.36) it can be checked that Brezis, Peletier and Terman's method applies and that the asymptotic formulas (2.3.38) and (2.3.39) are also valid for ψ and ψ_y / ψ respectively. From the maximum principle the function ψ is not increasing and its value at 0 is its limit. Taking $C > \max(1, \psi(0)/f(0))$ and setting $G(y) = y^{N-1}e^{y^2/4}$, we multiply the equation for f by G/f, the equation for ψ by G/ψ. Substracting the resulting identities, we multiply by $(\psi^2 - C^2 f^2)^+$ and integrate over (0,T), for $T > 0$. Therefore we obtain

$$\frac{(G\psi_y)_y}{\psi} - \frac{(GCf_y)_y}{Cf} - (\psi^{p-1} - (Cf)^{p-1}) \geq 0, \tag{2.3.40}$$

and finally

$$G(T)\left(\frac{\psi_y(T)}{\psi(T)} - \frac{f_y(T)}{f(T)}\right)\left(\psi^2(T) - C^2 f^2(T)\right)^+$$

$$- \int_0^T \left((\psi)^{p-1} - (Cf)^{p-1}\right)\left(\psi^2 - C^2 f^2\right)^+ dy$$

$$= \int_0^T \left[\left\{\psi_y\left(\psi_y - 2\frac{C^2 f f_y}{\psi} + \frac{C^2 f^2 \psi_y}{\psi^2}\right) + Cf_y\left(Cf_y - 2\frac{\psi\psi_y}{Cf} + \frac{Cf_y \psi^2}{C^2 f^2}\right)\right\}\right. \tag{2.3.41}$$

$$\left. \mathrm{sgn}^+(\psi - Cf)G\right] dy$$

But the right-hand side of (2.3.41) can be written under the form

$$\int_0^T \left\{\left(\psi_y - \frac{f_y \psi}{f}\right)^2 + \left(Cf_y - \frac{C\psi_y f}{\psi}\right)^2\right\}\mathrm{sgn}^+(\psi - Cf)Gdy. \tag{2.3.42}$$

Therefore, from estimates (2.3.35)-(2.3.36) on f and ψ, we deduce that

$$\limsup_{T\to\infty} \int_0^T (\psi^{p-1} - (Cf)^{p-1})(\psi^2 - C^2 f^2)^+ Gdy \leq 0. \tag{2.3.43}$$

Consequently $\psi \leq Cf$ on $[0,\infty)$ and the proof is complete.

Remark 6.4. Assume $\lambda > 0, p > 1$ and $w \in L_K^{p+1} \cap H_K^1$ is a solution of

$$K^{-1}\mathrm{div}(K\nabla w) + \lambda w - |w|^{p-1}w = 0 \tag{2.3.44}$$

in $\mathbb{R}^N$. Then w belongs $L_{K^{1/2}}^\infty \cap H_K^3$ and its bound in this space depends only on N, p and λ. This can be seen by an easy adaptation of Moser's iterative scheme. In particular $\mathcal{E}_{1/(p-1)}$ is a bounded subset of $L_{K^{1/2}}^\infty \cap H_K^3$.

6-2-4 Strong singularities

If u is a solution of (2.3.1)-(2.3.2), we consider now the following change of variable and unknown

$$w(y, \tau) = t^{1/(p-1)} u(x, t), \qquad (2.4.1)$$

where $\tau = \ln t$ and $y = \eta / \sqrt{t}$; then w satisfies

$$w_\tau - \Delta w - \frac{1}{2} <y, \nabla w> + w|w|^{p-1} - \frac{1}{p-1} w = 0 \qquad (2.4.2)$$

in $\mathbb{R}^N \times (-\infty, \ln T)$. The description of the isolated singularities of u at the origin is contained in the following result due to Moutoussamy and Véron [MV2]:

THEOREM 6.13. *Let* $1 < p < (N+2)/N$, *u be a solution of* (2.3.1)-(2.3.2) *and* w *be defined by* (2.4.1). *Then there exists a connected component* $\mathcal{E}^*$ *of* $\mathcal{E}_{1/(p-1)}$ *such that*

$$\lim_{\tau \to -\infty} \mathrm{dist}_\tau(w(., \tau), \mathcal{E}^*) = 0, \qquad (2.4.3)$$

where

$$\mathrm{dist}_\tau(w(., \tau), \mathcal{E}^*) = \inf \left\{ \|\omega - w(., \tau)\|_{H_K^1} + \|\omega - w(., \tau)\|_{L_{K^{1/2}}^-} : \omega \in \mathcal{E}^* \right\}. \qquad (2.4.4)$$

This theorem extends a previous one due to Oswald [Os] dealing only with nonnegative solutions. In fact the description of $\mathcal{E}_{1/(p-1)}$ allows to give a much more precise limit as the one in (2.4.3).

LEMMA 6.3. *Under the hypotheses of Theorem 6.13, we have the following :*

$$w \in L^\infty(-\infty, \ln T; H_K^1 \cap L_K^q \cap L_{K^\theta}^\infty) \cap L_{loc}^2(-\infty, \ln T; H_K^2) \qquad (2.4.5)$$

for any $\theta \in (0,1)$ *and any* $q \in (1, \infty)$.

Proof. Step 1. From estimate (2.4.43) and Proposition 6.6, we have

$$f(y) \leq C / K^\theta; \qquad (2.4.6)$$

therefore

$$w \in L^\infty(-\infty, \ln T; L_{K^\theta}^\infty \cap L_K^q) \qquad (2.4.7)$$

for any $\theta \in (0,1)$ and any $q \in (1, \infty)$. In the Hilbert space L_K^2 we define the following functional

$$\Phi(w) = \frac{1}{2} \int_{\mathbb{R}^N} |\nabla w|^2 K \, dy. \qquad (2.4.8)$$

It is clear that Φ is convex and lower semi continuous with domain H^1_K. Therefore its sub-differential $\partial\Phi$ is a maximal monotone operator in L^2_K. Let h be $\frac{1}{p-1}w - w|w|^{p-1}$, then h belongs to $L^2_{loc}(-\infty, \ln T; L^2_K)$ and the problem is to prove that w is a weak solution of

$$w_\tau + \partial\Phi(w) = h, \qquad\qquad (2.4.9)$$

in the sense of L^2_K and then to apply Brezis' regularity theory for example (see [Br3]).

Step 2. Let τ_0 be any real number and $\tilde{w}$ the weak solution of

$$\tilde{w}_\tau + \partial\Phi(\tilde{w}) = h \quad\text{on}\quad (\tau_0, \ln T), \text{ with } \tilde{w}(\tau_0) = w(\tau_0), \qquad (2.4.10)$$

and set $z = \tilde{w} - w$. Then the function z satisfies

$$z_\tau - K^{-1}\mathrm{div}(K\nabla z) = 0 \qquad\text{on}\quad (\tau_0, \ln T), \qquad\qquad (2.4.11)$$

with initial value 0. If n is any integer we define a sequence of smooth functions $\{\xi_n\}$ such that

$$\begin{cases} \text{(i)} & \sup p.(\xi_n) \subset B(0, 2n), \\ \text{(ii)} & \xi_n = 1 \quad\text{on}\quad B(O, n), \\ \text{(iii)} & \xi_n \geq 0. \end{cases} \qquad\qquad (2.4.12)$$

Multiplying (2.4.11) by $\xi_n K z$ and integrating over $\mathbb{R}^N$ yields

$$\frac{1}{2}\frac{d}{d\tau}\|\xi_n z\|^2_{L^2_K} + \|\xi_n\nabla z\|^2_{L^2_K} + 2\int z\xi_n\nabla z.\nabla\xi_n K d\eta = 0. \qquad (2.4.13)$$

But

$$2\int z\xi_n\nabla z.\nabla\xi_n K d\eta \geq -\frac{1}{2}\|\xi_n\nabla z\|^2_{L^2_K} - 2\|z\nabla\xi_n\|^2_{L^2_K}. \qquad (2.4.14)$$

Integrating (2.4.15) on (α, β) where $\tau_0 < \alpha < \beta$ and using the fact that $\nabla\xi_n$ is bounded uniformly in n, implies

$$\left[\|\xi_n z\|^2_{L^2_K}\right]^\beta_\alpha + \int_\alpha^\beta\|\xi_n\nabla z\|^2_{L^2_K}d\tau \leq C\int_\alpha^\beta\|z\|^2_{L^2_K}d\tau, \qquad (2.4.15)$$

for some constant $C > 0$. Letting n go to infinity and using Fatou's lemma yields with $\alpha = \tau_0$

$$\|z(\beta,.)\|^2_{L^2_K} \le C \int_{\tau_0}^{\beta} \|z(\tau,.)\|^2_{L^2_K} \, d\tau, \qquad (2.4.16)$$

which implies z=0 by Gronwall's inequality and w = $\tilde{w}$ on $[\tau_0, \ln T)$.

Step 3. As a consequence of Brezis' results, the function w satifies equation (2.4.10) a.e. and the following relations hold

$$\|w_\tau\|^2_{L^2_K} + \frac{d}{d\tau}\Phi(w) = \int hw_\tau K d\eta \qquad \text{a.e. on } (\tau_0, \ln T), \qquad (2.4.17)$$

$$\left(\int_{\tau_0}^{\tau_1} \Phi(w(\tau)d\tau\right)^{1/2} \le C\left(\|w(0)\|_{L^2_K} + \int_{\tau_0}^{\tau_1} \|h(\tau)\|_{L^2_K} d\tau\right), \qquad (2.4.18)$$

$$\sqrt{\tau - \tau_0}\, w_\tau \in L^2_{loc}([\tau_0, \ln T); L^2_K), \qquad (2.4.19)$$

$$\left(\int_{\tau_0}^{\tau_1} \|w_\tau(\tau)\|^2_{L^2_K}(\tau - \tau_0)d\tau\right)^{1/2} \le \left(\int_{\tau_0}^{\tau_1} \|h(\tau)\|^2_{L^2_K}(\tau - \tau_0)d\tau\right)^{1/2}$$

$$\qquad (2.4.20)$$

$$+ C\int_{\tau_0}^{\tau_1} \|h(\tau)\|_{L^2_K} d\tau + C\|w(\tau_0)\|_{L^2_K}.$$

Therefore, from (2.4.17)-(2.4.19) and by using the mean value theorem, we see that w belongs to $L^\infty(-\infty, \ln T; H^1_K)$ and $\partial\Phi(w) = K^{-1}\text{div}(K\nabla w)$ remains bounded in $L^2(R, R+k; L^2_K)$ by a constant independent of R, but not of k. The proof will be completed with the next regularity result.

LEMMA 6.4. *For any integer* m, *the operator* $w \mapsto K^{-1}\text{div}(K\nabla w)$ *is an isomorphism between* H^{m+2}_K *and* H^m_K.

Proof. Step 1. Let w be an element of H^1_K and fix an integer i between 1 and N; because $K_{\eta_i} = \frac{1}{2}\eta_i K$, we obtain

$$\int_{\mathbb{R}^N} \eta_i^2 w^2 K d\eta = -2\int_{\mathbb{R}^N}(w^2 + 2y_i ww_{\eta_i})K d\eta$$

$$\le 4\left(\int_{\mathbb{R}^N} \eta_i^2 w^2 K d\eta\right)^{1/2}\left(\int_{\mathbb{R}^N} w_{\eta_i}^2 K d\eta\right)^{1/2}, \qquad (2.4.21)$$

which implies

$$\big\| \eta |w| \big\|_{L_K^2} \le 4 \| \nabla w \|_{L_K^2}. \tag{2.4.22}$$

Step 2. Suppose $w \in H_K^1$ is a solution of

$$K^{-1} \mathrm{div}(K \nabla w) = F \qquad \text{in } \mathbb{R}^N, \tag{2.4.23}$$

with $F \in L_K^2$. Then we still know that

$$\| w \|_{H_K^1} \le \sqrt{1 + 2/N} \, \| F \|_{L_K^2}, \tag{2.4.24}$$

since $N/2$ is the first eigenvalue of $K^{-1}\mathrm{div}(K\nabla(.))$ in H_K^1. Differentiating (2.4.23) with respect to η_i, and using the fact that $w \in H_{loc}^2(\mathbb{R}^N)$ from the classical local regularity theory yields

$$K^{-1} \mathrm{div}\big(K \nabla w_{\eta_i} \big) = F_{\eta_i} - \frac{1}{2} w_{\eta_i}. \tag{2.4.25}$$

If we multiply (2.4.25) by $\xi_n^2 K w_{\eta_i}$ where ξ_n is as in Lemma 6.3, we deduce

$$\int_{\mathbb{R}^N} \big\| \nabla w_{\eta_i} \big\|^2 \xi_n^2 K d\eta + 2 \int_{\mathbb{R}^N} \xi_n w_{\eta_i} \nabla \xi_n . \nabla w_{\eta_i} K d\eta$$
$$= \int_{\mathbb{R}^N} F_{\eta_i} \xi_n^2 w_{\eta_i} K d\eta - \frac{1}{2} \int_{\mathbb{R}^N} (w_{\eta_i})^2 \xi_n^2 K d\eta. \tag{2.4.26}$$

By integration by parts the right-hand side of (2.4.26) is equal to

$$-\int_{\mathbb{R}^N} \xi_n^2 F w_{\eta_i \eta_i} K d\eta - 2 \int_{\mathbb{R}^N} \xi_n (\xi_n)_{\eta_i} F w_{\eta_i} K d\eta$$
$$-\frac{1}{2} \int_{\mathbb{R}^N} \xi_n^2 w_{\eta_i} F \eta_i K d\eta - \frac{1}{2} \int_{\mathbb{R}^N} (w_{\eta_i})^2 \xi_n^2 K d\eta. \tag{2.4.27}$$

Therefore, by (2.4.14) and similar applications of Schwarz inequality, and by Step 1, we obtain

$$\frac{1}{2} \big\| \xi_n \nabla w_{\eta_i} \big\|_{L_K^2}^2 - 4 \big\| w_{\eta_i} \nabla \xi_n \big\|_{L_K^2}^2$$
$$\le \big\| \xi_n w_{\eta_i \eta_i} \big\|_{L_K^2} \big\| \xi_n F \big\|_{L_K^2} + 2 \big\| \xi_n w_{\eta_i} \big\|_{L_K^2} \big\| F(\xi_n)_{\eta_i} \big\|_{L_K^2} + 2 \big\| \xi_n \nabla w_{\eta_i} \big\|_{L_K^2} \big\| \xi_n F K \big\|_{L_K^2}. \tag{2.4.28}$$

Letting n go to infinity yields

$$\| w \|_{H_K^2} \le C \| F \|_{L_K^2}, \tag{2.4.29}$$

for some $C > 0$. In the same way, we prove the following estimate: suppose $m \in \mathbb{N}$, then there exists some $C > 0$ such that, if $(w, F) \in H_K^{m+2} \times H_K^m$ satisfy (2.4.24), there holds

332

$$\|w\|_{H_K^{m+2}} \leq C\|f\|_{H_K^m}. \tag{2.4.30}$$

We complete the proof from (2.4.30) and Lax-Milgram Theorem .

Let us introduce now $\Sigma_\tau = \exp(-\tau L)$ as being the semi-group of contractions of L_K^2 generated generated by the operator $-L$ where $L = -K^{-1}\mathrm{div}(K\nabla.)$. This semi-group has the following regularizing effect:

LEMMA 6.5. *There exist a constant* $C = C(N)$ *such that*

$$\left\|\Sigma_t\right\|_{\mathcal{L}\left(L_K^2, L_{K^{1/2}}^\infty\right)} \leq Ct^{-N/4}e^{-Nt/2} \tag{2.4.31}$$

for any $t > 0$ *where* $\mathcal{L}\left(L_K^2, L_{K^{1/2}}^\infty\right)$ *is the usual norm for linear operator between* L_K^2 *and* $L_{K^{1/2}}^\infty$.

Proof. Step 1. It is essentially an application of Nash-Moser iterative scheme. Let $v(.,\tau) = \Sigma_\tau(v_0)$ be the solution of

$$v_\tau - K^{-1}\mathrm{div}(K\nabla v) = 0, \tag{2.4.32}$$

for $\tau > 0$. Assuming that v belongs to $L_{K^\beta}^q$, we multiply (2.3.34) by $|v|^{q-1}v\xi_n^2 K^\beta$ and we obtain the relation below by an easy approximation argument

$$\frac{1}{q}\frac{d}{d\tau}\int_{\mathbb{R}^N}|v|^q K^\beta d\eta + \frac{4}{q}\left(\frac{1}{\beta} - \frac{1}{q}\right)\int_{\mathbb{R}^N}|\nabla|w|^{q/2}|K^\beta d\eta \leq 0. \tag{2.4.33}$$

This implies that $\tau \mapsto \|v(.,\tau)\|_{L_{K^\beta}^q}$ is non-increasing, and the following inequality follows

$$\left\|v(.,\tau)K^{\beta/q}\right\|_{L^{\alpha q}} \leq \left(C^2(1+\beta)^2\frac{\beta q}{4(q-\beta)}\right)^{1/q}(\tau - \sigma)^{-1/q}\left\|v(.,\sigma)K^{\beta/q}\right\|_{L^q}. \tag{2.4.34}$$

Here we have assumed that $N > 2$, but the proof in the cases $N = 1$ or 2 is similar. Taking now $q = q_n = 2^n\alpha$, $\beta = q_n/2$ and $\tau = \tau_n = 2^{-n}t$, $\sigma = \tau_{n-1}$, we obtain

$$\left\|\Sigma_t\right\|_{L\left(L_K^2, L_{K^{1/2}}^\infty\right)} \leq Ct^{-N/4}. \tag{2.4.35}$$

Step 2. We decompose the space H_K^1 as

$$H_K^1 = \bigoplus_{k=1}^\infty N(L - \lambda_k I) \tag{2.4.36}$$

and let H' be $\overset{\infty}{\underset{k=2}{\oplus}} N(L - \lambda_k I)$ and Σ'_t the restriction of Σ_t to H'. We decompose v as $v_1 + v'$ where $v_1 \in N(L - \lambda_1 I)$ and $v' \in$ H', then we have

$$\Sigma_t v = e^{-\lambda_1 t} v_1 + \Sigma'_t v'. \tag{2.4.37}$$

Because the operator $L - \lambda_2 I$ is monotone on H', we have

$$\left\|\Sigma'_t v'\right\|_{L^2_K} \le e^{-\lambda_2 t} \|v'\|_{L^2_K}, \tag{2.4.38}$$

for any $t > 0$. Now we apply Step 1, (2.4.35), with $t = (1 - \lambda_1 / \lambda_2)\tau$ and (2.4.37) with $v' = \Sigma'_{(\lambda_1/\lambda_2)\tau} v'_0$. Therefore

$$\left\|\Sigma_\tau v'_0\right\|_{L^\infty_{K^{1/2}}} \le C((1 - \lambda_1 / \lambda_2)\tau)^{-N/4} e^{-\lambda_1 \tau} \|v'_0\|_{L^2_K}. \tag{2.4.39}$$

As for the first component, it is clear that

$$e^{-\lambda_1 t} \|v_1\|_{L^\infty_{K^{1/2}}} \le e^{-\lambda_1 t} \left(\left\|K^{-1}\right\|_{L^\infty_{K^{1/2}}} \Big/ \left\|K^{-1}\right\|_{L^2_K} \right) \|v_1\|_{L^2_K}. \tag{2.4.40}$$

Combining (2.4.37), (2.4.39) and (2.4.40) yields (2.4.31).

Proof of Theorem 6.13. We remember that the function w defined by (2.4.1) satisfies (2.4.2) on $\mathbb{R}^N \times (-\infty, \ln T)$. Let $E[w(.,\tau)] = E[w]$ be the energy function associated to (2.4.1) and defined by

$$E[w] = \frac{1}{2} \int_{\mathbb{R}^N} |\nabla w|^2 K d\eta - \frac{1}{2(p-1)} \int_{\mathbb{R}^N} w^2 K d\eta + \frac{1}{p-1} \int_{\mathbb{R}^N} |w|^{p+1} K d\eta. \tag{2.4.41}$$

Step 1. We claim that $\displaystyle\lim_{\tau \to -\infty} \|w_\tau(.,\tau)\|_{L^2_K} = 0$. The first thing to notice is that $w_\tau \in H^1\left(-\infty, 0; L^2_K\right)$. In fact, multiplying (2.4.2) by Kw and integrating over $\mathbb{R}^N$ gives

$$0 = \int w_\tau^2 K dy + \frac{d}{d\tau} E[w(.,\tau)]. \tag{2.4.42}$$

From Lemma 6.3 $E[w(.,\tau)]$ remains bounded uniformly on $\mathbb{R}^-$, which implies

$$\int_{-\infty}^0 \int_{\mathbb{R}^N} w_\tau^2 K d\eta d\tau = C < \infty. \tag{2.4.43}$$

Let z be w_τ, then differentiating (2.4.2) gives

334

$$z_\tau - \frac{1}{K}\operatorname{div}(K\nabla w) = \frac{1}{p-1}z - p|w|^{p-1}z. \tag{2.4.44}$$

Arguing as in the proof of Lemma 6.3, we see that z is the solution of

$$z_\tau + \partial\Phi(z) = g = \frac{1}{p-1}z + p|w|^{p-1}z \tag{2.4.45}$$

in L_K^2 and that

$$\int_{\mathbb{R}^N} z_\tau^2 K d\eta + \frac{d}{d\tau}\tilde{E}[z(.,\tau)] = \int_{\mathbb{R}^N}\left(\frac{1}{p-1} - p|w|^{p-1}\right)z^2 K d\eta, \tag{2.4.46}$$

where

$$\tilde{E}[z] = \frac{1}{2}\int_{\mathbb{R}^N}|\nabla z|^2 K d\eta - \frac{1}{2(p-1)}\int_{\mathbb{R}^N} z^2 K d\eta. \tag{2.4.47}$$

Moreover, as in Lemma 6.3, $\Phi(z)$ remains uniformly bounded on $\mathbb{R}^-$. Therefore, we deduce from (2.4.46) that

$$\int_{-\infty}^{0}\int_{\mathbb{R}^N} z_\tau^2 K dy d\tau = C < \infty. \tag{2.4.48}$$

This implies that $\tau \mapsto \int_{\mathbb{R}^N} w_\tau^2 K d\eta$ is uniformly continuous on $\mathbb{R}^-$ which in turn implies the claim.

Step 2. Completion of the proof. Let $\mathbb{T}$ be the negative trajectory of w, that is

$$\mathbb{T} = \{w(.,\tau): \tau \leq 0\}. \tag{2.4.49}$$

From Lemma 6.3, $K^{-1}\operatorname{div}(K\nabla w(.,\tau))$ is bounded in L_K^2 independently of τ. From Lemma 6.4, it implies that $\mathbb{T}$ is bounded in H_K^2. Therefore $\mathbb{T}$ is relatively compact in H_K^1. If we define the $\alpha-$limit set of the trajectory by

$$\Gamma^-(\mathbb{T}) = \bigcap_{\tau \leq 0}\operatorname{clo}_{H_K^1}\{w(.,t)\,/\,t \leq \tau\}, \tag{2.4.50}$$

then $\Gamma^-(\mathbb{T})$ is a non-empty compact connected subset of H_K^1 which is included in H_K^2. Now we prove that

$$\Gamma^-(\mathbb{T}) \subset \mathcal{E}_{1/(p-1)}. \tag{2.4.51}$$

In fact, we know from Remark 6.4, that $\mathcal{E}_{1/(p-1)}$ is included into $L^{\infty}_{K^{1/2}}$ and, from Proposition 6.4,

$$\mathcal{E}_{1/(p-1)} \subset L^{\infty}_{K^{\theta}} \tag{2.4.52}$$

for any $\theta < 1.$. Indeed, if ω is any element of $\Gamma^-(\mathbb{T})$ there exists a sequence $\{\tau_n\}_{n \leq 0}$ tending to $-\infty$ with n, such that

$$\lim_{\tau_n \to -\infty} \|w(.,\tau_n) - \omega(.)\|_{H^1_K} = 0 . \tag{2.4.53}$$

But, from Step 1,

$$\lim_{\tau_n \to -\infty} \|w_\tau(.,\tau_n)\|_{L^2_K} = 0, \tag{2.4.54}$$

and also

$$\lim_{\tau_n \to -\infty} w|w|^{p-1}(.,\tau_n) = \omega|\omega|^{p-1}(.) \tag{2.4.55}$$

in $L^{\sigma}_{K^{\alpha}}$ with $\alpha = N/(N-2)$ if $N > 2$ and $\alpha \in (0,\infty)$ if $N = 1$ or 2 , and $\sigma = 2\alpha/p$. Therefore we can go to the limit in (2.4.2) and find that ω belongs to $\mathcal{E}_{1/(p-1)}$. Finally, we deduce from (2.4.54) and Lemma 6.5 (since w is bounded) that

$$\lim_{\tau_n \to -\infty} \|w(.,\tau_n) - \omega(.)\|_{L^{\infty}_{K^{1/2}}} = 0, \tag{2.4.56}$$

which ends the proof.

Remark 6.5. In the particular case where $(N+3)/(N+1) \leq p < (N+2)/N$, the previous result reads as follows:

(i) either u can be extended to $\mathbb{R}^N \times [0,T)$ as a continuous solution,

(ii) or there exists a nonzero constant c such that $u(x,t) \approx cE(x,t)$ as t tends to 0, for x close enough to 0,

(iii) or $u(x,t) \approx \pm W(x,t)$ as tends to 0 f or $|x|$ small enough.

Moreover, if u is nonnegative, we just have to assume $1 < p < (N+2)/N$ and we have (ii) with $c > 0$, or (iii) with $u(x,t) \approx W(x,t)$. If u is a radial solution of (2.3.1)-(2.3.2), the previous assertion holds provided $(2N+2)/(N+2) < p < (N+2)/N$.

Remark 6.6. In their work on the single point blow-up of positive solutions of

$$u_t - \Delta u = u^p \tag{2.4.57}$$

(p > 1) Giga and Kohn [GK1] reduced the problem to the study of the behaviour at infinity of the positive and bounded solutions of

$$w_\tau - \Delta w + \frac{1}{2} y. \nabla w + \frac{1}{p-1} w - w^p = 0. \qquad (2.4.58)$$

In their case the energy space was $H^1_{K^{-1}}$, the ω-limit set of the positive trajectory of w was included in the set of bounded solutions of

$$-\Delta\omega + \frac{1}{2} y. \nabla\omega + \frac{1}{p-1}\omega - \omega^p = 0, \qquad (2.4.59)$$

and they proved that, if $1 < p < (N+2)/(N-2)$ -when $N > 2$ - this set of solutions was reduced to constant functions. In the isolated singularity case, the set $\mathcal{E} = \mathcal{E}_{1/(p-1)}$ is the set of self-similar solutions of (2.4.2), that is the set of solutions which are invariant under the scaling transformation

$$u \mapsto u_\lambda, \text{ with } u_\lambda(x,t) = \lambda^{1/(p-1)} u(\sqrt{\lambda} x, \lambda t) \qquad (2.4.60)$$

$(\lambda > 0)$. Actually, the scaling transformation is just a τ-translation for the function w.

6-2-5 Weak singularities

In the particular case where $\mathcal{E}^* = \{0\}$ we have the occurence of a weak singularity, and more precisely,

THEOREM 6.14. *Under the hypotheses of Theorem 6.13 and with the same notations, we assume that $\mathcal{E}^* = \{0\}$. We have the following:*
(i) if $(N+3)/(N+1) \leq p < (N+2)/N$, then there exists a constant c such that

$$\lim_{\tau \to -\infty} \left(\left\| e^{-\gamma\tau} w(.,\tau) - cK^{-1} \right\|_{H^1_K} + \left\| e^{-\gamma\tau} w(.,\tau) - cK^{-1} \right\|_{L^\infty_{\tilde{K}^{1/2}}} \right) = 0 \qquad (2.5.1)$$

where $\gamma = 1/(p-1) - N/2$; moreover, if $c = 0$, u can be extended as a continuous function in $\mathbb{R}^N \times [0, T)$.
(ii) if $1 < p < (N+3)/(N+1)$ and there exists   no integer ℓ such that $p = (N+\ell+2)/(N+\ell)$, let k_0 be the largest integer smaller than $\ell/(p-2) - N$, then either there exists an integer $k \in [0, k_0]$ and a nonzero linear combination of $D^\alpha(K^{-1})$ with $|\alpha| = k$, say ϕ, such that

$$\lim_{\tau \to -\infty} \left(\left\| e^{-\gamma_k\tau} w(.,\tau) - \phi \right\|_{H^1_K} + \left\| e^{-\gamma_k\tau} w(.,\tau) - \phi \right\|_{L^\infty_{\tilde{K}^{1/2}}} \right) = 0, \qquad (2.5.2)$$

337

where $\gamma_k = 1/(p-1) - (N+k)/2$, *or* u *can be extended to* $\mathbb{R}^N \times [0,T]$ *as a continuous function.*

LEMMA 6.6. *Let* w *be a solution of* (2.4.1)-(2.4.2) *such that*

$$\lim_{\tau \to -\infty} \|w(.,\tau)\|_{H_k^1} = 0. \tag{2.5.3}$$

Suppose also either $(N+3)/(N+1) \le p < (N+2)/N$, *or* $1 < p < (N+3)/(N+1)$ *and there exists no integer* ℓ *such that* $p = (N+\ell+2)/(N+\ell)$; *then there exists a positive constant* C *such that*

$$\|w(.,\tau)\|_{H_k^2} \le C e^{\gamma_k \tau} \tag{2.5.4}$$

for any $\tau \le 0$, *where*

$$\gamma_k = \frac{1}{p-1} - \frac{N+k-1}{2} = \frac{1}{p-1} - \lambda_k, \tag{2.5.5}$$

if $\dfrac{N+k+2}{N+k} < p < \dfrac{N+k+1}{N+k-1}$ *and* $k > 1$, *or*

$$\gamma_1 = \frac{1}{p-1} - N/2 = \frac{1}{p-1} - \lambda_1, \tag{2.5.6}$$

if $\dfrac{N+3}{N+1} \le p < 1 + \dfrac{2}{N}$.

Proof. Step 1. We claim that there exists a positive number ε such that

$$\limsup_{\tau \to -\infty} \left(e^{-\varepsilon\tau} \|w(.,\tau)\|_{H_k^1} \right) = 0. \tag{2.5.7}$$

In fact, if (2.5.7) is not true for some $\varepsilon > 0$, we recall that, from a result in [CMV], there exists a positive increasing function η such that

$$\begin{cases} \text{(i)} \quad \lim_{\tau \to -\infty} \eta(\tau) = 0, \\[2mm] \text{(ii)} \quad 0 < \limsup_{\tau \to -\infty} \|w(.,\tau)\|_{H_k^1} / \eta(\tau) < \infty, \\[2mm] \text{(iii)} \quad \lim_{\tau \to -\infty} \left(e^{-\varepsilon\tau} \eta(\tau) \right) = 0 \quad \text{for any } \varepsilon > 0, \\[2mm] \text{(iv)} \quad \lim_{\tau \to -\infty} \eta_\tau / \eta = 0 \, , \ \eta_\tau / \eta \in L^\infty(-\infty, 0) \\[2mm] \qquad \text{and } \left(\eta_\tau / \eta \right)_\tau \in L^1(-\infty, 0) \cap L^\infty(-\infty, 0). \end{cases} \tag{2.5.8}$$

Setting $\zeta = w / \eta$, then ς satsfies

338

$$\varsigma_\tau - K^{-1}\mathrm{div}(K\nabla\varsigma) - \frac{1}{p-1}\varsigma + \frac{\eta_\tau}{\eta}\varsigma + \eta^{p-1}|\varsigma|^{p-1}\varsigma = 0 \tag{2.5.9}$$

in $\mathbb{R}^N \times \mathbb{R}^-$. Let $\mathbb{T}^*$ be the negative trajectory of ς in H_K^1, then $\mathbb{T}^*$ is bounded, which implies that

$$\int_{-\infty}^{0}\int \varsigma_\tau^2 K dy d\tau < \infty. \tag{2.5.10}$$

Differentiating (2.5.9) and setting $\varsigma_\tau = \chi$, we see that χ satifies

$$\chi_\tau - K^{-1}\mathrm{div}(K\nabla\chi) - \left(\frac{1}{p-1} - \frac{\eta_\tau}{\eta} - p\eta^{p-1}|\varsigma|^{p-1}\right)\chi$$
$$+ \left(\frac{\eta_\tau}{\eta}\right)_\tau \varsigma + (p-1)\eta_\tau\eta^{p-1}|\varsigma|^{p-1}\varsigma = 0. \tag{2.5.11}$$

Multiplying by χ_τ and integrating over $\mathbb{R}^N$ yields

$$\int_{-\infty}^{0}\int_{\mathbb{R}^N} \chi_\tau^2 K dy d\tau < \infty, \tag{2.5.12}$$

which in turn implies that

$$\lim_{\tau\to-\infty}\left\|\varsigma_\tau\right\|_{L_K^2} = 0. \tag{2.5.13}$$

Because $\mathbb{T}^*$ is relatively compact in $L_K^2 \cap L_{K^\alpha}^{p+1}$, we deduce from (2.5.8)-(ii), that there exist a non-zero element b of H_K^1 and a sequence $\{\tau_n\}$ going to $-\infty$ with n such that $\varsigma(.,\tau_n)$ converges to b weakly in H_K^1 and strongly in $L_K^2 \cap L_{K^\alpha}^{p+1}$ as $\{\tau_n\}$ goes to $-\infty$. Clearly b is a H_K^1-solution of

$$K^{-1}\mathrm{div}(K\nabla b) + \frac{1}{p-1}b = 0 \tag{2.5.14}$$

in $\mathbb{R}^N$; but this is not possible since $1/(p-1)$ is not an eigenvalue of L in the case where $(N+k+2)/(N+k) < k < (N+k+1)/(N+k-1)$. Finally, when $p = (N+3)/(N+1)$, we can proceed as in the proof of Proposition 6.5 to see that w is radial, which is not compatible with the fact that b is a non zero linear combination of the derivatives of K^{-1} up to the order one.

Step 2. We assume that $(N+2)/N > p \geq (N+3)/(N+1)$ and we claim that

$$\|\mathrm{w}(.,\tau)\|_{H^1_K} \le Ce^{\gamma_1\tau}, \tag{2.5.15}$$

for some $C > 0$ and any $\tau \le 0$. Setting $\phi(.,\tau) = e^{-\varepsilon\tau}\mathrm{w}(.,\tau)$, then ϕ is bounded in H^1_K and satisfies

$$\phi_\tau - K^{-1}\mathrm{div}(K\nabla\phi) + \left(\varepsilon - \frac{1}{p-1}\right)\phi + e^{(p-1)\varepsilon\tau}|\phi|^{p-1}\phi = 0. \tag{2.5.16}$$

Let μ_ε be $1/(p-1) - \varepsilon$ and $H^1_K = \overset{\infty}{\underset{k=1}{\oplus}} N(L - \lambda_k I)$, $H' = \overset{\infty}{\underset{k=2}{\oplus}} N(L - \lambda_k I)$. Denote (ϕ_1, h_1) $(\mathrm{resp}(\phi', h'))$ the projection of $\left(\phi, |\phi|^{p-1}\phi\right)$ onto $N(L - \lambda_1 I)$ (resp. H'). For the first component, we have

$$(\phi_1)_\tau + (\lambda_1 - \mu_\varepsilon)\phi_1 + e^{(p-1)\varepsilon\tau}h_1 = 0, \tag{2.5.17}$$

which implies

$$\phi_1(0) = e^{(\lambda_1 - \mu_\varepsilon)\tau}\phi_1(\tau) - \int_\tau^0 e^{(\lambda_1 - \mu_\varepsilon + \varepsilon(p-1))t}h_1(t)dt. \tag{2.5.18}$$

Since h_1 is bounded, we get

$$|\phi_1(\tau)| \le e^{(\mu_\varepsilon - \lambda_1)\tau}\left(|\phi_1(0)| + C\right) + Ce^{\varepsilon(p-1)\tau} \tag{2.5.19}$$

(we can always assume that ε is such that $\lambda_1 - \mu_\varepsilon + \varepsilon(p-1)$ is non-zero). For the second component of ϕ we use the convolution formulation and we have

$$\phi'(\tau) = e^{\mu_\varepsilon(\tau-\sigma)}\Sigma'_{\tau-\sigma}\phi'(\sigma) - \int_\sigma^\tau e^{\mu_\varepsilon(\tau-t)}\Sigma'_{\sigma-t}e^{\varepsilon(p-1)t}h'(t)dt. \tag{2.5.20}$$

Using the decomposition of H', we also have

$$\|\nabla\Sigma'_\tau v\|_{L^2_K} \le e^{-\lambda_2\tau}\|\nabla v\|_{L^2_K}, \tag{2.5.21}$$

which, together with (2.5.20), implies

$$\|\phi'(\tau)\|_{H^1_K} \le e^{(\mu_\varepsilon - \lambda_2)(\tau-\sigma)}\|\phi'(\sigma)\|_{H^1_K} + \int_\sigma^\tau e^{(\mu_\varepsilon - \lambda_2)(\tau-t)}e^{\varepsilon(p-1)t}\|h'(t)\|_{H^1_K}dt, \tag{2.5.22}$$

for $\sigma < \tau < 0$. If we take $\tau = \sigma/n$, for some integer n, then

$$\left\|\phi'(\tau)\right\|_{H_K^1} \le C\left(e^{(\lambda_2-\mu_\varepsilon)(n-1)\tau} + \int_{n\tau}^{\tau} e^{(\mu_\varepsilon-\lambda_2)\tau}e^{(\lambda_2-\mu_\varepsilon+\varepsilon(p-1))t}\left\|h'(t)\right\|_{H_K^1}\right). \tag{2.5.23}$$

Since ϕ is bounded in $H_K^1 \cap L^\infty$, it is the same with $|\phi|^{p-1}\phi$ and with h'. Therefore

$$\left\|\phi'(\tau)\right\|_{H_K^1} \le C\left(e^{(\lambda_2-\mu_\varepsilon)(n-1)\tau} + C'\,e^{\varepsilon(p-1)\tau}\right). \tag{2.5.24}$$

From (2.5.19) and (2.5.24)) we obtain

$$\left\|w(.,\tau)\right\|_{H_K^1} \le C\left(e^{(1/(p-1)-\lambda_1)\tau} + e^{p\varepsilon\tau} + e^{(\lambda_2-1/(p-1))(n-1)\tau+n\varepsilon\tau}\right). \tag{2.5.25}$$

Since n is free and $\lambda_2 \ge 1/(p-1)$, we can choose it large enough in order to have

$$\left\|w(.,\tau)\right\|_{H_K^1} \le C_n\left(e^{(1/(p-1)-\lambda_1)\tau} + e^{p\varepsilon\tau}\right). \tag{2.5.26}$$

Iterating this process a finite number of times , we finally deduce that

$$\left\|w(.,\tau)\right\|_{H_K^1} \le Ce^{\gamma_1\tau} \tag{2.5.27}$$

holds, for some $C > 0$ and any $\tau \le 0$.

Step 3. We have in fact slightly more than (2.5.4) in assuming that (2.5.5)) holds for some $\varepsilon > 0$. If we assume that $\lambda_k < 1/(p-1) \le \lambda_{k+1}$ for some k > 1, then we prove in the same manner as in Step 2 that

$$\left\|w(.,\tau)\right\|_{H_K^1} \le Ce^{\gamma_k\tau}, \tag{2.5.28}$$

where $\gamma_k = 1/(p-1) - \lambda_k$.

Proof of Theorem 6.14. Let us assume that

$$\lim_{\tau\to-\infty}\left(\left\|w(.,\tau)\right\|_{H_K^1} + \left\|w(.,\tau)\right\|_{L_{K^{1/2}}^\infty}\right) = 0, \tag{2.5.29}$$

and either $(N+3)/(N+1) \le p < (N+2)/N$, or $1 < p < (N+3)/(N+1)$ and $1/(p-1)$ is not an eigenvalue of the operator L. From Lemmas 6.5- 6.6 there holds

$$\left\|w(.,\tau)\right\|_{H_K^1} + \left\|w(.,\tau)\right\|_{L_{K^{1/2}}^\infty} \le Ce^{\gamma_k\tau}, \tag{2.5.30}$$

for some $k \ge 1$. Defining the function ψ by $\psi = e^{-\gamma_k\tau}w$, then it is bounded in $H_K^1 \cap L_{K^{1/2}}^\infty$ (in fact, as in the proof of Theorem 6.13, it remains bounded in H_K^2) and satisfies

341

$$\psi_\tau - K^{-1}\mathrm{div}(K\nabla\psi) - \lambda_k\psi + e^{(p-1)\gamma_k\tau}|\psi|^{p-1}\psi = 0 \qquad \text{on } (-\infty, 0). \tag{2.5.31}$$

Step 1. We claim that there exists some ζ_k belonging to $N(L - \lambda_k I)$ such that

$$\lim_{\tau\to-\infty}\left(\left\|\psi(.,\tau) - \varsigma_k\right\|_{H^1_H} + \left\|\psi(.,\tau) - \varsigma_k\right\|_{L^\infty_{K^{1/2}}}\right) = 0. \tag{2.5.32}$$

In fact, if we call ψ_k the projection of ψ onto $N(L - \lambda_k I)$, we have

$$(\psi_k)_\tau - e^{(p-1)\gamma_k\tau}P_k\left(|\psi|^{p-1}\psi\right) = 0, \tag{2.5.33}$$

where P_k is the projection operator in H^1_K. Since $P_k\left(|\psi|^{p-1}\psi\right)$ is bounded, $(\psi_k)_\tau$ is integrable and $\psi_k(\tau)$ admits a limit when τ goes to $-\infty$. As in the proof of Theorem 6.13-Step 1, the negative trajectory $\mathbb{T}^k$ of ψ is relatively compact in $H^1_K \cap L^\infty_{K^{1/2}}$, and there holds

$$\lim_{\tau\to-\infty}\left\|\psi_\tau(.,\tau)\right\|_{H^1_K} = 0. \tag{2.5.34}$$

Therefore, going to the limit in (2.5.31) implies that the α-limit set of $\mathbb{T}^k$ in $H^1_K \cap L^\infty_{K^{1/2}}$ is non-empty and contained into $N(L - \lambda_k I)$. As a consequence (2.5.32) holds and the limit of ψ is just the limit of ψ_k.

Step 2. Assume that $\zeta_k = 0$. Then we claim that

$$\left\|w(.,\tau)\right\|_{H^1_K} \le Ce^{\gamma_{k-1}\tau}, \tag{2.5.35}$$

where $\gamma_{k-1} = 1/(p-1) - \lambda_{k-1}$ with $k > 1$. For proving (2.5.35), we can follow the proof of Lemma 6.6: we first have the existence of some $\varepsilon > 0$ such that

$$\limsup_{\tau\to-\infty}\left(e^{-\varepsilon\tau}\left\|\psi(.,\tau)\right\|_{H^1_K}\right) = 0. \tag{2.5.36}$$

For proving this assertion, we use the same function η and the fact that $1/(p-1)$ is not an eigenvalue of L. Proceeding as in Lemma 6.6-Step 2, we obtain (2.5.35) after a finite number of iterations.

Step 3. End of the proof. We first assume that $(N+3)/(N+1) \le p < (N+2)/N$. From Lemma 6.6-Steps 1-2, there exists some c such that

$$\lim_{\tau\to-\infty}\left(\left\|e^{-\gamma_1\tau}w(.,\tau) - cK^{-1}\right\|_{H^1_K} + \left\|e^{-\gamma_1\tau}w(.,\tau) - cK^{-1}\right\|_{L^\infty_{K^{1/2}}}\right) = 0. \tag{2.5.37}$$

If c = 0, then w is bounded. Actually this is a consequence of Step 2, but it can also be proved by taking a supersolution of (2.4.2) under the form $\varepsilon K^{-1} + C$ where $\varepsilon > 0$ is arbitrary and C is independant of ε. We complete the proof by using the classical regularity theory of parabolic equations. If $1 < p < (N+3)/(N+1)$ we apply the steps 1 and 2 and Lemma 6.6 and we conclude as above.

6-2-6 Miscellaneous

The previous techniques can be applied to describe the asymptotic behaviour of the solutions of (2.3.1) in $\mathbb{R}^N \times (0, \infty)$.

THEOREM 6.15. *Let $1 < p < (N+2)/N$ and u be a solution of (2.3.1) on $[0, \infty)$ with an initial value* u(x, 0) *satisfying*

$$\int |u(x,0)| e^{\alpha|x|^2} dx < \infty \tag{2.6.1}$$

for some $\alpha > 0$ and let w *be defined by (2.4.1)-(2.4.2). Then there exists a connected component $\mathcal{E}^*$ of $\mathcal{E}_{1/(p-1)}$ such that*

$$\lim_{\tau \to \infty} \mathrm{dist}_\tau(w(.,\tau), \mathcal{E}^*) = 0. \tag{2.6.2}$$

Moreover, if $(N+3)/(N+1) \leq p < (N+2)/N$, the set $\mathcal{E}^*$ is reduced to a single element which is either $\pm f$, where f *is the unique positive element of $\mathcal{E}_{1/(p-1)}$, or* 0 .

In the particular case where $\mathcal{E}^* = \{0\}$, the rate of decay is much stronger since we have:

THEOREM 6.16. *Under the hypotheses of Theorem 6.15 and with the same notations, we assume that $\mathcal{E}^* = \{0\}$. If there exists no integer ℓ such that $p = (N+\ell+2)/(N+\ell)$, and if k_0 is the smallest integer bigger than $2/(p-1) - N$, then either there exists an integer $k \in [k_0, \infty)$ and a non-zero linear combination, say ϕ, of $D^\alpha K^{-1}$ with $|\alpha| = k$ such that*

$$\lim_{\tau \to \infty} \left(\left\| e^{-\gamma_k \tau} w(.,\tau) - \phi \right\|_{H_K^1} + \left\| e^{-\gamma_k \tau} w(.,\tau) - \phi \right\|_{L^\infty_{K^{1/2}}} \right) = 0, \tag{2.6.3}$$

where $\gamma_k = 1/(p-1) - (N+k)/2$, or u *is identically zero.*

Remark 6.7. From the previous theorems it is possible to study some global singular solutions u of (2.3.1)-(2.3.2) in $\mathbb{R}^N \times (0, \infty)$. Defining the function $w(y, \tau)$ by (2.4.1)-(2.4.2), then the global singular solution w satisfies

343

$$w_\tau - K^{-1}\mathrm{div}(K\nabla w) - \frac{1}{p-1}w + w|w|^{p-1} = 0 \tag{2.6.4}$$

in $\mathbb{R}^N \times \mathbb{R}$. For such a solution we know that there exist ω_- and ω_+ belonging to $\mathcal{E}_{1/(p-1)}$ such that

$$\omega_- = \bigcap_{\tau<0} \mathrm{clo}_{H^1_K}\{w(.,t)\,/\,t < \tau\} \quad \text{and} \quad \omega_+ = \bigcap_{\tau>0} \mathrm{clo}_{H^1_K}\{w(.,t)\,/\,t > \tau\}. \tag{2.6.5}$$

Consequently we have the energy transition constraint

$$\int_{-\infty}^{\infty}\int w_\tau^2 K\,dy\,d\tau = E[\omega_-] - E[\omega_+], \tag{2.6.6}$$

where

$$E[\omega_\pm] = \frac{1}{2}\int_{\mathbb{R}^N}|\nabla\omega_\pm|^2 K\,d\eta - \frac{1}{2(p-1)}\int_{\mathbb{R}^N}\omega_\pm^2 K\,d\eta + \frac{1}{p+1}\int_{\mathbb{R}^N}|\omega_\pm|^{p+1}K\,d\eta. \tag{2.6.7}$$

As a consequence, either w is constant, or $E[\omega_-] > E[\omega_+]$.

It is worth noticing that the above techniques applies to symmetry properties of singular solutions of (2.3.1)-(2.3.2) in $\mathbb{R}^N \times (0,\infty)$. We give here a simple example (see [Ve4] for the proof).

THEOREM 6.17. *Let* $u \in C^{2,1}\big(\mathbb{R}^N \times (0,\infty)\big)$ *be a solution of* (2.3.1) *in* $\mathbb{R}^N \times (0,\infty)$ *such that*

$$\int_{\mathbb{R}^N} u^2(x,t)e^{|x|^2/4t}\,dx < \infty \qquad (\forall t > 0), \tag{2.6.8}$$

$$\liminf_{t\to 0} t^{1+N/2}\int_0^{\infty}\int_{S^{N-1}}|u(r,\sigma,t) - \bar{u}(r,t)|^2 e^{r^2/4t}r^{N-1}\,d\sigma\,dr = 0, \tag{2.6.9}$$

where (r,σ) *are the spherical coordinates in* $\mathbb{R}^N\setminus\{O\}$ *and* $\bar{u}(r,t)$ *is the spherical average of* $\sigma \mapsto u(r,\sigma,t)$. *Then* u *is is radially symmetric for any* $t > 0$.

6-3 Isolated singularities: the source case

6-3-1 Singularities of solutions of parabolic inequalities

The techniques of Section 3.1 dealing with elliptic inequalities can be adapted to treat some parabolic inequalities. This adaptation has been performed by Moutoussamy and Véron in

[MV1]. The parabolic version of the Brezis-Lions lemma is the following:

PROPOSITION 6.7. *Set* $Q = B(O,1) \times (0,T)$, $B^*(O,1) = B(O,1) \backslash \{O\}$, *and let* $\Psi \in L^1(Q)$ *and* $\omega \in C\big(\overline{Q} \backslash \{(O,0)\}\big)$ *be such that* $\omega_t - \Delta\omega \in L^1_{loc}\big(\overline{Q} \backslash \{(O,0)\}\big)$ *and*

$$\begin{cases} \omega \geq 0 & a.e.\ in\ Q, \\ \omega(x,0) = 0 & in\ B^*(O,1), \\ -\omega + \Delta\omega \leq \Psi. & a.e.\ in\ Q. \end{cases} \tag{3.1.1}$$

Then $\omega \in L^\infty\big((0,T); L^1(B(O,1))\big)$ *and there exist* $\beta \geq 0$ *and* $\Phi \in L^1(Q)$ *such that*

$$\begin{cases} -\omega + \Delta\omega = \Phi. & in\ Q, \\ \omega(x,0) = \beta\delta_O & in\ B^*(O,1). \end{cases} \tag{3.1.2}$$

Proof. Let φ_1 be the first eigenfunction of $-\Delta$ in $W_0^{1,2}(B(O,1))$ normalized by $\varphi(O) = 1$ and λ_1 the corresponding eigenvalue. For any $0 < s < t \leq T$, there holds

$$-\left[\int_{B(O,1)} \omega(x,\tau)\varphi_1(x)dx\right]_s^t - \lambda_1 \int_s^t \int_{B(O,1)} \omega\, dxd\tau - \int_s^t \int_{B(O,1)} \omega \frac{\partial\varphi_1}{\partial\nu} dSd\tau$$

$$\leq \int_s^t \int_{B(O,1)} \varphi_1 \Psi dxd\tau. \tag{3.1.3}$$

If we set $X(t) = \int_{B(O,1)} \omega(x,t)\varphi_1(x)dx$, then $\dfrac{d}{dt}\big[e^{\lambda_1 t}X(t)\big] \geq -e^{-\lambda_1 t}\int_{B(O,1)} |\Psi(x,t)|dx$, in the sense of distributions on $(0,T)$; consequently the function

$$t \mapsto e^{\lambda_1 t}\left[X(t) + \int_0^t \int_{B(O,1)} |\Psi(x,\tau)|dxd\tau\right] \tag{3.1.4}$$

is nondecreasing and there exists $A \geq 0$ such that

$$A = \lim_{t \to 0} e^{\lambda_1 t}\left[X(t) + \int_0^t \int_{B(O,1)} |\Psi(x,\tau)|dxd\tau\right] = \lim_{t \to 0} \int_{B(O,1)} \omega(x,t)\varphi_1(x)dx. \tag{3.1.5}$$

Therefore, there exist a positive measure μ on $B(O,1)$ and a sequence $\{t_n\}$ converging to 0 such that $\omega(.,t_n) \to \mu$ in the weak sense. In fact $\mu = A\delta_O$ since it is concentrated at O. Denote $h = -\omega_t + \Delta\omega - \Psi$, then h is nonnegative, and if $\phi \in C_0^\infty(B(O,1))$ with $0 \leq \phi \leq 1$ and value 1 in a neighborhood of O, we have

$$\int_0^T\!\!\int_{B(O,1)} h\phi\,dxdt = \int_0^T\!\!\int_{B(O,1)} \omega\Delta\phi\,dxdt$$

$$-\int_0^T\!\!\int_{B(O,1)} \Psi\phi\,dxdt + A\phi(O) - \int_{B(O,1)} \omega(x,T)\phi\,dx, \tag{3.1.6}$$

by integration on (t,T) and letting t go to O. Consequently, $-\omega_t + \Delta\omega = \Phi$ is integrable in Q and (3.1.2) follows.

The extension of Richard-Véron [RV] results to parabolic inequalities is the following:

THEOREM 6.18. *Let g be any continuous nondecreasing real valued function defined on $\mathbb{R}^+$, vanishing at 0 and such that*

$$g(E_{\delta_o}^{B(O,1)}) \in L^1(Q), \tag{3.1.7}$$

with the notations of section 6.2.2. Let $u \in C^{2,1}(\overline{Q} \setminus \{(O,0)\})$ satisfy

$$\begin{cases} u \geq 0 & a.e. \ \ in \ \ Q, \\ u(x,0) = 0 & in \ \ B^*(O,1), \\ -u + \Delta u \leq g(u). & in \ \ \overline{Q} - \{(O,0)\}. \end{cases} \tag{3.1.8}$$

Then we have the following alternative :

(i) either there exists $\gamma \geq 0$ such that $k^{N/2}\tilde{u}(\sqrt{k}x,kt)$ converges to $\gamma E_{\delta_o}^{B(O,1)}$ locally in measure in $\mathbb{R}^N \times \mathbb{R}^+$, where $\tilde{u}$ is the extension of u by 0 outside $\overline{Q}$, or

$$(ii) \quad \lim_{t\to 0} t^{N/2}u(x,t) = \infty, \tag{3.1.9}$$

uniformly on $E_\alpha \cap Q$, where $E_\alpha = \left\{(x,t) \in \mathbb{R}^N \times \mathbb{R}^+ : |x| \leq \alpha\sqrt{t}\right\}$, $\alpha > 0$.

Proof. Since (3.1.7) holds, for any $\lambda > 0$ we can define v_λ as being the solution of

$$\begin{cases} (v_\lambda)_t - \Delta v_\lambda + g(v_\lambda) = 0 & in \ \ Q, \\ v_\lambda(x,t) = \lambda\delta_o & in \ \ B(O,1), \\ v_\lambda(x,t) = 0 & on \ \ \partial B(O,1) \times (0,T), \end{cases} \tag{3.1.10}$$

and, for $\delta > 0$, set

$$p_\delta(t) = p(t) = \begin{cases} |t| - \delta/2 & if \ \ |t| \geq \delta, \\ t^2/2\delta & if \ \ |t| \leq \delta. \end{cases} \tag{3.1.11}$$

If $\omega_\delta = \dfrac{1}{2}(u + v_\lambda - p(u - v_\lambda))$ and $\omega^\lambda = \inf(u, v_\lambda) = \dfrac{1}{2}(u + v_\lambda - |u - v_\lambda|)$, then clearly

$$0 \leq \omega^\lambda \leq \omega_\delta \leq \omega^\lambda + \delta / 4. \qquad (3.1.12)$$

Step 1. We claim that there exist $\beta = \beta(\lambda) \geq 0$ and $\Phi \in L^1(Q)$ such that

$$\begin{cases} (\omega_\delta)_t - \Delta \omega_\delta = \Phi & \text{in } Q, \\ \omega_\delta(x, 0) = \beta \delta_o & \text{in } B(O, 1). \end{cases} \qquad (3.1.13)$$

By a direct computation

$$\begin{aligned}
-(\omega_\delta)_t + \Delta \omega_\delta &= -\frac{1}{2}\left[u_t + (v_\lambda)_t - p'(u - v_\lambda)(u_t - (v_\lambda)_t) \right] + \frac{1}{2}\Delta(u + v_\lambda) \\
&\quad - \frac{1}{2}p'(u - v_\lambda)\Delta(u - v_\lambda) - \frac{1}{2}p''(u - v_\lambda)|\nabla(u - v_\lambda)|^2, \\
&\leq -\frac{1}{2}(u_t + (v_\lambda)_t) + \frac{1}{2}\Delta(u + v_\lambda) \qquad (3.1.14) \\
&\quad + \frac{1}{2}p'(u - v_\lambda)\left[u_t - (v_\lambda)_t - \Delta(u - v_\lambda) \right] \\
&= F.
\end{aligned}$$

Defining the sets Q_1 by

$$Q_1 = \left\{ (x, t) \in \overline{Q} \setminus \{(O, 0)\} : (u - v_\lambda)(x, t) > \delta \right\},$$

$$Q_2 = \left\{ (x, t) \in \overline{Q} \setminus \{(O, 0)\} : (u - v_\lambda)(x, t) < -\delta \right\},$$

$$Q_3 = \left\{ (x, t) \in \overline{Q} \setminus \{(O, 0)\} : |(u - v_\lambda)(x, t)| \leq \delta \right\},$$

then

$$F = -(v_\lambda)_t + \Delta v_\lambda = g(v_\lambda) \leq g(v_\lambda + \delta) = g(\omega_\delta + 3\delta / 4) \qquad (3.1.15)$$

on Q_1 and

$$F = -u_t + \Delta u \leq g(u) \leq g(\omega_\delta + 3\delta / 4), \qquad (3.1.16)$$

on Q_2. On Q_3 we have $p'(u - v_\lambda) = (u - v_\lambda) / \delta$ and

$$F = -\frac{1}{2}\left[u_t\left(1 - \frac{u - v_\lambda}{\delta}\right) + (v_\lambda)_t\left(1 + \frac{u - v_\lambda}{\delta}\right)\right]$$

$$+ \frac{1}{2}\left(1 - \frac{u - v_\lambda}{\delta}\right)\Delta u + \frac{1}{2}\left(1 + \frac{u - v_\lambda}{\delta}\right)\Delta v_\lambda,$$

$$= \frac{1}{2}\left(1 - \frac{u - v_\lambda}{\delta}\right)(\Delta u - u_t) + \frac{1}{2}\left(1 + \frac{u - v_\lambda}{\delta}\right)(\Delta v_\lambda - (v_\lambda)_t), \tag{3.1.17}$$

$$\leq \frac{1}{2}\left(1 - \frac{u - v_\lambda}{\delta}\right)g(u) + \frac{1}{2}\left(1 + \frac{u - v_\lambda}{\delta}\right)g(v_\lambda).$$

By the mean value theorem and the monotonicity of g, the right-hand side of (3.1.17) is smaller than $g(\omega_\delta + \delta)$. Consequently the inequality

$$-(\omega_\delta)_t + \Delta\omega_\delta \leq g(\omega_\delta + \delta) \tag{3.1.18}$$

holds in $\overline{Q}\setminus\{(O,0)\}$. Since $g(\omega_\delta + \delta)$ is integrable in Q, Step 1 follows from Proposition 6.7.

Step 2. If $\overline{\omega}^\lambda$ is the extension of ω^λ by 0 outside $\overline{Q}$, then we shall prove that

$$\lim_{k \to 0} k^{N/2}\tilde{\omega}^\lambda(\sqrt{k}x, kt) = \beta(\lambda)E(x,t) \tag{3.1.19}$$

holds in $L^\infty_{loc}(0,\infty; L^1(\mathbb{R}^N))$, where $E(x,t)$ is the heat kernel in $\mathbb{R}^N$.

From (3.1.12), we can replace ω^λ by ω_δ and, with the maximum principle, we also have

$$E^{B(O,1)}_{\delta_\circ}(x,t) \leq E(x,t) \leq E^{B(O,1)}_{\delta_\circ}(x,t) + \max_{0 < \tau \leq t}\left((4\pi\,\tau)^{-N/2}e^{-1/4\tau}\right). \tag{3.1.20}$$

If ψ_δ denotes $\omega_\delta - \beta(\lambda)E^{B(O,1)}_{\delta_\circ}$, we have

$$\lim_{t \to 0}\int_{B(O,1)}|\psi_\delta(x,t)|dx = 0, \tag{3.1.21}$$

since Φ is integrable, or equivalently

$$\lim_{k \to 0}\int_{B(O,1/k)}|\psi_\delta(\sqrt{k}x, kt)|dx = 0. \tag{3.1.22}$$

Because $k^{N/2}\tilde{\omega}_\delta(\sqrt{k}x, kt) = \beta\, k^{N/2}\tilde{E}^{B(O,1)}_{\delta_\circ}(\sqrt{k}x, kt) + k^{N/2}\tilde{\psi}_\delta(\sqrt{k}x, kt)$, we get

$$\lim_{k \to 0}\left(\max_{0 < \tau \leq t}\left\| k^{N/2}\tilde{\omega}_\delta(\sqrt{k}\cdot, k\tau) - \beta\, k^{N/2}E(\sqrt{k}\cdot, k\tau)\right\|_{L^1(\mathbb{R}^N)}\right) = 0 \tag{3.1.23}$$

with the same extension notations outside $\overline{Q}$ as above. This means (3.1.19).

Step 3. End of the proof. Because $g(v_\lambda)$ is integrable in Q, the above technique applies also and

$$\lim_{k \to 0} k^{N/2} \tilde{v}_\lambda(\sqrt{k}x, kt) = \lambda E(x,t) \qquad (3.1.24)$$

holds in the topology of $L_{loc}^\infty(0, \infty; L^1(\mathbb{R}^N))$, where $\tilde{v}_\lambda(x,t)$ is the extension by 0 outside $\overline{Q}$, and this convergence is uniform on $E_\alpha \cap (\mathbb{R}^N \times [0,S])$ for any $S > 0$. Moreover $\lambda \mapsto v_\lambda$ is nondecreasing and it is the same with $\lambda \mapsto \omega_\lambda$ and $\lambda \mapsto \beta(\lambda)$. Therefore we are left with two possibilities:

Case 1. $\lim_{\lambda \to \infty} \beta(\lambda) = \gamma \in [0, \infty)$. In that case, we choose $\lambda > \gamma$ and from any sequence $\{k_n\}$ converging to 0, we can extract a subsequence such that

$$\lim_{k_{n_j} \to 0} k_{n_j}^{N/2} \tilde{\omega}_\lambda(\sqrt{k_{n_j}}x, k_{n_j}t) = \beta(\lambda)E(x,t) \qquad (3.1.25)$$

a.e. in $\mathbb{R}^N \times \mathbb{R}^+$. Since $\beta(\lambda) \le \gamma < \lambda$, we deduce from (3.1.24) that

$$\lim_{k_{n_j} \to 0} k_{n_j}^{N/2} \tilde{u}(\sqrt{k_{n_j}}x, k_{n_j}t) = \beta(\lambda)E(x,t). \qquad (3.1.26)$$

Choosing $\lambda' \ne \lambda$ and larger than γ, we have

$$\lim_{k_{n'_j} \to 0} k_{n'_j}^{N/2} \tilde{\omega}_{\lambda'}(\sqrt{k_{n'_j}}x, k_{n'_j}t) = \beta(\lambda')E(x,t) \qquad (3.1.27)$$

for another subsequence $\{k_{n'_j}\}$ extracted from $\{k_{n_j}\}$. Therefore $\beta(\lambda) = \beta(\lambda')$ and β is constant on (γ, ∞), with value γ. In that case, (i) follows.

Case 2. $\lim_{\lambda \to \infty} \beta(\lambda) = \infty$. In that case we fix $\mu > 0$; we set $\lambda > 0$ such that $\beta(\lambda) \ge \mu$ and for $\sigma > 0$, we define $q(r) = \min(1, r^+ / \sigma)$ and $j(r) = \int_0^r q(t)dt$ and deduce from (3.1.10)-(3.1.18) that

$$\left[\int_{B(O,1)} j(v_\mu - \omega_\delta)dx\right]_s^t + \int_s^t \int_{B(O,1)} q'(v_\mu - \omega_\delta)|\nabla(v_\mu - \omega_\delta)|^2 dxd\tau$$

$$+ \int_s^t \int_{B(O,1)} q(v_\mu - \omega_\delta)\big(g(v_\mu) - g(\omega_\delta + \delta)\big)dxd\tau \le 0, \qquad (3.1.28)$$

holds for $0 < s < t \le T$. Letting δ, and then σ go to 0 implies that the function $t \mapsto \int_{B(O,1)}(v_\mu - \omega^\lambda)(x,t)dx$ is nonincreasing. From Step 2 there exists a sequence $\{k_n\}$ going to 0 when n tends to infinity such that

349

$$\lim_{n\to\infty} k_n^{N/2}\left(\tilde{v}_\mu - \tilde{\omega}^\lambda\right)(\sqrt{k_n}\,x, k_n t) = (\mu - \beta(\lambda))E(x,t) \le 0, \tag{3.1.29}$$

a.e. in $\mathbb{R}^N \times \mathbb{R}^+$. We deduce from Lebesgue's theorem that

$$\lim_{n\to\infty} k_n^{N/2} \int_{\mathbb{R}^N} \left(\tilde{v}_\mu - \tilde{\omega}^\lambda\right)^+ (\sqrt{k_n}\,x, k_n t)\,dx$$
$$= \lim_{n\to\infty} \int_{B(O,1)} \left(v_\mu - \omega^\lambda\right)^+ (x, k_n t)\,dx = 0 \tag{3.1.30}$$

Consequently $v_\mu \le \omega_\lambda$. Since $\lambda E(t,.)$ and $v_\lambda(t,.)$ are both radial and radially decreasing, it is the same with the solution w of

$$\begin{cases} w_t - \Delta w = g(u) & \text{in } Q, \\ w(x,t) = (4\pi t)^{-N/2} e^{-1/4t} & \text{for } |x| = 1, \\ w(x,0) = 0 & \text{for } |x| \le 1, \end{cases} \tag{3.1.31}$$

and w is nonnegative. Therefore (3.1.24) and the following estimate

$$\int_{B(O,1)} \left(\lambda E(x,t) - k^{N/2} v_\lambda(\sqrt{k}\,x, kt)\right) dx$$
$$\ge \int_{|x|\le \varepsilon\sqrt{t}} \left(\lambda E(x,t) - k^{N/2} v_\lambda(\sqrt{k}\,x, kt)\right) dx, \tag{3.1.32}$$
$$\ge \text{meas}(B(O,\varepsilon\sqrt{t})) \max_{|x|\ge \varepsilon\sqrt{t}} \left(\lambda E(x,t) - k^{N/2} v_\lambda(\sqrt{k}\,x, kt)\right),$$

implies that the right-hand side of (3.1.32) tends to 0 when k goes to 0. Because $v_\mu \le \omega_\lambda \le u$, we obtain (3.1.9).

As an application we give an isotropy result for any nonnegative solution of

$$u_t - \Delta u + g(u) = 0. \tag{3.1.33}$$

COROLLARY 6.2. *Let Q and g be as in Theorem 6.18 and* $u \in C^{2,1}\left(\overline{Q}\setminus \{(O,0)\}\right)$ *be a nonnegative solution of (3.1.34) in* $\overline{Q}\setminus \{(O,0)\}$, *vanishing on* $\overline{B}(O,1)\times\{0\}\setminus \{(O,0)\}$. *Then*
(i) either $\lim_{t\to 0} t^{N/2} u(x,t) = \infty$, *uniformly on the sets* $E_\alpha \cap Q$, $\alpha > 0$,
(ii) or there exists $\gamma \ge 0$ *such that* $t^{N/2}|u(x,t) - \gamma E(x,t)|$ *converges to* 0, *uniformly on the set* $E_\alpha^c \cap Q$ *for any* $\alpha > 0$ *when* t *tends to 0. Moreover* g(u) *is integrable in Q and* u *satisfies (3.1.33) with initial data* $\gamma \delta_O$.

In the particular case where $g(r) = r^p$ with $1 < p < (N+2)/N$, then Oswald result [Os] follows.

6-3-2 Isolated singularities of source type equations

In this section we consider positive solutions of the following type of equations

$$u_t - \Delta u = g(u), \tag{3.2.1}$$

where g is a continuous real valued function vanishing at 0 and taking nonnegative values on $\mathbb{R}^+$. In the general result no assumption of monotonicity is made on g.

THEOREM 6.19. *Assume that* u *belongs to* $C^{2,1}\left(\overline{Q}\setminus\{(O,0)\}\right)$ *and is a nonnegative solution of* (3.2.1) *in* $\overline{Q}\setminus\{(O,0)\}$ *vanishing on* $\overline{B}(O,1)\setminus\{O\}$ *and let* $\tilde{u}$ *be the extension of* u *by* 0 *outside* $\overline{Q}$. *Then there exists* $\gamma \geq 0$ *such that*

$$\lim_{k\to 0} k^{N/2}\tilde{u}(\sqrt{k}x, kt) = \gamma E(x,t) \tag{3.2.2}$$

holds in $L^\infty_{loc}(0,\infty; L^1(\mathbb{R}^N))$. *Moreover* g(u) *is integrable in* Q *and* u *solves* (3.2.1) *with* $\gamma\delta_O$ *as an initial data.*

Proof. From Theorem 6.18, $u \in L^\infty(0,T; L^1(\mathbb{R}^N))$ and there exist $\gamma \geq 0$ and $\Phi \in L^1(Q)$ such that

$$\begin{cases} -u_t + \Delta u = \Phi & \text{in } Q, \\ u(x,0) = \gamma\delta_O & \text{in } B(O,1). \end{cases} \tag{3.2.3}$$

Therefore $\Phi = -g(u)$ and if G^γ is the solution of

$$\begin{cases} G^\gamma_t - \Delta G^\gamma = 0 & \text{in } Q, \\ G^\gamma(x,t) = u(x,t) & \text{on } \partial B(O,1) \times [O,T], \\ G^\gamma(x,0) = \gamma\delta_O & \text{in } B(O,1), \end{cases} \tag{3.2.4}$$

then

$$\gamma E_{\delta_O}^{B(O,1)}(x,t) \leq G^\gamma(x,t) \leq \gamma E_{\delta_O}^{B(O,1)}(x,t) + \sup_{0\leq\tau\leq t}\|u(.,\tau)\|_{L^\infty(\partial B(O,1))}, \tag{3.2.5}$$

and

$$\lim_{k\to 0} k^{N/2}G^\gamma(\sqrt{k}x, kt) = \gamma E(x,t). \tag{3.2.6}$$

Setting $w(x,t) = u(x,t - G^\gamma(x,t)$, then

$$\lim_{t\to 0} \int_{B(O,1)} |w(x,t)|dx = 0 \tag{3.2.7}$$

which implies

$$\lim_{k \to 0} k^{N/2} \tilde{u}(\sqrt{k}x, kt) = \gamma E(x,t) \tag{3.2.8}$$

in $L^{\infty}_{loc}(0, \infty; L^1(\mathbb{R}^N))$.

Remark 6.8. If u is radially decreasing for any $t > 0$, then (2.2.2) is improved by improved by

$$\lim_{t \to 0} t^{N/2}(u(x,t) - \gamma E(x,t)) = 0, \tag{3.2.9}$$

holds uniformly on $E^c_\alpha \cap Q$ for any $\alpha > 0$. It is also worth noticing that if g satisfies

$$\int_0^T \int_{B(O,1)} g(E) dx dt = \infty, \tag{3.2.10}$$

then $\gamma = 0$.

In the particular case of a subcritical power growth, we have

THEOREM 6.20. *Let g be a C^1 nondecreasing function on $\mathbb{R}^+$, vanishing at 0. We assume that it satisfies the following relation*

$$\int_0^T \left(\int_{B(O,1)} \sup_{0 \leq \theta \leq 1} |g'(\theta\phi + (1-\theta)\psi|^p dx \right)^{q/p} dt < \infty, \tag{3.2.11}$$

for any nonnegative measurable functions ϕ and ψ such that $g(\phi)$ and $g(\psi)$ are integrable in Q, in which formula r, $q > 1$ are linked by the relations

$$\frac{N}{2r} + \frac{1}{q} < 1. \tag{3.2.12}$$

If $u \in C^{2,1}(\overline{Q} \setminus \{(O,0)\})$ is a nonnegative solution of (3.2.1) in $\overline{Q} \setminus \{(O,0)\}$ which vanishes on $\overline{B}(O,1) \setminus \{O\}$, there exists $\gamma \geq 0$ such that

$$\lim_{t \to 0} t^{N/2} \left| u(x,t) - \gamma E^{B(O,1)}_{\delta_o} \right| = 0 \tag{3.2.13}$$

holds uniformly on any set $E^c_\alpha \cap Q$. Moreover, if $\gamma = 0$, u is bounded in $\overline{Q}$.

Proof. The nonnegative number γ exists from Theorem 6.20 and g(u) is integrable in Q.
Step 1. Suppose $\gamma > 0$, then there exists a solution u_γ of

$$\begin{cases} u_{\gamma_t} - \Delta u_\gamma = g(u_\gamma) & \text{in } Q, \\ u_\gamma(x,t) = 0 & \text{on } \partial B(O,1) \times (0,T), \\ u_\gamma(x,0) = \gamma \delta_O & \text{in } B(O,1). \end{cases} \qquad (3.2.14)$$

This solution is obtained via the iterative scheme

$$\begin{cases} u_{\gamma_t}^n - \Delta u_\gamma^n = g(u_\gamma^{n-1}) & \text{in } Q, \\ u_\gamma^n(x,t) = 0 & \text{on } \partial B(O,1) \times (0,T), \\ u_\gamma^n(x,0) = \gamma \delta_O & \text{in } B(O,1). \end{cases} \qquad (3.2.15)$$

for $n \in \mathbb{N}$ with $u_\gamma^0 = 0$. From the monotonicity of g and the integrability of $g(u)$, there holds

$$0 \le u_\gamma^0 = \gamma E_{\delta_o}^{B(O,1)} \le u_\gamma^1 \le \ldots \le u_\gamma^n \le u_\gamma^{n+1} \le u. \qquad (3.2.16)$$

The limit of the sequence $\left\{ u_\gamma^n \right\}$ is some function u_γ which solves (3.2.1), is radial and radially decreasing.

Step 2. Proof in the case $\gamma > 0$. Setting $w = u - u_\gamma$, $d = \big(g(u) - g(u_\gamma)\big)\big/(u - u_\gamma)$, the relation

$$\int_0^T \int_{B(O,1)} \left[w\big(-\phi_t - \Delta\phi\big) - dw \right] dx dt = 0 \qquad (3.2.17)$$

is satisfied for any test function $\phi \in C^{2,1}(\overline{B}(O,1) \times [0,T])$ with compact support in $B(O,1) \times [0,T]$. Moreover we can write d in the following way:

$$d(x,t) = g'\big(\theta u(x,t) + (1-\theta)u_\gamma(x,t)\big), \text{ with } \theta = \theta(x,t) \in (0,1). \qquad (3.2.18)$$

Because

$$\int_0^T \left(\int_{B(O,1)} |d(x,t)|^r dx \right)^{q/r} dt < \infty, \qquad (3.2.19)$$

with p and q satisfying (3.2.12) we deduce from Theorem 6.3 that w remains bounded in $\overline{Q}$, and (3.2.13) follows from Remark 6.8 (the sign of the integrable nonlinear term is not important).

Step 3. The case $\gamma = 0$. We just write (3.2.1) under the form

$$u_t - \Delta u - \frac{g(u)}{u}u = 0, \qquad\qquad (3.2.20)$$

and we have $\int_0^T \left(\int_{B(O,1)} |g(u)/u|^r dx \right)^{q/r} dt < \infty$. Therefore u remains bounded and (3.2.13) follows.

The existence proof of Step 1 can be adapted in order to have the following type of existence result:

PROPOSITION 6.8. *Let g be a nondecreasing real valued fonction defined on $\mathbb{R}^+$, vanishing at 0, $T > 0$, $\Lambda > 0$ are real numbers and $\phi \in C^0\left(\overline{B}(O,1) \times [0,T] \backslash \{(O,0)\}\right)$ is a nonnegative function which satisfies*

$$\begin{cases} g(\phi) \in L^1(B(O,1) \times (0,T)), \\ \phi(x,t) \geq \Lambda E_{\delta_o}^{B(O,1)} & in \ \overline{B}(O,1) \times (0,T], \\ \phi_t - \Delta\phi \geq g(\phi) & in \ \mathcal{D}'\left(\overline{B}(O,1) \times (0,T]\right). \end{cases} \qquad (3.2.21)$$

Then for any $\lambda \in [0,\Lambda]$ there exists at least one nonnegative function $u \in C^0\left(\overline{B}(O,1) \times [0,T]\right)$, such that $g(u)$ is integrable in Q satisfying

$$\begin{cases} u_t - \Delta u = g(u) & in \ Q, \\ u(x,t) = 0 & on \ \partial B(O,1) \times (0,T), \\ u(x,0) = \lambda\delta_o & in \ \mathcal{D}'(B(O,1)). \end{cases} \qquad (3.2.22)$$

Remark 6.9. It is not always easy to construct such a ϕ, however, if $0 \leq g(r) \leq Cr^p$ for $r > 0$ with $1 < p < (N+2)/N$, it can be checked (see [BPT]) that there exists $D > 0$ and $\sigma = \sigma(N,p) > 0$ such that

$$\left| \int_0^t S^*(t-s)g(kE_{\delta_o}^{B(O,1)}(.,s))(x)ds \right| \leq Dk^p t^\sigma E_{\delta_o}^{B(O,1)}(x,t). \qquad (3.2.23)$$

Here $\left(S^*(t)\right)_{t \geq 0}$ is the semi-group associated to the linear heat equation in $B(O,1)$ with Dirichlet boundary data. Consequently, for any $\lambda > 0$, there exists $T > 0$ such that the function ϕ defined by the formula

$$\phi(x,t) = \lambda E_{\delta_o}^{B(O,1)}(x,t) + \int_0^t S^*(t-s)g(2\lambda E_{\delta_o}^{B(O,1)}(.,s))(x)ds, \qquad (3.2.24)$$

satisfies (3.2.21). A similar estimate is used in Remark 4.15.

COROLLARY 6.3. *Let* $1 < p < (N+2)/N$ *and* $u \in C^{2,1}(\overline{Q} \setminus \{(O,0)\})$ *is a nonnegative solution of*

$$u_t - \Delta u = u^p \tag{3.2.25}$$

in $\overline{Q} \setminus \{(O,0)\}$ *vanishing on* $\overline{B}(O,1) \setminus \{O\}$; *then there exists* $\gamma \geq 0$ *such that the relation*

$$\lim_{t \to 0} \left(u(x,t) - \gamma E_{\delta_o}^{B(O,1)} \right) = 0, \tag{3.2.26}$$

holds uniformly in $\overline{B}(O,1)$.

Proof. Since $1 < p < (N+2)/N$, the relation (3.2.11) is verified and we can apply Theorem 6.20 with $r = q = p/(p-1)$. We have the existence of $\gamma \geq 0$ such that (3.2.13) holds. From Remark 6.9 there exists $T^* \in (0,T]$ and u_γ such that $\phi = (\gamma+1)E_{\delta_o}^{B(O,1)}$ satisfies (3.2.21) on $(0,T^*] \times B(O,1)$ and (3.2.14) holds, with T^* replacing T. This function u_γ is obtained via the increasing scheme (3.2.15) with the following estimate

$$0 \leq u_\gamma^0 = \gamma E_{\delta_o}^{B(O,1)} \leq u_\gamma^1 \leq \ldots \leq u_\gamma^n \leq u_\gamma^{n+1} \leq (\gamma+1)E_{\delta_o}^{B(O,1)}. \tag{3.2.27}$$

Consequently, we have uniformly in $\overline{B}(O,1) \times (0,T^*]$,

$$0 < u_\gamma(x,t) - \gamma E_{\delta_o}^{B(O,1)}(x,t) \leq D' \gamma^p t^\sigma E_{\delta_o}^{B(O,1)}(x,t), \tag{3.2.28}$$

and from Theorem 6.3

$$0 \leq (u - u_\gamma)(x,t) \leq C \max_{0 \leq \tau \leq t} \| u(.,\tau) \|_{L^\infty(\partial B(O,1))}. \tag{3.2.29}$$

holds. Estimate (3.2.26) follows from (3.2.28)-(3.2.29).

6-3-3 Rapidly decaying solutions

If we look for singular solutions of

$$u_t - \Delta u = |u|^{p-1} u \qquad (\, p > 1 \,), \tag{3.3.1}$$

in $\mathbb{R}^N \times \mathbb{R}^+ \setminus \{(O,0)\}$ under the form

$$u(x,t) = t^{-1/(p-1)} f(x/\sqrt{t}), \tag{3.3.2}$$

then f satisfies (with $x/\sqrt{t} = \eta$)

$$\Delta f + \frac{1}{2} x . \nabla f + \frac{1}{p-1} f + |f|^{p-1} f = 0 \tag{3.3.3}$$

in $\mathbb{R}^N$. If we suppose also that $\lim_{t \to 0} u(x,t) = 0$ for any $x \neq O$ and uniformly with respect to $|x|$, then $\lim_{|\eta| \to \infty} |\eta|^{2/(p-1)} f(\eta) = 0$. This type of rapidly decaying solutions have been intensively studied by Weissler ([We1], [We2]) from an ODE point of view and by Escobedo and Kavian ([EK]) by variational methods. In fact, if we consider the more general equation

$$\Delta \omega + \frac{1}{2} x . \nabla \omega + \lambda \omega + |\omega|^{p-1} \omega = 0, \tag{3.3.4}$$

we can summarise below the existence results (we recall that $K(\eta) = e^{|\eta|^2/4}$):

THEOREM 6.21. *Let* $1 < p < (N+2)/(N-2)$, *then there exist infinitely many solutions of (3.3.4) in* $H^1_K(\mathbb{R}^N)$ *and any of these solution is* C^2 *and satisfies*

$$|\omega(\eta)| + |\nabla \omega(\eta)| \le C e^{-|\eta|^2/8} \tag{3.3.5}$$

for some $C > 0$. *Moreover, if* $\lambda < N/2$, *one of these solutions is positive.*

The proof of the existence of infinitely many solutions is obtained in [EK] via Ljusternik-Schnirelman theory. In [We1], [We2], there are infinitely many radial solutions which are constructed via nodal or variational methods. Very recently, an interesting uniqueness result has be proved by Yanagida [Ya] in the study of radial solutions of (3.3.3) which are solutions $\rho \mapsto f(\rho)$ of

$$\begin{cases} f'' + \left(\dfrac{N-1}{\rho} + \dfrac{\rho}{2} \right) f' + \dfrac{1}{p-1} f + |f|^{p-1} f = 0 & \text{for} \quad \rho > 0, \\ f(0) = \alpha, \quad f'(0) = 0, \end{cases} \tag{3.3.6}$$

and are obtained via the classical shooting method. Any local solution f of (3.3.6) can be constructed as a fixed point of the operator $f \mapsto F(f)$ for $f \in C([0,T])$, where

$$F(f)(\rho) = \alpha + \int_0^\rho s^{1-N} e^{-s^2/4} \int_0^s \sigma^{N-1} e^{\sigma^2/4} \left(|f|^{p-1} f(\sigma) + \frac{1}{p-1} f(\sigma) \right) d\sigma ds. \tag{3.3.7}$$

The local solution is necessarily a global since (3.3.6) implies that it remains bounded on $(0, \infty)$. Let $\rho \to f(\alpha, \rho)$ be the solution of (3.3.6), then Yanagida's theorem reads as follows

THEOREM 6.22. *Let* $1 < p \leq N/(N-2)$, *then for each nonnegative integer* m *there exists at most one* $\alpha > 0$ *such that* $\rho \mapsto f(\alpha, \rho)$ *is a solution of* (3.3.6) *in* $H_K^1(\mathbb{R}^N)$ *with exactly* m *zeroes on* $(0, \infty)$.

An immediate consequence is the uniqueness of the positive solution in $H_K^1(\mathbb{R}^N)$ obtained through Theorem 6.21. An interesting case deals with the critical case $p = (N+2)/(N-2)$. The following result proved in [EK] extends a well-known result of Brezis and Nirenberg [BN].

THEOREM 6.23. I- *Suppose* $N > 3$, *then for any* $\lambda \in (N/4, N/2)$ *there is at least one positive solution of* (3.3.4) *in* $H_K^1(\mathbb{R}^N)$. *If* $\lambda \notin (N/4, N/2)$, *no such solution exists.*

        II- *Suppose* $N = 3$, *then for* $\lambda \in (1, 3/2)$ *the previous existence result holds. If* $\lambda \leq 3/4$ *or* $\lambda \geq 3/2$, *no such solution exists.*

6-4 Some quasilinear equations

6-4-1 The porous media and the p-laplacian flows

The density u of a gaz through a porous medium obeys Darcy's law which implies the following equation in $\mathbb{R}^N \times (0, \infty)$, called the porous medium equation,

$$u_t - \Delta u^m = 0, \tag{4.1.1}$$

with m > 1. This equation possesses a self-similar singular solution found by Barenblatt [Ba], Pattle [Pa] and Zel'dovich [ZK], namely

$$W_M(x, t) = t^{-k}\left(C - \frac{(m-1)k}{2mN}\frac{|x|^2}{t^{2k/N}}\right)_+^{1/(m-1)} \tag{4.1.2}$$

where $k = (m - 1 + 2/N)^{-1}$ and C is determined by the mass M of the solution W

$$M = \int_{\mathbb{R}^N} W(x, t)dx, \tag{4.1.3}$$

via $C = a(m, N)M^\ell$ with $\ell = 2(m-1)k/N$. Therefore W_M satisfies (4.1.1) with initial data $M\delta_0$. One of the most remarkable property of the porous medium equation is the finite speed of propagation of the support of integrable solutions. Another equation with the same finite speed of propagation property and called the p-Laplacian evolution equation or p-Laplacian flow is the following

$$u_t - \text{div}\left(|\nabla u|^{p-2}\nabla u\right) = 0, \tag{4.1.4}$$

with p > 2. Such an equation also admits a singular solution (see [Ba])

$$Z_M(x,t) = t^{-h}\left(D - \frac{p-2}{p}\left(\frac{h}{N}\right)^{1/(p-1)}\left(\frac{|x|}{t^{h/N}}\right)^{p/(p-1)}\right)_+^{(p-1)/((p-2)}$$

(4.1.5)

where $h = (p - 2 + p / N)^{-1}$ and $D = b(p,N)M^\alpha$ with $\alpha = p(p-2)h / N(p-1)$ where M is the time-invariant mass of Z_M. As above, Z solves (4.1.4) with $M\delta_0$ as an initial condition.

6-4-2 The porous media equation with absorption

If we consider the following equation

$$u_t - \Delta u^m + |u|^{p-1}u = 0$$

(4.2.1)

in $\mathbb{R}^N \times (0,\infty)$, with m, p > 1, then the study of solutions with an isolated singularty at (O,0) gives rise to a situation similar to the one of the heat equation with absoption. The following result is essentially due to Kamin and Peletier ([KP2]), but we shall give a more complete proof adapted from Gmira ([Gm]).

THEOREM 6.24. I- *Suppose* $p \geq m + 2 / N$, *then any function* u *which is continuous in* $\mathbb{R}^N \times [0,T]\setminus \{(O,0)\}$, *vanishes on* $\mathbb{R}^N \times \{0\}\setminus \{(O,0)\}$ *and satisfies* (4.2.1) *in* $\mathbb{R}^N \times (0,T)$ *is identically* 0.

    II- *Suppose* $1 < p < m + 2 / N$, *then for any* $c > 0$ *there exists a solution* u *of the singular initial value problem*

$$\begin{cases} u_t - \Delta u^m + |u|^{p-1}u = 0 & in \ \mathbb{R}^N \times (0,\infty), \\ u(x,0) = c\delta_0 & in \ \mathbb{R}^N. \end{cases}$$

(4.2.2)

Proof of I. Step 1. As in Lemma 6.2 an *a priori* estimate of Osserman type exists: there exists $C = C(N,m,p) > 0$ such that, for any (x,t) in $\mathbb{R}^N \times [0,T]\setminus \{(O,0)\}$, there holds

$$|u(x,t)| \leq \frac{C}{t^{1/(p-1)} + |x|^{2/(p-m)}}.$$

(4.2.3)

This is proved by considering a well-chosen supersolution.

Step 2. Every solution u vanishes outside the set

$$P = \left\{(x,t) : |x| \leq \rho t^{(p-m)/2(p-1)}\right\}$$

(4.2.4)

$\rho = \rho(N,m,p) > 0$. In fact, from (4.2.3), every solution is uniformly bounded by some D > 0 depending on N, p, q, r T for $(x,t) \in \partial B(O,r) \times [0,T]$. Therefore we can choose a function W_M given by (4.1.2) with M large enough so that $W_M(x,t) \geq D$ on $\partial B(O,r) \times [1, T+1]$. By the maximum principle

$$|u(x,t)| \leq W_M(x,t+1), \tag{4.2.5}$$

if $(x,t) \in B(O,r) \times [0,T]$. Consequently $u(.,t)$ has compact support for every $t \in (0,T]$. In particular there exists $\rho > 0$ such that $u(x,1) = 0$ for $|x| \geq \rho$, and it important to notice that the solution u is arbitrary. Finally, for $0 < \tau \leq T$, we can consider the rescaled function $\tilde{u}$ defined by

$$\tilde{u}(x,t) = \tau^{1/(p-1)} u(\tau^{(p-m)/2(p-1)} x, \tau t). \tag{4.2.6}$$

It satisfies the same equation in $\mathbb{R}^N \times (0, T/\tau)$. Since $T/\tau \geq 1$, we have $\tilde{u}(y,1) = 0$ for any $|y| \geq \rho$, which implies that $u(x,\tau) = 0$ if $|x| \geq \rho \tau^{(p-m)/2(p-1)}$.

Step 3. Since, as in the semilinear case, we always have

$$|u(x,t)| \leq \left(\frac{1}{t(p-1)}\right)^{1/(p-1)}, \tag{4.2.7}$$

we deduce

$$\int_{\mathbb{R}^N} |u(x,t)| dx \leq \left(\frac{1}{t(p-1)}\right)^{1/(p-1)} \mathrm{meas}\left(B(O, \rho t^{(p-m)/2(p-1)})\right). \tag{4.2.8}$$

From (4.2.2) we also have the mass decay that is

$$\int_{\mathbb{R}^N} |u(x,t)| dx \geq \int_{\mathbb{R}^N} |u(x,s)| dx \tag{4.2.9}$$

for $0 < t \leq s \leq T$.

If $p > m + 2/N$, the right-hand side of (4.2.8) goes to 0 with t, which implies that u is identically 0.

If $p = m + 2/N$, we can always assume without any loss of generality that u is nonnegative. Therefore there exists $K = K(n,m,p) \geq 0$ such that

$$\lim_{t \to 0} \int_{\mathbb{R}^N} u(x,t) dx = C_K \leq K. \tag{4.2.10}$$

Let us assume that $C_K > 0$ and take $K_1 > K$. For some $a > 1$ we define

$$\tilde{u}(x,t) = a^{-2/(m-1)} u(x/a, t) \tag{4.2.11}$$

and

$$\tilde{u}_t - \Delta\tilde{u}^m + a^{-2(p-1)/(m-1)}\tilde{u}^p = 0 \tag{4.2.12}$$

with

$$\lim_{t\to 0}\int_{\mathbb{R}^N}\tilde{u}(x,t)dx = a^{N+2/(m-1)}C_K. \tag{4.2.13}$$

Now take $\tau > 0$ and introduce u^τ as being the solution of (4.2.1) in $\mathbb{R}^N\times(0,\infty)$ with initial data $\tilde{u}(x,\tau)$. Since $a > 1$, $\tilde{u}$ is a supersolution and

$$0 \le u^\tau(x,t) \le \tilde{u}(x,t+\tau). \tag{4.2.14}$$

When τ tends to 0, the L^1-contractivity property of the equation implies that u^τ converges to some u^0 which is also a solution of (4.2.1) in $\mathbb{R}^N\times(0,\infty)$, and from (4.2.13)(4.2.14) there holds

$$\lim_{t\to 0}\int_{\mathbb{R}^N}u^0(x,t)dx \le a^{N+2/(m-1)}C_K. \tag{4.2.15}$$

In order to find a contradiction we want to prove that $\lim_{t\to 0}\int_{\mathbb{R}^N}u^0(x,t)dx = a^{N+2/(m-1)}C_K$. For that we fix $0 < s < 2s < t < T$, $\chi \in C_0^\infty(\mathbb{R}^N)$ and $\eta \in C_0^\infty(0,T)$ with $0 \le \eta \le 1$ and

$$\eta(\sigma) = \begin{cases} 0 & \text{if } 0 < \sigma \le s, \\ 1 & \text{if } 2s < \sigma < t-2s, \\ 0 & \text{if } t-s < \sigma < t. \end{cases} \tag{4.2.16}$$

$$\chi(x) = \begin{cases} 1 & \text{if } |x| \le \rho T^{(p-m)/2(p-1)}, \\ 0 & \text{if } |x| \ge 2\rho T^{(p-m)/2(p-1)}. \end{cases} \tag{4.2.17}$$

Since u^τ is a solution of (4.2.1), we have

$$\int_0^T\int_{\mathbb{R}^N}\left(-u^\tau\chi\eta_t - u^{\tau m}\eta\Delta\chi + u^{\tau p}\chi\eta\right)dxdt = 0. \tag{4.2.18}$$

Letting s go to 0 implies

$$\left|\int_{\mathbb{R}^N}u^\tau(x,0)\chi(x)dx - \int_{\mathbb{R}^N}u^\tau(x,t)\chi(x)dx\right|$$
$$\le \int_0^t\int_{\mathbb{R}^N}u^{\tau p}(x,\sigma)dxd\sigma \le \int_t^{t+\tau}\int_{\mathbb{R}^N}\tilde{u}^p(x,\sigma)dxd\sigma. \tag{4.2.19}$$

Letting τ tend to 0 implies, with (4.2.13),

$$\left| C_K a^{N+2/(m-1)} - \int_{\mathbb{R}^N} u^0(x,t)dx \right| \le \int_0^t \int_{\mathbb{R}^N} \tilde{u}^p(x,\sigma)dxd\sigma, \tag{4.2.20}$$

and this contradicts (4.2.10) since a is arbitrary. Therefore, for any solution u,

$$\lim_{t \to 0} \int_{\mathbb{R}^N} u(x,t)dx = 0, \tag{4.2.21}$$

and we complete the proof with (4.2.9).

Proof of II. Step 4. The existence follows some ideas in the one of Theorem 6.12. For a positive integer ℓ, set u_ℓ the solution of

$$\begin{cases} u_{\ell t} - \Delta u_\ell^m + u_\ell^p = 0 & \text{in } \mathbb{R}^N \times (0,\infty), \\ u_\ell(x,0) = W_c(x,1/\ell) & \text{in } \mathbb{R}^N. \end{cases} \tag{4.2.22}$$

The choice of c means that $\int_{\mathbb{R}^N} u_\ell(x,0)dx = c$. From the maximum principle

$$0 < u_\ell(x,t) < W_c(x,t+1/\ell) \tag{4.2.23}$$

and $x \mapsto u_\ell(x,t) = 0$ vanishes outside some ball $B(O, K(1+t^{k/N}))$, where K is independent of ℓ. From estimate (4.2.23), the following estimate holds

$$\int_0^t \int_{\mathbb{R}^N} u_\ell^p(x,t)dxdt \le C' t^{1-kp}, \tag{4.2.24}$$

where the constant is again independent of ℓ. By DiBenedetto regularity result [DiB], the sequence $\{u_\ell\}$ is equicontinuous in K for any compact subset $K \subset \mathbb{R}^N \times (0,\infty)$. As a consequence there exist a function u belonging to $C(0,\infty;L^1(\mathbb{R}^N)) \cap L^\infty(\mathbb{R}^N \times (\tau,\infty))$ for every $\tau > 0$ and a subsequence $\{u_{\ell_n}\}$ such that $\lim_{n \to \infty} u_{\ell_n} = u$ in the $C_{loc}(\mathbb{R}^N \times (0,\infty))$-topology. It is clear that u is a weak solution of (4.2.1) in $\mathbb{R}^N \times (0,\infty)$. Set $\zeta \in C_0^\infty(\mathbb{R}^N)$, the last problem is to prove that

$$\lim_{t \to 0} \int_{\mathbb{R}^N} u(x,t)\zeta(x)dx = c\zeta(O). \tag{4.2.25}$$

Since u belongs to $C(0,\infty;L^1(\mathbb{R}^N))$, the same approxition argument as above proves that

$$\left| \int_{\mathbb{R}^N} u_\ell(t,x)\zeta(x)dx - \int_{\mathbb{R}^N} u_\ell(t',x)\zeta(x)dx \right| \le$$
$$A\int_{t'}^t \int_{\mathbb{R}^N} u_\ell^p(s,x)dxds + B\int_{t'}^t \int_{\mathbb{R}^N} u_\ell^m(s,x)dxds, \tag{4.2.26}$$

for $0 < t' < t$, where $A = \|\zeta\|_{L^\infty}$ and $B = \|\Delta\zeta\|_{L^\infty}$. When t' goes to 0, the limit of the left-hand side of (4.2.26) is $\left|\int_{\mathbb{R}^N} u_\ell(t,x)\zeta(x)dx - \int_{\mathbb{R}^N} W_c(\ell^{-1},x)\zeta(x)dx\right|$, as for the right-hand side, it is dominated by $L\left(t^{1-kp} + t^{1-km}\right)$ for some constant $L > 0$, and this from (4.2.24) and Hölder's inequality. Letting $\{\ell_n\}$ go to 0 yields

$$\left|\int_{\mathbb{R}^N} u(t,x)\zeta(x)dx - c\zeta(O)\right| \leq L\left(t^{1-kp} + t^{1-km}\right), \tag{4.2.27}$$

which in turn implies (4.2.27).

Remark 6.10. In the range of values $1 < m < p < m + 2/N$, it is proved in [KPV] that, for any $c > 0$, the solution u_c of (4.2.2) is unique, nonnegative and radially decreasing. When c goes to infinity, u_c converges to some $u_\infty(x,t)$ which has the following form

$$u_\infty(x,t) = t^{-1/(p-1)}f\left(|x| / t^{-(p-m)/2(p-1)}\right). \tag{4.2.28}$$

The function f is nonnegative on $\mathbb{R}^+$ where it satisfies

$$\frac{d^2f^m}{d\eta^2} + \frac{N-1}{\eta}\frac{df^m}{d\eta} + \frac{p-m}{2(p-1)}\eta\frac{df}{d\eta} + \frac{1}{p-1}f - f^p = 0, \tag{4.2.29}$$

with the limit conditions

$$f'(0) = 0, \qquad \lim_{\eta\to\infty}\eta^{2/(p-m)}f(\eta) = 0. \tag{4.2.30}$$

The existence of a nonnegative solution of (4.2.29)-(4.2.30) has been first proved by Peletier and Terman [PT] by using ODE techniques. In fact, it not difficult to see that the function u_∞ satisfies the integral blow-up estimate

$$\lim_{t\to 0}\int_{B(O,r)} u_\infty(x,t)dx = \infty \tag{4.2.31}$$

for any $r > 0$. A general existence and uniqueness result for this class of solutions has been proved by Kamin and Véron ([KV]).

THEOREM 6.25. *Let $N \geq 1$ and $1 < m < p < m + 2/N$. Then there exists a unique nonnegative function U continuous in $\mathbb{R}^N \times [0,\infty) \backslash \{(O,0)\}$, vanishing on the set $\mathbb{R}^N \times \{0\} \backslash \{(O,0)\}$, satisfying (4.2.1) in $\mathbb{R}^N \times (0,\infty)$ and such that*

$$\lim_{t\to 0}\int_{B(O,r)} U(x,t)dx = \infty \tag{4.2.32}$$

for any $r > 0$.

Thanks to this theorem, Kamin, Peletier and Vazquez[KPV] obtained a complete classification of the positive solutions of (4.2.1) with an isolated singularity at $t = 0$.

THEOREM 6.26. *Under the assumption of Theorem 6.25, any nonnegative function* u, *continuous in* $\mathbb{R}^N \times [0,\infty)\setminus \{(O,0)\}$, *vanishing on* $\mathbb{R}^N \times \{0\}\setminus \{(O,0)\}$ *and satisfying* (4.2.1) *in* $\mathbb{R}^N \times (0,\infty)$ *is either the zero function, either the solution* u_c *of* (4.2.2) *for some* $c > 0$, *or the very singular solution* U.

Remark 6.11. If $1 < p \le m$, the absorption is not strong enough to allow the existence of the very singular solution. Therefore the only singular solution of (4.2.1) with an isolated singularity located at $(O,0)$ are the solutions u_c of (4.2.2); moreover for any $c > 0$ such a solution u_c exists.

6-4-3 The p-Laplacian evolution equation with absorption

The main results concerning the porous mediium equation with absorption have been extended to the p-Laplacian evolution equation with absorption

$$u_t - \operatorname{div}\left(|\nabla u|^{p-2}\nabla u\right) + |u|^{q-1}u = 0, \tag{4.3.1}$$

in $\mathbb{R}^N \times (0,\infty)$, when $p > 2$. The following has been proved by Gmira[Gm]

THEOREM 6.27. I- *Suppose* $q \ge p-1+p/N$, *then any function* u *which is continuous in* $\mathbb{R}^N \times [0,T]\setminus \{(O,0)\}$, *vanishes on* $\mathbb{R}^N \times \{0\}\setminus \{(O,0)\}$ *and satisfie* (4.3.1) *in* $\mathbb{R}^N \times (0,T)$ *is identically* 0.

II- *Suppose* $0 < q < p-1+2/N$, *then for any* $c > 0$ *there exists a solution* u *of the singular initial value problem*

$$\begin{cases} u_t - \operatorname{div}\left(|\nabla u|^{p-2}\nabla u\right) + |u|^{q-1}u = 0 & \textit{in } \mathbb{R}^N \times (0,\infty), \\ u(x,0) = c\delta_O & \textit{in } \mathbb{R}^N. \end{cases} \tag{4.3.2}$$

The proof is the same as the one of Theorem 6.24. Moreover it is proved in [KVa] that the solution of (4.3.2) is unique. In the same way as for the porous medium equation with absorption, in the range of values $p > 2$ and $0 < q < p-1+2/N$, Peletier and Wang [PW] proved the existence of a radial very singular solution of (4.3.1) under the form

$$V(x,t) = t^{-1/(q-1)}f(t^{-p(q-1)/((q+1-p)}|x|) = t^{-1/(q-1)}f(\eta). \tag{4.3.3}$$

In that case f is nonnegative and satisfies

$$\frac{d}{d\eta}\left(\left|\frac{df}{d\eta}\right|^{p-2}\frac{df}{d\eta}\right) + \frac{N-1}{\eta}\left|\frac{df}{d\eta}\right|^{p-2}\frac{df}{d\eta} + \frac{q+1-p}{p(q-1)}\eta\frac{df}{d\eta} + \frac{1}{q-1}f - f^q = 0 \qquad (4.3.4)$$

on $(0,\infty)$ with the following limit conditions

$$f'(0) = 0, \qquad \lim_{\eta \to \infty}\eta^{p/(q+1-p)}f(\eta) = 0. \qquad (4.3.5)$$

Later on, Kamin and Vazquez [KVa] proved the analogous of theorems 6.25.

THEOREM 6.28. *Let* $N \geq 1, p > 2$ *and* $0 < q < p-1+2/N$. *Then there exists a unique nonnegative function* V *continuous in* $\mathbb{R}^N \times [0,\infty) \setminus \{(O,0)\}$, *vanishing on the set* $\mathbb{R}^N \times \{0\} \setminus \{(O,0)\}$, *satisfying* (4.3.1) *in* $\mathbb{R}^N \times (0,\infty)$ *and such that*

$$\lim_{t \to 0}\int_{B(O,r)} U(x,t)dx = \infty, \qquad (4.3.6)$$

for any $r > 0$. *The function* V *is defined by* (4.3.3)-(4.3.5).

It is likely that a classification theorem similar to Theorem 6.25 holds, however it has not yet been proved.

References

[Ad] R. A. Adams, **Sobolev Spaces**, Academic Press, New-York (1975).

[AP] D.R. Adams & J.C. Polking, The equivalence of two definitions of capacity, **Proc. Amer. Math. Soc. 37**, 529-534 (1973).

[Ar1] D.G. Aronson, Removable singularities for linear parabolic equations, **Arch. Rat. Mech. Anal. 17**, 79-84 (1965).

[Ar2] D.G. Aronson, Isolated singularities of solutions of second order linear parabolic equations, **Arch. Rat. Mech. Anal. 19**, 231-238 (1965).

[Ar3] D.G. Aronson, On the Green's function for second order parabolic differential equations with discontinuous coefficients, **Bull. Amer. Math. Soc. 69**, 841-847 (1963).

[AS] D.G. Aronson & J. Serrin, Local behaviour of quasilinear parabolic equations, **Arch. Rat. Mech. Anal. 25**, 81-122 (1967).

[Au1] Th. Aubin, Equations différentielles non-linéaires et problème de Yamabe concernant la courbure scalaire, **Jl. Math. Pures Appl. 55**, 269-293 (1976).

[Au2] Th. Aubin, **Nonlinear Analysis on Manifolds. Monge-Ampère Equations**, Springer-Verlag, New-York, Heidelberg, Berlin (1982).

[Av1] P. Aviles, On isolated singularities in some nonlinear partial differential equations, **Indiana Univ. Math. Jl. 32**, 773-791 (1983).

[Av2] P. Aviles, A study of the singularities of solutions of a class of nonlinear partial differential equations, **Comm. Part. Diff. Equ. 7**, 609-643 (1982)

[Av3] P. Aviles, Local behaviour of solutions of some nonlinear elliptic equations, **Comm. Math. Phys. 108**, 177-192 (1987).

[BE] C. Bandle & M. Essen, On positive solutions of Emden equations in cone-like domain, **Arch. Rat. Mech. Anal. 112**, 319-338 (1990).

[BM] C. Bandle & M. Marcus, Large solutions of semilinear elliptic equations: existence, uniqueness and asymptotic behaviour, **Jl. Anal. Math. 58**, 9-24 (1992).

[Ba] G.I. Barenblatt, On some unsteady motions of a liquid and a gas in a porous medium, **Prikladnaja Matematika i Mechanika 16**, 67-78 (1952).

[BHV] P. Baras, J.C. Hassan & L. Véron, Compacité de l'opérateur définissant la solution d'une équation d'évolution non homogène, **C.R. Acad. Sci. Paris t. 284 Ser. I**, 799-802 (1977)

[BP1]   P. Baras & M. Pierre, Singularités éliminables pour des équations semi-linéaires, **Ann. Inst. Fourier, Grenoble 34**, 185-206 (1984).

[BP2]   P. Baras & M. Pierre, Problèmes paraboliques semi-linéaires avec données mesures, **Applicable Anal. 18**, 111 149 (1984).

[BP3]   P. Baras & M. Pierre, Critères d'existence de solutions positives pour des équations semi-linéaires non monotones, **Ann. Inst. H. Poincaré, Anal. Non-Linéaire 2,** 185-212 (1985).

[BBL]   R. Benguria, H. Brezis & E.H. Lieb, The Thomas-Fermi-Von Weizsacker theory of atoms and molecules, **Comm. Math. Phys. 79**, 167-180 (1980).

[BB]   Ph. Bénilan & H. Brezis, Nonlinear problems related to the Thomas-Fermi equation, unpublished work (see also [Br1]).

[BBC]   Ph. Bénilan, H. Brezis & M. Crandall, A semilinear elliptic equation in $L^1(\mathbb{R}^N)$, **Ann. Scu. Norm. Sup. Pisa Cl. Sci. 2**, 523-555 (1975).

[BGM]   M. Berger, P. Gauduchon & E. Mazet, **Le Spectre d'une Variété Riemannienne,** Springer-Verlag, Lecture Notes in Math. 194 (19771).

[Be]   L. Bers, Isolated singularities of minimal surfaces, **Ann. Math. 53**, 364-386 (1951).

[B·V1] M.F. Bidaut-Véron, Local and global behaviour of solutions of quasilinear elliptic equations of Emden-Fowler type, **Arch. Rat. Mech. Anal. 107**, 293-324 (1989).

[B·V2] M.F. Bidaut-Véron, Global existence and uniqueness results for singular solutions of the capillarity equation, **Pacific Jl. Math. 124**, 317-333 (1986).

[B·V3] M.F. Bidaut-Véron, Rotationnally hypersurfaces with prescribed mean curvature, **Pacific Jl. Math.** (to appear).

[B·VR] M.F. Bidaut-Véron & Th. Raoux, Asymptotics of solutions of some nonlinear elliptic systems, **Comm. Part. Diff. Equ.** (to appear).

[B·VV] M.F. Bidaut-Véron & L. Véron, Nonlinear elliptic equations on compact Riemannian manifolds and asymptotics of Emden equations, **Invent. Math. 106**, 489-539 (1991).

[Bô]   M. Bôcher, Singular points of functions which satisfy partial differential equations
of   elliptic type, **Bull. Amer. Math. Soc. 9**, 455-465 (1903).

[BV]   M.Bouhar & L. Véron, Integral representation of solutions of semilinear elliptic equations in cylinders and applications, **Nonlinear Anal. 23**, 275-296 (1994).

[Br1]   H. Brezis, Some variational problem of the Thomas-Fermi type, in **Variational Inequalities**, eds R.W. Cottle, F. Gianessi & J.L. Lions, Wiley, Chichester, 53-73 (1980).

[Br2]   H. Brezis, Une équation semi-linéaire avec conditions aux limites dans L^1, unpublished work.

[Br3]   H. Brezis, **Opérateurs Maximaux Monotones et Semi-Groupes de Contractions dans les Espaces de Hilbert**, North-Holland, Amsterdam (1973).

[BF]   H. Brezis & A. Friedman, Nonlinear parabolic equationsq involving measure as initial conditions, **Jl. Math. Pures Appl. 62**, 73-97 (1983).

[BLb]   H. Brezis & E.H. Lieb, Long range atomic potential in Thomas-Fermi theory, **Comm. Math. Phys. 65**, 231-346 (1979).

[BLs]   H. Brezis & P.L. Lions, A note on isolated singularities for linear elliptic equations, **Math. Anal. Appl. Adv., Suppl. Stud. 7A**, 263-266 (1981).

[BBT]   H. Brezis, L.A. Peletier & D. Terman, A very singular solution of the heat equation with absorption, **Arch. Rat. Mech. Anal. 95**, 185-209 (1986).

[BS]   H. Brezis & W.A. Strauss, Semilinear second order elliptic equations in L^1, **Jl. Math. Soc. Japan 25**, 565-590 (1973).

[BV]   H. Brezis & L. Véron, Removable singularities of some nonlinear equations, **Arch.Rat. Mech. Anal. 75**, 1-6 (1980).

[CS]   L.A. Caffarelli & J. Spruck, Variational problems with critical Sobolev growth and positive Dirichlet data, **Indiana Univ. Math. Jl. 39**, 1-18 (1990).

[CGS]   L.A. Caffarelli, B. Gidas & J. Spruck, Asymptotic symmetry and local behaviour of semilineaar elliptic equations with critical Sobolev growth, **Comm. Pure Appl. Math. 42**, 271-297 (1989).

[Ch]   S. Chandrasekhar, **An Introduction to the Study of Stellar Structure**, Dover Publ. Inc. (1967)

[CMV]   X.Y. Chen, H. Matano & L. Véron, Anisotropic singularities of nonlinear elliptic equations in $\mathbb{R}^2$, **Jl. Funct. Anal. 83**, 50-97 (1989).

[CL1]   W.X. Chen & C.M. Li, *A priori* estimates for solutions to nonlinear elliptic equations, **Arch. Rat. Mech. Anal. 122**, 145-157 (1993).

[CL2]   W.X. Chen & C.M. Li, Classification of solutions of some nonlinear elliptic equations, **Duke Math. Jl. 63**, 615-622 (1991).

[CL]   C.C. Chen & C.S. Lin, Local behavior of singular positive solutions of semilinear elliptic equations with Sobolev exponents, preprint.

[CF1]   P. Concus & R. Finn, A singular solution of the capillarity equation I. Existence, **Invent. Math. 29**, 143-148 (1975).

[CF2]   P. Concus & R. Finn, A singular solution of the capillarity equation I. Uniqueness, **Invent. Math. 29**, 149-160 (1975).

[CF3]   P. Concus & R. Finn, The shape of a pendent liquid drop, **Philos. Trans. Roy. Soc. London A 292**, 307-340 (1979).

[CR]   R. Courant & D. Hilbert, **Methods of Mathematical Physics**, Vol. I-II, Interscience, New-York (1962).

[DiB]   E. DiBenedetto, Continuity of weak solutions to a general porous media equation, **Indiana Univ. Math. Jl. 32**, 83-118 (1983).

[Dy]   E.B. Dynkin, A probabilistic approach to one class of nonlinear differential equations, **Prob. Th. Rel. Fields 89**, 89-115 (1991).

[DK1]   E.B. Dynkin & S.E. Kuznetsov, Superdiffusion and removable singularities for quasilinear partial differential equations, **Comm. Pure Appl. Math. Vol. XLIX**, 125-176 (1996).

[DK2]   E.B. Dynkin & S.E. Kuznetsov, Linear additive functionals of superdiffusions and related nonlinear P.D.E. (to appear).

[Em]   V.R. Emden, **Gaskugeln**, Teubner, Leipzig (1897).

[EK]   M. Escobedo & O. Kavian, Variational problems related to self-similar solutions of the heat equation, **Nonlinear Analysis, T.M.A. 10**, 1103-1133 (1987).

[Fa]   J. Fabbri, **Problèmes elliptiques non linéaires singuliers au bord dans des ouverts non réguliers**, PhD Thesis, Univ. Tours (1994).

[FV1]   J. Fabbri & L. Véron, Equations elliptiques non linéaires singulières au bord dans des domaines non réguliers, **C.R. Acad. Sci. Paris t. 320 Ser. I**, 941-946 (1995).

[FV2]   J. Fabbri & L. Véron, Singular boundary value problems for nonlinear elliptic equations in non smooth domains, **Adv. in Diff. Equ.** (to appear).

[FLN]   D. de Figuereido, P.L. Lions & R.D. Nussbaum, A priori estimatesand existence of positive solutions of semilinear elliptic equations, **Jl. Math. Pures Appl. 61**, 41-63 (1982).

[Fo]   R.H. Fowler, Further studies on Emden's and similar differential equations, **Quart. Jl. Math. 2**, 259-288 (1931).

[Fd]   A. Friedman, **Partial Differential Equations of Parabolic Type**, Englewood Cliffs, New-Jersey, Prentice-Wood (1964).

[FV]   A. Friedman & L. Véron, Singular solutions of some quasilinear elliptic equations, **Arch. Rat. Mech. Anal. 96**, 359-387 (1986).

[Fr]   J. Frehse, Capacity methods in the theory of partial differential equations, **Jber. d. Dt. Math.-Verein 84**, 1-44 (1982).

[GKS]   V.A. Galaktionov, S.P. Kurdyumov & A.A. Samarskii, On asymptotics "eigenfunctions" of the Cauchy problem for a nonlinear parabolic equation, **Mat. Sbornik 126**, 435-472 (1985). English transl.: **Math. USSR Sbornik 54**, 421-455 (1986).

[Ge1]   M. Gevrey, Sur certaines propriétés des fonctions harmoniques et leur extension aux équations aux dérivées partielles, **C.R. Acad. Sci. Paris 183 Ser I**, 546-548 (1926).

[Ge2]   M. Gevrey, Sur une généralisation du principe des singularités positives de M. Picard, **C.R. Acad. Sci. Paris 211 Ser I**, 581-584 (1940).

[Gi]   B. Gidas, Symmetry properties and isolated singularities of positive solutions of nonlinear elliptic equations, in **Nonlinear Differential Equations in Engeneering and Applied Sciences**, R.L. Sternberg ed., Marcel   Dekker , Inc. (1980).

[GNN]   B. Gidas, W.M. NI & L. Nirenberg, Symmetry and related propertries via the maximum principle, **Comm. Math. Phys. 68**, 209-243 (1979).

[GSk1]   B. Gidas & J. Spruck, Local and global behaviour of positive solutions of nonlinear elliptic equations, **Comm. Pure Appl. Math. 34**, 525-598 (1980).

[GSk2]   B. Gidas & J. Spruck, A priori bounds for positive solutions of nonlinear elliptic equations, **Comm. Part. Diff. Equ. 8**, 883-901 (1981).

[GK1]   Y. Giga & R.V. kohn, Asymptotic self-similar blow-up of semilinear heat equations, **Comm. Pure Appl. Math. 38**, 297-319 (1985).

[GK2]   Y. Giga & R.V. kohn, Characterizing blowup using similarity variables, **Indiana Univ. Math. Jl. 36**, 1-39 (1987).

[GS]   D. Gilbarg & J. Serrin, On isolated singularities of second order elliptic differential equations, **Jl Anal. Math. 4**, 309-340 (1955-56).

[GT]   D. Gilbarg & N.S. Trudinger, **Elliptic Partial Differential Equations of Second Order**, 2nd Ed., Springer-Verlag, Berlin/New-York (1983).

[Gm]   A. Gmira, On quasilinear parabolic equations involving measure data, **Asymptotic Anal. 3**, 43-56 (1990).

[GV]   A. Gmira & L. Véron, Boundary singularities of solutions of some nonlinear elliptic equations, **Duke Math. Jl. 64**, 271-324 (1991).

[Gr1]   M. Grillot, **Sur la construction de solutions d'équations elliptiques non-linéaires singulières sur une sous variété**, PhD Thesis, Univ. Tours (1996).

[Gr2]   M. Grillot,Solutions d'équations elliptiques non-linéaires singulières sur une sous variété, **C.R. Acad. Sci. Paris 322**, 49-54 (1996).

[GhV]   B. Guerch & L. Véron, Local properties of stationnary solutions of some nonlinear singular Schrödinger equation, **Rev. Mat. Iberoamericana 7**, 65-114 (1991).

[GaV]   M. Guedda & L. Véron, Local and global properties of solutions of quasilinear elliptic equations, **Jl. Diff. Equ. 76**, 159-189 (1988).

[Ha]   H. Hamza, Private communication.

[Hi1]   E. Hille, Some aspects of the Thomas-Fermi equation, **Jl. Analyse Math. 23**, 147-170 (1970).

[Hi2]   E. Hille, Aspects of Emden's equations, **Jl. Fac. Sci. Tokyo Sect. I-17**, 12-30 (1970).

[HM]   Th. Horsin-Molinaro, Construction of some solutions of a nonlinear partial differential equation having a non punctual prescribed singular set, **Comm. P.D.E.**

[Is]   I. Iscoe, On the support of measure-valued critical critical branching Brownian motion, **Ann. Prob. 16**, 200-221 (1988).

[KP1]   S. Kamin & L.A. Peletier, Singular solutions of the heat equation with absorption, **Proc. Amer. Math. Soc. 95**, 205-210 (1985).

[KP2]   S. Kamin & L.A. Peletier, Source type solutions of degenerate diffusion equations with absorption, **Israel Jl. Math. 50**, 219-230 (1985).

[KPV1] S. Kamin, L.A. Peletier & J.L. Vazquez, Classification des solutions singulières à l'origine pour une équation de la chaleur non linéaire, **C.R. Acad. Sci. Paris 305 Ser. I**, 595-598 (1987).

[KPV2] S. Kamin, L.A. Peletier & J.L. Vazquez, Classification of singular solutions of a nonlinear heat equation, **Duke Math. Jl. 58**, 601-615 (1989).

[KVa]   S. Kamin & J.L. Vazquez, Singular solutions of some nonlinear parabolic equations, **Jl. Analyse Math. 59**, 51-74 (1992).

[KV]   S. Kamin & L. Véron, Existence and uniqueness of the very singular solution of the porous media equation with absorption, **Jl. Anal. Math. 51**, 245-258 (1988).

[Ke]   J.B. Keller, On the solutions of $\Delta u = f(u)$, **Comm. Pure Appl. Math. 10**, 503-510 (1957).

[KV]   S. Kichenassamy & L. Véron, Singular solutions of the p-Laplace equation, **Math. Ann. 275**, 599-615 (1986).

[KL]   V. Kondratiev & E. Landis, On the qualitaive properties of the solutions of a nonlinear second order equation, **Math. Sbornik 135**, 346-360 (1988) ⌊in Russian⌋.

[KN]   V. Kondratiev & A. Nikishkin, On positive solutions of singular boundary problems for the equation $\Delta u = u^k$, **Russian Jl. Math. Phys. 1**, 123-138 (1993).

[KM]   I.N. Kroll & V.G. Mazja, The lack of continuity and Hölder continuity of the solutions of a certain quasilinear equation, **Sem. Math. V.A. Steklov Math. Inst. Leningrad 14**, 44-45 (1969).

[LL]   J.M. Lasry & P.L. Lions, Nonlinear elliptic equations with singular boundary conditions and stochastic control with state constraints, **Math. Annal. 283**, 583-630 (1989).

[LP]   J.M. Lee & T.H. Parker, The Yamabe problem, **Bull. Amer. Math. Soc. 17**, 37-91 (1987).

[LG1]   J.F. Le Gall, A path-valued Markov process and its connections with partial differential equations, **Proc. 1st European Congress Math. Vol 2**, Birkhäuser Boston, Basel, 185-212 (1994).

[LG2]   J.F. Le Gall, Solutions positives de $\Delta u = u^2$ dans le disque unité, **C.R. Acad. Sci. 317 Ser. I**, 873-878 (1993).

[Le1]   J.L. Lewis, Smoothness of certain degenerate elliptic equations, **Proc. Amer. Math. Soc. 80**, 259-265 (1980).

[Le2]   J.L. Lewis, Regularity of the derivatives of solutions to certain degenerate elliptic equations, **Indiana Univ. Math. Jl. 32**, 849-858 (1983).

[Li1]   C.M. Li, local asymptotic symmetry of singular solutions to nonlinear elliptic equations, preprint.

[Li2]   C.M. Li, Monotonicity and symmetry of solutions of fully nonlinear elliptic equations on bounded domains, **Comm. Part. Diff. Equ. 16**, 491-526 (1991).

[Lz]   A. Lichnerowicz, **Géométrie des Groupes de Transformations**, Dunod, Paris (1958).

[Li]   J.R. Licois, Asymptotiques d'un système dynamique conservatif associé à une équation elliptique non linéaire, **Jl. Anal. Math.** (to appear).

[LV1]   J.R. Licois & L. Véron, Un théorème d'annulation pour des équations elliptiques nonlinéaires sur des variétés riemanniennes compact, **C.R. Acad. Sci. 320 Ser. I**, 1337-1342 (1995).

[LV2]   J.R. Licois & L. Véron, A class of nonlinear conservative elliptic equations in cylinders, **Proc. London Math. Soc.** (submitted).

[LM]   P. Lindqvist & O. Martio, Regularity and polar sets for supersolutions of certain degenerate elliptic equations, **Jl. d'Anal. Math. 50**, 1-17 (1988).

[Ls]   P.L. Lions, Isolated singularities in semilinear problems, **Jl. Diff. Equ. 38**, 441-450 (1980).

[LN]   C. Loewner & L. Nirenberg, Partial differential equations invariant under conformal or projective transformations, in **Contributions to Analysis**, 245-272, Academic Press, Orlando Flo, (1974).

[Lo]   S. Lojaciewicz, Ensembles semi-analytiques, **Inst. Hautes Etud. Sci. Notes** (1965).

[MV1]   M. Marcus & L. Véron, Uniqueness and asymptotic behaviour of solutions with boundary blowup for a class of nonlinear elliptic equations, **Ann. Inst. H. Poincaré, Anal. Non-Linéaire** (à paraître).

[MV2]   M. Marcus & L. Véron, Trace au bord des solutions positives d'équations elliptiques non linéaires, **C.R. Acad. Sci. Paris t. 231, Ser. I**, (1995).

[MV3] M. Marcus & L. Véron, Boundary trace of positive solutions of nonlinear partial differential eaquations, **Ann. Math.** (submitted).

[MV4] M. Marcus & L. Véron, Trace au bord des solutions positives d'équations elliptiques et paraboliques non linéaires. Résultats d'existence et d'unicité, **C.R. Acad. Sci.** (à paraître).

[Ma1] V.G. Maz'ya, **Sobolev spaces**, Springer-Verlag (1985).

[Ma2] V.G. Maz'ya, Beurling's theorem on a minimum principle for positive harmonic functions, **Jl. Soviet. Math. 4**, 367-379 (1973).

[MS] R. Mazzeo & N. Smale, Conformally flat metrics of constant positive scalar curvature, **Jl. Diff. Geom. 34**, 581-621 (1991).

[MP] R. Mazzeo & F. Pacard, A construction of singular solutions for a semilinear elliptic equation using asymptotic analysis (to appear).

[Me] N.G. Meyers, A theory of capacities for potentials ofg functions in Lebesgues classcs, **Math. Scand. 26**, 255-293 (1970).

[MV1] I. Moutoussamy & L. Véron, Source type positive solutions of nonlinear parabolic inequalities, **Ann. Scu. Norm. Pisa Ser IV 16**, 527-555 (1989).

[MV2] I. Moutoussamy & L. Véron, Isolated singularities and asymptotic behaviour of the solutions of a semilinear heat equation, **Asymptotic Anal. 9**, 259-289 (1994).

[Ng] L. Nirenberg, On non-linear elliptic partial differential equations and Hölder continuity, **Comm. Pure Appl. Math. 6**, 103-156 (1953).

[Ni] J. Nitsche, Uber die isolierten singularitäten der lösungen von $\Delta u = e^{u}$, **Math. Z, 59**, 316-324 (1957).

[NS] W.M. Ni & J. Serrin, Nonexistence theorem for singular solutions of quasilinear partial differential equations, **Comm. Pure Appl. Math. 39**, 379-399 (1986).

[Ob] M. Obata, The conjecture on conformal transformations of Riemannian manifolds, **Jl. Diff. Geom. 6**, 247-258 (1971).

[Pc1] F. Pacard, Solutions de $\Delta u = -\lambda e^{u}$ ayant des singularites ponctuelles prescrites, **C.R. Acad. Sci. 311 Ser. I**, 317-320 (1990).

[Pc2] F. Pacard, Existence et convergence de solutions faibles positives de $-\Delta u = u^{n/(n-2)}$ dans des ouverts bornés de $\mathbb{R}^{n}$, $n \geq 3$, **C.R. Acad. Sci. 314 Ser. I**, 729-734 (1992).

[Pc3]   F. Pacard, Existence de solutions faibles positives de $-\Delta u = u^\alpha$ dans des ouverts bornés de $\mathbb{R}^n$, **C.R. Acad. Sci. 315 Ser. I**, 793-798 (1992).

[Pc4]   F. Pacard, Solutions with high dimensional singular set, to a conformally invariant elliptic equation in $\mathbb{R}^4$ and in $\mathbb{R}^6$, **Comm. Math. Phys.** (to appear).

[Pc5]   F. Pacard, the Yamabe problem on subdomains of even dimensional spheres, **Jl. Diff. Geom.** (to appear).

[Pa]   R.E. Pattle, Diffusion from an instantaneous point source with concentration dependent coefficient, **Quart. Jl. Mech. Anal. 12**, 407-409 (1959).

[PT]   L.A. Peletier & D. Terman, A very singular solution of the porous media equation with absorption, **Jl. Diff. Equ. 65**, 396-410 (1986).

[PW]   L.A. Peletier & J. Wang, A very singular solution of a quasilkinear degenerate diffusion equation with absorption, **Trans. Amer. Math. Soc. 307**, 813-826 (1988).

[Os]   R.Osserman, On the inequality $\Delta u \geq f(u)$, **Pacific Jl. Math. 7**, 1641-1647 (1957).

[Os]   L. Oswald, Isolated positive singularies of the nonlinear heat equation, **Houston Jl. Math. 14**, 543-572 (1988).

[RRV]   A. Ratto, M. Rigoli & L. Véron, Scalar curvature and conformal deformation of hyperbolic space, **Jl. Funct. Anal. 121**, 15-77 (1994).

[Re]   Y.G. Reshetniak, Mapping with bounded deformation as extremals of Dirichlet type integral, **Sibirsk. Mat. Zh.9**, 652-666 (1966).

[RV]   Y. Richard & L. Véron, Isotropic singularities of solutions of nonlinear elliptic inequalities, **Ann. Inst. H. Poincaré, Anal. Non-Linéaire 6**, 37-72 (1989).

[Sc]   R. Schoen, The existence of weak solutions with prescribed singular behaviour for a conformally invariant scalar equation, **Comm. Pure Appl. Math. 41**, 317-392 (1988).

[SY]   R. Schoen & S.T. Yau, Conformally flat manifolds, Kleinian group and scalar curvature, **Invent. Math. 92**, 47-71 (1988).

[Se1]   J. Serrin, Local behaviour of solutions of quasilinear equations, **Acta Math. 111**, 247-302 (1964).

[Se2]   J. Serrin, Isolated singularities of solutions of quasilinear equations, **Acta Math. 113**, 219-240 (1965).

[Se3]   J. Serrin, Singularities of solutions of nonlinear equations, **Proc. Symp. in Appl. Math. 17**, 68-88 (1965).

[Se4]   J. Serrin, Removable singularities of solutions of elliptic equations, **Arch. Rat. Mech. Anal. 17**, 67-76 (1964).

[Se5]   J. Serrin, Removable singularities of solutions of elliptic equations. II, **Arch. Rat. Mech. Anal. 20**, 163-169 (1965).

[Sh]   Y.C. Sheu, Removable boundary singularities for solutions of some nonlinear differential equations, **Duke Math. Jl. 74**, 701-711 (1994).

[Si1]   L. Simon, Asymptotics for a class of nonlinear evolution equations with applications to geometric problem, **Ann. Math. 118**, 525-571 (1983).

[Si2]   L. Simon, Isolated singularities of extrema of geometric variational problems, in **Harmonic Mappings and Minimal Immersions**, E. Giusti ed., Springer-Verlag Lecture Notes in Math. **1161** (1985).

[Sm1]   N. Smale, A bridge principle for minimal surfaces and constant mean curvature submanifolds of $\mathbb{R}^N$, **Invent. Math. 90**, 505-549 (1987).

[Sm2]   N. Smale, Minimal hypersurfaces with many isolated singularities, **Ann. Math.130**, 603-642 (1989).

[So]   A. Sommerfeld, Asymptotishe integration der differentialgleichung des Thomas-Fermishen atom, **Z für Phys. 78** 283-308 (1932).

[St1]   E. Stein, **Boundary behaviour of holomorphic functions of several complex variables**, Math. Notes 9, Princeton Univ. Press, Princeton (1972).

[St2]   E. Stein, **Topics in Harmonic analysis**, Annals of Math. Studies, Princeton Univ. Press, Princeton (1970).

[Ta]   S.D. Taliafero, Asymptotic integration of $u'' = \Phi(t)u^\lambda$, **Jl. Math. Anal. Appl. 66**, 95-134 (1978).

[To1]   P. Tolksdorf, Regularity for a more general class of quasilinear elliptic equations, **Jl. Diff. Equ. 51**, 126-150 (1984).

[To2]   P. Tolksdorf, On the Dirichlet problem for quasilinear equations in domains with conical boundary points, **Comm. Part. Diff. Equ. 8**, 773-817 (1983).

[Tr]   N.S. Trudinger, On Harnack type inequality and their applications to quasilinear elliptic equations, **Comm. Pure Appl. Math. 20**, 721-747 (1967).

[Ua]   N.N. Ural'Ceva, Degenerate quasilinear systems, **Sem. Math. V.A. Steklov, Math. Inst. Leningrad 7**, 83-99 (1968).

[Va1]   J.L. Vazquez, An *a priori* interior estimate for the solution of a nonlinear problem representing weak diffusion, **Nonlinear Anal. 5**, 95-105 (1981).

[Va2]   J.L. Vazquez, On a semilinear equation in $\mathbb{R}^2$ involving bounded measures, **Proc. Roy. Soc. Edinburgh 95A**, 181-202 (1983).

[VV1]   J.L. Vazquez & L. Véron, Removable singularities of some strongly nonlinear elliptic equations, **Manuscripta Math. 33**, 129-144 (1980).

[VV2]   J.L. Vazquez & L. Véron, Singularities of elliptic equations with an exponential nonlinearity, **Math. Ann. 269**, 119-135 (1984).

[VV3]   J.L. Vazquez & L. Véron, Isolated singularities of some nonlinear elliptic equations, **Jl. Diff. Equ. 60**, 301-321 (1985).

[Ve1]   L. Véron, Singular solutions of some nonlinear elliptic equations, **Nonlinear Anal. 5**, 225-242 (1981).

[Ve2]   L. Véron, Comportement asymptotique des solutions d'équations elliptiques semilinéaires dans $\mathbb{R}^N$, **Ann. Mat. Pura Appl. 127**, 25-50 (1981).

[Ve3]   L. Véron, Global behaviour and symmetry properties of singular solutions of nonlinear elliptic equations, **Ann. fac. Sci. Toulouse 6,** 1-31 (1984).

[Ve4]   L. Véron, Geometric invariance of singular solutions of some nonlinear partial differential equations, **Indiana Univ. Math. Jl. 38**, 75-100 (1989).

[Ve5]   L. Véron, Equations d'évolution semi-linéaires du second ordre dans L^1, **Rev. Roum. Math. Pures Appl. 27**, 95-123 (1982).

[Ve6]   L. Véron, Singularities of some quasilinear equtions, in **Nonlinear Diffusion Equations and their Equilibrium States Vol. II**, W.N. Ni ed., Sringer-Verlag New-York 333-365 (1986).

[Ve7]   L. Véron, Conformal asymptotics of the isothermal gas spheres equation, in **Nonlinear Diffusion Equations and their Equilibrium States Vol. 3**, eds N.G. Llyod, W.M. Ni, L.A. Peletier & J. Serrin, Birkhäuser Boston, Basel, 537-559 (1992).

[Ve8]   L. Véron, Semilinear elliptic equations with uniform blow-up on the boundary, **Jl. Analyse Math. 59**, 231-250 (1992).

[Ve9]   L. Véron, Weak and strong singularities of nonlinear elliptic equations, **Proc.Symp. Pure Math. 45**, 599-615 (1986).

[We1]   F.B. Weissler, Asymptotic analysis of an ordinary differential equation and non-uniqueness for a semilinear partial differential equation, **Arch. Rat. Mech. Anal. 91**, 231-245 (1986).

[We2]   F.B. Weissler, Rapidly decaying solutions of an ordinary differential equation with applications to semilinear elliptic and parabolic partial differential equations, **Arch. Rat. Mech. Anal. 91**, 247-266 (1986).

[Ya]   E. Yanagida, Uniqueness of rapidly decaying solution of the Haraux-Weissler equation, to appear.

[ZK]   Y.B. Zel'dovich & A.S. Kompanets, **On the theory of heat propagation where thermal conductivity depends on temperature**, Collection published on the occasion of the seventieth birthday of Academician A.F. Ioffe, Akad. Nauk. SSSR (Russian), Moscow (1950).